LINEAR PROGRAMMING AND NETWORK FLOWS

Board of Advisors, Engineering

LINEAR PROGRAMMING AND NETWORK FLOWS

Mokhtar S. Bazaraa

John J. Jarvis

School of Industrial and Systems Engineering
Georgia Institute of Technology, Atlanta, Georgia

JOHN WILEY & SONS
New York • Chichester • Brisbane • Toronto • Singapore

To Alice and Sadek

Library of Congress Cataloging in Publication Data:

Bazarra, M S 1943-
Linear programming and network flows.

Includes bibliographies and index.
1. Linear programming. 2. Network analysis (Planning) I. Jarvis, John J., joint author. II. Title.
T57.74.B39 519.7'2 76-42241
ISBN 0-471-06015-1

Printed in the United States of America

20 19

PREFACE

Linear Programming deals with the problem of minimizing or maximizing a linear function in the presence of linear inequalities. Since the development of the simplex method by George B. Dantzig in 1947, linear programming has been extensively used in the military, industrial, governmental, and urban planning fields, among others. The popularity of linear programming can be attributed to many factors including its ability to model large and complex problems, and the ability of the users to solve large problems in a reasonable amount of time by the use of the simplex method and computers.

During and after World War II it became evident that planning and coordination among various projects and the efficient utilization of scarce resources were essential. Intensive work by the U. S. Air Force team SCOOP (Scientific Computation of Optimum Programs) began in June 1947. As a result, the simplex method was developed by George B. Dantzig by the end of summer 1947. Interest in linear programming spread quickly among economists, mathematicians, statisticians, and government institutions. In the summer of 1949 a conference on linear programming was held under the sponsorship of the Cowles Commission for Research in Economics. The papers presented at that conference were later collected in 1951 by T. C. Koopmans into the book *Activity Analysis of Production and Allocation.*

Since the development of the simplex method many people have contributed to the growth of linear programming by developing its mathematical theory, devising efficient computational methods and codes, exploring new applications, and by their use of linear programming as an aiding tool for solving more complex problems, for instance, discrete programs, nonlinear programs, combinatorial problems, stochastic programming problems, and problems of optimal control.

This book addresses the subjects of linear programming and network flows. The simplex method represents the backbone of most of the techniques presented in the book. Whenever possible, the simplex method is specialized to take advantage of problem structure. Throughout we have attempted first to present the techniques, to illustrate them by numerical examples, and then to provide detailed mathematical analysis and an argument showing convergence to an optimal solution. Rigorous proofs of the results are given without the theorem-proof format. Even though this may bother some readers, we believe that the format and mathematical level adopted in this book will provide an adequate and smooth study for those who wish to learn the techniques and the know-how to use them, and for those who wish to study the algorithms at a more rigorous level.

The book can be used both as a reference and as a textbook for advanced undergraduate students and first-year graduate students in the fields of industrial engineering, management, operations research, computer science, mathematics, and other engineering disciplines that deal with the subjects of linear programming and network flows. Even though the book's material requires some mathematical maturity, the only prerequisite is linear algebra. For convenience of the reader, pertinent results from linear algebra and convex analysis are summarized in Chapter two. In a few places in the book, the notion of differentiation would be helpful. These, however, can be omitted without loss of understanding or continuity.

This book can be used in several ways. It can be used in a two-course sequence on linear programming and network flows, in which case all of its material could be easily covered. The book can also be utilized in a one-semester course on linear programming and network flows. The instructor may have to omit some topics at his discretion. The book can also be used as a text for a course on either linear programming or network flows.

Following the introductory first chapter and the second chapter on linear algebra and convex analysis, the book is organized into two parts: linear programming and networks flows. The linear programming part consists of Chapters three to seven. In Chapter three the simplex method is developed in detail, and in Chapter four the initiation of the simplex method by the use of artificial variables and the problem of degeneracy are discussed. Chapter five deals with some specializations of the simplex method and the development of optimality criteria in linear programming. In Chapter six we consider the dual problem, develop several computational procedures based on duality, and discuss sensitivity and parametric analysis. Chapter seven introduces the reader to the decomposition principle and to large-scale programming. The part on network flows consists of Chapters eight to eleven. Many of the procedures in this part are presented as a direct simplification of the simplex method. In Chapter eight the transportation problem and the assignment problem are both examined. Chapter nine considers the minimal cost network flow problem from the simplex method point of view. In Chapter ten we present the out-of-kilter algorithm for solving the same problem. Finally, Chapter eleven covers the special topics of the maximal flow problem, the shortest path problem, and the multicommodity minimal cost flow problem.

We thank the graduate students at the School of Industrial and Systems Engineering at the Georgia Institute of Technology who suffered through two earlier drafts of this manuscript and who offered many constructive criticisms. We express our appreciation to Gene Ramsay, Dr. Jeff Kennington, Dr. Michael Todd, and Dr. Ron Rardin for their many fine suggestions. We are especially grateful to Süleyman Tüfekçi for preparing the solutions manual and to Carl H. Wohlers for preparing the bibliography. We also thank Dr. Robert N. Lehrer, director of the School of Industrial and Systems Engineering at the Georgia Institute of Technology, for his support during all phases of the preparation of the manuscript. Special thanks are due to Mrs. Alice Jarvis, who typed the first and third drafts of the manuscript; and to Mrs. Carolyn Piersma, Mrs. Amelia Williams, and Miss Kaye Watkins, who typed portions of the second draft.

Mokhtar S. Bazaraa
John J. Jarvis

CONTENTS

ONE: INTRODUCTION

In 1949 George B. Dantzig published the "simplex method" for solving linear programs. Since that time a number of individuals have contributed to the field of linear programming in many different ways including theoretical development, computational aspects, and exploration of new applications of the subject. The simplex method of linear programming enjoys wide acceptance because of (1) its ability to model important and complex management decision problems and (2) its capability for producing solutions in a reasonable amount of time. In subsequent chapters of this text we shall consider the simplex method and its variants, with emphasis on the understanding of the methods.

In this chapter the linear programming problem is introduced. The following topics are discussed: basic definitions in linear programming, assumptions leading to linear models, manipulation of the problem, examples of linear problems, and geometric solution in the feasible region space and the requirement space. This chapter is elementary and may be skipped if the reader has previous knowledge of linear programming.

1.1 THE LINEAR PROGRAMMING PROBLEM

A linear programming problem is a problem of minimizing or maximizing a linear function in the presence of linear constraints of the inequality and/or the equality type. In this section the linear programming problem is formulated.

Basic Definitions

Consider the following linear programming problem.

$$
\begin{array}{lrcl}
\text{Minimize} & c_1x_1 + c_2x_2 + \cdots + c_nx_n & & \\
\text{Subject to} & a_{11}x_1 + a_{12}x_2 + \cdots + a_{1n}x_n & \geqslant & b_1 \\
& a_{21}x_1 + a_{22}x_2 + \cdots + a_{2n}x_n & \geqslant & b_2 \\
& \vdots \qquad\quad \vdots \qquad\qquad\quad \vdots & & \vdots \\
& a_{m1}x_1 + a_{m2}x_2 + \cdots + a_{mn}x_n & \geqslant & b_m \\
& x_1, \quad x_2, \quad \ldots, \quad x_n & \geqslant & 0
\end{array}
$$

Here $c_1x_1 + c_2x_2 +, \ldots, + c_nx_n$ is the *objective function* (or criterion function) to be minimized and will be denoted by z. The coefficients $c_1, c_2, \ldots, c_n$ are the (known) *cost coefficients* and $x_1, x_2, \ldots, x_n$ are the *decision variables* (variables, or activity levels) to be determined. The inequality $\sum_{j=1}^{n} a_{ij}x_j \geqslant b_i$ denotes the ith *constraint* (or restriction). The coefficients a_{ij} for $i = 1, 2, \ldots, m, j = 1, 2, \ldots, n$ are called the *technological coefficients*. These technological coefficients form the *constraint matrix* **A** given below.

$$
\mathbf{A} = \begin{bmatrix} a_{11} & a_{12} & \cdots & a_{1n} \\ a_{21} & a_{22} & \cdots & a_{2n} \\ \vdots & \vdots & & \vdots \\ a_{m1} & a_{m2} & \cdots & a_{mn} \end{bmatrix}
$$

The column vector whose ith component is b_i, which is referred to as the *right-hand-side vector*, represents the minimal requirements to be satisfied. The constraints $x_1, x_2, \ldots, x_n \geqslant 0$ are the *nonnegativity constraints*. A set of variables $x_1, \ldots, x_n$ satisfying all the constraints is called a *feasible point* or a *feasible vector*. The set of all such points constitutes the *feasible region* or the *feasible space*.

Using the foregoing terminology, the linear programming problem can be stated as follows: Among all feasible vectors, find that which minimizes (or maximizes) the objective function.

Example 1.1

Consider the following linear problem.

$$\text{Minimize} \quad 2x_1 + 5x_2$$

$$\begin{aligned} \text{Subject to} \quad x_1 + x_2 &\geq 6 \\ -x_1 - 2x_2 &\geq -18 \\ x_1, \; x_2 &\geq 0 \end{aligned}$$

In this case we have two decision variables x_1 and x_2. The objective function to be minimized is $2x_1 + 5x_2$. The constraints and the feasible region are illustrated in Figure 1.1. The optimization problem is thus to find a point in the feasible region with the smallest possible objective.

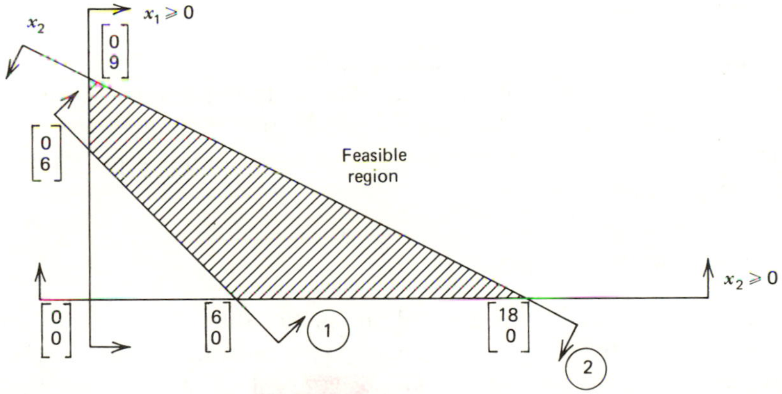

Figure 1.1. **Illustration of the feasible region.**

Assumptions of Linear Programming

In order to represent an optimization problem as a linear program, several assumptions that are implicit in the linear programming formulation discussed above are needed. A brief discussion of these assumptions is given below.

1. *Proportionality.* Given a variable x_j, its contribution to cost is $c_j x_j$ and its contribution to the ith constraint is $a_{ij} x_j$. This means that if x_j is doubled, say, so is its contribution to cost and to each of the constraints. To illustrate, suppose that x_j is the amount of activity j used. For instance, if

$x_j = 10$, then the cost of this activity is $10c_j$. If $x_j = 20$, then the cost is $20c_j$, and so on. This means that no savings (or extra costs) are realized by using more of activity j. Also no setup cost for starting the activity is realized.

2. *Additivity*. This assumption guarantees that the total cost is the sum of the individual costs, and that the total contribution to the ith restriction is the sum of the individual contributions of the individual activities.
3. *Divisibility*. This assumption ensures that the decision variables can be divided into any fractional levels so that noninteger values for the decision variables are permitted.

To summarize, an optimization problem can be cast as a linear program only if the aforementioned assumptions hold. This precludes situations where economies of scale exist; for example, when the unit cost decreases as the amount produced is increased. In these situations one must resort to nonlinear programs. It should also be noted that the parameters c_j, a_{ij}, and b_i must be known or estimated.

Problem Manipulation

Recall that a linear program is a problem of minimizing or maximizing a linear function in the presence of linear inequality and/or equality constraints. By simple manipulations the problem can be transformed from one form to another equivalent form. These manipulations are most useful in linear programming, as will be seen throughout the text.

INEQUALITIES AND EQUATIONS

An inequality can be easily transformed into an equation. To illustrate, consider the constraint given by $\sum_{j=1}^{n} a_{ij}x_j \geqslant b_i$. This constraint can be put in an equation form by subtracting the nonnegative *slack variable* x_{n+i} (sometimes denoted by s_i) leading to $\sum_{j=1}^{n} a_{ij}x_j - x_{n+i} = b_i$ and $x_{n+i} \geqslant 0$. Similarly the constraint $\sum_{j=1}^{n} a_{ij}x_j \leqslant b_i$ is equivalent to $\sum_{j=1}^{n} a_{ij}x_j + x_{n+i} = b_i$ and $x_{n+i} \geqslant 0$. Also an equation of the form $\sum_{j=1}^{n} a_{ij}x_j = b_i$ can be transformed into the two inequalities $\sum_{j=1}^{n} a_{ij}x_j \leqslant b_i$ and $\sum_{j=1}^{n} a_{ij}x_j \geqslant b_i$.

NONNEGATIVITY OF THE VARIABLES

For most practical problems the variables represent physical quantities and hence must be nonnegative. The simplex method is designed to solve linear programs where the variables are nonnegative. If a variable x_j is unrestricted in sign, then it can be replaced by $x_j' - x_j''$ where $x_j' \geqslant 0$ and $x_j'' \geqslant 0$. If $x_j \geqslant l_j$, then the new variable $x_j' = x_j - l_j$ is automatically nonnegative. Also if a variable x_j is restricted such that $x_j \leqslant u_j$ where $u_j \leqslant 0$, then the substitution $x_j' = u_j - x_j$ produces a nonnegative variable x_j'.

MINIMIZATION AND MAXIMIZATION PROBLEMS

Another problem manipulation is to convert a maximization problem into a minimization problem and conversely. Note that over any region

$$\text{Maximum} \sum_{j=1}^{n} c_j x_j = -\text{minimum} \sum_{j=1}^{n} -c_j x_j$$

So a maximization (minimization) problem can be converted into a minimization (maximization) problem by multiplying the coefficients of the objective function by -1. After the optimization of the new problem is completed, the objective of the old problem is -1 times the optimal objective of the new problem.

Standard and Canonical Formats

From the foregoing discussion we see that a given linear program can be put in different equivalent forms by suitable manipulations. Two forms in particular will be useful. These are the standard and the canonical forms. A linear program is said to be in *standard format* if all restrictions are equalities and all variables are nonnegative. The simplex method is designed to be applied only after the problem is put in the standard form. The canonical form is also useful especially in exploiting duality relationships. A minimization problem is in *canonical form* if all variables are nonnegative and all the constraints are of the $\geqslant$ type. A maximization problem is in canonical format if all the variables are nonnegative and all the constraints are of the $\leqslant$ type. The standard and canonical forms are summarized in Table 1.1.

Linear Programming in Matrix Notation

A linear programming problem can be stated in a more convenient form using matrix notation. To illustrate, consider the following problem.

$$\begin{aligned} &\text{Minimize} && \sum_{j=1}^{n} c_j x_j \\ &\text{Subject to} && \sum_{j=1}^{n} a_{ij} x_j = b_i \quad i = 1, 2, \ldots, m \\ &&& x_j \geqslant 0 \quad j = 1, 2, \ldots, n \end{aligned}$$

Table 1.1 Standard and Canonical Forms

	MINIMIZATION PROBLEM			MAXIMIZATION PROBLEM		
Standard Form	Minimize	$\sum_{j=1}^{n} c_j x_j$		Maximize	$\sum_{j=1}^{n} c_j x_j$	
	Subject to	$\sum_{j=1}^{n} a_{ij} x_j = b_i$	$i = 1, \ldots, m$	Subject to	$\sum_{j=1}^{n} a_{ij} x_j = b_i$	$i = 1, \ldots, m$
		$x_j \geqslant 0$	$j = 1, \ldots, n$		$x_j \geqslant 0$	$j = 1, \ldots, n$
Canonical Form	Minimize	$\sum_{j=1}^{n} c_j x_j$		Maximize	$\sum_{j=1}^{n} c_j x_j$	
	Subject to	$\sum_{j=1}^{n} a_{ij} x_j \geqslant b_i$	$i = 1, \ldots, m$	Subject to	$\sum_{j=1}^{n} a_{ij} x_j \leqslant b_i$	$i = 1, \ldots, m$
		$x_j \geqslant 0$	$j = 1, \ldots, n$		$x_j \geqslant 0$	$j = 1, \ldots, n$

Denote the row vector $(c_1, c_2, \ldots, c_n)$ by $\mathbf{c}$, and consider the following column vectors $\mathbf{x}$ and $\mathbf{b}$, and the $m \times n$ matrix $\mathbf{A}$.

$$\mathbf{x} = \begin{bmatrix} x_1 \\ x_2 \\ \vdots \\ x_n \end{bmatrix} \qquad \mathbf{b} = \begin{bmatrix} b_1 \\ b_2 \\ \vdots \\ b_m \end{bmatrix} \qquad \mathbf{A} = \begin{bmatrix} a_{11} & a_{12} & \cdots & a_{1n} \\ a_{21} & a_{22} & \cdots & a_{2n} \\ \vdots & \vdots & & \vdots \\ a_{m1} & a_{m2} & \cdots & a_{mn} \end{bmatrix}$$

Then the above problem can be written as follows.

$$\begin{aligned} \text{Minimize} \quad & \mathbf{cx} \\ \text{Subject to} \quad & \mathbf{Ax} = \mathbf{b} \\ & \mathbf{x} \geqslant \mathbf{0} \end{aligned}$$

The problem can also be conveniently represented via the columns of $\mathbf{A}$. Denoting $\mathbf{A}$ by $[\mathbf{a}_1, \mathbf{a}_2, \ldots, \mathbf{a}_n]$ where $\mathbf{a}_j$ is the jth column of $\mathbf{A}$, the problem can be formulated as follows.

$$\begin{aligned} \text{Minimize} \quad & \sum_{j=1}^{n} c_j x_j \\ \text{Subject to} \quad & \sum_{j=1}^{n} \mathbf{a}_j x_j = \mathbf{b} \\ & x_j \geqslant 0 \qquad j = 1, 2, \ldots, n \end{aligned}$$

1.2 EXAMPLES OF LINEAR PROBLEMS

In this section we describe several problems that can be formulated as linear programs. The purpose is to show the varieties of problems that can be recognized and expressed in precise mathematical terms as linear programs.

Feed Mix Problem

An agricultural mill manufactures feed for chickens. This is done by mixing several ingredients, such as corn, limestone, or alfalfa. The mixing is to be done in such a way that the feed meets certain levels for different types of nutrients, such as protein, calcium, carbohydrates, and vitamins. To be more specific, suppose that n ingredients $j = 1, 2, \ldots, n$ and m nutrients $i = 1, 2, \ldots, m$ are considered. Let the unit cost of ingredient j be c_j and let the amount of ingredient j to be used be x_j. The cost is therefore $\Sigma_{j=1}^{n} c_j x_j$. If the amount of the final product needed is b, then we must have $\Sigma_{j=1}^{n} x_j = b$. Further suppose that a_{ij} is the amount of nutrient i present in a unit of ingredient j, and that the acceptable lower and upper limits of nutrient i in a unit of the chicken feed are l_i' and u_i' respectively. Therefore we must have the constraints $l_i' b \leqslant \Sigma_{j=1}^{n} a_{ij} x_j \leqslant u_i' b$ for $i = 1, 2, \ldots, m$. Finally, because of shortages, suppose that the mill cannot acquire more than u_j units of ingredient j. The problem of mixing the ingredients such that the cost is minimized and the above restrictions are met, can be formulated as follows.

$$
\begin{array}{llcl}
\text{Minimize} & & c_1 x_1 + c_2 x_2 \cdots + c_n x_n & \\
\text{Subject to} & & x_1 + x_2 \cdots + x_n = b & \\
& bl_1' \leqslant & a_{11} x_1 + a_{12} x_2 \cdots + a_{1n} x_n & \leqslant bu_1' \\
& bl_2' \leqslant & a_{21} x_1 + a_{22} x_2 \cdots + a_{2n} x_n & \leqslant bu_2' \\
& \vdots & \vdots \qquad \vdots \qquad\qquad \vdots & \vdots \\
& bl_m' \leqslant & a_{m1} x_1 + a_{m2} x_2 \cdots + a_{mn} x_n & \leqslant bu_m' \\
& & 0 \leqslant x_1 \leqslant u_1 & \\
& & 0 \leqslant x_2 \leqslant u_2 & \\
& & \vdots & \\
& & 0 \leqslant x_n \leqslant u_n &
\end{array}
$$

Production Scheduling: An Optimal Control Problem

A company wishes to determine the production rate over the planning horizon of the next T weeks such that the known demand is satisfied and the total production and inventory cost is minimized. Let the known demand rate at time t be $g(t)$, and similarly denote the production rate and inventory at t by $x(t)$ and $y(t)$. Further suppose that the initial inventory at time 0 is y_0 and that the desired inventory at the end of the planning horizon is y_T. Suppose that the inventory cost is proportional to the units in storage, so that the inventory cost is given by $c_1 \int_0^T y(t)\, dt$ where $c_1 > 0$ is known. Also suppose that the production cost is proportional to the rate of production, and so is given by $c_2 \int_0^T x(t)\, dt$. Then the total cost is $\int_0^T [c_1 y(t) + c_2 x(t)]\, dt$. Also note that the inventory at any time is given according to the relationship

$$
y(t) = y_0 + \int_0^t [x(\tau) - g(\tau)]\, d\tau \qquad t \in [0, T]
$$

Suppose that no backlogs are allowed; that is, all demand must be satisfied. Further suppose that the present manufacturing capacity restricts the production rate so that it does not exceed b_1 at any time. Also the available storage restricts the maximum inventory to be less than or equal to b_2. Hence the production scheduling problem can be stated as follows.

$$\begin{array}{lll} \text{Minimize} & \int_0^T [c_1 y(t) + c_2 x(t)]\, dt & \\ \text{Subject to} & y(t) = y_0 + \int_0^t [x(\tau) - g(\tau)]\, d\tau & t \in [0, T] \\ & y(T) = y_T & \\ & 0 \leqslant x(t) \leqslant b_1 & t \in [0, T] \\ & 0 \leqslant y(t) \leqslant b_2 & t \in [0, T] \end{array}$$

The foregoing model is a linear control problem, where the *control variable* is the production rate $x(t)$ and the *state variable* is the inventory level $y(t)$. The problem can be approximated by a linear program by discretizing the continuous variables x and y. First the planning horizon $[0, T]$ is divided into n smaller periods $[0, \Delta], [\Delta, 2\Delta], \ldots, [(n-1)\Delta, n\Delta]$ where $n\Delta = T$. The production rate, the inventory, and the demand rate are assumed constant over each period. In particular let the production rate, the inventory, and the demand rate on period j be x_j, y_j, and g_j respectively. Then the production scheduling problem above can be approximated by the following linear program (why?).

$$\begin{array}{lll} \text{Minimize} & \sum_{j=1}^{n} (c_1\Delta) y_j + \sum_{j=1}^{n} (c_2\Delta) x_j & \\ \text{Subject to} & y_j = y_{j-1} + (x_j - g_j)\Delta & j = 1, 2, \ldots, n \\ & y_n = y_T & \\ & 0 \leqslant x_j \leqslant b_1 & j = 1, 2, \ldots, n \\ & 0 \leqslant y_j \leqslant b_2 & j = 1, 2, \ldots, n \end{array}$$

Cutting Stock Problem

A manufacturer of metal sheets produces rolls of standard fixed width w and of standard length l. A large order is placed by a customer who needs sheets of width w and varying lengths. In particular, b_i sheets with length l_i and width w for $i = 1, 2, \ldots, m$ are ordered. The manufacturer would like to cut the standard rolls in such a way as to satisfy the order and to minimize the waste. Since scrap pieces are useless to the manufacturer, the objective is to minimize the number of rolls needed to satisfy the order. Given a standard sheet of length l, there are many ways of cutting it. Each such way is called a *cutting pattern*.

The jth cutting pattern is characterized by the column vector $\mathbf{a}_j$ where the ith component of $\mathbf{a}_j$, namely a_{ij}, is a nonnegative integer denoting the number of sheets of length l_i in the jth pattern. For instance, suppose that the standard sheets have length $l = 10$ meters and that sheets of lengths 1.5, 2.5, 3.0, and 4.0 meters are needed. The following are typical cutting patterns:

$$\mathbf{a}_1 = \begin{bmatrix} 3 \\ 2 \\ 0 \\ 0 \end{bmatrix}, \quad \mathbf{a}_2 = \begin{bmatrix} 0 \\ 4 \\ 0 \\ 0 \end{bmatrix}, \quad \mathbf{a}_3 = \begin{bmatrix} 0 \\ 0 \\ 3 \\ 0 \end{bmatrix}, \ldots$$

Note that the vector $\mathbf{a}_j$ represents a cutting pattern if and only if $\sum_{i=1}^{n} a_{ij} l_i \leq l$ and each a_{ij} is a nonnegative integer. The number of cutting patterns n is finite. If we let x_j be the number of standard rolls cut according to the jth pattern, the problem can be formulated as follows.

$$\begin{array}{lll} \text{Minimize} & \sum_{j=1}^{n} x_j & \\ \text{Subject to} & \sum_{j=1}^{n} a_{ij} x_j \geq b_i & i = 1, 2, \ldots, m \\ & x_j \geq 0 & j = 1, 2, \ldots, n \\ & x_j \ \ \text{integer} & j = 1, 2, \ldots, n \end{array}$$

If the integrality requirement on the x_j's is dropped, the abovementioned problem is a linear program. Of course the difficulty of this problem is that the number of possible cutting patterns n is very large, and also it is not computationally feasible to enumerate each cutting pattern and its column $\mathbf{a}_j$ beforehand. The decomposition algorithm of Chapter 7 is particularly suited to solve the preceding problem where a new cutting pattern is generated at each iteration (see also Exercise 7.25). In Section 6.7 we suggest a method for handling the integrality requirements.

The Transportation Problem

The Brazilian coffee company processes coffee beans into coffee at m plants. The coffee is then shipped every week to n warehouses in major cities for retail, distribution, and exporting. Suppose that the unit shipping cost from plant i to warehouse j is c_{ij}. Further suppose that the production capacity at plant i is a_i and that the demand at warehouse j is b_j. It is desired to find the production-shipping pattern that minimizes the overall shipping cost. This is the well-known transportation problem. The essential elements of the problem are shown in the

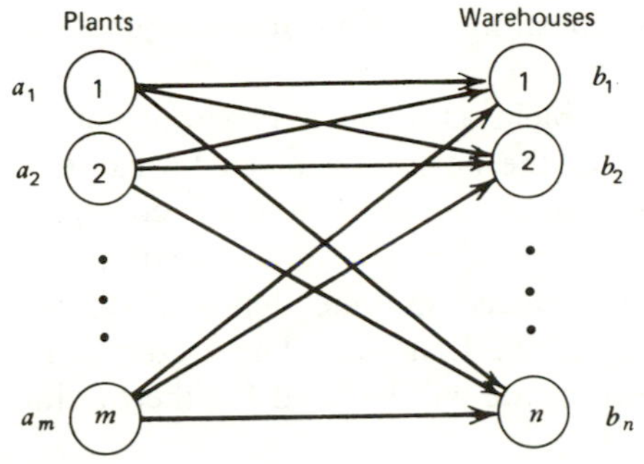

Figure 1.2. The transportation problem.

network of Figure 1.2. The transportation problem discussed above can be formulated as the following linear program.

$$\text{Minimize} \quad \sum_{i=1}^{m} \sum_{j=1}^{n} c_{ij} x_{ij}$$

$$\text{Subject to} \quad \sum_{j=1}^{n} x_{ij} \leqslant a_i \qquad i = 1, 2, \ldots, m$$

$$\sum_{i=1}^{m} x_{ij} = b_j \qquad j = 1, 2, \ldots, n$$

$$x_{ij} \geqslant 0 \qquad i = 1, 2, \ldots, m \qquad j = 1, 2, \ldots, n$$

Capital Budgeting Problem

A municipal construction project has funding requirements over the next four years of \$2 million, \$4 million, \$8 million, and \$5 million respectively. Assume that all of the money for a given year is required at the beginning of the year. The city intends to sell exactly enough long-term bonds to cover the project funding requirements, and all of these bonds, regardless of when they are sold, will be paid off (*mature*) on the same date in a distant future year. The long-term bond market interest rates (that is, the costs of selling bonds) for the next four years are projected to be 7 percent, 6 percent, 6.5 percent, and 7.5 percent respectively. Bond interest paid will commence one year after the project is completed and will continue for 20 years, after which the bonds will be paid off. During the same period the short-term interest rates on time deposits (that is, what the city can earn on deposits) are projected to be 6 percent, 5.5 percent, and 4.5 percent respectively (the city will clearly not invest money in short-term deposits during the fourth year). What is the city's optimal strategy for selling bonds and depositing funds in time accounts in order to complete the construction project?

To formulate this problem as a linear program, let x_j, $j = 1, \ldots, 4$ be the amount of bonds sold at the beginning of each year j. When bonds are sold, some of the money will immediately be used for construction and some money will be placed in short-term deposits to be used in later years. Let y_j, $j = 1, \ldots, 3$ be the money placed in time deposits at the beginning of year j. Consider the beginning of the first year. The amount of bonds sold minus the amount of time deposits made will be used for the funding requirement at that year. Thus we may write

$$x_1 - y_1 = 2$$

We could have expressed the constraint as $\geqslant$. However, it is clear in this case that any excess funds will be deposited so that $=$ is also acceptable.

Consider the beginning of the second year. In addition to bonds sold and time deposits made, we also have time deposits plus interest becoming available from the previous year. Thus we have

$$1.06y_1 + x_2 - y_2 = 4$$

The third and fourth constraints are constructed in a similar manner.

Ignoring the fact that the amounts occur in different years (that is, the time value of money), the unit cost of selling bonds is 20 times the interest rate. Thus for bonds sold at the beginning of the first year we have $c_1 = 20(0.07)$. The other cost coefficients are computed similarly.

Finally, the linear programming model becomes as follows.

$$\begin{array}{ll} \text{Minimize} & 20(0.07)x_1 + 20(0.06)x_2 + 20(0.065)x_3 + 20(0.075)x_4 \\ \text{Subject to} & \begin{array}{rl} x_1 - y_1 & = 2 \\ 1.06y_1 + x_2 - y_2 & = 4 \\ 1.055y_2 + x_3 - y_3 & = 8 \\ 1.045y_3 + x_4 & = 5 \\ x_1, x_2, x_3, x_4, y_1, y_2, y_3 & \geqslant 0 \end{array} \end{array}$$

Tanker Scheduling Problem

A shipline company requires a fleet of ships to service requirements for carrying cargo between six cities. There are four specific routes that must be served daily. These routes and the number of ships required for each route are as follows.

ROUTE #	ORIGIN	DESTINATION	NUMBER OF SHIPS PER DAY NEEDED
1	Dhahran	New York	3
2	Marseilles	Istanbul	2
3	Naples	Bombay	1
4	New York	Marseilles	1

All cargo are compatible, and therefore only one type of ship is needed. The travel time matrix between the various cities is shown below.

	NAPLES	MARSEILLES	ISTANBUL	NEW YORK	DHAHRAN	BOMBAY	
Naples	0	1	2	14	7	7	
Marseilles	1	0	3	13	8	8	
Istanbul	2	3	0	15	5	5	t_{ij} matrix (days)
New York	14	13	15	0	17	20	
Dhahran	7	8	5	17	0	3	
Bombay	7	8	5	20	3	0	

It takes one day to off-load and one day to on-load each ship. How many ships must the shipline company purchase?

In addition to nonnegativity there are two types of constraints that must be maintained in this problem. First, we must ensure that ships coming off of some route get assigned to some (other) route. Second, we must ensure that each route gets its required number of ships per day. Let x_{ij} be the number of ships per day coming off of route i and assigned to route j. Let b_i represent the number of ships per day required on route i.

To ensure that ships from a given route get assigned to other routes we write the constraint

$$\sum_{j=1}^{4} x_{ij} = b_i \qquad i = 1, \ldots, 4$$

To ensure that a given route gets its required number of ships we write the constraint

$$\sum_{k=1}^{4} x_{ki} = b_i \qquad i = 1, \ldots, 4$$

Computing the cost coefficients is a bit more involved. Since the objective is to minimize the total number of ships, let c_{ij} be the number of ships required to ensure a continuous daily flow of one ship coming off of route i and assigned to route j. To illustrate the computation of the c_{ij}'s, consider c_{23}. It takes one day to load a ship at Marseilles, three days to travel from Marseilles to Istanbul, one day to unload the cargo at Istanbul, and two days to head from Istanbul to Naples—a total of seven days. This implies that seven ships are needed to ensure that one ship will be assigned daily from route 2 to route 3 (why?). In particular, one ship will be on-loading at Marseilles, three ships on route from Marseilles to Istanbul, one ship off-loading at Istanbul, and two ships on route from Istanbul to Naples.

In general c_{ij} is given as follows:

$$\begin{aligned} c_{ij} = {} & \text{one day for on-loading} + \text{number of days for transit on route } i \\ & + \text{one day for off-loading} \\ & + \text{number of days for travel from the destination of} \\ & \quad \text{route } i \text{ to the origin of route } j \end{aligned}$$

Therefore the tanker scheduling problem becomes as follows.

$$\begin{aligned} \text{Minimize} \quad & 36x_{11} + 32x_{12} + 33x_{13} + 19x_{14} + 10x_{21} + 8x_{22} + 7x_{23} \\ & + 20x_{24} + 12x_{31} + 17x_{32} + 16x_{33} + 29x_{34} + 23x_{41} \\ & + 15x_{42} + 16x_{43} + 28x_{44} \\ \text{Subject to} \quad & \sum_{j=1}^{4} x_{ij} = b_i \qquad i = 1, 2, 3, 4 \\ & \sum_{k=1}^{4} x_{ki} = b_i \qquad i = 1, 2, 3, 4 \\ & x_{ij} \geqslant 0 \qquad i, j = 1, 2, 3, 4 \end{aligned}$$

where $b_1 = 3$, $b_2 = 2$, $b_3 = 1$, and $b_4 = 1$.

It can be easily seen that this is another application of the transportation problem (it will be instructive for the reader to form the origins and destinations of the corresponding transportation problem).

1.3 GEOMETRIC SOLUTION

We describe here a geometric procedure for solving a linear programming problem. Even though this method is only suitable for very small problems, it provides a great deal of insight into the linear programming problem. To be more specific, consider the following problem.

$$\begin{aligned} \text{Minimize} \quad & \mathbf{cx} \\ \text{Subject to} \quad & \mathbf{Ax} \geqslant \mathbf{b} \\ & \mathbf{x} \geqslant \mathbf{0} \end{aligned}$$

Note that the feasible region consists of all vectors $\mathbf{x}$ satisfying $\mathbf{Ax} \geqslant \mathbf{b}$ and $\mathbf{x} \geqslant \mathbf{0}$. Among all such points we wish to find a point with minimal $\mathbf{cx}$ value. Note that points with the same objective z satisfy the equation $\mathbf{cx} = z$, that is, $\sum_{j=1}^{n} c_j x_j = z$. Since z is to be minimized, then the plane (line in a two-dimensional space) $\sum_{j=1}^{n} c_j x_j = z$ must be moved parallel to itself in the direction that minimizes the objective most. This direction is $-\mathbf{c}$, and hence the plane is moved in the direction $-\mathbf{c}$ as much as possible. This process is illustrated in Figure 1.3. Note that as the optimal point $\mathbf{x}^*$ is reached, the line $c_1x_1 + c_2x_2 = z^*$, where $z^* = c_1x_1^* + c_2x_2^*$, cannot be moved farther in the direction $-\mathbf{c} = (-c_1, -c_2)$ because this will lead to only points outside the feasible region. We therefore conclude that $\mathbf{x}^*$ is indeed the optimal solution. Needless to say, for a maximization problem, the plane $\mathbf{cx} = z$ must be moved as much as possible in the direction $\mathbf{c}$.

The foregoing process is convenient for problems with two variables and is obviously impractical for problems with more than three variables. It is worth

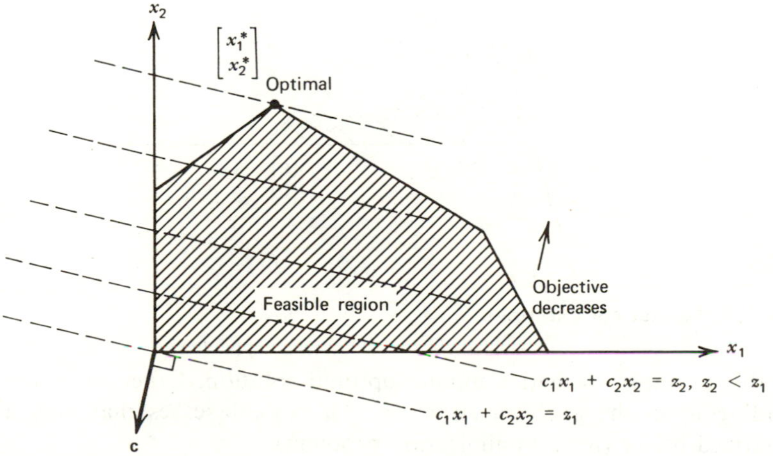

Figure 1.3. Geometric solution.

noting that the optimal point $\mathbf{x}^*$ in Figure 1.3 is one of the five corner points that are called *extreme points*. We shall show in Section 3.1 that if a linear program has a finite optimal solution, then it has an optimal corner (or extreme) solution.

Example 1.2

$$\begin{array}{ll} \text{Minimize} & -x_1 - 3x_2 \\ \text{Subject to} & x_1 + x_2 \leqslant 6 \\ & -x_1 + 2x_2 \leqslant 8 \\ & x_1, \quad x_2 \geqslant 0 \end{array}$$

The feasible region is illustrated in Figure 1.4. The first and second constraints represent points "below" lines 1 and 2 respectively. The nonnegativity constraints restrict the points to be in the first quadrant. The equations $-x_1 - 3x_2 = z$ are called the *objective contours* and are represented by dotted lines in Figure 1.4. In particular the contour $-x_1 - 3x_2 = z = 0$ passes through the origin. The contours are moved in the direction $-\mathbf{c} = (1, 3)$ as much as possible until the optimal point $(4/3, 14/3)$ is reached.

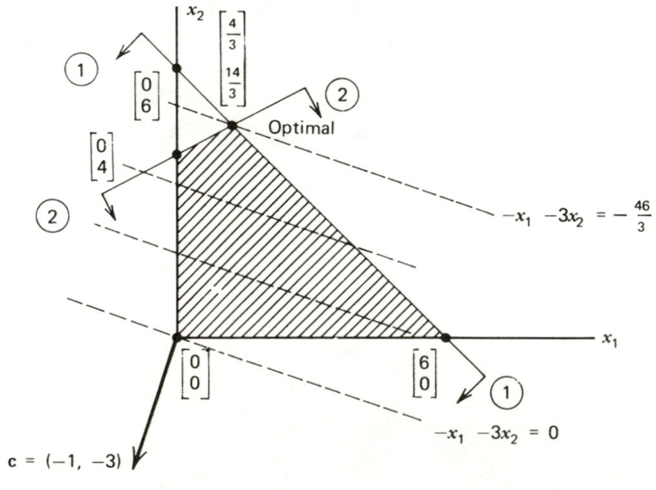

Figure 1.4. **Numerical example.**

In this example we had a unique optimal solution. Other cases may occur depending upon the problem structure. All possible cases that may arise are summarized below (for a minimization problem).

1. *Unique Finite Optimal Solution*. If the optimal finite solution is unique, then it occurs at an extreme point. Figures 1.5*a* and *b* show a unique

optimal solution. In Figure 1.5*a* the feasible region is bounded; that is, there is a ball around the origin that contains the feasible region. In Figure 1.5*b* the feasible region is not bounded. In each case, however, the unique optimal solution is finite.

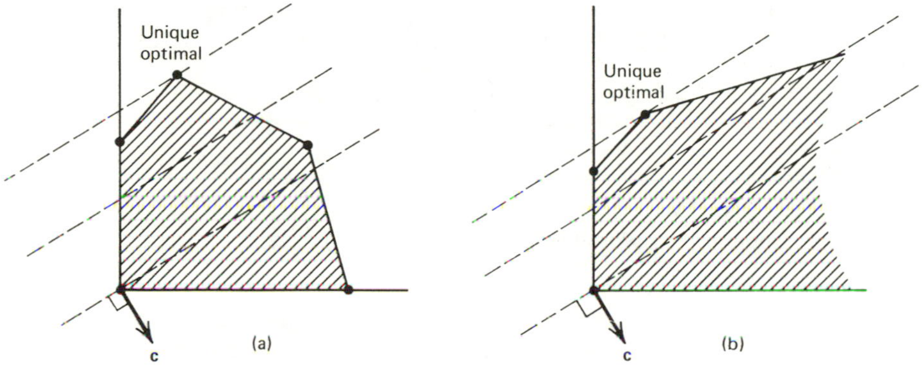

Figure 1.5. Unique finite optimal solution: (*a*) Bounded region. (*b*) Unbounded Region.

2. *Alternative Finite Optimal Solutions*. This case is illustrated in Figure 1.6. Note that in Figure 1.6*a* the feasible region is bounded. The two corner points $\mathbf{x}_1^*$ and $\mathbf{x}_2^*$ are optimal, and also any point on the line segment joining them. In Figure 1.6*b* the feasible region is unbounded but the optimal objective is finite. Any point on the "ray" with vertex $\mathbf{x}^*$ in Figure 1.6*b* is optimal.

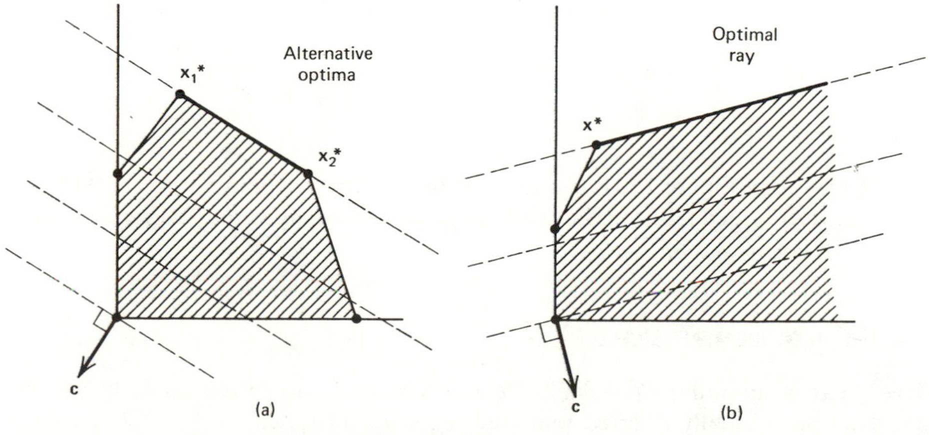

Figure 1.6. Alternative finite optima: (*a*) Bounded Region. (*b*) Unbounded Region.

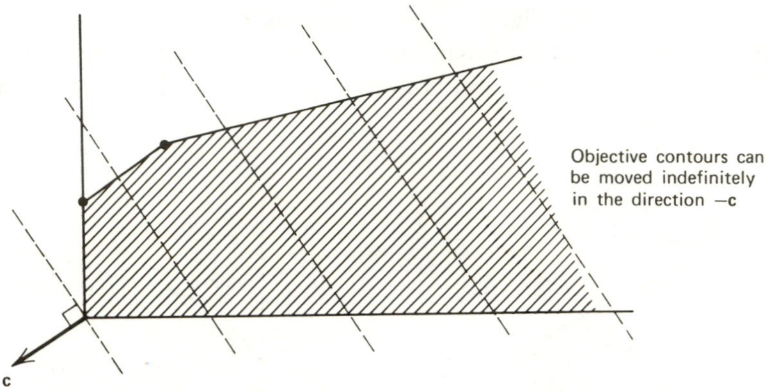
Objective contours can be moved indefinitely in the direction −c
c

Figure 1.7. **Unbounded optimal solution.**

3. *Unbounded Optimal Solution.* This case is illustrated in Figure 1.7 where both the feasible region and the optimal solution are unbounded. For a minimization problem the plane $\mathbf{cx} = z$ can be moved in the direction $-\mathbf{c}$ indefinitely while always intersecting with the feasible region. In this case the optimal objective is unbounded with value $-\infty$.
4. *Empty Feasible Region.* In this case the system of equations and/or inequalities defining the feasible region is *inconsistent*. To illustrate, consider the following problem.

$$\begin{aligned} \text{Minimize} \quad & -2x_1 + 3x_2 \\ \text{Subject to} \quad & -x_1 + 2x_2 \leq 2 \\ & 2x_1 - x_2 \leq 3 \\ & x_2 \geq 4 \\ & x_1, \quad x_2 \geq 0 \end{aligned}$$

Examining Figure 1.8, it is clear that there exists no point (x_1, x_2) satisfying the above inequalities. The problem is said to be *infeasible*, *inconsistent*, or with *empty feasible region*.

1.4 THE REQUIREMENT SPACE

The linear programming problem can be interpreted and solved geometrically in another space usually referred to as the requirement space.

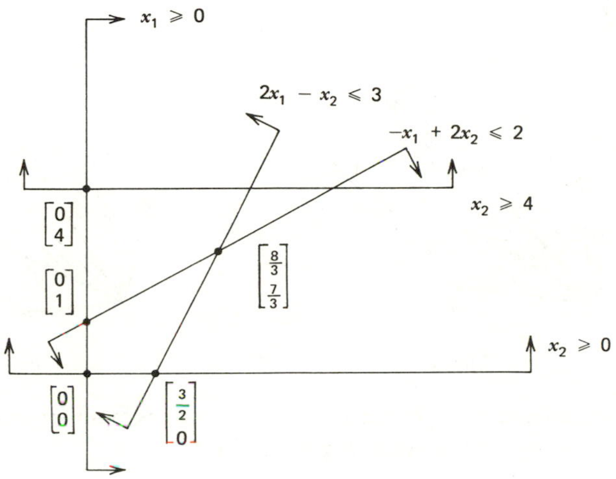

Figure 1.8. **An example of an empty feasible region.**

Interpretation of Feasibility

Consider the following linear programming problem in standard form.

$$\begin{aligned} \text{Minimize} \quad & \mathbf{cx} \\ \text{Subject to} \quad & \mathbf{Ax} = \mathbf{b} \\ & \mathbf{x} \geqslant \mathbf{0} \end{aligned}$$

where $\mathbf{A}$ is an $m \times n$ matrix whose jth column is denoted by $\mathbf{a}_j$. The problem can be rewritten as follows.

$$\begin{aligned} \text{Minimize} \quad & \sum_{j=1}^{n} c_j x_j \\ \text{Subject to} \quad & \sum_{j=1}^{n} \mathbf{a}_j x_j = \mathbf{b} \\ & x_j \geqslant 0 \qquad j = 1, 2, \ldots, n \end{aligned}$$

Given the vectors $\mathbf{a}_1, \mathbf{a}_2, \ldots, \mathbf{a}_n$, we wish to find nonnegative scalars $x_1, x_2, \ldots, x_n$ such that $\sum_{j=1}^{n} \mathbf{a}_j x_j = \mathbf{b}$ and such that $\sum_{j=1}^{n} c_j x_j$ is minimized. Note, however, that the collection of vectors of the form $\sum_{j=1}^{n} \mathbf{a}_j x_j$, where $x_1, x_2, \ldots, x_n \geqslant 0$, is the cone generated by $\mathbf{a}_1, \mathbf{a}_2, \ldots, \mathbf{a}_n$ (see Figure 1.9).

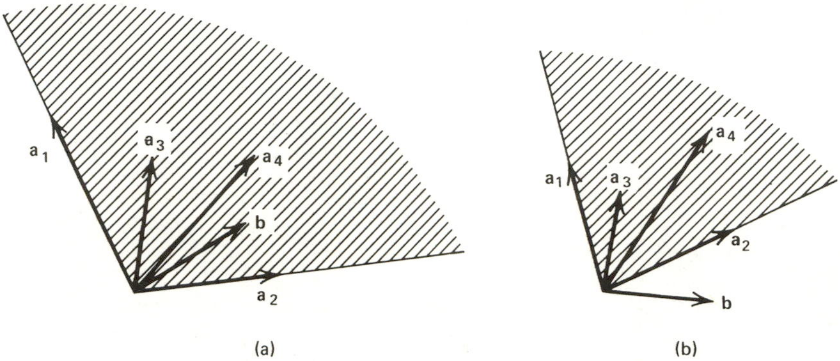

Figure 1.9. **Interpretation of feasibility in the requirement space: (*a*) Feasible region is not empty. (*b*) Feasible region is empty.**

Thus the problem has a feasible solution if and only if the vector **b** belongs to this cone. Since the vector **b** usually reflects requirements to be satisfied, Figure 1.9 above is usually referred to as the requirement space.

Example 1.3

Consider the following two systems.

System 1

$$\begin{aligned} 2x_1 + x_2 + x_3 \quad\quad &= 2 \\ -x_1 + 3x_2 \quad\quad + x_4 &= 3 \\ x_1, \; x_2, \; x_3, \; x_4 &\geqslant 0 \end{aligned}$$

System 2

$$\begin{aligned} 2x_1 + x_2 + x_3 \quad\quad &= -1 \\ -x_1 + 3x_2 \quad\quad + x_4 &= 2 \\ x_1, \; x_2, \; x_3, \; x_4 &\geqslant 0 \end{aligned}$$

Figure 1.10 shows the requirement space of both systems. For System 1 the vector **b** belongs to the cone generated by the vectors $\begin{bmatrix} 2 \\ -1 \end{bmatrix}$, $\begin{bmatrix} 1 \\ 3 \end{bmatrix}$, $\begin{bmatrix} 1 \\ 0 \end{bmatrix}$, and $\begin{bmatrix} 0 \\ 1 \end{bmatrix}$ and hence admits feasible solutions. For the second system, **b** does not belong to the cone and the system is hence inconsistent.

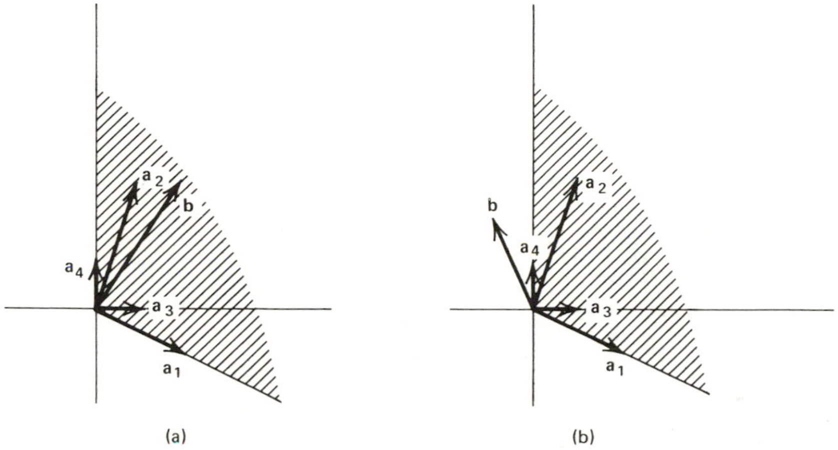

Figure 1.10. **Illustration of the requirement space: (*a*) System 1 is feasible. (*b*) System 2 is inconsistent.**

The Requirement Space and Inequality Constraints

We now illustrate the interpretation of feasibility for the inequality case. Consider the following inequality system:

$$\sum_{j=1}^{n} \mathbf{a}_j x_j \leqslant \mathbf{b}$$
$$x_j \geqslant 0 \qquad j = 1, 2, \ldots, n$$

Note that the collection of vectors $\Sigma_{j=1}^{n} \mathbf{a}_j x_j$ where $x_j \geqslant 0$ for $j = 1, 2, \ldots, n$ is the cone generated by $\mathbf{a}_1, \mathbf{a}_2, \ldots, \mathbf{a}_n$. If a feasible solution exists, then this cone must intersect the collection of vectors that are less than or equal to the requirement vector **b**. Figure 1.11 shows both a feasible system and an infeasible system.

Optimality

We have seen above that the system $\Sigma_{j=1}^{n} \mathbf{a}_j x_j = \mathbf{b}$ and $x_j \geqslant 0$ for $j = 1, 2, \ldots, n$ is feasible if and only if **b** belongs to the cone generated by $\mathbf{a}_1, \mathbf{a}_2, \ldots, \mathbf{a}_n$. The variables $x_1, x_2, \ldots, x_n$ must be chosen so that feasibility is satisfied and also $\Sigma_{j=1}^{n} c_j x_j$ is minimized. Therefore the linear programming problem can be stated as follows. Find nonnegative $x_1, x_2, \ldots, x_n$ such that

$$\begin{bmatrix} c_1 \\ \mathbf{a}_1 \end{bmatrix} x_1 + \begin{bmatrix} c_2 \\ \mathbf{a}_2 \end{bmatrix} x_2 + \cdots + \begin{bmatrix} c_n \\ \mathbf{a}_n \end{bmatrix} x_n = \begin{bmatrix} z \\ \mathbf{b} \end{bmatrix}$$

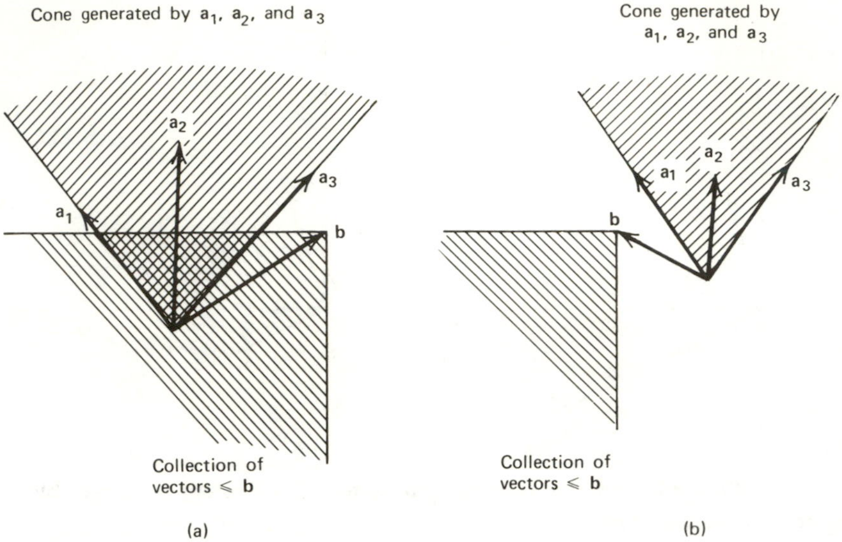

Figure 1.11. **Requirement space and inequality constraints: (*a*) System is feasible. (*b*) System is infeasible.**

where the objective z is to be minimized. In other words we seek to represent the vector $\begin{bmatrix} z \\ \mathbf{b} \end{bmatrix}$ for the smallest possible z, in the cone spanned by the vectors $\begin{bmatrix} c_1 \\ \mathbf{a}_1 \end{bmatrix}$, $\begin{bmatrix} c_2 \\ \mathbf{a}_2 \end{bmatrix}, \ldots,$ and $\begin{bmatrix} c_n \\ \mathbf{a}_n \end{bmatrix}$. The reader should note that the price we must pay for including the objective function explicitly in the requirement space is to increase the dimensionality from m to $m + 1$.

Example 1.4

$$\begin{array}{ll} \text{Minimize} & -2x_1 - 3x_2 \\ \text{Subject to} & x_1 + 2x_2 \leqslant 2 \\ & x_1,\quad x_2 \geqslant 0 \end{array}$$

Add the slack variable $x_3 \geqslant 0$. The problem is then to choose x_1, x_2, $x_3 \geqslant 0$ such that

$$\begin{bmatrix} -2 \\ 1 \end{bmatrix} x_1 + \begin{bmatrix} -3 \\ 2 \end{bmatrix} x_2 + \begin{bmatrix} 0 \\ 1 \end{bmatrix} x_3 = \begin{bmatrix} z \\ 2 \end{bmatrix}$$

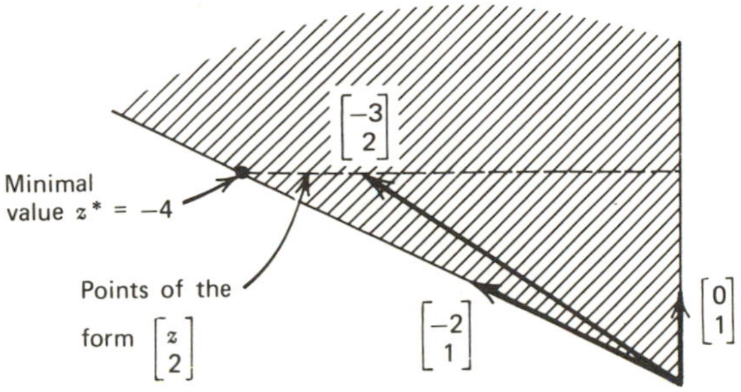

Figure 1.12. **Optimal bounded objective in the requirement space.**

where z is to be minimized. The cone generated by the vectors $\begin{bmatrix} -2 \\ 1 \end{bmatrix}$, $\begin{bmatrix} -3 \\ 2 \end{bmatrix}$, and $\begin{bmatrix} 0 \\ 1 \end{bmatrix}$ is shown in Figure 1.12. We want to choose a vector $\begin{bmatrix} z \\ 2 \end{bmatrix}$ in this cone with minimal z. This gives the optimal solution $z^* = -4$ with $x_1^* = 2$ and $x_2^* = x_3^* = 0$.

Example 1.5

Minimize $-2x_1 - 3x_2$

Subject to $x_1 + 2x_2 \geqslant 2$

$x_1, \quad x_2 \geqslant 0$

Obviously the optimal solution is unbounded. We illustrate this fact in the requirement space. Subtracting the slack variable $x_3 \geqslant 0$, the problem can be restated as follows: Find $x_1, x_2, x_3 \geqslant 0$ such that

$$\begin{bmatrix} -2 \\ 1 \end{bmatrix} x_1 + \begin{bmatrix} -3 \\ 2 \end{bmatrix} x_2 + \begin{bmatrix} 0 \\ -1 \end{bmatrix} x_3 = \begin{bmatrix} z \\ 2 \end{bmatrix}$$

such that z is minimized. The cone generated by $\begin{bmatrix} -2 \\ 1 \end{bmatrix}$, $\begin{bmatrix} -3 \\ 2 \end{bmatrix}$, and $\begin{bmatrix} 0 \\ -1 \end{bmatrix}$ is shown in Figure 1.13. We want to choose $\begin{bmatrix} z \\ 2 \end{bmatrix}$ in this cone with smallest possible z. Note that we can find points of the form $\begin{bmatrix} z \\ 2 \end{bmatrix}$ in the cone with arbitrarily small z. Therefore the optimal solution is unbounded with value $-\infty$.

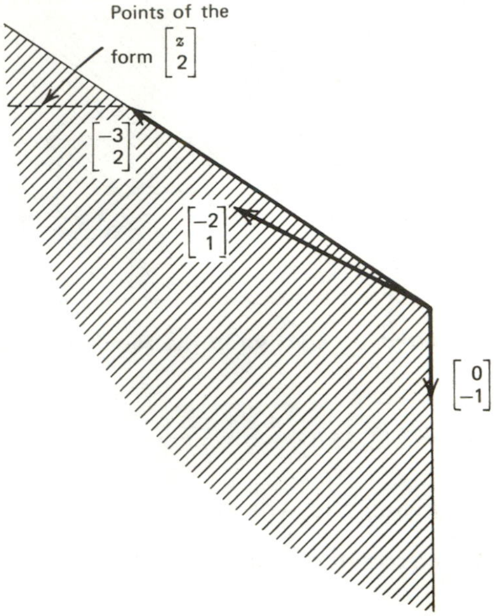

Figure 1.13. **Optimal unbounded objective in the requirement space.**

1.5 NOTATION

Throughout the text we shall utilize notation that is insofar as possible consistent with generally accepted standards for the field of mathematics and operations research. In this section we indicate some of the notation that may require special attention either because of its infrequency of use in linear programming literature or else because of the possibility of confusion with other terms.

In Chapter 2 we shall present review material on vectors and matrices. We indicate vectors by lowercase, boldface Greek or Roman letters or numerals, such as $\mathbf{a}$, $\mathbf{b}$, $\mathbf{x}$, $\mathbf{1}$, $\boldsymbol{\lambda}$; matrices by uppercase, boldface Greek or Roman letters, such as $\mathbf{A}$, $\mathbf{B}$, $\mathbf{N}$, $\boldsymbol{\Phi}$; and all scalars by Greek or Roman letters or numerals that are not boldface, such as a, b, 1, ϵ. Column vectors are generally denoted by subscripts, such as $\mathbf{a}_j$, unless clear in the context. When special emphasis is required, row vectors are indicated by superscripts, such as $\mathbf{a}^i$.

In calculus, the partial derivative, indicated by $\partial z / \partial x$, represents the expected rate of change in the variable z with respect to a unit change in the variable x. We shall also utilize the symbol $\partial z / \partial \mathbf{x}$ to indicate the vector of partial derivatives of z with respect to each element of the vector $\mathbf{x}$. That is, if $\mathbf{x} = (x_1, x_2, \ldots, x_n)$, then

$$\frac{\partial z}{\partial \mathbf{x}} = \left(\frac{\partial z}{\partial x_1}, \frac{\partial z}{\partial x_2}, \ldots, \frac{\partial z}{\partial x_n} \right)$$

Also, we shall sometimes consider the partial derivative of one vector with respect to another vector, such as $\partial\mathbf{y}/\partial\mathbf{x}$. If $\mathbf{y} = (y_1, y_2, \ldots, y_m)$ and $\mathbf{x} = (x_1, x_2, \ldots, x_n)$ then

$$\frac{\partial\mathbf{y}}{\partial\mathbf{x}} = \begin{bmatrix} \frac{\partial y_1}{\partial x_1} & \frac{\partial y_1}{\partial x_2} & \cdots & \frac{\partial y_1}{\partial x_n} \\ \frac{\partial y_2}{\partial x_1} & \frac{\partial y_2}{\partial x_2} & \cdots & \frac{\partial y_2}{\partial x_n} \\ \vdots & \vdots & & \vdots \\ \frac{\partial y_m}{\partial x_1} & \frac{\partial y_m}{\partial x_2} & \cdots & \frac{\partial y_m}{\partial x_n} \end{bmatrix}$$

Note that if z is a function of the vector $\mathbf{x} = (x_1, x_2, \ldots, x_n)$, then $\partial z/\partial\mathbf{x}$ is called the *gradient*.

We shall, when necessary, use (a, b) to refer to the *open interval* $a < x < b$, and $[a, b]$ to refer to the *closed interval* $a \leqslant x \leqslant b$. Finally we shall utilize the standard set operators $\cup$, $\cap$, $\subset$, and $\in$ to refer to union, intersection, set inclusion, and set membership respectively.

EXERCISES

1.1 An agricultural mill manufactures feed for cattle, sheep, and chickens. This is done by mixing the following main ingredients: corn, limestone, soybeans, and fish meal. These ingredients contain the following nutrients: vitamins, protein, calcium, and crude fat. The contents of the nutrients in each kilogram of the ingredients is summarized below.

	NUTRIENT			
INGREDIENT	VITAMINS	PROTEIN	CALCIUM	CRUDE FAT
Corn	8	10	6	8
Limestone	6	5	10	6
Soybeans	10	12	6	6
Fish meal	4	8	6	9

The mill contracted to produce 10, 6, and 8 (metric) tons of cattle feed, sheep feed, and chicken feed. Because of shortages, a limited amount of the ingredients is available—namely, 6 tons of corn, 10 tons of limestone, 4 tons of soybeans, and 5 tons of fish meal. The price per kilogram of these ingredients is respectively \$0.20, \$0.12, \$0.24, and \$0.12. The minimal and maximal units of the various nutrients that are permitted is summarized below for a kilogram of the cattle feed, the sheep feed, and the chicken feed.

	NUTRIENT							
	VITAMINS		PROTEIN		CALCIUM		CRUDE FAT	
PRODUCT	MIN	MAX	MIN	MAX	MIN	MAX	MIN	MAX
Cattle feed	6	∞	6	∞	7	∞	4	8
Sheep feed	6	∞	6	∞	6	∞	4	6
Chicken feed	4	6	6	∞	6	∞	4	6

Formulate this feed-mix problem so that the total cost is minimized.

1.2 The technical staff of a hospital wishes to develop a computerized menu-planning system. To start with, a lunch menu is sought. The menu is divided into three major categories: vegetables, meat, and dessert. At least one equivalent serving of each category is desired. The cost per serving of some suggested items as well as their content of carbohydrates, vitamins, protein, and fats is summarized below.

	CARBOHYDRATES	VITAMINS	PROTEIN	FATS	COST IN \$/SERVING
Vegetables					
Peas	1	3	1	0	0.10
Green beans	1	5	2	0	0.12
Okra	1	5	1	0	0.13
Corn	2	6	1	2	0.09
Macaroni	4	2	1	1	0.10
Rice	5	1	1	1	0.07
Meat					
Chicken	2	1	3	1	0.70
Beef	3	8	5	2	1.20
Fish	3	6	6	1	0.63
Dessert					
Orange	1	3	1	0	0.28
Apple	1	2	0	0	0.42
Pudding	1	0	0	0	0.15
Jello	1	0	0	0	0.12

Suppose that the minimal requirements of carbohydrates, vitamins, protein, and fats per meal are respectively 5, 10, 10, and 2.

a. Formulate the menu-planning problem as a linear program.

b. Many practical aspects have been left out in the foregoing model. These include planning the breakfast, lunch, and supper menus together, weekly planning so that different varieties of food are used, and special menus for patients on particular diets. Discuss in detail how can these aspects be incorporated in a comprehensive menu-planning system.

1.3 Consider the problem of locating a new machine to an existing layout consisting of four machines. These machines are located at the following x_1 and x_2 coordinates: $\begin{pmatrix} 3 \\ 0 \end{pmatrix}$, $\begin{pmatrix} 0 \\ -3 \end{pmatrix}$, $\begin{pmatrix} -2 \\ 1 \end{pmatrix}$, and $\begin{pmatrix} 1 \\ 4 \end{pmatrix}$. Let the coordinates of the new machine be $\begin{pmatrix} x_1 \\ x_2 \end{pmatrix}$. Formulate the problem of finding an optimal location as a *linear program* for each of the following cases.

a. The sum of the distances from the new machine to the four machines is minimized. Use the street distance; for example, the distance from $\begin{pmatrix} x_1 \\ x_2 \end{pmatrix}$ to the first machine located at $\begin{pmatrix} 3 \\ 0 \end{pmatrix}$ is $|x_1 - 3| + |x_2|$.

b. Because of various amounts of flow between the new machine and the existing machines, reformulate the problem where the sum of the weighted distances is minimized, where the weights corresponding to the four machines are 5, 7, 3, and 1 respectively.

c. In order to avoid congestion, suppose that the new machine must be located in the square $\{(x_1, x_2) : -1 \leqslant x_1 \leqslant 2, 0 \leqslant x_2 \leqslant 1\}$. Formulate (a) and (b) with this added restriction.

d. Suppose that the new machine must be located so that its distance from the first machine does not exceed 3/2. Formulate the problem with this added restriction.

1.4 Consider the following problem of launching a rocket to a fixed altitude b in a given time T, while expending a minimum amount of fuel. Let $u(t)$ be the acceleration force exerted at time t and let $y(t)$ be the rocket altitude at time t. The problem can be formulated as follows.

$$\begin{array}{ll} \text{Minimize} & \displaystyle\int_0^T |u(t)|\, dt \\ \text{Subject to} & \ddot{y}(t) = u(t) - g \\ & y(T) = b \\ & y(t) \geqslant 0 \qquad t \in [0, T] \end{array}$$

where g is the gravitational force and $\ddot{y}$ is the second derivative of the altitude y. Discretize the problem and reformulate it as a linear programming problem. In particular formulate the problem where $T = 10$, $b = 15$, and $g = 32$. (*Hint*. Replace the integration by proper summation and the differentiation by difference equations. Make the change of variable $|u_j| = x_j$ and note that $x_j \geqslant u_j$ and $x_j \geqslant -u_j$).

1.5 A company wishes to plan its production of two items with seasonal demands over a 12-month period. The monthly demand of item 1 is 100,000 units during the months of October, November, and December; 10,000 units during the months of January, February, March, and April; and 30,000 units during the remaining months. The demand of item 2 is 50,000 during the months of October through February and 15,000 during the remaining months. Suppose that the unit product cost of items 1 and 2 is \$5.00 and \$8.00 respectively, provided that these were manufactured prior to June. After June, the unit costs are reduced to \$4.50 and \$7.00 because of the installation of an improved manufacturing system. The total units of items 1 and 2 that can be manufactured during any particular month cannot exceed 120,000. Furthermore, each unit of item 1 occupies 2 cubic feet and each unit of item 2 occupies 4 cubic feet of inventory. Suppose that the maximum inventory space allocated to these items is 150,000 cubic feet and that the holding cost per cubic foot during any month is \$0.10. Formulate the production scheduling problem so that the total production and inventory costs are minimized.

1.6 Fred has \$2200 to invest over the next five years. At the beginning of each year he can invest money in one- or two-year time deposits. The bank pays 8 percent interest on one-year time deposits and 17 percent (total) on two-year time deposits. In addition, West World Limited will offer three-year certificates at the beginning of the second year. These certificates will return 27 percent (total). If Fred reinvests his money available every year, formulate a linear program to show him how to maximize his total cash on hand at the end of the fifth year.

1.7 A steel manufacturer produces four sizes of I beams: small, medium, large, and extra large. These beams can be produced on any one of three machine types: A, B, and C. The lengths in feet of the I beams that can be produced on the machines per hour are summarized below.

	MACHINE		
BEAM	A	B	C
Small	300	600	800
Medium	250	400	700
Large	200	350	600
Extra large	100	200	300

Assume that each machine can be used up to 50 hours per week and that the hourly operating costs of these machines are respectively \$30.00, \$50.00, and \$80.00. Further suppose that 10,000, 8,000, 6,000, and 6,000 feet of the different-size I beams are required weekly. Formulate the machine scheduling problem as a linear program.

1.8 A cheese firm produces two types of cheese: Swiss cheese and sharp cheese. The firm has 60 experienced workers and would like to increase its working force to 90 workers during the next eight weeks. Each experienced worker can train 3 new employees in a period of two weeks during which the workers involved virtually produce nothing. It takes one hour to produce 10 pounds of Swiss cheese and one hour to produce 6 pounds of sharp cheese. A work week is 40 hours. The weekly demands (in 1000 pounds) are summarized below.

	WEEK							
CHEESE TYPE	1	2	3	4	5	6	7	8
Swiss cheese	12	12	12	16	16	20	20	20
Sharp cheese	8	8	10	10	12	12	12	12

Suppose that a trainee receives full salary as an experienced worker. Further suppose that overaging destroys the flavor of the cheese so that inventory is limited to one week. How should the company hire and train its new employees so that the labor cost is minimized? Formulate the problem as a linear program.

1.9 A lathe is used to reduce the diameter of a steel shaft whose length is 36 in. from 14 in. to 12 in. The speed x_1 (in revolutions per minute), the depth feed x_2 (in inches per minute), and the length feed x_3 (in inches per minute) must be determined. The duration of the cut is given by $36/x_2x_3$. The compression and side stresses exerted on the cutting tool are given by $30x_1 + 4000x_2$ and $40x_1 + 6000x_2 + 6000x_3$ pounds per square inch respectively. The temperature (in degrees Fahrenheit) at the tip of the cutting tool is $200 + 0.5x_1 + 150(x_2 + x_3)$. The maximum compression stress, side stress, and temperature allowed are 150,000 psi, 100,000 psi, and 800°F. It is desired to determine the speed (which must be in the range from 600 to 800 rpm), the depth feed, and the length feed such that the duration of the cut is minimized. In order to use a linear model the following approximation is made. Since $36/x_2x_3$ is minimized if and only if x_2x_3 is maximized, it was decided to replace the objective by the maximization of the minimum of x_2 and x_3. Formulate the problem as a linear model and comment on the validity of the approximation used in the objective function. (We ask the reader to solve this problem in Exercise 3.22.)

1.10 An oil refinery can buy two types of oil: light crude oil and heavy crude oil. The cost per barrel of these types is respectively \$11 and \$9. The following quantities of gasoline, kerosene, and jet fuel are produced per barrel of each type of oil.

	GASOLINE	KEROSENE	JET FUEL
Light crude oil	0.4	0.2	0.35
Heavy crude oil	0.32	0.4	0.2

Note that 5% and 8% of the crude are lost respectively during the refining process. The refinery has contracted to deliver 1 million barrels of gasoline, 400,000 barrels of kerosene, and 250,000 barrels of jet fuel. Formulate the problem of finding the number of barrels of each crude oil that satisfy the demand and minimize the total cost as a linear program. (We ask the reader to solve this problem in Exercise 3.23.)

1.11 A company manufactures an assembly consisting of a frame, a shaft, and a ball bearing. The company manufactures the shafts and frames but purchases the ball bearings from a ball bearing manufacturer. Each shaft must be processed on a forging machine, a lathe, and a grinder. These operations require 0.5 hours, 0.2 hours, and 0.3 hours per shaft respectively. Each frame requires 0.8 hours on a forging machine, 0.1 hours on a drilling machine, 0.3 hours on a milling machine, and 0.5 hours on a grinder. The company has 5 lathes, 10 grinders, 20 forging machines, 3 drillers, and 6 millers. Assume that each machine operates a maximum of 2400 hours per year. Formulate the problem of finding the maximum number of assembled components that can be produced as a linear program. (We ask the reader to solve this problem as Exercise 3.40.)

1.12 A television set manufacturing firm has to decide on the mix of color and black-and-white TV's to be produced. A market research indicates that at most 1000 units and 4000 units of color and black-and-white TV's can be sold per month. The maximum number of man-hours available is 50,000 per month. A color TV requires 20 man-hours and a black-and-white TV requires 15 man-hours. The unit profits of the color and black-and-white TV's are \$60 and \$30 respectively. It is desired to find the number of units of each TV type that the firm must produce in order to maximize its profit. Formulate the problem. (We ask the reader to solve this problem in Exercise 3.41.)

1.13 A manufacturer of plastics is planning to blend a new product from four chemical compounds. These compounds are mainly composed of three elements A, B, and C. The composition and unit cost of these chemicals are shown below.

CHEMICAL COMPOUND	1	2	3	4
Percentage of A	30	20	40	20
Percentage of B	20	60	30	40
Percentage of C	40	15	25	30
Cost/kilogram	20	30	20	15

The new product consists of 20% element A, at least 30% element B, and at least 20% element C. Owing to side effects of compounds 1 and 2, they must not exceed 30% and 40% of the content of the new product. Formulate the problem of finding the least costly way of blending as a linear program. (We ask the reader to solve this problem in Exercise 5.25.)

1.14 A production manager is planning the scheduling of three products on four machines. Each product can be manufactured on each of the machines. The unit production costs (in \$) are summarized below.

	MACHINE			
PRODUCT	1	2	3	4
1	4	4	5	7
2	6	7	5	6
3	12	10	8	11

The time (in hours) required to produce each unit product on each of the machines is summarized below.

	MACHINE			
PRODUCT	1	2	3	4
1	0.3	0.25	0.2	0.2
2	0.2	0.3	0.2	0.25
3	0.8	0.6	0.6	0.5

Suppose that 4000, 5000, and 3000 units of the products are required, and that the available machine-hours are 1500, 1200, 1500, and 2000 respectively. Formulate the scheduling problem as a linear program. (We ask the reader to solve this problem in Exercise 8.19.)

1.15 A furniture manufacturer has three plants, which need 500, 700, and 600 tons of lumber weekly. The manufacturer may purchase the lumber from three lumber companies. The first two lumber manufacturers virtually have unlimited supply, and because of other commitments the third manufacturer cannot ship more than 500 tons weekly. The first lumber manufacturer uses rail for transportation and there is no limit on the tonnage that can be shipped to the furniture facilities. On the other hand, the last two lumber companies use trucks that limit the maximum tonnage that can be shipped to any of the furniture companies to 200 tons. The following table gives the transportation cost from the lumber companies to the furniture manufacturers ($ per ton).

LUMBER COMPANY	FURNITURE FACILITY 1	2	3
1	2	3	5
2	2.5	4	4.8
3	3	3.6	3.2

Formulate the problem as a linear program. (We ask the reader to solve this problem in Exercise 8.32.)

1.16 A corporation has $30 million available for the coming year to allocate to its three subsidiaries. Because of commitments to stability of personnel employment and for other reasons, the corporation has established a minimal level of funding for each subsidiary. These funding levels are $3 million, $5 million, and $8 million respectively. Owing to the nature of its operation, subsidiary 2 cannot utilize more than $17 million without major new capital expansion. The corporation is unwilling to undertake such expansion at this time. Each subsidiary has the opportunity to conduct various projects with the funds it receives. A rate of return (as a percent of investment) has been established for each project. In addition, certain of the projects permit only limited investment. The data for each project are given below.

SUBSIDIARY	PROJECT	RATE OF RETURN	UPPER LIMIT OF INVESTMENT
1	1	8%	$6 million
	2	6%	$5 million
	3	7%	$9 million
2	4	5%	$7 million
	5	8%	$10 million
	6	9%	$4 million
3	7	10%	$6 million
	8	6%	$3 million

Formulate this problem as a linear program. (We ask the reader to solve this problem in Exercise 9.46.)

1.17 A ten-acre slum in New York City is to be cleared. The officials of the city must decide on the redevelopment plan. Two housing plans are to be considered: low-income housing and middle-income housing. The types of housing can be developed at 20 and 15 units per acre respectively. The unit costs of the low- and middle-income housing are \$13,000 and \$18,000. The lower and upper limits set by the officials on the number of low-income housing units are 60 and 100. Similarly, the number of middle-income housing units must lie between 30 and 70. The combined maximum housing market potential is estimated to be 150 (which is less than the sum of the individual market limits due to the overlap between the two markets). The total mortgage committed to the renewal plan is not to exceed \$2 million. Finally, it was suggested by the architectural adviser that the number of low-income housing units be at least 50 units greater than one-half the number of middle-income housing units.

a. Formulate the minimal cost renewal planning problem as a linear program and solve it graphically.

b. Resolve the problem if the objective is to maximize the number of houses being constructed.

1.18 A region is divided into m residential and central business districts. Each district is represented by a node and the nodes are inter-connected by links representing major routes. People living in the various districts go to their business in the same and/or at other districts so that each node attracts and/or generates a number of trips. In particular, let a_{ij} be the number of trips generated at node i with final destination at node j and let b_{ij} be the time to travel from node i to node j. It is desired to determine the routes to be taken by the people living in the region.

a. Illustrate the problem by a suitable network.

b. Develop some measures of effectiveness for this *traffic assignment problem* and for each devise a suitable model.

1.19 Consider the problem of scheduling court hearings over a planning horizon consisting of n periods. Let b_j be the available judge-hours at period j, h_{ij} be the number of hearings of class i arriving at period j, and a_i be the number of judge-hours required to process a hearing of class i. It is desired to determine the number of hearings x_{ij} of class i processed at period j.

a. Formulate the problem as a linear program.

b. Modify the model in part a so that hearings would not be delayed for too long.

1.20 Suppose that there are m sources which generate waste and n disposal sites. The amount of waste generated at source i is a_i and the capacity of site j is b_j. It is desired to select appropriate transfer facilities from among K candidate facilities. Potential transfer facility k has fixed cost f_k, capacity

q_k, and unit processing cost α_k per ton of waste. Let c_{ik} and $\bar{c}_{kj}$ be the unit shipping costs from source i to transfer station k and from transfer station k to disposal site j respectively. The problem is to choose the transfer facilities and the shipping pattern that minimize the total capital and operating costs of the transfer stations plus the transportation costs. Formulate this *distribution* problem.
(*Hint*. Let y_k be 1 if transfer station k is selected and 0 otherwise.)

1.21 A governmental planning agency wishes to determine the sources of purchase of fuel for use by n depots from among m bidders. Suppose that the maximum quantity offered by bidder i is a_i gallons and that the demand of depot j is b_j gallons. Let c_{ij} be the unit delivery cost of bidder i to the jth depot.

a. Formulate the problem of minimizing the total purchasing cost as a linear program.
b. Suppose that a discount in the unit delivery cost is offered by bidder i if the ordered quantity exceeds the level α_i. How would you incorporate this modification in the model developed in part a?

1.22 The quality of air in an industrial region largely depends on the effluent emission from n plants. Each plant can use m different types of fuel. Suppose that the total energy needed at plant j is b_j british thermal units per day and that c_{ij} is the effluent emission per ton of fuel type i at plant j. Further suppose that fuel type i costs c_i dollars per ton and that each ton of this fuel type generates α_{ij} british thermal units at plant j. The level of air pollution in the region is not to exceed b micro-grams per cubic meter. Finally, let γ_j be a meteorological parameter relating emissions at plant j to air quality at the region.

a. Formulate the problem of determining the mix of fuels to be used at each plant.
b. How would you incorporate technology constraints that prohibit the use of certain mixes of fuel at certain plants?
c. How could you ensure equity among the plants?

1.23 Consider the following linear programming problem.

$$
\begin{aligned}
\text{Minimize} \quad & x_1 - 2x_2 - 3x_3 \\
\text{Subject to} \quad & x_1 + x_2 + x_3 \leqslant 6 \\
& x_1 + 2x_2 + 4x_3 \geqslant 12 \\
& x_1 - x_2 + x_3 \geqslant 2 \\
& x_1, \; x_2, \; x_3 \quad \text{unrestricted}
\end{aligned}
$$

a. Reformulate the problem so that it is in standard format.
b. Reformulate the problem so that it is in canonical format.
c. Convert the problem into a maximization problem.

1.24 Consider the following problem.

$$\begin{array}{ll} \text{Maximize} & 2x_1+5x_2 \\ \text{Subject to} & x_1+2x_2\leqslant 16 \\ & 2x_1+\ x_2\leqslant 12 \\ & x_1,\ x_2\geqslant 0 \end{array}$$

a. Sketch the feasible region in the (x_1, x_2) space.
b. Identify the regions in the (x_1, x_2) space where the slack variables x_3 and x_4 are equal to zero.
c. Solve the problem geometrically.
d. Draw the requirement space and interpret feasibility.

1.25 Sketch the feasible region of the set $\{\mathbf{x} : \mathbf{Ax} \leqslant \mathbf{b}\}$ where $\mathbf{A}$ and $\mathbf{b}$ are given below. In each case state whether the feasible region is empty or not, and whether it is bounded or not.

a. $\mathbf{A} = \begin{bmatrix} 1 & 1 \\ 2 & -1 \\ 0 & 1 \end{bmatrix} \qquad \mathbf{b} = \begin{bmatrix} 6 \\ 6 \\ 2 \end{bmatrix}$

b. $\mathbf{A} = \begin{bmatrix} -1 & 0 \\ 0 & -1 \\ 2 & 3 \\ 1 & -1 \end{bmatrix} \qquad \mathbf{b} = \begin{bmatrix} 0 \\ 0 \\ 12 \\ 5 \end{bmatrix}$

c. $\mathbf{A} = \begin{bmatrix} 1 & 1 \\ -1 & -2 \\ -1 & 0 \end{bmatrix} \qquad \mathbf{b} = \begin{bmatrix} 4 \\ -12 \\ 0 \end{bmatrix}$

1.26 Consider the following problem.

$$\begin{array}{ll} \text{Maximize} & 2x_1+3x_2 \\ \text{Subject to} & x_1+\ x_2\leqslant 2 \\ & 4x_1+6x_2\leqslant 9 \\ & x_1,\ x_2\geqslant 0 \end{array}$$

a. Sketch the feasible region.
b. Find two alternative optimal extreme (corner) points.
c. Find an infinite class of optimal solutions.

1.27 Consider the following problem.

$$\begin{aligned} \text{Maximize} \quad & 3x_1 + x_2 \\ \text{Subject to} \quad & -x_1 + 2x_2 \leqslant 6 \\ & x_2 \leqslant 4 \end{aligned}$$

a. Sketch the feasible region.
b. Verify that the problem has an unbounded optimal solution.

1.28 Consider the following problem.

$$\begin{aligned} \text{Minimize} \quad & -x_1 - x_2 + 2x_3 + x_4 \\ \text{Subject to} \quad & x_1 + x_2 + x_3 + x_4 \geqslant 6 \\ & x_1 - x_2 - 2x_3 + x_4 \leqslant 4 \\ & x_1, x_2, x_3, x_4 \geqslant 0 \end{aligned}$$

a. Introduce slack variables and draw the requirement space.
b. Interpret feasibility in the requirement space.
c. You are told that an optimal solution can be obtained by having at most two positive variables while all other variables are set at zero. Utilize this statement and the requirement space to find an optimal solution.

1.29 Consider the problem: Minimize $\mathbf{cx}$ subject to $\mathbf{Ax} \geqslant \mathbf{b}$, $\mathbf{x} \geqslant \mathbf{0}$. Suppose that one component of the vector $\mathbf{b}$, say b_i, is increased by one unit to $b_i + 1$.
a. What happens to the feasible region?
b. What happens to the optimal objective?

1.30 From the results of the previous problem, assuming $\partial z^*/\partial b_i$ exists, is it $\leqslant 0$, $= 0$, or $\geqslant 0$?

1.31 Solve Exercises 1.29 and 1.30 above if the restrictions $\mathbf{Ax} \geqslant \mathbf{b}$ are replaced by $\mathbf{Ax} \leqslant \mathbf{b}$.

1.32 Consider the problem: Minimize $\mathbf{cx}$ subject to $\mathbf{Ax} \geqslant \mathbf{b}$, $\mathbf{x} \geqslant \mathbf{0}$. Suppose that a new constraint, $m + 1$, is added to the problem.
a. What happens to the feasible region?
b. What happens to the optimal objective z^*?

1.33 Consider the problem: Minimize $\mathbf{cx}$ subject to $\mathbf{Ax} \geqslant \mathbf{b}$, $\mathbf{x} \geqslant \mathbf{0}$. Suppose that a new variable, $n + 1$, is added to the problem.
a. What happens to the feasible region?
b. What happens to the optimal objective z^*?

1.34 Consider the problem: Minimize $\mathbf{cx}$ subject to $\mathbf{Ax} \geqslant \mathbf{b}$, $\mathbf{x} \geqslant \mathbf{0}$. Suppose that a constraint, say constraint i, is deleted from the problem.
   a. What happens to the feasible region?
   b. What happens to the optimal objective z^*?

1.35 Consider the problem: Minimize $\mathbf{cx}$ subject to $\mathbf{Ax} \geqslant \mathbf{b}$, $\mathbf{x} \geqslant \mathbf{0}$. Suppose that a variable, say x_k, is deleted from the problem.
   a. What happens to the feasible region?
   b. What happens to the optimal objective z^*?

NOTES AND REFERENCES

1. Linear programming and the simplex method were developed by Dantzig in 1947 in connection with planning of the military. A great deal of work has influenced the development of linear programming, including World War II developments and the need for scheduling of supply and maintenance operations as well as the need for training of flying personnel, Leontief's input-output model [311], Von Neumann's Equilibrium Model [451], Koopman's Model of Transportation [290], the Hitchcock transportation problem [240], the work of Kantorovich [271], Von Neumann-Morgenstern game theory [454], and the rapid progress in electronic computing machines.
2. Linear programming has found numerous applications in the military, the government, industry, and urban engineering.
3. Linear programming is also frequently used as a part of general computational schemes for solving nonlinear programming problems, discrete programs, combinatorial problems, problems of optimal control, and programming under uncertainty.

TWO: RESULTS FROM LINEAR ALGEBRA AND CONVEX ANALYSIS

In this chapter we review some basic results from linear algebra and convex analysis. These results will be used throughout the book. The reader may skip any sections of this chapter, according to his familiarity with the subject material. Sections 2.1 and 2.2 review some elementary results from vector and matrix algebra. In Section 2.3 we discuss solvability of a system of linear equations and introduce the important notion of basic solutions. The remaining sections discuss results from convex analysis, including the notions of convex sets, convex and concave functions, convex cones, hyperplanes, and polyhedral sets. The sections on polyhedral sets and their representation in terms of extreme points and extreme directions are very important in linear programming, and hence they deserve a thorough study. The last section treats Farkas's Theorem, which will be used to prove the optimality criteria in linear programming.

2.1 VECTORS

An n vector is a row or a column array of n numbers. For example, $\mathbf{a} = (1, 2, 3, -1, 4)$ is a row vector of size $n = 5$, and $\mathbf{a} = \begin{pmatrix} 1 \\ -1 \end{pmatrix}$ is a column vector of size $n = 2$. Row and column vectors are denoted by lowercase boldface letters, such as **a**, **b**, **c**. Whether a vector is a row or a column vector will be clear from the context. Figure 2.1 shows some vectors in a two-dimensional space. Each vector can be represented by a point or by a line from the origin to the point, with an arrowhead at the end point of the line.

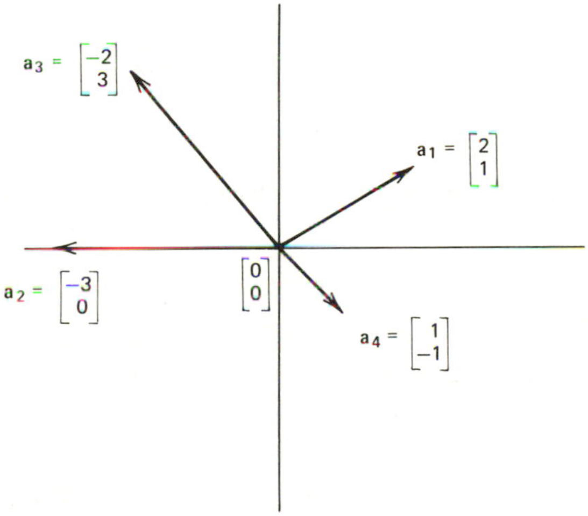

Figure 2.1. **Some examples of vectors.**

Special Vectors

ZERO VECTOR

The *zero vector*, denoted by **0**, is a vector with all components equal to zero. This vector is also referred to as the origin.

*i*TH UNIT VECTOR

This is a vector with zero components, except for a 1 in the ith position. This vector is denoted by $\mathbf{e}_i$ and is sometimes called the *ith coordinate vector*.

$$\mathbf{e}_i = (0, 0, \ldots, \underset{\substack{\uparrow \\ i\text{th position}}}{1}, \ldots, 0, 0)$$

SUM VECTOR

This is a vector with each component equal to one. The sum vector is denoted by $\mathbf{1}$.

Addition and Multiplication of Vectors

ADDITION

Two vectors of the same size can be added, where addition is performed componentwise. To illustrate, let $\mathbf{a}_1$ and $\mathbf{a}_2$ be the following two n vectors:

$$\mathbf{a}_1 = (a_{11}, a_{21}, \ldots, a_{n1})$$

$$\mathbf{a}_2 = (a_{12}, a_{22}, \ldots, a_{n2})$$

Then the addition of $\mathbf{a}_1$ and $\mathbf{a}_2$, denoted by $\mathbf{a}_1 + \mathbf{a}_2$, is the following vector:

$$\mathbf{a}_1 + \mathbf{a}_2 = (a_{11} + a_{12}, a_{21} + a_{22}, \ldots, a_{n1} + a_{n2})$$

The operation of vector addition is illustrated in Figure 2.2. Note that $\mathbf{a}_1 + \mathbf{a}_2$ is the diagonal of the parallelogram with sides $\mathbf{a}_1$ and $\mathbf{a}_2$.

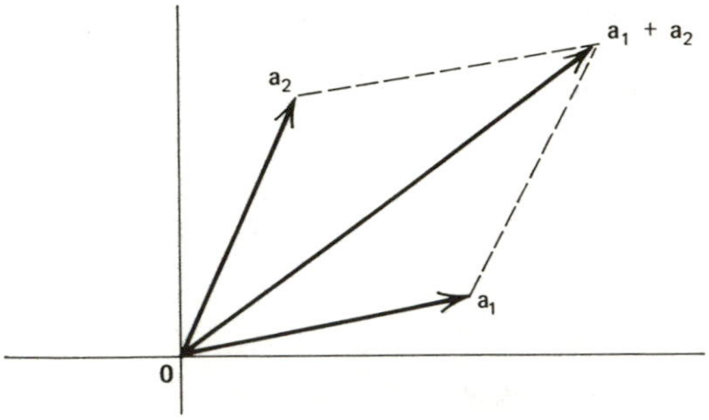

Figure 2.2. **Vector addition.**

SCALAR MULTIPLICATION

The operation of multiplying a vector $\mathbf{a}$ with a scalar k is performed componentwise. If $\mathbf{a} = (a_1, a_2, \ldots, a_n)$, then the vector $k\mathbf{a} = (ka_1, ka_2, \ldots, ka_n)$. This operation is shown in Figure 2.3. If $k > 0$, then $k\mathbf{a}$ points in the same direction as $\mathbf{a}$. On the other hand if $k < 0$, then $k\mathbf{a}$ points in the opposite direction.

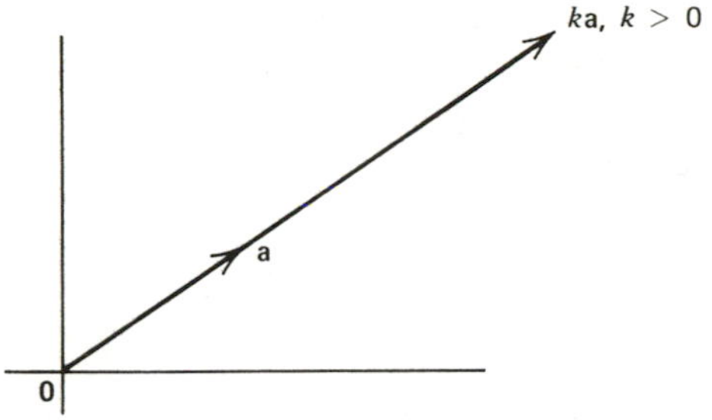

Figure 2.3. **Scalar multiplication.**

Inner Product

Any two n vectors **a** and **b** can be multiplied. The result of this multiplication is a real number called the *inner product* of the two vectors. It is defined below:

$$\mathbf{ab} = a_1b_1 + a_2b_2 \cdots + a_nb_n = \sum_{j=1}^{n} a_jb_j$$

For example, if $\mathbf{a} = (1, -1)$ and $\mathbf{b} = \begin{pmatrix} -2 \\ 1 \end{pmatrix}$, then $\mathbf{ab} = -3$.

Norm of a Vector

Various norms (measures of size) of a vector can be used. We shall use here the *Euclidean norm*. This norm is the square root of the inner product of the vector and itself. In other words, the norm of a vector **a**, denoted by $\|\mathbf{a}\|$, is given by $\sqrt{\sum_{j=1}^{n} a_j^2}$. Note that

$$\|\mathbf{a} + \mathbf{b}\|^2 = \|\mathbf{a}\|^2 + \|\mathbf{b}\|^2 + 2\mathbf{ab}$$

for any two vectors **a** and **b** of the same size.

Schwartz Inequality

Given two vectors **a** and **b** of the same size, the following inequality, called the *Schwartz inequality*, holds:

$$|\mathbf{ab}| \leqslant \|\mathbf{a}\| \, \|\mathbf{b}\|$$

To illustrate, let $\mathbf{a} = (0, 2)$ and $\mathbf{b} = (3, 4)$. Then $\mathbf{ab} = 8$, whereas $\|\mathbf{a}\| = 2$ and

$\|\mathbf{b}\| = 5$. Clearly $8 \leqslant 2 \times 5$. In fact, the "discrepancy" between the inner product $\mathbf{ab}$ and $\|\mathbf{a}\|\ \|\mathbf{b}\|$ measures the angle θ between the two vectors. In particular, $\cos\theta = \mathbf{ab}/\|\mathbf{a}\|\ \|\mathbf{b}\|$. Of course, if $\mathbf{ab} = 0$, then $\cos\theta = 0$; that is, the two vectors $\mathbf{a}$ and $\mathbf{b}$ are *orthogonal*. Figure 2.4 shows two orthogonal vectors $\mathbf{a}$ and $\mathbf{b}$.

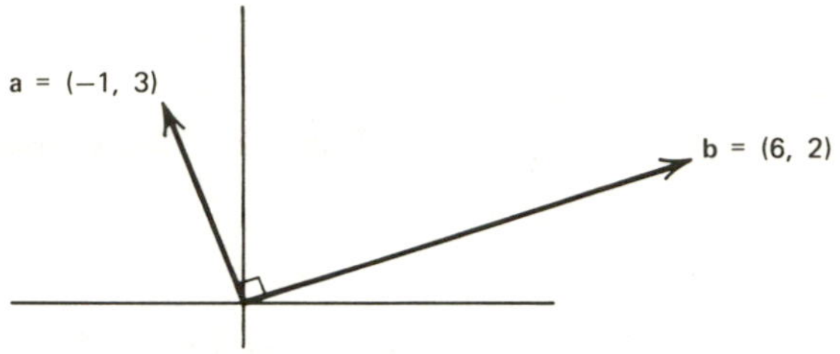

Figure 2.4. **Orthogonal vectors.**

Euclidean Space

An n-dimensional *Euclidean* space, denoted by E^n, is the collection of all vectors of dimension n. Addition and scalar multiplication of vectors in E^n are defined above. Also, associated with any vector in E^n is its norm, and associated with any two vectors in E^n is their inner product, defined above.

Linear Combination

A vector $\mathbf{b}$ in E^n is said to be a *linear combination* of $\mathbf{a}_1, \mathbf{a}_2, \ldots, \mathbf{a}_k$ in E^n, if $\mathbf{b} = \sum_{j=1}^{k} \lambda_j \mathbf{a}_j$, where $\lambda_1, \lambda_2, \ldots, \lambda_k$ are real numbers.

Linear Independence

A collection of vectors $\mathbf{a}_1, \mathbf{a}_2, \ldots, \mathbf{a}_k$ of dimension n is called *linearly independent* if

$$\sum_{j=1}^{k} \lambda_j \mathbf{a}_j = \mathbf{0} \text{ implies that } \lambda_j = 0 \text{ for } j = 1, 2, \ldots, k$$

For example, let $\mathbf{a}_1 = (1, 2)$ and $\mathbf{a}_2 = (-1, 1)$. These two vectors are linearly independent because $\lambda_1(1, 2) + \lambda_2(-1, 1) = (0, 0)$ implies that $\lambda_1 = \lambda_2 = 0$.

A collection of vectors is called *linearly dependent* if they are not linearly independent. Therefore $\mathbf{a}_1, \mathbf{a}_2, \ldots, \mathbf{a}_k$ are linearly dependent if there exist $\lambda_1, \lambda_2, \ldots, \lambda_k$, *not all zero*, such that $\sum_{j=1}^{k} \lambda_j \mathbf{a}_j = \mathbf{0}$. For example, let $\mathbf{a}_1 = (1, 2, 3)$, $\mathbf{a}_2 = (-1, 1, -1)$, and $\mathbf{a}_3 = (0, 3, 2)$. These three vectors are linearly dependent because $\lambda_1\mathbf{a}_1 + \lambda_2\mathbf{a}_2 + \lambda_3\mathbf{a}_3 = \mathbf{0}$ for $\lambda_1 = \lambda_2 = 1$ and $\lambda_3 = -1$.

Spanning Set

A collection of vectors $\mathbf{a}_1, \mathbf{a}_2, \ldots, \mathbf{a}_k$ in E^n is said to *span* E^n if any vector in E^n can be represented as a linear combination of $\mathbf{a}_1, \mathbf{a}_2, \ldots, \mathbf{a}_k$. In other words, given any vector $\mathbf{b}$ in E^n, we must be able to find scalars $\lambda_1, \lambda_2, \ldots, \lambda_k$ such that $\mathbf{b} = \sum_{j=1}^{k} \lambda_j \mathbf{a}_j$.

To illustrate, let $n = 2$, and consider $\mathbf{a}_1 = (1, 0)$, $\mathbf{a}_2 = (-1, 3)$, and $\mathbf{a}_3 = (2, 1)$. The vectors $\mathbf{a}_1, \mathbf{a}_2, \mathbf{a}_3$ span E^2 since any vector $\mathbf{b}$ in E^2 can be represented as a linear combination of these vectors. For example $\mathbf{b} = (b_1, b_2)$ can be represented as $\lambda_1\mathbf{a}_1 + \lambda_2\mathbf{a}_2 + \lambda_3\mathbf{a}_3$ where $\lambda_1 = b_1 + \frac{1}{3}b_2$, $\lambda_2 = \frac{1}{3}b_2$, and $\lambda_3 = 0$. In this case the representation is not unique. Another representation is given by letting $\lambda_1 = b_1 - 2b_2$, $\lambda_2 = 0$, and $\lambda_3 = b_2$.

Basis

A collection of vectors $\mathbf{a}_1, \mathbf{a}_2, \ldots, \mathbf{a}_k$ forms a *basis* of E^n if the following conditions hold:

1. $\mathbf{a}_1, \mathbf{a}_2, \ldots, \mathbf{a}_k$ span E^n.
2. If any of the vectors is removed, the remaining collection of vectors does not span E^n.

It can be shown that the foregoing conditions are equivalent to the following two requirements: $k = n$ and $\mathbf{a}_1, \mathbf{a}_2, \ldots, \mathbf{a}_n$ are linearly independent. To illustrate, consider the two vectors $\mathbf{a}_1 = \begin{pmatrix} 1 \\ 1 \end{pmatrix}$ and $\mathbf{a}_2 = \begin{pmatrix} 0 \\ 1 \end{pmatrix}$ in E^2. These two vectors form a basis of E^2 since $k = n = 2$, and $\mathbf{a}_1$ and $\mathbf{a}_2$ are linearly independent.

Given a basis of E^n, say $\mathbf{a}_1, \mathbf{a}_2, \ldots, \mathbf{a}_n$, any vector $\mathbf{b}$ in E^n is uniquely represented in terms of this basis. If $\mathbf{b} = \sum_{j=1}^{n} \lambda_j \mathbf{a}_j$ and also $\mathbf{b} = \sum_{j=1}^{n} \lambda_j' \mathbf{a}_j$ then $\sum_{j=1}^{n} (\lambda_j - \lambda_j')\mathbf{a}_j = \mathbf{0}$, which implies that $\lambda_j = \lambda_j'$ for each j, since otherwise we would violate the linear independence of $\mathbf{a}_1, \mathbf{a}_2, \ldots, \mathbf{a}_n$.

Since a basis in E^n must always have n vectors, then the dimension of a basis is unique, namely n. But a basis itself is not unique, since any set of n vectors that are linearly independent will form a basis.

Replacing a Vector in the Basis by Another Vector

In the simplex method, different bases will be generated, where one vector from the last basis is replaced by another vector. We have to be careful in choosing the vectors entering and leaving the basis, because otherwise the new vectors may not be linearly independent, and hence will not form a basis. To illustrate, the vectors $\mathbf{a}_1 = (1, 2, 1)$, $\mathbf{a}_2 = (3, 0, 1)$, and $\mathbf{a}_3 = (2, -2, 1)$ are linearly independent, and hence form a basis of E^3. We cannot replace $\mathbf{a}_3$ by $(2, -2, 0)$, because $\mathbf{a}_1$, $\mathbf{a}_2$, and (2, − 2, 0) are linearly dependent and do not form a basis.

This leads to the following natural question: If we have a basis of E^n, what is the condition that will guarantee that if a vector of the basis, say $\mathbf{a}_j$, is replaced by another vector, say $\mathbf{a}$, then the new set of vectors still forms a basis?

Let $\mathbf{a}_1, \mathbf{a}_2, \ldots, \mathbf{a}_n$ form a basis of E^n. We want to replace $\mathbf{a}_j$ by $\mathbf{a}$. Since $\mathbf{a}_1, \ldots, \mathbf{a}_n$ form a basis, then $\mathbf{a}$ can be represented as a linear combination of these vectors, that is,

$$\mathbf{a} = \sum_{i=1}^{n} \lambda_i \mathbf{a}_i$$

Suppose that $\lambda_j \neq 0$. We shall show that the vectors $\mathbf{a}_1, \mathbf{a}_2, \ldots, \mathbf{a}_{j-1}, \mathbf{a}, \mathbf{a}_{j+1}, \ldots, \mathbf{a}_n$ are linearly independent, and hence form a basis. Suppose that there exist μ and $\mu_i (i \neq j)$, such that

$$\sum_{i \neq j} \mu_i \mathbf{a}_i + \mu \mathbf{a} = \mathbf{0}$$

Substituting $\mathbf{a} = \sum_{i=1}^{n} \lambda_i \mathbf{a}_i$, we get

$$\sum_{i \neq j} \mu_i \mathbf{a}_i + \mu \sum_{i=1}^{n} \lambda_i \mathbf{a}_i = \mathbf{0}$$

$$\sum_{i \neq j} (\mu_i + \mu\lambda_i)\mathbf{a}_i + \mu\lambda_j \mathbf{a}_j = \mathbf{0}$$

But since $\mathbf{a}_1, \mathbf{a}_2, \ldots, \mathbf{a}_j, \ldots, \mathbf{a}_n$ are linearly independent, then $\mu\lambda_j = 0$, and $\mu_i + \mu\lambda_i = 0$ for $i \neq j$. Since $\lambda_j \neq 0$ by assumption, then $\mu = 0$. But this implies that $\mu_i = 0$ for $i \neq j$. In other words, $\sum_{i \neq j} \mu_i \mathbf{a}_i + \mu \mathbf{a} = \mathbf{0}$ is only possible if $\mu = 0$ and $\mu_i = 0$ for $i \neq j$, and hence $\mathbf{a}$ and $\mathbf{a}_i (i \neq j)$ are linearly independent and must form a basis. From this discussion it is obvious that the condition $\lambda_j \neq 0$ is sufficient for the new set of vectors to be linearly independent. Obviously the condition is also necessary, because if λ_j were zero, then $\mathbf{a} - \sum_{i \neq j} \lambda_i \mathbf{a}_i = \mathbf{0}$, and hence $\mathbf{a}$, and $\mathbf{a}_i (i \neq j)$ would be linearly dependent.

2.2 MATRICES

A *matrix* is a rectangular array of numbers. If the matrix has m rows and n columns, it is called an $m \times n$ matrix (reads "m by n"). Matrices will be denoted by capital boldface letters, such as **A**, **B**, **C**. An example of a 3×2 matrix is given below.

$$\mathbf{A} = \begin{bmatrix} 1 & -1 \\ 2 & 2 \\ 3 & 1 \end{bmatrix}$$

The entry in row i and column j is denoted by a_{ij}; for example, $a_{12} = -1$ and $a_{31} = 3$. An $m \times n$ matrix $\mathbf{A}$ can be represented by its columns or by its rows. If we denote the columns of $\mathbf{A}$ by $\mathbf{a}_1, \mathbf{a}_2, \ldots, \mathbf{a}_n$, then $\mathbf{A} = [\mathbf{a}_1, \mathbf{a}_2, \ldots, \mathbf{a}_n]$. Similarly, $\mathbf{A}$ can be represented as

$$\begin{bmatrix} \mathbf{a}^1 \\ \mathbf{a}^2 \\ \vdots \\ \mathbf{a}^m \end{bmatrix}$$

where $\mathbf{a}^1, \mathbf{a}^2, \ldots, \mathbf{a}^m$ are the rows of $\mathbf{A}$. Note that every vector is a matrix, but not every matrix is a vector.

Addition of Matrices

The addition of two matrices of the same dimension is defined componentwise; that is, if $\mathbf{A}$ and $\mathbf{B}$ are $m \times n$ matrices, then $\mathbf{C} = \mathbf{A} + \mathbf{B}$ is defined by letting $c_{ij} = a_{ij} + b_{ij}$ for $i = 1, 2, \ldots, m$ and $j = 1, 2, \ldots, n$.

Multiplication by a Scalar

Let $\mathbf{A}$ be an $m \times n$ matrix and let k be a scalar. Then $k\mathbf{A}$ is an $m \times n$ matrix whose ij entry is ka_{ij}.

Matrix Multiplication

Let $\mathbf{A}$ be an $m \times n$ matrix and $\mathbf{B}$ be an $n \times p$ matrix. Then the product $\mathbf{AB}$ is defined to be the $m \times p$ matrix $\mathbf{C}$ with

$$c_{ij} = \sum_{k=1}^{n} a_{ik} b_{kj} \qquad \text{for } i = 1, \ldots, m \quad j = 1, \ldots, p$$

In other words, the ij entry of $\mathbf{C}$ is determined as the inner product of the ith row of $\mathbf{A}$ and the jth column of $\mathbf{B}$. Let

$$\mathbf{A} = \begin{bmatrix} 1 & -1 & 1 \\ 4 & -2 & 5 \\ 2 & 0 & 1 \end{bmatrix} \qquad \text{and} \quad \mathbf{B} = \begin{bmatrix} 5 & 0 \\ 3 & 0 \\ 1 & 1 \end{bmatrix}$$

Then

$$\mathbf{C} = \mathbf{AB} = \begin{bmatrix} 1 & -1 & 1 \\ 4 & -2 & 5 \\ 2 & 0 & 1 \end{bmatrix} \begin{bmatrix} 5 & 0 \\ 3 & 0 \\ 1 & 1 \end{bmatrix} = \begin{bmatrix} 3 & 1 \\ 19 & 5 \\ 11 & 1 \end{bmatrix}$$

The following points need to be emphasized. If **A** is an $m \times n$ matrix and **B** is a $p \times q$ matrix, then

1. **AB** is defined only if $n = p$. **AB** is then an $m \times q$ matrix.
2. **BA** is defined only if $q = m$. **BA** is then an $p \times n$ matrix.
3. Even if **AB** and **BA** are both defined (if $m = n = p = q$), then **AB** is not necessarily equal to **BA**. Note that **AB** in the foregoing example is defined but **BA** is not defined.

Special Matrices

ZERO MATRIX

An $m \times n$ matrix is called the *zero* matrix if each entry in the matrix is zero.

IDENTITY MATRIX

A square $n \times n$ matrix is called the *identity* matrix, denoted by **I** (sometimes the notation $\mathbf{I}_n$ is used to denote the size), if it has entries equal to one on the diagonal and zero entries everywhere else. Note that $\mathbf{AI}_n = \mathbf{A}$ and $\mathbf{I}_m\mathbf{A} = \mathbf{A}$ for any $m \times n$ matrix **A**.

TRIANGULAR MATRIX

A square $n \times n$ matrix is called an *upper triangular* matrix if all the entries below the diagonal are zeros. Similarly an $n \times n$ matrix is called a *lower triangular* matrix if all elements above the diagonal are zeros.

Transposition

Given an $m \times n$ matrix **A** with a_{ij} as its ij entry, the *transpose* of **A**, denoted by $\mathbf{A}^t$, is an $n \times m$ matrix whose ij entry is a_{ji}. In other words, $\mathbf{A}^t$ is formed by letting the jth column of **A** be the jth row of $\mathbf{A}^t$ (similarly by letting the jth row of **A** be the jth column of $\mathbf{A}^t$). A square matrix **A** is called *symmetric* if $\mathbf{A} = \mathbf{A}^t$ and *skew-symmetric* if $\mathbf{A} = -\mathbf{A}^t$. The following results are obvious.

1. $(\mathbf{A}^t)^t = \mathbf{A}$.
2. If **A** and **B** have the same dimension, then $(\mathbf{A} + \mathbf{B})^t = \mathbf{A}^t + \mathbf{B}^t$.
3. If **AB** is defined, then $(\mathbf{AB})^t = \mathbf{B}^t\mathbf{A}^t$.

Partitioned Matrices

Given an $m \times n$ matrix **A**, we can obtain a submatrix of **A** by deleting certain rows and/or columns of **A**. Hence we can think of a matrix **A** as being *partitioned* into submatrices. For example, consider

$$\mathbf{A} = \left[\begin{array}{cc|cc} a_{11} & a_{12} & a_{13} & a_{14} \\ a_{21} & a_{22} & a_{23} & a_{24} \\ a_{31} & a_{32} & a_{33} & a_{34} \\ \hline a_{41} & a_{42} & a_{43} & a_{44} \end{array}\right]$$

Here $\mathbf{A}$ has been partitioned into four submatrices, $\mathbf{A}_{11}$, $\mathbf{A}_{12}$, $\mathbf{A}_{21}$, and $\mathbf{A}_{22}$; therefore

$$\mathbf{A} = \left[\begin{array}{c|c} \mathbf{A}_{11} & \mathbf{A}_{12} \\ \hline \mathbf{A}_{21} & \mathbf{A}_{22} \end{array}\right]$$

where

$$\mathbf{A}_{11} = \begin{bmatrix} a_{11} & a_{12} \\ a_{21} & a_{22} \\ a_{31} & a_{32} \end{bmatrix} \qquad \mathbf{A}_{12} = \begin{bmatrix} a_{13} & a_{14} \\ a_{23} & a_{24} \\ a_{33} & a_{34} \end{bmatrix}$$

$$\mathbf{A}_{21} = \begin{bmatrix} a_{41} & a_{42} \end{bmatrix} \qquad \mathbf{A}_{22} = \begin{bmatrix} a_{43} & a_{44} \end{bmatrix}$$

Now suppose that $\mathbf{A}$ and $\mathbf{B}$ are partitioned as follows:

$$\begin{array}{c} \quad\; n_1 \quad\;\; n_2 \\ \mathbf{A} = \left[\begin{array}{c|c} \mathbf{A}_{11} & \mathbf{A}_{12} \\ \hline \mathbf{A}_{21} & \mathbf{A}_{22} \end{array}\right] \begin{array}{c} m_1 \\ m_2 \end{array} \end{array} \qquad \begin{array}{c} \quad\; q_1 \quad\;\; q_2 \quad\;\; q_3 \\ \mathbf{B} = \left[\begin{array}{c|c|c} \mathbf{B}_{11} & \mathbf{B}_{12} & \mathbf{B}_{13} \\ \hline \mathbf{B}_{21} & \mathbf{B}_{22} & \mathbf{B}_{23} \end{array}\right] \begin{array}{c} p_1 \\ p_2 \end{array} \end{array}$$

Then $\mathbf{AB}$ is defined by

$$\begin{aligned} \mathbf{AB} &= \left[\begin{array}{c|c} \mathbf{A}_{11} & \mathbf{A}_{12} \\ \hline \mathbf{A}_{21} & \mathbf{A}_{22} \end{array}\right] \left[\begin{array}{c|c|c} \mathbf{B}_{11} & \mathbf{B}_{12} & \mathbf{B}_{13} \\ \hline \mathbf{B}_{21} & \mathbf{B}_{22} & \mathbf{B}_{23} \end{array}\right] \\ &= \left[\begin{array}{c|c|c} \mathbf{A}_{11}\mathbf{B}_{11} + \mathbf{A}_{12}\mathbf{B}_{21} & \mathbf{A}_{11}\mathbf{B}_{12} + \mathbf{A}_{12}\mathbf{B}_{22} & \mathbf{A}_{11}\mathbf{B}_{13} + \mathbf{A}_{12}\mathbf{B}_{23} \\ \hline \mathbf{A}_{21}\mathbf{B}_{11} + \mathbf{A}_{22}\mathbf{B}_{21} & \mathbf{A}_{21}\mathbf{B}_{12} + \mathbf{A}_{22}\mathbf{B}_{22} & \mathbf{A}_{21}\mathbf{B}_{13} + \mathbf{A}_{22}\mathbf{B}_{23} \end{array}\right] \end{aligned}$$

Note that we must have $n_1 = p_1$ and $n_2 = p_2$, so that the product of the submatrices is well defined.

Elementary Matrix Operations

Given an $m \times n$ matrix $\mathbf{A}$, we can perform some elementary row and column operations. These operations are most helpful in solving a system of linear equations and in finding the inverse of a matrix (to be defined later).

An *elementary row operation* on a matrix **A** is one of the following operations:

1. Row i and row j of **A** are interchanged.
2. Row i is multiplied by a nonzero scalar k.
3. Row i is replaced by row i plus k times row j.

Elementary row operations on a matrix **A** are equivalent to premultiplying **A** by a specific matrix. *Elementary column operations* are defined similarly. Elementary column operations on **A** are equivalent to postmultiplying **A** by a specific matrix.

Example 2.1

Let $\mathbf{A} = \begin{bmatrix} 2 & 1 & 1 & 10 \\ -1 & 2 & 1 & 8 \\ 1 & -1 & 2 & 2 \end{bmatrix}$.

We shall perform the following elementary operations on **A**. Divide row 1 by 2, add the new row 1 to row 2, and subtract it from row 3. This gives

$$\begin{bmatrix} 1 & \frac{1}{2} & \frac{1}{2} & 5 \\ 0 & \frac{5}{2} & \frac{3}{2} & 13 \\ 0 & -\frac{3}{2} & \frac{3}{2} & -3 \end{bmatrix}$$

Now multiply row 2 by $\frac{2}{5}$, multiply the new row 2 by $\frac{3}{2}$ and add to row 3. This gives

$$\begin{bmatrix} 1 & \frac{1}{2} & \frac{1}{2} & 5 \\ 0 & 1 & \frac{3}{5} & \frac{26}{5} \\ 0 & 0 & \frac{24}{10} & \frac{24}{5} \end{bmatrix}$$

Divide row 3 by $\frac{24}{10}$. This gives

$$\begin{bmatrix} 1 & \frac{1}{2} & \frac{1}{2} & 5 \\ 0 & 1 & \frac{3}{5} & \frac{26}{5} \\ 0 & 0 & 1 & 2 \end{bmatrix}$$

Note that the matrix **A** is reduced to the foregoing matrix through elementary row operations.

Solving a System of Linear Equations by Elementary Matrix Operations

Consider the system $\mathbf{Ax} = \mathbf{b}$ of m equations in n unknown, where **A** is an $m \times n$ matrix, **b** is an m vector, and **x** is an n vector of variables. The following fact is helpful in solving this system: $\mathbf{Ax} = \mathbf{b}$ if and only if $\mathbf{A}'\mathbf{x} = \mathbf{b}'$, where $(\mathbf{A}', \mathbf{b}')$ is obtained from $(\mathbf{A}, \mathbf{b})$ by a finite number of elementary row operations. To illustrate, consider the following system:

$$\begin{aligned} 2x_1 + x_2 + x_3 &= 10 \\ -x_1 + 2x_2 + x_3 &= 8 \\ x_1 - x_2 + 2x_3 &= 2 \end{aligned}$$

$$(\mathbf{A}, \mathbf{b}) = \begin{bmatrix} 2 & 1 & 1 & 10 \\ -1 & 2 & 1 & 8 \\ 1 & -1 & 2 & 2 \end{bmatrix}$$

This matrix was reduced in Example 2.1 above through elementary row operations to

$$\begin{bmatrix} 1 & \frac{1}{2} & \frac{1}{2} & 5 \\ 0 & 1 & \frac{3}{5} & \frac{26}{5} \\ 0 & 0 & 1 & 2 \end{bmatrix}$$

Therefore $\mathbf{x}$ solves the original system if and only if it solves the following system.

$$\begin{aligned} x_1 + \tfrac{1}{2}x_2 + \tfrac{1}{2}x_3 &= 5 \\ x_2 + \tfrac{3}{5}x_3 &= \tfrac{26}{5} \\ x_3 &= 2 \end{aligned}$$

Note that $\mathbf{A}'$ is upper triangular, and we can solve the system by *back substitution*. From the third equation $x_3 = 2$, and from the second equation $x_2 = 4$, and from the first equation $x_1 = 2$. The process of reducing $\mathbf{A}$ into an upper triangular matrix with ones on the diagonal is called *Gaussian reduction* of the system.

Matrix Inversion

Let $\mathbf{A}$ be a square $n \times n$ matrix. If $\mathbf{B}$ is an $n \times n$ matrix such that $\mathbf{AB} = \mathbf{I}$ and $\mathbf{BA} = \mathbf{I}$, then $\mathbf{B}$ is called the *inverse* of $\mathbf{A}$. The inverse matrix, if it exists, is unique and is denoted by $\mathbf{A}^{-1}$. If $\mathbf{A}$ has an inverse, $\mathbf{A}$ is called *nonsingular*; otherwise $\mathbf{A}$ is called *singular*.

CONDITION FOR EXISTENCE OF THE INVERSE

Given an $n \times n$ matrix $\mathbf{A}$, it has an inverse if and only if the rows of $\mathbf{A}$ are linearly independent or, equivalently, if the columns of $\mathbf{A}$ are linearly independent.

CALCULATION OF THE INVERSE

The inverse matrix, if it exists, can be obtained through a finite number of elementary row operations. This can be done by noting that if a sequence of elementary row operations reduce **A** to the identity, then the same sequence of operations will reduce (**A**, **I**) to (**I**, $\mathbf{A}^{-1}$). In fact, this is equivalent to premultiplying the system by $\mathbf{A}^{-1}$. Further, if (**A**, **B**) is reduced to (**I**, **F**) by elementary row operations, then $\mathbf{F} = \mathbf{A}^{-1}\mathbf{B}$.

In order to calculate the inverse, we adjoin the identity to **A**. The matrix **A** is reduced to the identity by elementary row operations. This will result in reducing the identity to $\mathbf{A}^{-1}$. Of course, if $\mathbf{A}^{-1}$ does not exist, then the elementary row operations will fail to produce the identity. This discussion is made clear by the following two examples.

Example 2.2 ($\mathbf{A}^{-1}$ exists)

Consider the matrix **A** below.

$$\mathbf{A} = \begin{bmatrix} 2 & 1 & 1 \\ -1 & 2 & 1 \\ 1 & -1 & 2 \end{bmatrix}$$

To find the inverse, form the augmented matrix (**A**, **I**). Reduce **A** by elementary row operations to the identity. The matrix in place of **I** is $\mathbf{A}^{-1}$.

$$\left[\begin{array}{rrr|rrr} 2 & 1 & 1 & 1 & 0 & 0 \\ -1 & 2 & 1 & 0 & 1 & 0 \\ 1 & -1 & 2 & 0 & 0 & 1 \end{array}\right]$$

Divide the first row by 2. Add the new first row to the second row and subtract it from the third row.

$$\left[\begin{array}{rrr|rrr} 1 & \frac{1}{2} & \frac{1}{2} & \frac{1}{2} & 0 & 0 \\ 0 & \frac{5}{2} & \frac{3}{2} & \frac{1}{2} & 1 & 0 \\ 0 & -\frac{3}{2} & \frac{3}{2} & -\frac{1}{2} & 0 & 1 \end{array}\right]$$

Multiply the second row by $\frac{2}{5}$. Multiply the new second row by $-\frac{1}{2}$ and add to the first row, and multiply the new second row by $\frac{3}{2}$ and add to the third row.

$$\left[\begin{array}{rrr|rrr} 1 & 0 & \frac{1}{5} & \frac{2}{5} & -\frac{1}{5} & 0 \\ 0 & 1 & \frac{3}{5} & \frac{1}{5} & \frac{2}{5} & 0 \\ 0 & 0 & \frac{12}{5} & -\frac{1}{5} & \frac{3}{5} & 1 \end{array}\right]$$

Multiply the third row by $\frac{5}{12}$. Multiply the new third row by $-\frac{3}{5}$ and add to the second row, and multiply the new third row by $-\frac{1}{5}$ and add to the first row.

$$\left[\begin{array}{ccc|ccc} 1 & 0 & 0 & \frac{5}{12} & -\frac{3}{12} & -\frac{1}{12} \\ 0 & 1 & 0 & \frac{3}{12} & \frac{3}{12} & -\frac{3}{12} \\ 0 & 0 & 1 & -\frac{1}{12} & \frac{3}{12} & \frac{5}{12} \end{array}\right]$$

Therefore the inverse of **A** exists and is given by

$$\mathbf{A}^{-1} = \frac{1}{12}\begin{bmatrix} 5 & -3 & -1 \\ 3 & 3 & -3 \\ -1 & 3 & 5 \end{bmatrix}$$

Example 2.3 ($\mathbf{A}^{-1}$ does not exist)

Consider the matrix **A** below.

$$\mathbf{A} = \begin{bmatrix} 1 & 1 & 2 \\ 2 & -1 & 1 \\ 1 & 2 & 3 \end{bmatrix}$$

The inverse does not exist since $\mathbf{a}_3 = \mathbf{a}_1 + \mathbf{a}_2$. If we use the foregoing procedure, the elementary matrix operations will fail to produce the identity.

$$\left[\begin{array}{ccc|ccc} 1 & 1 & 2 & 1 & 0 & 0 \\ 2 & -1 & 1 & 0 & 1 & 0 \\ 1 & 2 & 3 & 0 & 0 & 1 \end{array}\right]$$

Multiply the first row by -2 and add to the second row, and multiply the first row by -1 and add to the third row.

$$\left[\begin{array}{ccc|ccc} 1 & 1 & 2 & 1 & 0 & 0 \\ 0 & -3 & -3 & -2 & 1 & 0 \\ 0 & 1 & 1 & -1 & 0 & 1 \end{array}\right]$$

Multiply the second row by $-\frac{1}{3}$. Then multiply the new second row by -1 and add to the first row, and multiply the new second row by -1 and add to the third row.

$$\left[\begin{array}{ccc|ccc} 1 & 0 & 1 & \frac{1}{3} & \frac{1}{3} & 0 \\ 0 & 1 & 1 & \frac{2}{3} & -\frac{1}{3} & 0 \\ 0 & 0 & 0 & -\frac{5}{3} & \frac{1}{3} & 1 \end{array}\right]$$

There is no way that the left-hand-side matrix can be transformed into the identity matrix by elementary row operations, and hence the matrix **A** has no inverse.

The following facts about matrix inversion are useful.

1. If **A** is nonsingular, then $\mathbf{A}^t$ is also nonsingular, and $(\mathbf{A}^t)^{-1} = (\mathbf{A}^{-1})^t$.
2. If **A** and **B** are both $n \times n$ nonsingular matrices, then **AB** is nonsingular, and $(\mathbf{AB})^{-1} = \mathbf{B}^{-1}\mathbf{A}^{-1}$.
3. A triangular matrix (either lower or upper triangular) with nonzero diagonal elements has an inverse. This can be easily established by noting that such a matrix can be reduced to the identity by a finite number of elementary row operations.
4. Let **A** be partitioned as follows, where **D** is nonsingular.

$$\mathbf{A} = \overset{\begin{matrix} n_1 & n_2 \end{matrix}}{\left[\begin{array}{c|c} \mathbf{I} & \mathbf{C} \\ \hline \mathbf{0} & \mathbf{D} \end{array}\right]} \begin{matrix} n_1 \\ n_2 \end{matrix}$$

Then **A** is nonsingular, and

$$\mathbf{A}^{-1} = \left[\begin{array}{c|c} \mathbf{I} & -\mathbf{C}\mathbf{D}^{-1} \\ \hline \mathbf{0} & \mathbf{D}^{-1} \end{array}\right]$$

Determinant of a Matrix

Associated with each square $n \times n$ matrix is a real number, called the *determinant* of the matrix. Let **A** be an $n \times n$ matrix whose ij element is a_{ij}. The determinant of **A**, denoted by det **A**, is defined as follows:

$$\det \mathbf{A} = \sum_{i=1}^{n} a_{i1} A_{i1}$$

where A_{i1} is the *cofactor* of a_{i1} defined as $(-1)^{i+1}$ times the determinant of the submatrix of **A** obtained by deleting the ith row and first column. The determinant of a 1×1 matrix is just the element. To illustrate, consider the following example.

$$\det\begin{bmatrix} 1 & 0 & 1 \\ 2 & 1 & -3 \\ -3 & 2 & 1 \end{bmatrix} = 1A_{11} + 2A_{21} - 3A_{31}$$

$$= \det\begin{bmatrix} 1 & -3 \\ 2 & 1 \end{bmatrix} - 2\det\begin{bmatrix} 0 & 1 \\ 2 & 1 \end{bmatrix} - 3\det\begin{bmatrix} 0 & 1 \\ 1 & -3 \end{bmatrix}$$

Note that the foregoing definition reduces the calculation of a determinant of an $n \times n$ matrix to n determinants of $(n-1) \times (n-1)$ matrices. The same definition can be used to reduce a determinant of an $(n-1) \times (n-1)$ matrix to determinants of $(n-2) \times (n-2)$ matrices, and so forth. Obviously, by the above definition, the determinant of a 2×2 matrix, say $\mathbf{A}' = \begin{bmatrix} a_{11} & a_{12} \\ a_{21} & a_{22} \end{bmatrix}$, is simply $a_{11}a_{22} - a_{21}a_{12}$.

To summarize, the determinant of an $n \times n$ matrix can be calculated by successively applying the foregoing definition. The determinant of **A** above is therefore given by $1(1 + 6) - 2(0 - 2) - 3(0 - 1) = 14$. We summarize below some important facts about determinants of square matrices.

1. In the definition above, the first column was used as a reference in calculating det **A**. Any column or row can be used as a reference in the calculations; that is,

$$\det \mathbf{A} = \sum_{i=1}^{n} a_{ij}A_{ij} \qquad \text{for any} \quad j = 1, 2, \ldots, n$$

and similarly

$$\det \mathbf{A} = \sum_{j=1}^{n} a_{ij}A_{ij} \qquad \text{for any} \quad i = 1, 2, \ldots, n$$

where A_{ij} is the cofactor of a_{ij} given as $(-1)^{i+j}$ times the determinant of the submatrix obtained from **A** by deleting the ith row and jth column.
2. $\det \mathbf{A} = \det \mathbf{A}^t$.
3. Let **B** be obtained from **A** by interchanging two rows (or columns). Then $\det \mathbf{B} = -\det \mathbf{A}$.
4. Let **B** be obtained from **A** by adding to one row (column) a constant times another row (column). Then $\det \mathbf{B} = \det \mathbf{A}$.
5. Let **B** be obtained from **A** by multiplying a row (or column) by a scalar k. Then $\det \mathbf{B} = k \det \mathbf{A}$.
6. Let **A** be partitioned as follows, where **B** and **C** are square.

$$\mathbf{A} = \left[\begin{array}{c|c} \mathbf{B} & \mathbf{0} \\ \hline \mathbf{D} & \mathbf{C} \end{array}\right]$$

Then $\det \mathbf{A} = \det \mathbf{B} \cdot \det \mathbf{C}$.
7. Let **A** and **B** be $n \times n$ matrices. Then $\det(\mathbf{AB}) = \det \mathbf{A} \cdot \det \mathbf{B}$.
8. $\det \mathbf{A} \neq 0$ if and only if the columns (and rows) of **A** are linearly independent. Equivalently, $\det \mathbf{A} = 0$ if and only if the rows (columns) of **A** are linearly dependent. Therefore a square matrix **A** has an inverse if and only if its determinant is not zero.

9. Let $\mathbf{A}$ be an $n \times n$ matrix whose determinant is not zero. Then $\mathbf{A}^{-1}$ exists and is given by

$$\mathbf{A}^{-1} = \frac{\mathbf{B}}{\det \mathbf{A}}$$

where $\mathbf{B}$ is the transpose of the matrix whose ij entry is A_{ij}, the cofactor of a_{ij}. Here $\mathbf{B}$ is called the *adjoint matrix* of $\mathbf{A}$.

10. Consider the system $\mathbf{Ax} = \mathbf{b}$ where $\mathbf{A}$ is $n \times n$, $\mathbf{b}$ is an n vector, and $\mathbf{x}$ is an n vector of unknowns. If $\mathbf{A}$ has an inverse (that is, if $\det \mathbf{A} \neq 0$), then the unique solution to this system is given by

$$x_j = \frac{\det \mathbf{A}_j}{\det \mathbf{A}} \qquad \text{for} \quad j = 1, 2, \ldots, n$$

where $\mathbf{A}_j$ is obtained from $\mathbf{A}$ by replacing the jth column of $\mathbf{A}$ by $\mathbf{b}$. This method for solving the system is called *Cramer's rule*.

11. The determinant of a triangular matrix is the product of the diagonal entries.

The Rank of a Matrix

Let $\mathbf{A}$ be an $m \times n$ matrix. The *row rank* of the matrix is equal to the maximum number of linearly independent rows of $\mathbf{A}$. The *column rank* of $\mathbf{A}$ is the maximum number of linearly independent columns of $\mathbf{A}$.

It can be shown that the row rank of a matrix is always equal to its column rank, and hence the *rank* of the matrix is equal to the maximum number of linearly independent rows (or columns) of $\mathbf{A}$. Thus it is clear that rank $(\mathbf{A}) \leqslant$ minimum (m, n). If rank $(\mathbf{A})$ = minimum (m, n), $\mathbf{A}$ is said to be of *full rank*. It can be shown that the rank of $\mathbf{A}$ is k if and only if $\mathbf{A}$ can be reduced to $\left[\begin{array}{c|c} \mathbf{I}_k & \mathbf{Q} \\ \hline \mathbf{0} & \mathbf{0} \end{array}\right]$ through a finite sequence of elementary matrix operations.

2.3 SIMULTANEOUS LINEAR EQUATIONS

Consider the system $\mathbf{Ax} = \mathbf{b}$ and the augmented matrix $(\mathbf{A}, \mathbf{b})$ with m rows and $n + 1$ columns. If the rank of $(\mathbf{A}, \mathbf{b})$ is greater than the rank of $\mathbf{A}$, then $\mathbf{b}$ cannot be represented as a linear combination of $\mathbf{a}_1, \mathbf{a}_2, \ldots, \mathbf{a}_n$, and hence there is no solution to the system $\mathbf{Ax} = \mathbf{b}$ (and in particular there is no solution to the system $\mathbf{Ax} = \mathbf{b}, \mathbf{x} \geqslant \mathbf{0}$).

Now let us suppose that rank $(\mathbf{A})$ = rank $(\mathbf{A}, \mathbf{b}) = k$. Possibly after rearranging the rows of $(\mathbf{A}, \mathbf{b})$, let

$$(\mathbf{A}, \mathbf{b}) = \begin{bmatrix} \mathbf{A}_1 & \mathbf{b}_1 \\ \mathbf{A}_2 & \mathbf{b}_2 \end{bmatrix}$$

where $\mathbf{A}_1$ is $k \times n$, $\mathbf{b}_1$ is a k vector, $\mathbf{A}_2$ is an $(m - k) \times n$ matrix, $\mathbf{b}_2$ is an $m - k$ vector, and rank $(\mathbf{A}_1)$ = rank $(\mathbf{A}_1, \mathbf{b}_1) = k$.

Note that if a vector $\mathbf{x}$ satisfies $\mathbf{A}_1\mathbf{x} = \mathbf{b}_1$, then it satisfies $\mathbf{A}_2\mathbf{x} = \mathbf{b}_2$ automatically. Thus we can throw away the "redundant" or "dependent" constraints $\mathbf{A}_2\mathbf{x} = \mathbf{b}_2$, and keep the independent constraints $\mathbf{A}_1\mathbf{x} = \mathbf{b}_1$. Since rank $(\mathbf{A}_1) = k$, we can pick k linearly independent columns of $\mathbf{A}_1$. Possibly after rearranging the columns of $\mathbf{A}_1$, let $\mathbf{A}_1 = (\mathbf{B}, \mathbf{N})$, where $\mathbf{B}$ is a $k \times k$ nonsingular matrix, and $\mathbf{N}$ is $k \times (n - k)$. Note that such a matrix $\mathbf{B}$ exists since $\mathbf{A}_1$ has rank k. Here $\mathbf{B}$ is called the *basic matrix* (since the columns of $\mathbf{B}$ form a basis of E^k) and $\mathbf{N}$ is called the *nonbasic matrix*. Let us decompose $\mathbf{x}$ accordingly into $\mathbf{x}_B$ and $\mathbf{x}_N$, where $\mathbf{x}_B$ is composed of $x_1, x_2, \ldots, x_k$ and $\mathbf{x}_N$ is composed of $x_{k+1}, \ldots, x_n$. Now $\mathbf{A}_1\mathbf{x} = \mathbf{b}_1$ means that $(\mathbf{B}, \mathbf{N})\begin{bmatrix} \mathbf{x}_B \\ \mathbf{x}_N \end{bmatrix} = \mathbf{b}_1$; that is, $\mathbf{Bx}_B + \mathbf{Nx}_N = \mathbf{b}_1$. Since $\mathbf{B}$ has an inverse, then we can solve $\mathbf{x}_B$ in terms of $\mathbf{x}_N$ by premultiplying by $\mathbf{B}^{-1}$, and we get

$$\mathbf{x}_B = \mathbf{B}^{-1}\mathbf{b}_1 - \mathbf{B}^{-1}\mathbf{N}\mathbf{x}_N$$

In the case $k = n$, $\mathbf{N}$ is vacuous, and we have a unique solution to the system $\mathbf{A}_1\mathbf{x} = \mathbf{b}_1$, namely $\mathbf{x}_B = \mathbf{B}^{-1}\mathbf{b}_1 = \mathbf{A}_1^{-1}\mathbf{b}_1$. On the other hand, if $n > k$, then by assigning arbitrary values to the vector $\mathbf{x}_N$, we can uniquely solve for $\mathbf{x}_B$, by the equation $\mathbf{x}_B = \mathbf{B}^{-1}\mathbf{b}_1 - \mathbf{B}^{-1}\mathbf{N}\mathbf{x}_N$, to obtain a solution $\begin{bmatrix} \mathbf{x}_B \\ \mathbf{x}_N \end{bmatrix}$ to the system $\mathbf{A}_1\mathbf{x} = \mathbf{b}_1$. In this case we have an infinite number of solutions to the system $\mathbf{A}_1\mathbf{x} = \mathbf{b}_1$ (and hence to the system $\mathbf{Ax} = \mathbf{b}$). Note that the notion of decomposing $\mathbf{A}_1$ into $\mathbf{B}$ and $\mathbf{N}$ and solving $\mathbf{x}_B = \mathbf{B}^{-1}\mathbf{b}_1 - \mathbf{B}^{-1}\mathbf{N}\mathbf{x}_N$ can be interpreted as follows. We have a system of k equations in n unknowns. Assign arbitrary values to $n - k$ of the variables, corresponding to $\mathbf{x}_N$, and then solve for the remaining system of k equations in k unknowns. This is done such that the k equations in k unknowns have a *unique solution*, and that is why we require $\mathbf{B}$ to have an inverse. Such a solution obtained by letting $\mathbf{x}_N = \mathbf{0}$ and $\mathbf{x}_B = \mathbf{B}^{-1}\mathbf{b}_1$ is called a *basic solution* of the system $\mathbf{A}_1\mathbf{x} = \mathbf{b}_1$. Let us now summarize the different possible cases that may arise:

1. Rank $(\mathbf{A}, \mathbf{b}) >$ rank $(\mathbf{A})$ and hence $\mathbf{Ax} = \mathbf{b}$ has no solution.
2. Rank $(\mathbf{A}, \mathbf{b}) =$ rank $(\mathbf{A}) = k = n$, and there exists a unique solution to the system $\mathbf{Ax} = \mathbf{b}$.
3. Rank $(\mathbf{A}, \mathbf{b}) =$ rank $(\mathbf{A}) = k < n$, and we have an infinite number of solutions to the system $\mathbf{Ax} = \mathbf{b}$.

Example 2.4

Consider the following system:

$$\begin{aligned} x_1 + 2x_2 + x_3 - 2x_4 &= 10 \\ -x_1 + 2x_2 - x_3 + x_4 &= 6 \\ x_2 + x_3 \qquad &= 2 \end{aligned}$$

We shall solve this system by matrix inversion and Gaussian reduction.

1. *Matrix Inversion.* Reduce 3 columns of $\mathbf{A}$ to the identity [this is possible since rank $(\mathbf{A}) = 3$]

$$\left[\begin{array}{cccc|c} 1 & 2 & 1 & -2 & 10 \\ -1 & 2 & -1 & 1 & 6 \\ 0 & 1 & 1 & 0 & 2 \end{array}\right]$$

Add the first row to the second row.

$$\left[\begin{array}{cccc|c} 1 & 2 & 1 & -2 & 10 \\ 0 & 4 & 0 & -1 & 16 \\ 0 & 1 & 1 & 0 & 2 \end{array}\right]$$

Divide the second row by 4. Multiply the new second row by -2 and add to the first row. Multiply the new second row by -1 and add to the third row.

$$\left[\begin{array}{cccc|c} 1 & 0 & 1 & -\frac{3}{2} & 2 \\ 0 & 1 & 0 & -\frac{1}{4} & 4 \\ 0 & 0 & 1 & \frac{1}{4} & -2 \end{array}\right]$$

Multiply the third row by -1 and add to the first row.

$$\begin{array}{cccc|c} x_1 & x_2 & x_3 & x_4 & \\ \hline 1 & 0 & 0 & -\frac{7}{4} & 4 \\ 0 & 1 & 0 & -\frac{1}{4} & 4 \\ 0 & 0 & 1 & \frac{1}{4} & -2 \end{array}$$

The original system has been reduced to the system above. Equivalence of the two systems is assured since the new system is obtained from the original system after performing a finite number of elementary row operations. The solution to the system is as follows. Assign x_4 arbitrarily, say $x_4 = \lambda$. Then $x_1 = 4 + \frac{7}{4}\lambda$, $x_2 = 4 + \frac{1}{4}\lambda$, and $x_3 = -2 - \frac{1}{4}\lambda$.

2. *Gaussian Reduction.*

$$\left[\begin{array}{cccc|c} 1 & 2 & 1 & -2 & 10 \\ -1 & 2 & -1 & 1 & 6 \\ 0 & 1 & 1 & 0 & 2 \end{array}\right]$$

Add the first row to the second row.

$$\left[\begin{array}{cccc|c} 1 & 2 & 1 & -2 & 10 \\ 0 & 4 & 0 & -1 & 16 \\ 0 & 1 & 1 & 0 & 2 \end{array}\right]$$

Divide the second row by 4. Subtract the new second row from the third row.

$$\begin{array}{cccc|c} x_1 & x_2 & x_3 & x_4 & \\ \hline 1 & 2 & 1 & -2 & 10 \\ 0 & 1 & 0 & -\frac{1}{4} & 4 \\ 0 & 0 & 1 & \frac{1}{4} & -2 \end{array}$$

The foregoing matrix has an upper triangular submatrix. Let x_4 be equal to an arbitrary value λ. Then $x_3 = -2 - \frac{1}{4}\lambda$, $x_2 = 4 + \frac{1}{4}\lambda$, and $x_1 = 10 + 2\lambda - 2x_2 - x_3 = 4 + \frac{7}{4}\lambda$. This gives the same general solution obtained earlier.

2.4 CONVEX SETS AND CONVEX FUNCTIONS

In this section we consider some basic properties of convex sets, convex functions, and concave functions.

Convex Sets

A set X in E^n is called a *convex set* if given any two points $\mathbf{x}_1$ and $\mathbf{x}_2$ in X, then $\lambda\mathbf{x}_1 + (1 - \lambda)\mathbf{x}_2 \in X$ for each $\lambda \in [0, 1]$.

Note that $\lambda\mathbf{x}_1 + (1 - \lambda)\mathbf{x}_2$ for λ in the interval $[0, 1]$ represents a point on the line segment joining $\mathbf{x}_1$ and $\mathbf{x}_2$. Any point of the form $\lambda\mathbf{x}_1 + (1 - \lambda)\mathbf{x}_2$ where $0 \leqslant \lambda \leqslant 1$, is called a *convex combination* (or weighted average) of $\mathbf{x}_1$ and $\mathbf{x}_2$. If $\lambda \in (0, 1)$, then the convex combination is called *strict*. Hence convexity of X can be interpreted geometrically as follows. For each pair of points $\mathbf{x}_1$ and $\mathbf{x}_2$ in X, the line segment joining them, or the convex combinations of the two points, must belong to X.

Figure 2.5 below shows an example of a convex set and an example of a nonconvex set. In the latter case, we see that not all convex combinations of $\mathbf{x}_1$ and $\mathbf{x}_2$ belong to X. The following are some examples of convex sets.

1. $\{(x_1, x_2) : x_1^2 + x_2^2 \leqslant 1\}$.

2. $\{\mathbf{x} : \mathbf{A}\mathbf{x} = \mathbf{b}\}$, where $\mathbf{A}$ is an $m \times n$ matrix and $\mathbf{b}$ is an m vector,

3. $\{\mathbf{x} : \mathbf{A}\mathbf{x} = \mathbf{b}, \mathbf{x} \geqslant \mathbf{0}\}$, where $\mathbf{A}$ is an $m \times n$ matrix and $\mathbf{b}$ is an m vector.

4.
$$\left\{\mathbf{x} : \mathbf{x} = \lambda_1 \begin{bmatrix} 1 \\ 0 \\ 0 \end{bmatrix} + \lambda_2 \begin{bmatrix} 1 \\ 2 \\ 1 \end{bmatrix} + \lambda_3 \begin{bmatrix} -1 \\ 2 \\ -3 \end{bmatrix}, \quad \lambda_1 + \lambda_2 + \lambda_3 = 1, \quad \lambda_1, \lambda_2, \lambda_3 \geqslant 0 \right\}.$$

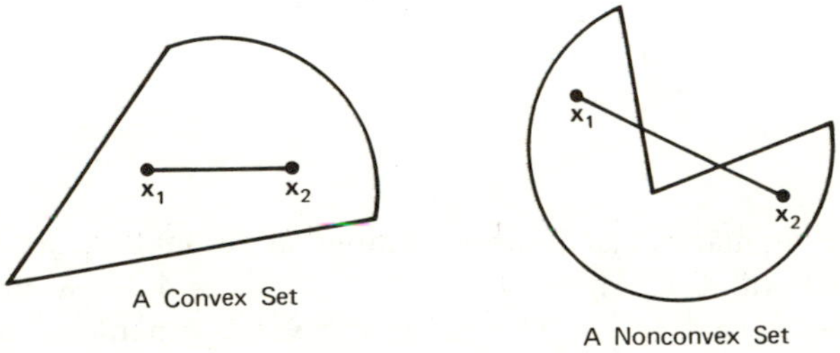

Figure 2.5. Example of convex and nonconvex sets.

Extreme Points

The notion of extreme points plays an especially important role in the theory of linear programming. A point $\mathbf{x}$ in a convex set X is called an *extreme point* of X, if $\mathbf{x}$ cannot be represented as a strict convex combination of two distinct points in X. In other words, if $\mathbf{x} = \lambda \mathbf{x}_1 + (1 - \lambda)\mathbf{x}_2$ with $\lambda \in (0, 1)$ and $\mathbf{x}_1, \mathbf{x}_2 \in X$, then $\mathbf{x} = \mathbf{x}_1 = \mathbf{x}_2$.

Figure 2.6 shows some examples of extreme and nonextreme points of convex sets. Note that $\mathbf{x}_1$ is an extreme point of X whereas $\mathbf{x}_2$ and $\mathbf{x}_3$ are not.

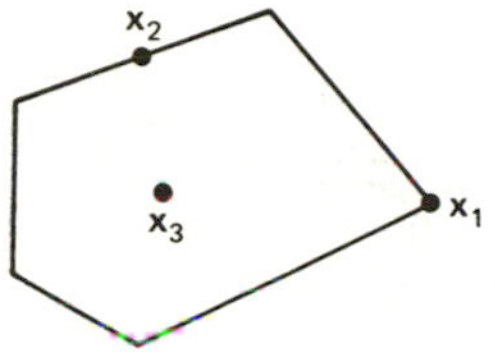

Figure 2.6. **Extreme and nonextreme points.**

Hyperplanes and Halfspaces

A hyperplane in E^n generalizes the notion of a straight line in E^2 and the notion of a plane in E^3. A *hyperplane* H in E^n is a set of the form $\{\mathbf{x} : \mathbf{px} = k\}$ where $\mathbf{p}$ is a nonzero vector in E^n, and k is a scalar. Here $\mathbf{p}$ is usually called the *normal* to the hyperplane.

Equivalently, a hyperplane consists of all points $\mathbf{x} = (x_1, x_2, \ldots, x_n)$ satisfying the equation $\sum_{j=1}^{n} p_j x_j = k$. The constant k can be eliminated by referring to a fixed point $\mathbf{x}_0$ on the hyperplane. If $\mathbf{x}_0 \in H$, then $\mathbf{px}_0 = k$, and for any $\mathbf{x} \in H$, we have $\mathbf{px} = k$. Upon subtraction we get $\mathbf{p}(\mathbf{x} - \mathbf{x}_0) = 0$. In other words, H can be represented as the collection of points satisfying $\mathbf{p}(\mathbf{x} - \mathbf{x}_0) = 0$, where $\mathbf{x}_0$ is any fixed point in H. A hyperplane is a convex set.

Figure 2.7 shows a hyperplane and its normal vector $\mathbf{p}$. Note that $\mathbf{p}$ is orthogonal to $\mathbf{x} - \mathbf{x}_0$ for each $\mathbf{x}$ in the hyperplane H.

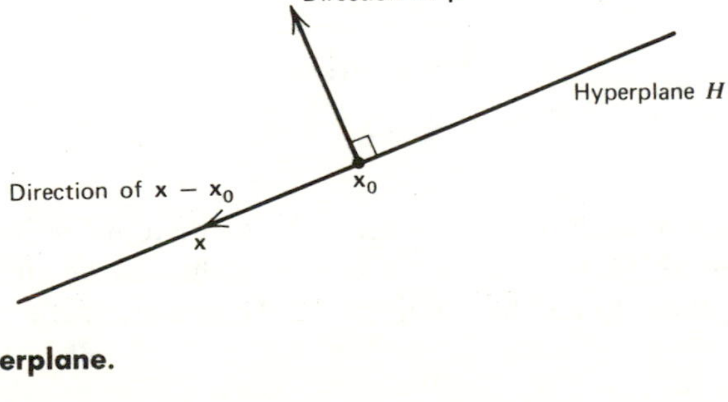

Figure 2.7. **Hyperplane.**

A hyperplane divides E^n into two regions, called halfspaces. Hence a *halfspace* is a collection of points of the form $\{\mathbf{x} : \mathbf{px} \geqslant k\}$, where $\mathbf{p}$ is a nonzero vector in E^n and k is a scalar. A halfspace can also be represented as a set of points of the form $\{\mathbf{x} : \mathbf{px} \leqslant k\}$. The union of the two halfspaces $\{\mathbf{x} : \mathbf{px} \geqslant k\}$ and $\{\mathbf{x} : \mathbf{px} \leqslant k\}$ is E^n. Referring to a fixed point $\mathbf{x}_0$ in the hyperplane defining the halfspace, the latter can be represented as $\{\mathbf{x} : \mathbf{p}(\mathbf{x} - \mathbf{x}_0) \geqslant 0\}$ or as $\{\mathbf{x} : \mathbf{p}(\mathbf{x} - \mathbf{x}_0) \leqslant 0\}$ as shown in Figure 2.8.

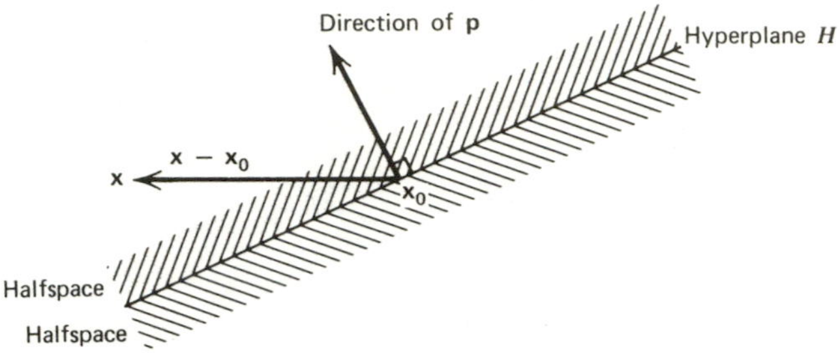

Figure 2.8. **Halfspaces.**

Rays and Directions

Another example of a convex set is a ray. A *ray* is a collection of points of the form $\{\mathbf{x}_0 + \lambda\mathbf{d} : \lambda \geqslant 0\}$, where $\mathbf{d}$ is a nonzero vector. Here $\mathbf{x}_0$ is called the *vertex* of the ray, and $\mathbf{d}$ is the *direction of the ray*.

Directions of a Convex Set

Given a convex set, a nonzero vector $\mathbf{d}$ is called a *direction of the set*, if for each $\mathbf{x}_0$ in the set, the ray $\{\mathbf{x}_0 + \lambda\mathbf{d} : \lambda \geqslant 0\}$ also belongs to the set. Clearly if the set is bounded, then it has no directions.

Consider the nonempty polyhedral set $X = \{\mathbf{x} : \mathbf{Ax} = \mathbf{b},\ \mathbf{x} \geqslant \mathbf{0}\}$. Then a nonzero $\mathbf{d}$ is a direction of X if and only if

$$\mathbf{A}(\mathbf{x} + \lambda\mathbf{d}) = \mathbf{b}$$

$$\mathbf{x} + \lambda\mathbf{d} \geqslant \mathbf{0}$$

for each $\lambda \geqslant 0$ and each $\mathbf{x} \in X$. Since $\mathbf{x} \in X$, then $\mathbf{Ax} = \mathbf{b}$ and the above equation reduces to $\mathbf{Ad} = \mathbf{0}$. Also, since $\mathbf{x} + \lambda\mathbf{d}$ has to be nonnegative for λ arbitrarily large, then $\mathbf{d}$ must be nonnegative. To summarize, $\mathbf{d}$ is a direction of

X if and only if

$$\mathbf{d} \geqslant \mathbf{0}, \mathbf{d} \neq \mathbf{0}, \qquad \text{and} \qquad \mathbf{A}\mathbf{d} = \mathbf{0}$$

Similarly, it can be shown (see Exercise 2.42) that $\mathbf{d}$ is a direction of the nonempty set $X = \{\mathbf{x} : \mathbf{A}\mathbf{x} \geqslant \mathbf{b}, \mathbf{x} \geqslant \mathbf{0}\}$ if and only if $\mathbf{d} \neq \mathbf{0}$, $\mathbf{d} \geqslant \mathbf{0}$, and $\mathbf{A}\mathbf{d} \geqslant \mathbf{0}$. The set of directions forms a convex set.

Example 2.5

Consider the set $X = \{(x_1, x_2) : x_1 - 2x_2 \geqslant -6,\ x_1 - x_2 \geqslant -2,\ x_1 \geqslant 0,\ x_2 \geqslant 1\}$ depicted in Figure 2.9. Let $\mathbf{x}_0 = \begin{pmatrix} x_1 \\ x_2 \end{pmatrix}$ be an arbitrary fixed feasible point. Then $\mathbf{d} = \begin{pmatrix} d_1 \\ d_2 \end{pmatrix}$ is a direction of X if and only if $\begin{pmatrix} d_1 \\ d_2 \end{pmatrix} \neq \begin{pmatrix} 0 \\ 0 \end{pmatrix}$ and $\mathbf{x}_0 + \lambda\mathbf{d} = \begin{pmatrix} x_1 + \lambda d_1 \\ x_2 + \lambda d_2 \end{pmatrix}$ belongs to X for all $\lambda \geqslant 0$. Therefore

$$\begin{aligned} x_1 - 2x_2 + \lambda(d_1 - 2d_2) &\geqslant -6 \\ x_1 - x_2 + \lambda(d_1 - d_2) &\geqslant -2 \\ x_1 + \lambda d_1 &\geqslant 0 \\ x_2 + \lambda d_2 &\geqslant 1 \end{aligned}$$

for all $\lambda \geqslant 0$. Since the last two inequalities must hold for the fixed x_1 and x_2 and for all $\lambda \geqslant 0$, we conclude that d_1 and $d_2 \geqslant 0$ (why?). Similarly, from the first two inequalities we conclude that $d_1 - 2d_2 \geqslant 0$ and $d_1 - d_2 \geqslant 0$ (why?). Since d_1 and $d_2 \geqslant 0$, then $d_1 \geqslant 2d_2$ implies that $d_1 \geqslant d_2$. Therefore $\begin{pmatrix} d_1 \\ d_2 \end{pmatrix}$ is a direction of X if and only if

$$\begin{pmatrix} d_1 \\ d_2 \end{pmatrix} \neq \begin{pmatrix} 0 \\ 0 \end{pmatrix}$$

$$d_1 \geqslant 0, \quad d_2 \geqslant 0$$
$$d_1 \geqslant 2d_2$$

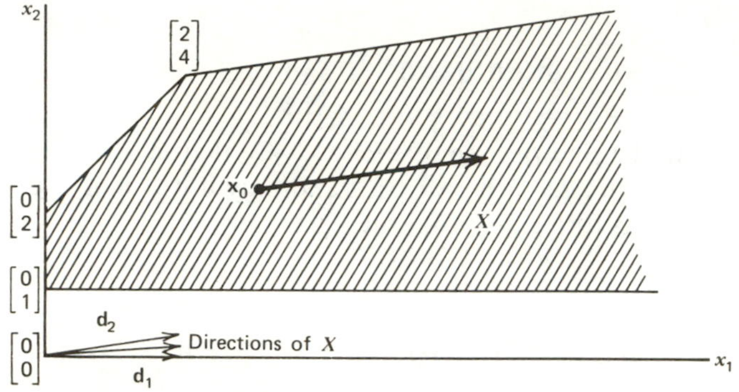

Figure 2.9. **Directions of convex sets.**

This collection of vectors is shown in Figure 2.9 and can be *normalized* such that each direction has norm (or length) equal to 1.

EXTREME DIRECTIONS OF A CONVEX SET

The notion of *extreme directions* is similar to the notion of extreme points. An extreme direction of a convex set is a direction of the set that cannot be represented as a positive combination of two distinct directions of the set. Two vectors, $\mathbf{d}_1$ and $\mathbf{d}_2$, are said to be *distinct* or not equivalent if $\mathbf{d}_1$ cannot be represented as a positive multiple of $\mathbf{d}_2$. In the foregoing example, after normalization, we have two extreme directions $\mathbf{d}_1 = (1, 0)$ and $\mathbf{d}_2 = (2/\sqrt{5}, 1/\sqrt{5})$. Any other direction of the set, which is not a multiple of $\mathbf{d}_1$ or $\mathbf{d}_2$, can be represented as $\lambda_1 \mathbf{d}_1 + \lambda_2 \mathbf{d}_2$ where $\lambda_1, \lambda_2 > 0$. Any ray that is contained in the convex set, and whose direction is an extreme direction, is called an *extreme ray*.

Convex Cones

A special important class of convex sets is convex cones. A *convex cone* C is a convex set with the additional property that $\lambda\mathbf{x} \in C$ for each $\mathbf{x} \in C$, and for each $\lambda \geqslant 0$. Note that a convex cone always contains the origin by letting $\lambda = 0$, and also that given any point $\mathbf{x} \in C$, the ray or *halfline* $\{\lambda\mathbf{x} : \lambda \geqslant 0\}$ belongs to C. Hence a convex cone is a convex set that consists entirely of rays emanating from the origin. Figure 2.10 shows some examples of convex cones.

Since a convex cone is formed by its rays, then a convex cone can be entirely characterized by its directions. In fact, not all directions are needed, since a nonextreme direction can be represented as a positive combination of extreme directions. In other words, a convex cone is fully characterized by its extreme directions.

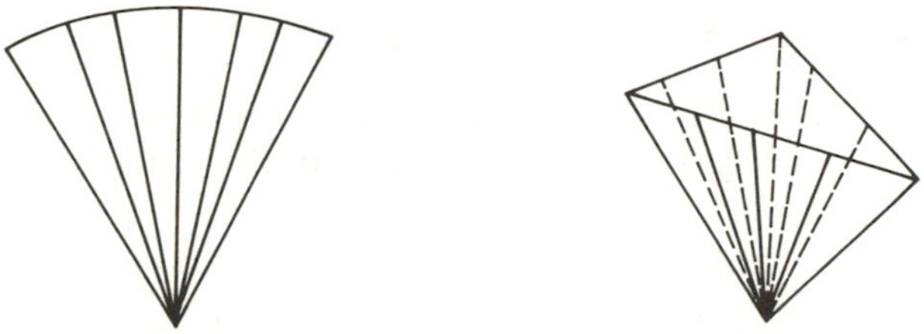

Figure 2.10. **Some examples of convex cones.**

As an example, consider the convex cone whose extreme directions are (1, 1) and (0, 1). From Figure 2.11 it is clear that the convex cone must be the set $\{(x_1, x_2) : x_1 \geqslant 0, x_1 \leqslant x_2\}$.

Given a set of vectors $\mathbf{a}_1, \mathbf{a}_2, \ldots, \mathbf{a}_k$, we can form the convex cone C generated by these vectors. This cone consists of all nonnegative combinations of $\mathbf{a}_1, \mathbf{a}_2, \ldots, \mathbf{a}_k$, that is,

$$C = \left\{ \sum_{j=1}^{k} \lambda_j \mathbf{a}_j : \lambda_j \geqslant 0 \quad \text{for} \quad j = 1, 2, \ldots, k \right\}$$

Figure 2.11 shows the convex cone generated by the points (0, 1) and (1, 1).

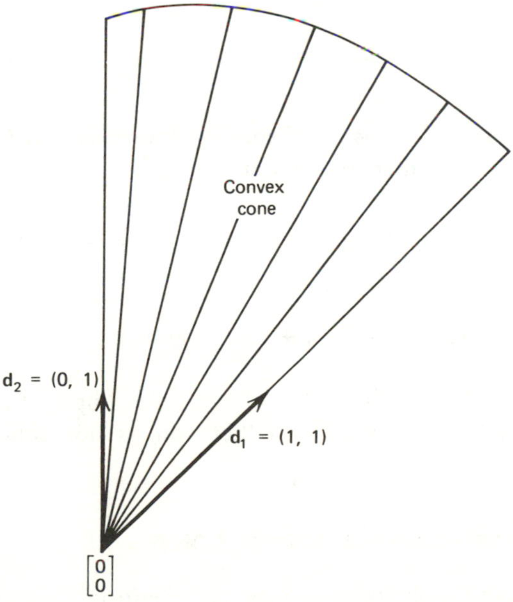

Figure 2.11. **Characterization of convex cones in terms of extreme directions.**

Convex and Concave Functions

Convex and concave functions play an important role in optimization problems. These functions naturally arise in linear optimization problems when dealing with parametric analysis.

A function f of the vector $(x_1, x_2, \ldots, x_n)$ is said to be *convex* if the following inequality holds for any two vectors $\mathbf{x}_1$ and $\mathbf{x}_2$:

$$f(\lambda \mathbf{x}_1 + (1 - \lambda)\mathbf{x}_2) \leq \lambda f(\mathbf{x}_1) + (1 - \lambda) f(\mathbf{x}_2) \qquad \text{for all} \quad \lambda \in [0, 1]$$

Figure 2.12*a* below shows an example of a convex function. Note that the foregoing inequality can be interpreted as follows: $\lambda f(\mathbf{x}_1) + (1 - \lambda) f(\mathbf{x}_2)$ where $\lambda \in [0, 1]$ represents the height of the *chord* joining $(\mathbf{x}_1, f(\mathbf{x}_1))$ and $(\mathbf{x}_2, f(\mathbf{x}_2))$ at the point $\lambda \mathbf{x}_1 + (1 - \lambda)\mathbf{x}_2$. Since $\lambda f(\mathbf{x}_1) + (1 - \lambda) f(\mathbf{x}_2) \geq f(\lambda \mathbf{x}_1 + (1 - \lambda)\mathbf{x}_2)$, then the height of the chord is at least as large as the height of the function itself.

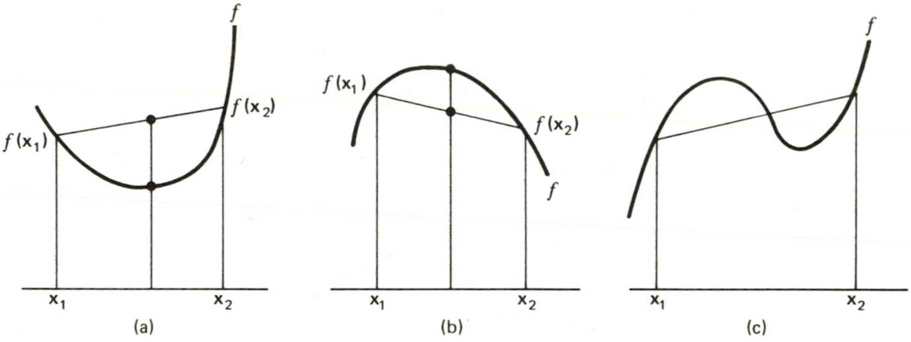

Figure 2.12. Examples of convex and concave functions: (*a*) Convex function. (*b*) Concave function. (*c*) Neither convex nor concave.

A function f is *concave* if and only if $-f$ is convex. This can be restated as follows:

$$f(\lambda \mathbf{x}_1 + (1 - \lambda)\mathbf{x}_2) \geq \lambda f(\mathbf{x}_1) + (1 - \lambda) f(\mathbf{x}_2) \qquad \text{for all} \quad \lambda \in [0, 1]$$

for any given $\mathbf{x}_1$ and $\mathbf{x}_2$. Figure 2.12*b* shows an example of a concave function. An example of a function that is neither convex nor concave is depicted in Figure 2.12*c*.

2.5 POLYHEDRAL SETS AND POLYHEDRAL CONES

Polyhedral sets and polyhedral cones represent important special cases of convex sets and convex cones. A *polyhedral set* is the intersection of a finite number of halfspaces. Since a halfspace can be represented by an inequality of the type $\mathbf{a}^i\mathbf{x} \leqslant b_i$, then a polyhedral set can be represented by the system $\mathbf{a}^i\mathbf{x} \leqslant b_i$ for $i = 1, \ldots, m$. Hence a polyhedral set can be represented by $\{\mathbf{x} : \mathbf{A}\mathbf{x} \leqslant \mathbf{b}\}$ where $\mathbf{A}$ is an $m \times n$ matrix whose ith row is $\mathbf{a}^i$ and $\mathbf{b}$ an m vector. Since an equation can be written as two inequalities, a polyhedral set can be represented by a finite number of linear inequalities and/or equations. As an example, consider the polyhedral set defined by the following inequalities:

$$
\begin{aligned}
-2x_1+x_2 &\leqslant 4\\
x_1+x_2 &\leqslant 3\\
x_1 \quad &\leqslant 2\\
x_1 \quad &\geqslant 0\\
x_2 &\geqslant 0
\end{aligned}
$$

The intersection of these five halfspaces gives the shaded set of Figure 2.13. Clearly the set is a convex set. We can see a distinct difference between the first inequality and the remaining inequalities. If the first inequality is disregarded, the polyhedral set is not affected. To differentiate between the first inequality and the remaining inequalities, we say that the hyperplanes corresponding to the

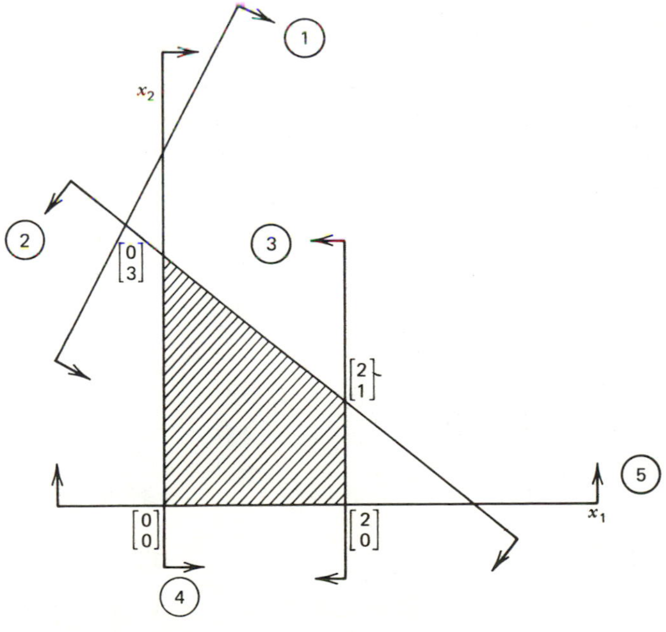

Figure 2.13. **Polyhedral set.**

second, third, fourth, and fifth inequalities, namely $\{(x_1, x_2) : x_1 + x_2 = 3\}$, $\{(x_1, x_2) : x_1 = 2\}$, $\{(x_1, x_2) : x_1 = 0\}$, and $\{(x_1, x_2) : x_2 = 0\}$, are *faces* of the polyhedral set.

A special class of polyhedral sets is *polyhedral cones*. A polyhedral cone is the intersection of a finite number of halfspaces, whose hyperplanes pass through the origin. In other words, C is a polyhedral cone if it can be represented as $\{\mathbf{x} : \mathbf{Ax} \leqslant \mathbf{0}\}$, where $\mathbf{A}$ is an $m \times n$ matrix. Note that the ith row of the matrix $\mathbf{A}$ is the normal vector to the hyperplane defining the ith halfspace. Figure 2.11 shows an example of a polyhedral cone.

2.6 REPRESENTATION OF POLYHEDRAL SETS

In this section we discuss the representation of a polyhedral set in terms of extreme points and extreme directions. This alternative representation will prove very useful throughout the book. The proof of the representation theorem is given in the Appendix. The reader may want to delay studying the Appendix until he develops more confidence with the notions of extreme points and extreme directions by the end of Chapter 3.

Bounded Polyhedral Sets

Consider the bounded polyhedral set (recall that a set is bounded if there is a number k such that $\|\mathbf{x}\| < k$ for each point $\mathbf{x}$ in the set) of Figure 2.14, which is formed as the intersection of five halfspaces. We have five extreme points, namely $\mathbf{x}_1$, $\mathbf{x}_2$, $\mathbf{x}_3$, $\mathbf{x}_4$, and $\mathbf{x}_5$. Note that any point in the set can be represented as a convex combination, or a weighted average, of these five extreme points. To illustrate, choose the point $\mathbf{x}$ shown in Figure 2.14. Note that $\mathbf{x}$ can be represented as a convex combination of $\mathbf{y}$ and $\mathbf{x}_4$, that is,

$$\mathbf{x} = \lambda \mathbf{y} + (1 - \lambda)\mathbf{x}_4 \qquad \text{where} \quad \lambda \in (0, 1)$$

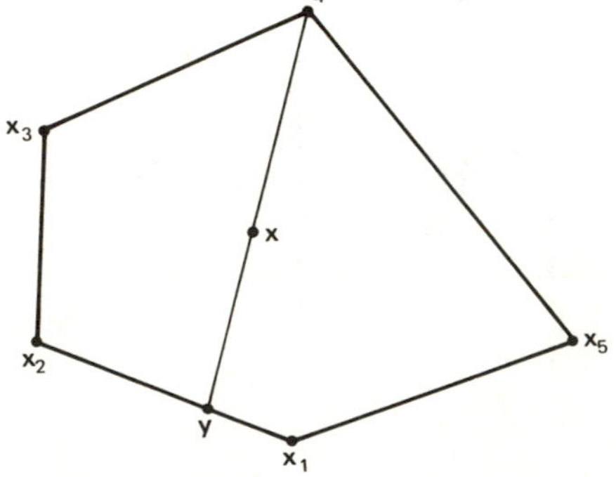

Figure 2.14. **Representation in terms of extreme points.**

But $\mathbf{y}$ can itself be represented as a convex combination of $\mathbf{x}_1$ and $\mathbf{x}_2$, that is,

$$\mathbf{y} = \mu\mathbf{x}_1 + (1 - \mu)\mathbf{x}_2 \qquad \text{where} \quad \mu \in (0, 1)$$

Substituting, we get

$$\mathbf{x} = \lambda\mu\mathbf{x}_1 + \lambda(1 - \mu)\mathbf{x}_2 + (1 - \lambda)\mathbf{x}_4$$

Since $\lambda \in (0, 1)$ and $\mu \in (0, 1)$, then $\lambda\mu$, $\lambda(1 - \mu)$, and $(1 - \lambda) \in (0, 1)$. Also $\lambda\mu + \lambda(1 - \mu) + (1 - \lambda) = 1$. In other words, $\mathbf{x}$ can be represented as a convex combination of the extreme points $\mathbf{x}_1$, $\mathbf{x}_2$, and $\mathbf{x}_4$. In general, any point in a bounded polyhedral set can be represented as a convex combination of its extreme points.

The above discussion is made more precise by the following theorem. The theorem is a special case of a more general result that will be stated later in the section.

***Theorem* 1** (Representation Theorem for the Bounded Case)

Let $X = \{\mathbf{x} : \mathbf{A}\mathbf{x} = \mathbf{b},\ \mathbf{x} \geqslant \mathbf{0}\}$ be a nonempty bounded (polyhedral) set. Then the set of extreme points is not empty and has a finite number of points, say $\mathbf{x}_1$, $\mathbf{x}_2, \ldots, \mathbf{x}_k$. Furthermore, $\mathbf{x} \in X$ if and only if $\mathbf{x}$ can be represented as a convex combination of $\mathbf{x}_1, \ldots, \mathbf{x}_k$, that is,

$$\mathbf{x} = \sum_{j=1}^{k} \lambda_j \mathbf{x}_j$$

$$\sum_{j=1}^{k} \lambda_j = 1$$

$$\lambda_j \geqslant 0 \qquad j = 1, 2, \ldots, k$$

Unbounded Polyhedral Sets

Let us now consider the case of an unbounded polyhedral set. An example is shown in Figure 2.15. We see that the set has three extreme points $\mathbf{x}_1$, $\mathbf{x}_2$, and $\mathbf{x}_3$, as well as two extreme directions $\mathbf{d}_1$ and $\mathbf{d}_2$. From Figure 2.15 it is clear that in general we can represent every point in the set as a convex combination of the extreme points, plus a nonnegative linear combination of the extreme directions. To illustrate, consider the point $\mathbf{x}$ in Figure 2.15. The point $\mathbf{x}$ can be represented as $\mathbf{y}$ plus a positive multiple of the extreme direction $\mathbf{d}_2$. Note that $\mathbf{x} - \mathbf{y}$ points in the direction $\mathbf{d}_2$. But $\mathbf{y}$ itself is a convex combination of the extreme points $\mathbf{x}_1$

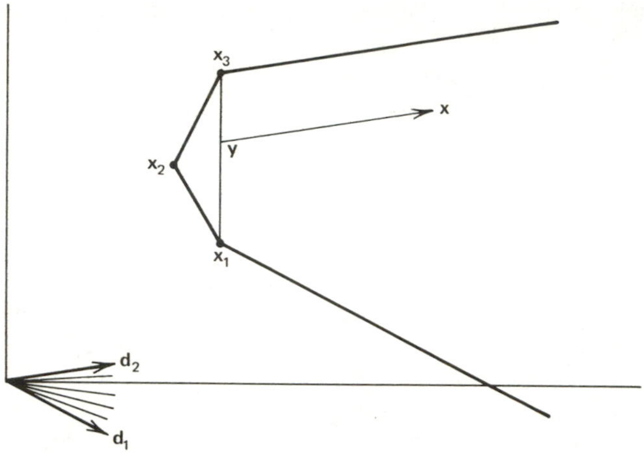

Figure 2.15. **Representation of polyhedral sets in terms of extreme points and extreme directions.**

and $\mathbf{x}_3$, and hence

$$\begin{aligned}\mathbf{x} &= \mathbf{y} + \mu \mathbf{d}_2 \\ &= \lambda \mathbf{x}_1 + (1 - \lambda)\mathbf{x}_3 + \mu \mathbf{d}_2\end{aligned}$$

where $\lambda \in (0, 1)$, and $\mu > 0$.

This discussion is made more precise by the following theorem, a proof of which is provided in the Appendix. The theorem applies whether the set is bounded or not. Of course if the set is bounded, then it has no directions, and the theorem reduces to the previous representation theorem.

***Theorem* 2** (Representation Theorem for the General Case)

Let $X = \{\mathbf{x} : \mathbf{A}\mathbf{x} = \mathbf{b}, \mathbf{x} \geqslant \mathbf{0}\}$ be a nonempty (polyhedral) set. Then the set of extreme points is not empty and has a finite number of points, say $\mathbf{x}_1, \mathbf{x}_2, \ldots, \mathbf{x}_k$. Furthermore, the set of extreme directions is empty if and only if X is bounded. If X is not bounded, then the set of extreme directions is nonempty and has a finite number of vectors, say $\mathbf{d}_1, \mathbf{d}_2, \ldots, \mathbf{d}_l$. Furthermore, $\mathbf{x} \in X$ if and only if it can be represented as a convex combination of $\mathbf{x}_1, \ldots, \mathbf{x}_k$ plus a nonnegative linear combination of $\mathbf{d}_1, \ldots, \mathbf{d}_l$, that is,

$$\mathbf{x} = \sum_{j=1}^{k} \lambda_j \mathbf{x}_j + \sum_{j=1}^{l} \mu_j \mathbf{d}_j$$

$$\sum_{j=1}^{k} \lambda_j = 1$$

$$\lambda_j \geqslant 0 \qquad j = 1, 2, \ldots, k$$

$$\mu_j \geqslant 0 \qquad j = 1, 2, \ldots, l$$

Example 2.6

Consider the polyhedral set formed by the following inequalities:

$$
\begin{aligned}
-3x_1 + x_2 &\leqslant -2 \\
-x_1 + x_2 &\leqslant 2 \\
-x_1 + 2x_2 &\leqslant 8 \\
-x_2 &\leqslant -2
\end{aligned}
$$

The set is illustrated in Figure 2.16. The extreme points and extreme directions are given below:

$$\mathbf{x}_1 = \begin{pmatrix} \frac{4}{3} \\ 2 \end{pmatrix} \qquad \mathbf{x}_2 = \begin{pmatrix} 2 \\ 4 \end{pmatrix} \qquad \mathbf{x}_3 = \begin{pmatrix} 4 \\ 6 \end{pmatrix}$$

$$\mathbf{d}_1 = \begin{pmatrix} 1 \\ 0 \end{pmatrix} \qquad \mathbf{d}_2 = \begin{pmatrix} 2 \\ 1 \end{pmatrix}$$

Let $\mathbf{x} = \begin{pmatrix} 4 \\ 3 \end{pmatrix}$ and note that $\mathbf{x}$ belongs to the above polyhedral set. Then $\mathbf{x}$ can be represented as follows:

$$\begin{pmatrix} 4 \\ 3 \end{pmatrix} = \lambda_1 \begin{pmatrix} \frac{4}{3} \\ 2 \end{pmatrix} + \lambda_2 \begin{pmatrix} 2 \\ 4 \end{pmatrix} + \lambda_3 \begin{pmatrix} 4 \\ 6 \end{pmatrix} + \mu_1 \begin{pmatrix} 1 \\ 0 \end{pmatrix} + \mu_2 \begin{pmatrix} 2 \\ 1 \end{pmatrix}$$

where $\lambda_1 = \lambda_2 = \frac{1}{2}$, $\lambda_3 = 0$, $\mu_1 = \frac{7}{3}$, and $\mu_2 = 0$. Note that the representation is not unique. By letting $\lambda_1 = \frac{3}{4}$, $\lambda_2 = 0$, $\lambda_3 = \frac{1}{4}$, $\mu_1 = 2$, and $\mu_2 = 0$, we get another representation of $\mathbf{x}$.

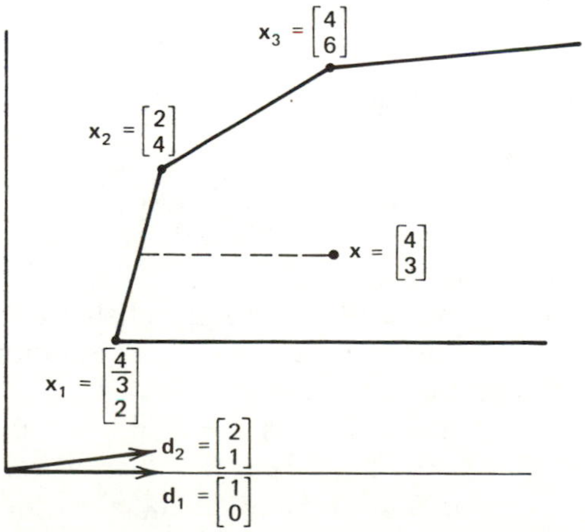

Figure 2.16. Numerical example.

2.7 FARKAS'S THEOREM

Farkas's theorem represents an important result that will be used to develop the Kuhn-Tucker optimality conditions for linear programs. In this section we state the theorem and discuss its geometric interpretation. Farkas's theorem deals with solvability of two systems of equations and inequalities.

***Theorem* 3** (Farkas's Theorem)

One and only one of the following two systems has a solution.

$$\text{System 1: } \mathbf{Ax} \leqslant \mathbf{0} \quad \text{and} \quad \mathbf{cx} > 0$$
$$\text{System 2: } \mathbf{wA} = \mathbf{c} \quad \text{and} \quad \mathbf{w} \geqslant \mathbf{0}$$

where $\mathbf{A}$ is a given $m \times n$ matrix, and $\mathbf{c}$ a given n vector.

Proof

The variables in the two systems are $\mathbf{x}$ and $\mathbf{w}$ respectively. The theorem can be restated as follows. If there exists an $\mathbf{x}$ with $\mathbf{Ax} \leqslant \mathbf{0}$ and $\mathbf{cx} > 0$, then there is no $\mathbf{w} \geqslant \mathbf{0}$ with $\mathbf{wA} = \mathbf{c}$. Conversely, if there exists no $\mathbf{x}$ with $\mathbf{Ax} \leqslant \mathbf{0}$ and $\mathbf{cx} > 0$, then there exists a $\mathbf{w} \geqslant \mathbf{0}$ such that $\mathbf{wA} = \mathbf{c}$.

Suppose that System 2 has a solution $\mathbf{w}$ such that $\mathbf{wA} = \mathbf{c}$ and $\mathbf{w} \geqslant \mathbf{0}$. Let $\mathbf{x}$ be such that $\mathbf{Ax} \leqslant \mathbf{0}$. Then $\mathbf{cx} = \mathbf{wAx} \leqslant 0$ since $\mathbf{w} \geqslant \mathbf{0}$ and $\mathbf{Ax} \leqslant \mathbf{0}$. This shows that $\mathbf{cx}$ cannot be positive and so System 1 has no solution. Now suppose that System 2 has no solution. This means that $\mathbf{c} \notin S = \{\mathbf{wA} : \mathbf{w} \geqslant \mathbf{0}\}$. Note that S is a closed convex set (why?). Applying Lemma 1 of the Appendix, we conclude that there exists an $\mathbf{x}$ such that $\mathbf{cx} > \mathbf{wAx}$ for all $\mathbf{w} \geqslant \mathbf{0}$. By letting $\mathbf{w} = \mathbf{0}$, we conclude that $\mathbf{cx} > 0$. Furthermore, since $\mathbf{w}$ can be chosen arbitrarily large, then we must have $\mathbf{Ax} \leqslant \mathbf{0}$ (why?). This shows that System 1 has a solution and Farkas's theorem is proved.

Geometric Interpretation of the Theorem

Denote the ith row of $\mathbf{A}$ by $\mathbf{a}^i$, $i = 1, 2, \ldots, m$. Let us consider System 1, in which $\mathbf{Ax} \leqslant \mathbf{0}$ means that $\mathbf{a}^i\mathbf{x} \leqslant 0$ for each i. That is, the "angle" between $\mathbf{x}$ and each row vector $\mathbf{a}^i$ is greater than or equal to 90°. Then $\mathbf{cx} > 0$ requires that the angle between $\mathbf{x}$ and $\mathbf{c}$ be less than 90°. Therefore System 1 has a solution if the intersection of the cone $\{\mathbf{x} : \mathbf{Ax} \leqslant \mathbf{0}\}$ and the open halfspace $\{\mathbf{x} : \mathbf{cx} > 0\}$ is not empty. Figure 2.17 shows a case where System 1 has a solution, with any $\mathbf{x}$ in the shaded area as a solution of System 1.

Now let us consider System 2. Here $\mathbf{wA} = \mathbf{c}$ and $\mathbf{w} \geqslant \mathbf{0}$ simply means that $\mathbf{c} = \sum_{i=1}^{m} w_i \mathbf{a}^i$, $w_i \geqslant 0$ for $i = 1, 2, \ldots, m$. In other words, System 2 has a solution if and only if $\mathbf{c}$ belongs to the cone generated by the rows of $\mathbf{A}$, namely the vectors $\mathbf{a}^1, \mathbf{a}^2, \ldots, \mathbf{a}^m$.

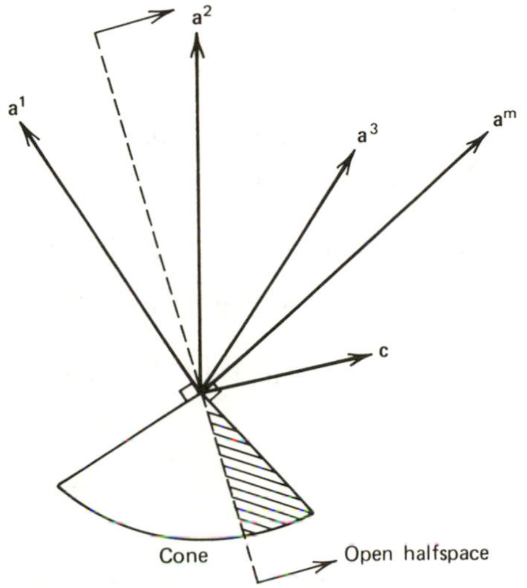

Figure 2.17. System 1 has a solution.

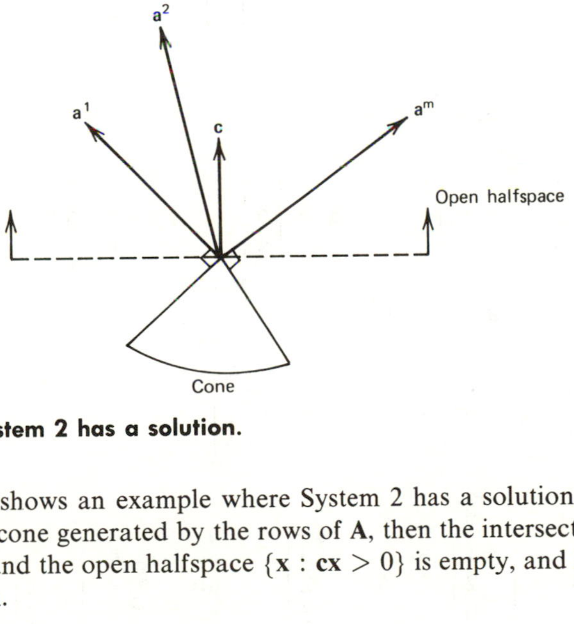

Figure 2.18. System 2 has a solution.

Figure 2.18 shows an example where System 2 has a solution. Note that if $\mathbf{c}$ belongs to the cone generated by the rows of $\mathbf{A}$, then the intersection of the cone $\{\mathbf{x} : \mathbf{Ax} \leqslant \mathbf{0}\}$ and the open halfspace $\{\mathbf{x} : \mathbf{cx} > 0\}$ is empty, and hence System 1 has no solution.

Other Forms of Farkas's Theorem

Farkas's theorem can be presented in various other forms. We present here a different form of Farkas's theorem, and we shall ask the reader to state and prove some other forms of the theorem in Exercise 2.52.

Corollary 1 (Alternate Form of Farkas's Theorem)

One and only one of the following two systems has a solution.

$$\text{System 1: } \mathbf{Ax} \leqslant \mathbf{0},\ \mathbf{x} \leqslant \mathbf{0} \text{ and } \mathbf{cx} > 0$$
$$\text{System 2: } \mathbf{wA} \leqslant \mathbf{c} \text{ and } \mathbf{w} \geqslant \mathbf{0}.$$

This form can be easily deduced from the format above by changing System 2 into equality form. Since $\mathbf{wA} \leqslant \mathbf{c}$ and $\mathbf{w} \geqslant \mathbf{0}$ is equivalent to $(\mathbf{w}, \mathbf{v})\begin{pmatrix}\mathbf{A}\\ \mathbf{I}\end{pmatrix} = \mathbf{c}$ and $(\mathbf{w}, \mathbf{v}) \geqslant (\mathbf{0}, \mathbf{0})$, then System 1 must read $\begin{pmatrix}\mathbf{A}\\ \mathbf{I}\end{pmatrix}\mathbf{x} \leqslant \begin{pmatrix}\mathbf{0}\\ \mathbf{0}\end{pmatrix}$ and $\mathbf{cx} > 0$, that is, $\mathbf{Ax} \leqslant \mathbf{0}$, $\mathbf{x} \leqslant \mathbf{0}$, and $\mathbf{cx} > 0$.

EXERCISES

2.1 Which of the following collection of vectors form a basis of E^3, span E^3, or neither?

a. $\mathbf{a}_1 = \begin{bmatrix}1\\2\\1\end{bmatrix}$, $\mathbf{a}_2 = \begin{bmatrix}-1\\0\\-1\end{bmatrix}$, $\mathbf{a}_3 = \begin{bmatrix}0\\0\\1\end{bmatrix}$

b. $\mathbf{a}_1 = \begin{bmatrix}1\\3\\2\end{bmatrix}$, $\mathbf{a}_2 = \begin{bmatrix}1\\0\\5\end{bmatrix}$

c. $\mathbf{a}_1 = \begin{bmatrix}-1\\2\\3\end{bmatrix}$, $\mathbf{a}_2 = \begin{bmatrix}0\\1\\0\end{bmatrix}$, $\mathbf{a}_3 = \begin{bmatrix}1\\2\\3\end{bmatrix}$, $\mathbf{a}_4 = \begin{bmatrix}-3\\2\\4\end{bmatrix}$

d. $\mathbf{a}_1 = \begin{bmatrix}1\\2\\1\end{bmatrix}$, $\mathbf{a}_2 = \begin{bmatrix}-3\\-1\\2\end{bmatrix}$, $\mathbf{a}_3 = \begin{bmatrix}-5\\0\\5\end{bmatrix}$

e. $\mathbf{a}_1 = \begin{bmatrix}1\\4\\2\end{bmatrix}$, $\mathbf{a}_2 = \begin{bmatrix}-1\\-4\\-1\end{bmatrix}$, $\mathbf{a}_3 = \begin{bmatrix}0\\1\\0\end{bmatrix}$

2.2 Let

$$\mathbf{a}_1 = \begin{bmatrix}-1\\2\\0\end{bmatrix}, \quad \mathbf{a}_2 = \begin{bmatrix}3\\2\\5\end{bmatrix}, \quad \text{and} \quad \mathbf{a}_3 = \begin{bmatrix}5\\-2\\5\end{bmatrix}$$

Are these vectors linearly independent? Do they span E^3?

2.3 Let $\mathbf{a}_1, \mathbf{a}_2, \ldots, \mathbf{a}_k$ form a basis for E^n. Show that $\mathbf{a}_1, \mathbf{a}_2, \ldots, \mathbf{a}_k$ are linearly independent. Also show that $k = n$.

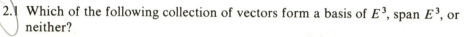

2.4 Show that the vectors

$$\mathbf{a}_1 = \begin{bmatrix} 1 \\ 0 \\ 0 \end{bmatrix}, \quad \mathbf{a}_2 = \begin{bmatrix} 0 \\ 1 \\ 0 \end{bmatrix}, \quad \text{and} \quad \mathbf{a}_3 = \begin{bmatrix} 1 \\ 5 \\ 3 \end{bmatrix}$$

form a basis for E^3. Supposing that $\mathbf{a}_2$ is replaced by $\begin{bmatrix} 1 \\ 1 \\ 0 \end{bmatrix}$, indicate whether the new set of vectors still forms a basis of E^3.

2.5 Suppose that $\mathbf{a}_1, \mathbf{a}_2, \ldots, \mathbf{a}_n$ form a basis of E^n and $\mathbf{y} = \lambda_1\mathbf{a}_1 + \lambda_2\mathbf{a}_2 + \cdots + \lambda_n\mathbf{a}_n$ with $\lambda_j = 0$. Prove that $\mathbf{a}_1, \ldots, \mathbf{a}_{j-1}, \mathbf{y}, \mathbf{a}_{j+1}, \ldots, \mathbf{a}_n$ do not form a basis of E^n.

2.6 Let $\mathbf{B}$ be an invertible matrix. Show that $\mathbf{B}^{-1}$ is unique.

2.7 Let

$$\mathbf{A} = \left[\begin{array}{c|c} \mathbf{B} & \mathbf{0} \\ \hline \mathbf{T} & \mathbf{I} \end{array}\right]$$

where $\mathbf{B}$ is an $m \times m$ invertible matrix, $\mathbf{I}$ is a $k \times k$ identity matrix, $\mathbf{0}$ is an $m \times k$ zero matrix, and $\mathbf{T}$ is an arbitrary $k \times m$ matrix. Show that $\mathbf{A}$ has an inverse and that

$$\mathbf{A}^{-1} = \left[\begin{array}{c|c} \mathbf{B}^{-1} & \mathbf{0} \\ \hline -\mathbf{T}\mathbf{B}^{-1} & \mathbf{I} \end{array}\right]$$

2.8 Find the inverse of the following triangular matrix.

$$\mathbf{A} = \begin{bmatrix} 1 & 4 & -3 & -4 \\ 0 & 5 & 1 & 2 \\ 0 & 0 & 2 & 5 \\ 0 & 0 & 0 & -1 \end{bmatrix}$$

2.9 Does the following matrix have an inverse? If the answer is yes, find $\mathbf{A}^{-1}$.

$$\mathbf{A} = \begin{bmatrix} 1 & -1 & 2 & 0 \\ 2 & 5 & 0 & 1 \\ 0 & 2 & 0 & 1 \\ 1 & 3 & 1 & 2 \end{bmatrix}$$

2.10 Let $\mathbf{A}$ be an $n \times n$ invertible matrix. Show that $\mathbf{A}^t$ has an inverse and that $(\mathbf{A}^t)^{-1} = (\mathbf{A}^{-1})^t$.

2.11 Show that if **A** and **B** are $n \times n$ matrices that are both invertible, then $(\mathbf{AB})^{-1} = \mathbf{B}^{-1}\mathbf{A}^{-1}$.

2.12 Let $\mathbf{A} = (\mathbf{a}_1, \mathbf{a}_2, \ldots, \mathbf{a}_j, \ldots, \mathbf{a}_m)$ be an invertible $m \times m$ matrix. Show that $\mathbf{A}^{-1}\mathbf{a}_j = \mathbf{e}_j$, where $\mathbf{e}_j$ is a vector of zeros except for a 1 at position j.

2.13 If the ith row of a square nonsingular matrix **B** is multiplied by a scalar $\lambda \neq 0$, what changes would result in $\mathbf{B}^{-1}$?

2.14 If the ith column of a square nonsingular matrix **B** is multiplied by a scalar $\lambda \neq 0$, what changes would result in $\mathbf{B}^{-1}$?

2.15 Let **B** be an invertible matrix with nonnegative entries. Show that every row of $\mathbf{B}^{-1}$ has at least one positive entry.

2.16 Let **A** be an $n \times n$ matrix. Suppose that **B** is an $n \times n$ matrix such that $\mathbf{AB} = \mathbf{I}$. Is it necessarily true that **A** has an inverse? Is it necessarily true that $\mathbf{B} = \mathbf{A}^{-1}$?

2.17 Find the determinants of the following matrices.

a. $\mathbf{A} = \begin{bmatrix} 1 & 0 & 1 \\ 2 & 1 & -1 \\ 0 & 2 & 2 \end{bmatrix}$

b. $\mathbf{A} = \begin{bmatrix} 1 & 0 & -2 & 1 \\ 2 & 1 & -1 & 1 \\ -2 & 2 & 2 & 2 \\ 1 & 3 & 1 & 5 \end{bmatrix}$

c. $\mathbf{A} = \begin{bmatrix} 1 & -2 & 1 \\ 2 & 1 & 5 \\ 3 & -2 & 2 \end{bmatrix}$

2.18 Find the rank of the following matrices.

$$\mathbf{A} = \begin{bmatrix} 1 & 0 & 1 & 1 \\ 2 & 2 & 4 & -1 \\ 1 & 0 & 5 & 3 \end{bmatrix}$$

$$\mathbf{A} = \begin{bmatrix} -1 & 1 & 0 \\ 1 & 4 & 5 \\ 2 & 3 & 5 \end{bmatrix}$$

2.19 Show that the determinant of a square triangular matrix is the product of the diagonal entries.

2.20 Solve the following system by Cramer's rule.

$$2x_1 + x_2 = 6$$

$$5x_1 - 2x_2 = 4$$

2.21 Demonstrate by enumeration that every basis matrix of the following system is triangular.

$$\begin{aligned} x_1 + x_2 - x_3 \qquad\qquad &= 1 \\ -x_1 \qquad + x_3 + x_4 \qquad &= 3 \\ -x_2 \qquad - x_4 + x_5 &= 3 \end{aligned}$$

2.22 Solve the following system of equations.

$$\begin{aligned} x_1 + 2x_2 + x_3 &= 1 \\ -x_1 + x_2 - x_3 &= 5 \\ 2x_1 + 3x_2 + x_3 &= -4 \end{aligned}$$

Without resolving the system, what is the solution if the right-hand side of the first equation is changed from 1 to 2?

2.23 Construct a general solution of the system $\mathbf{Ax} = \mathbf{b}$ where $\mathbf{A}$ is an $m \times n$ matrix with rank m.

2.24 What is the general solution of the following system?

$$\begin{aligned} x_1 + 2x_2 + x_3 &= 3 \\ -x_1 + 5x_2 + x_3 &= 6 \end{aligned}$$

2.25 Find all basic solutions of the following system.

$$\begin{aligned} -x_1 + x_2 + x_3 + x_4 - 2x_5 &= 4 \\ x_1 - 2x_2 \qquad + x_4 - x_5 &= 3 \end{aligned}$$

2.26 Determine whether the following system possesses: (a) no solution, (b) a unique solution, or (c) many (how many?) solutions.

$$\begin{aligned} x_1 + 3x_2 + x_3 - x_4 &= 1 \\ 5x_2 - 6x_3 + x_4 &= 0 \\ x_1 - 2x_2 + 4x_3 \qquad &= 1 \end{aligned}$$

2.27 Consider the system $\mathbf{Ax} = \mathbf{b}$ where $\mathbf{A} = [\mathbf{a}_1, \mathbf{a}_2, \ldots, \mathbf{a}_n]$ is an $m \times n$ matrix of rank m. Let $\mathbf{x}$ be any solution of this system. Starting with $\mathbf{x}$, construct a basic solution. (*Hint*. Suppose that $x_1, \ldots, x_p \neq 0$ and

$x_{p+1}, \ldots, x_n = 0$. If $p > m$, represent one of the columns $\mathbf{a}_j$ for $j = 1, 2, \ldots, p$ as a linear combination of the remaining vectors. This results in a new solution with a smaller number of nonzero variables. Repeat the process.)

2.28 Which of the following sets are convex and which are not?
a. $\{(x_1, x_2) : x_1^2 + x_2^2 \leqslant 1\}$
b. $\{(x_1, x_2, x_3) : x_1 + x_2 \leqslant 1, x_1 - x_3 \leqslant 2\}$
c. $\{(x_1, x_2) : x_2 - x_1^2 = 0\}$
d. $\{(x_1, x_2, x_3) : x_2 \geqslant x_1^2, x_1 + x_2 + x_3 \leqslant 6\}$
e. $\{(x_1, x_2) : x_1 = 1, |x_2| \leqslant 4\}$
f. $\{(x_1, x_2, x_3) : x_3 = |x_2|, x_1 \leqslant 4\}$

2.29 Show that a hyperplane $H = \{\mathbf{x} : \mathbf{px} = k\}$ and a halfspace $H^+ = \{\mathbf{x} : \mathbf{px} \geqslant k\}$ are convex sets.

2.30 Consider the set $\{(x_1, x_2) : -x_1 + x_2 \leqslant 2,\ x_1 + 2x_2 \leqslant 8,\ x_1 \geqslant 0, x_2 \geqslant 0\}$. What is the minimum distance from (4, 4) to the set? What is the point in the set closest to (4, 4)?

2.31 Consider the set $X = \{(x_1, x_2) : x_1, x_2 \geqslant 0, x_1 + x_2 \geqslant 2, x_2 \leqslant 4\}$. Find a hyperplane H such that X and the point $(-2, 1)$ are on different sides of the hyperplane. Write the equation of the hyperplane.

2.32 Let $\mathbf{a}_1 = \begin{pmatrix} 1 \\ 0 \end{pmatrix}$, $\mathbf{a}_2 = \begin{pmatrix} 2 \\ 3 \end{pmatrix}$, $\mathbf{a}_3 = \begin{pmatrix} -1 \\ 4 \end{pmatrix}$, $\mathbf{a}_4 = \begin{pmatrix} 5 \\ 3 \end{pmatrix}$, and $\mathbf{a}_5 = \begin{pmatrix} -4 \\ 3 \end{pmatrix}$. Illustrate geometrically the collection of all convex combinations of these five points.

2.33 Show that the set of feasible solutions to the following linear program forms a convex set.

$$\begin{aligned} \text{Minimize } & \mathbf{cx} \\ \text{Subject to } & \mathbf{Ax} = \mathbf{b} \\ & \mathbf{x} \geqslant \mathbf{0} \end{aligned}$$

2.34 Which of the following functions are convex, concave, or neither?
a. $f(x) = x^2$
b. $f(x_1, x_2) = e^{-x_1 - x_2} + x_1^2 - 2x_1$
c. $f(x_1, x_2) = \text{Maximum } (f_1(x_1, x_2), f_2(x_1, x_2))$ where $f_1(x_1, x_2) = x_1^2 + x_2^2$ and $f_2(x_1, x_2) = 2x_1^2 - x_2$
d. $f(x_1, x_2, x_3) = -x_1^2 - 2x_2^2 - x_3^2 + 2x_1x_2 - x_2x_3 + 2x_1 + 5x_3$
e. $f(x_1, x_2) = x_1^2 + 2x_2^2 - 2x_1x_2 + x_2$

2.35 Show that f is convex if and only if its *epigraph* $= \{(\mathbf{x}, y) : \mathbf{x} \in E^n, y \in E^1, y \geqslant f(\mathbf{x})\}$ is a convex set. Similarly show that f is concave if and only if its *hypograph* $= \{(\mathbf{x}, y) : \mathbf{x} \in E^n, y \in E^1, y \leqslant f(\mathbf{x})\}$ is a convex set.

2.36 Show that a differentiable function f is convex if and only if the following inequality holds for each fixed point $\mathbf{x}_0$ in E^n: $f(\mathbf{x}) \geqslant f(\mathbf{x}_0) + \nabla f(\mathbf{x}_0)(\mathbf{x} - \mathbf{x}_0)$for all $\mathbf{x} \in E^n$, where $\nabla f(\mathbf{x}_0)$ is the gradient vector of f at $\mathbf{x}_0$ given by

$$\left(\frac{\partial f(\mathbf{x}_0)}{\partial x_1}, \quad \frac{\partial f(\mathbf{x}_0)}{\partial x_2}, \ldots, \frac{\partial f(\mathbf{x}_0)}{\partial x_n} \right).$$

2.37 If S is an *open set*, show that the problem

$$\begin{aligned} &\text{Maximize} \quad \mathbf{cx} \\ &\text{Subject to} \quad \mathbf{x} \in S \end{aligned}$$

where $\mathbf{c} \neq \mathbf{0}$ possesses no optimal point. (*Note*. S is *open* if for each $\mathbf{x}_0 \in S$ there is an $\epsilon > 0$ such that $\|\mathbf{x} - \mathbf{x}_0\| < \epsilon$ implies that $\mathbf{x} \in S$.)

2.38 Show that if C is a convex cone, then C has at most one extreme point, namely the origin.

2.39 Show that C is a convex cone if and only if $\mathbf{x}$ and $\mathbf{y} \in C$ imply that $\lambda\mathbf{x} + \mu\mathbf{y} \in C$ for all $\lambda \geqslant 0$ and $\mu \geqslant 0$.

2.40 Find all extreme points of the following polyhedral set.

$$X = \{(x_1, x_2, x_3) : x_1 + x_2 + x_3 \leqslant 1, \quad -x_1 + 2x_2 \leqslant 4, \quad x_1, x_2, x_3 \geqslant 0\}$$

2.41 Find the extreme points of the region defined by the following inequalities.

$$\begin{aligned} x_1 + x_2 + x_3 &\leqslant 5 \\ -x_1 + x_2 + 2x_3 &\leqslant 6 \\ x_1, \; x_2, \; x_3 &\geqslant 0 \end{aligned}$$

(Hint. Introduce slack variables and consider basic solutions to the resulting system.)

2.42 Consider the nonempty polyhedral set $X = \{\mathbf{x} : \mathbf{Ax} \leqslant \mathbf{b}, \ \mathbf{x} \geqslant \mathbf{0}\}$. Show that $\mathbf{d}$ is a direction of the set if and only if $\mathbf{d} \neq \mathbf{0}$, $\mathbf{Ad} \leqslant \mathbf{0}$, and $\mathbf{d} \geqslant \mathbf{0}$. Obtain analogous results if the inequality $\mathbf{Ax} \leqslant \mathbf{b}$ is replaced by $\mathbf{Ax} = \mathbf{b}$, and if it is replaced by $\mathbf{Ax} \geqslant \mathbf{b}$.

2.43 Let $X = \{\mathbf{x} : \mathbf{Ax} \leqslant \mathbf{b}\}$ and let $\mathbf{x}_0$ be such at $\mathbf{Ax}_0 < \mathbf{b}$. Show that $\mathbf{x}_0$ cannot be an extreme point of X.

2.44 Prove that a polyhedral set X is bounded if and only if it has no directions.

2.45 Does the following set have any directions? Why?

$$\begin{aligned} -x_1+x_2 \quad\quad &= 4 \\ x_1+x_2+x_3 &\leqslant 6 \\ x_3 &\geqslant 1 \\ x_1, x_2, x_3 &\geqslant 0 \end{aligned}$$

2.46 Let $X = \{\mathbf{x} : \mathbf{Ax} = \mathbf{b}, \mathbf{x} \geqslant \mathbf{0}\}$ where $\mathbf{A}$ is an $m \times n$ matrix with rank m. Show that $\mathbf{d}$ is an extreme direction of X if and only if $\mathbf{d}$ is a positive multiple of the vector $(-\mathbf{y}_j{}', 0, 0, \ldots, 1, 0, \ldots, 0)'$ where

$\mathbf{y}_j = \mathbf{B}^{-1}\mathbf{a}_j \leqslant \mathbf{0}$
$\mathbf{A} = [\mathbf{B}, \mathbf{N}]$ where $\mathbf{B}$ is an $m \times m$ *invertible matrix*
$\mathbf{a}_j$ = *a* column of $\mathbf{N}$
the 1 appears in position j

Illustrate by the following system.

$$\begin{aligned} -x_1+\ x_2+x_3 \quad\quad &= 2 \\ -x_1+2x_2 \quad\quad +x_4 &= 6 \\ x_1, x_2, x_3, x_4 &\geqslant 0 \end{aligned}$$

2.47 Show that an unbounded polyhedral set of the form $\{\mathbf{x} : \mathbf{Ax} = \mathbf{b}, \mathbf{x} \geqslant \mathbf{0}\}$ has at least one extreme direction. (*Hint*. Start with any direction and reduce it to the direction characterized in Exercise 2.46 above.)

2.48 You are given the following polyhedral set. Identify the faces, extreme points, extreme directions, and extreme rays of the set.

$$\begin{aligned} x_1-\ x_2+\ x_3 &\leqslant 10 \\ 2x_1-\ x_2+2x_3 &\leqslant 40 \\ 3x_1-2x_2+3x_3 &\leqslant 50 \\ x_1,\ x_2,\ x_3 &\geqslant 0 \end{aligned}$$

(Hint. Introduce slack variables and examine the basic solutions of the resulting system and the directions given in Exercise 2.46.)

2.49 Consider the polyhedral set $X = \{\mathbf{x} : \mathbf{px} = k\}$ where $\mathbf{p}$ is a nonzero vector and k is a scalar. Show that X has neither extreme points nor extreme rays. How do you explain this in terms of the general representation theorem?

2.50 Let $X = \{(x_1, x_2) : x_1 - x_2 \leqslant 3, 2x_1 + x_2 \leqslant 4, x_1 \geqslant -3\}$. Find all extreme points of X and represent $\mathbf{x} = (0, 1)$ has a convex combination of the extreme points.

2.51 Find all extreme points and extreme directions of the following polyhedral set:

$$X = \{(x_1, x_2, x_3, x_4) : x_1, x_2, x_3, x_4 \geqslant 0, -x_1 + x_2 + x_3 = 1, \; x_2 + x_4 = 2\}$$

Represent $x = (1, \frac{3}{2}, \frac{1}{2}, \frac{1}{2})$ as a convex combination of the extreme points of X plus a nonnegative combination of the extreme directions of X.

2.52 Suppose that the following system has no solution.

$$\mathbf{Ax} = \mathbf{0}, \qquad \mathbf{x} \geqslant \mathbf{0}, \qquad \text{and} \quad \mathbf{cx} > 0$$

Devise another system that must have a solution. (*Hint.* Use Farkas's theorem.)

2.53 Let $\mathbf{A} = \begin{bmatrix} 1 & 1 \\ 0 & 2 \\ -1 & 4 \end{bmatrix}$ and $\mathbf{c} = (1, 4)$. Which of the following two systems has a solution?

$$\text{System 1:} \quad \mathbf{Ax} \leqslant \mathbf{0} \quad \mathbf{cx} > 0$$
$$\text{System 2:} \quad \mathbf{wA} = \mathbf{c} \quad \mathbf{w} \geqslant \mathbf{0}$$

Illustrate geometrically.

2.54 Consider the following problem, where $\mathbf{A}$ is an $m \times n$ matrix.

$$\begin{aligned} \text{Minimize} \quad & \mathbf{cx} \\ \text{Subject to} \quad & \mathbf{Ax} = \mathbf{b} \\ & \mathbf{x} \geqslant \mathbf{0} \end{aligned}$$

Let $\mathbf{x}^*$ be an optimal solution. Show that there is a $\mathbf{w}$ such that

$$\mathbf{wA} \leqslant \mathbf{c}$$
$$(\mathbf{wA} - \mathbf{c})\mathbf{x}^* = 0$$

These are the *Kuhn-Tucker optimality conditions* of linear programming. (*Hint*. Suppose that $x_1^*, \ldots, x_p^* > 0$ and $x_j^* = 0$ for $j = p + 1, \ldots, n$.

Show that the system

$$\begin{aligned} \mathbf{Ax} &= \mathbf{0} \\ -\mathbf{cx} &> 0 \\ x_{p+1}, \ldots, x_n &\geqslant 0 \end{aligned}$$

has no solution. Then use Farkas's theorem.)

NOTES AND REFERENCES

1. Sections 2.1 through 2.3 present a quick review of some relevant results of vector and matrix algebra.
2. Sections 2.4 and 2.5 give some basic definitions and properties of convex sets, convex cones, and convex functions. For more details the reader may refer to Eggleston [134], Mangasarian [319], and the more advanced text of Rockafellar [377].
3. Correspondence between bases and extreme points is established in Section 3.2, and characterization of extreme directions is presented in Exercise 2.46 (also see the Appendix).
4. The representation theorem for polyhedral sets evolved from the work of Minkowski [335] and Goldman and Tucker [200]. The result is also true for (nonpolyhedral) convex sets which contain no lines. See Rockafellar [377] and Bazaraa and Shetty [26].
5. In Section 2.7 Farkas's theorem is presented. The theorem was published by Farkas [142] in 1902 and is used extensively in the literature of mathematical programming. More specifically, Farkas's theorem is used to establish optimality conditions, duality relationships, and other theorems of the alternative. The reader may refer to Kuhn and Tucker [295], and Mangasarian [319].

THREE: THE SIMPLEX METHOD

In this chapter the simplex method for solving a linear programming problem is developed. We first show that if an optimal solution exists, then an optimal extreme point also exists. Extreme points are then characterized in terms of basic feasible solutions. We then describe the simplex method for improving these solutions until optimality is reached, or else until we conclude that the optimal value is unbounded. The well-known tableau format of the simplex method is also discussed. This is a key chapter, fundamental to the development of many other chapters in the book.

3.1 EXTREME POINTS AND OPTIMALITY

We observed from Figure 1.3 that when an optimal solution of a linear programming problem exists, an optimal extreme point also exists. This observation is always true, as will be shown shortly.

Consider the following linear programming problem.

$$\begin{aligned} \text{Minimize} \quad & \mathbf{cx} \\ \text{Subject to} \quad & \mathbf{Ax} = \mathbf{b} \\ & \mathbf{x} \geqslant \mathbf{0} \end{aligned}$$

Let $\mathbf{x}_1, \mathbf{x}_2, \ldots, \mathbf{x}_k$ be the extreme points of the constraint set, and let $\mathbf{d}_1, \mathbf{d}_2, \ldots, \mathbf{d}_l$ be the extreme directions of the constraint set. Recall that any point $\mathbf{x}$ such that $\mathbf{Ax} = \mathbf{b}$ and $\mathbf{x} \geqslant \mathbf{0}$ can be represented as

$$\mathbf{x} = \sum_{j=1}^{k} \lambda_j \mathbf{x}_j + \sum_{j=1}^{l} \mu_j \mathbf{d}_j$$

where

$$\sum_{j=1}^{k} \lambda_j = 1$$

$$\lambda_j \geqslant 0 \quad j = 1, 2 \ldots, k$$

$$\mu_j \geqslant 0 \quad j = 1, 2 \ldots, l$$

Therefore the linear programming problem can be transformed into a problem in the variables $\lambda_1, \lambda_2, \ldots, \lambda_k, \mu_1, \mu_2, \ldots, \mu_l$, resulting in the following linear program.

$$\text{Minimize} \sum_{j=1}^{k} (\mathbf{cx}_j)\lambda_j + \sum_{j=1}^{l} (\mathbf{cd}_j)\,\mu_j$$

$$\text{Subject to} \sum_{j=1}^{k} \lambda_j = 1$$

$$\lambda_j \geqslant 0 \qquad j = 1, 2, \ldots, k$$

$$\mu_j \geqslant 0 \qquad j = 1, 2, \ldots, l$$

Since the μ_j's can be made arbitrarilv large, the minimum is $-\infty$ if $\mathbf{cd}_j < 0$ for some $j = 1, 2, \ldots, l$. If $\mathbf{cd}_j \geqslant 0$ for all $j = 1, 2, \ldots, l$, then the corresponding μ_j can be chosen as zero. Now in order to minimize $\Sigma_{j=1}^{k}(\mathbf{cx}_j)\,\lambda_j$ over $\lambda_1, \lambda_2, \ldots, \lambda_k$ satisfying $\lambda_j \geqslant 0$ for $j = 1, 2, \ldots, k$, and $\Sigma_{j=1}^{k}\lambda_j = 1$, we simply find the Minimum $\mathbf{cx}_j$, say $\mathbf{cx}_p$, let $\lambda_p = 1$, and all other λ_j's equal to zero.

To summarize, the optimal solution of the linear problem is finite if and only if $\mathbf{cd}_j \geqslant 0$ for all extreme directions. Furthermore, if this is the case, then we find the minimum point, by picking the minimum objective value among all extreme points. This shows that if an optimal solution exists, we must be able to find an optimal extreme point. Of course, if the Minimum $\mathbf{cx}_j$ occurs at more than one index, then each corresponding extreme point is an optimal point, and also each convex combination of these points is an optimal solution (why?).

Example 3.1

Consider the region defined by the following constraints:

$$\begin{aligned} -x_1 + x_2 &\leqslant 2 \\ -x_1 + 2x_2 &\leqslant 6 \\ x_1, \quad x_2 &\geqslant 0 \end{aligned}$$

Note that this region has three extreme points $\mathbf{x}_1$, $\mathbf{x}_2$, and $\mathbf{x}_3$, and two extreme directions $\mathbf{d}_1$ and $\mathbf{d}_2$ (see Figure 3.1). These are

$$\mathbf{x}_1 = \begin{bmatrix} 0 \\ 0 \end{bmatrix} \qquad \mathbf{x}_2 = \begin{bmatrix} 0 \\ 2 \end{bmatrix} \qquad \mathbf{x}_3 = \begin{bmatrix} 2 \\ 4 \end{bmatrix}$$

$$\mathbf{d}_1 = \begin{bmatrix} 1 \\ 0 \end{bmatrix} \qquad \mathbf{d}_2 = \begin{bmatrix} 2 \\ 1 \end{bmatrix}$$

Now suppose that we are minimizing $x_1 - 3x_2$ over the foregoing region. We see from Figure 3.1*a* that the optimal is unbounded and has value $-\infty$. In this case we have

$$\mathbf{cx}_1 = (1, -3)\begin{bmatrix} 0 \\ 0 \end{bmatrix} = 0$$

$$\mathbf{cx}_2 = (1, -3)\begin{bmatrix} 0 \\ 2 \end{bmatrix} = -6$$

$$\mathbf{cx}_3 = (1, -3)\begin{bmatrix} 2 \\ 4 \end{bmatrix} = -10$$

$$\mathbf{cd}_1 = (1, -3)\begin{bmatrix} 1 \\ 0 \end{bmatrix} = 1$$

$$\mathbf{cd}_2 = (1, -3)\begin{bmatrix} 2 \\ 1 \end{bmatrix} = -1$$

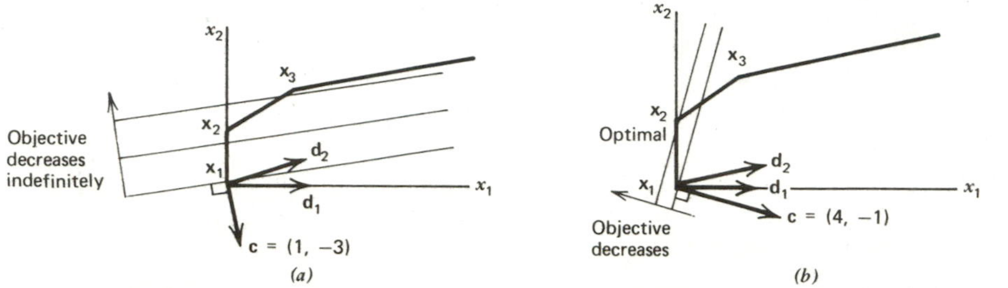

Figure 3.1. **Extreme directions and optimality:** (*a*) **Unbounded optimal solution.** (*b*) **Bounded optimal solution.**

The problem is equivalent to the following.

$$\text{Minimize } 0\lambda_1 - 6\lambda_2 - 10\lambda_3 + \mu_1 - \mu_2$$

$$\text{Subject to } \lambda_1 + \lambda_2 + \lambda_3 = 1$$

$$\lambda_1, \lambda_2, \lambda_3, \mu_1, \mu_2 \geqslant 0$$

Since $\mathbf{cd}_2 = -1 < 0$ and μ_2 can be made arbitrarily large without violating the foregoing constraints, the optimal is achieved by letting $\mu_2 = \infty$, giving an unbounded objective of $-\infty$. Then μ_1 can be chosen equal to zero. Any set of nonnegative λ_1, λ_2, λ_3 adding to 1 satisfies the foregoing constraints, for example, $\lambda_1 = 1$, $\lambda_2 = \lambda_3 = 0$. This illustrates the necessary and sufficient condition for unboundness, namely $\mathbf{cd} < 0$ for some extreme direction.

Now consider the problem of minimizing $4x_1 - x_2$ over the same region. From Figure 3.1*b* the optimal solution is the extreme point $\mathbf{x}_2 = \begin{bmatrix} 0 \\ 2 \end{bmatrix}$. In this case we have

$$\mathbf{cx}_1 = (4, -1)\begin{bmatrix} 0 \\ 0 \end{bmatrix} = 0$$

$$\mathbf{cx}_2 = (4, -1)\begin{bmatrix} 0 \\ 2 \end{bmatrix} = -2$$

$$\mathbf{cx}_3 = (4, -1)\begin{bmatrix} 2 \\ 4 \end{bmatrix} = 4$$

$$\mathbf{cd}_1 = (4, -1)\begin{bmatrix} 1 \\ 0 \end{bmatrix} = 4$$

$$\mathbf{cd}_2 = (4, -1)\begin{bmatrix} 2 \\ 1 \end{bmatrix} = 7$$

The problem is therefore equivalent to the following.

$$\text{Minimize} \quad 0\lambda_1 - 2\lambda_2 + 4\lambda_3 + 4\mu_1 + 7\mu_2$$

$$\begin{aligned} \text{Subject to} \quad & \lambda_1 + \lambda_2 + \lambda_3 \qquad\qquad = 1 \\ & \lambda_1, \quad \lambda_2, \quad \lambda_3, \quad \mu_1, \quad \mu_2 \geq 0 \end{aligned}$$

Since the coefficients of μ_1 and μ_2 in the objective function are positive, we let $\mu_1 = \mu_2 = 0$. In order to minimize the expression $0\lambda_1 - 2\lambda_2 + 4\lambda_3$ subject to $\lambda_1 + \lambda_2 + \lambda_3 = 1$ and $\lambda_1, \lambda_2, \lambda_3 \geq 0$, we let $\lambda_2 = 1$ and $\lambda_1 = \lambda_3 = 0$. This shows that the optimal solution is the extreme point $\mathbf{x}_2 = \begin{pmatrix} 0 \\ 2 \end{pmatrix}$.

Minimizing $\mathbf{cx}$ corresponds to moving the plane $\mathbf{cx}$ = constant in the direction $-\mathbf{c}$ as far as possible. When $\mathbf{c} = (1, -3)$ we can move the plane indefinitely while always intersecting the feasible region, and hence the optimal value is $-\infty$. When $\mathbf{c} = (4, -1)$ we cannot move the plane indefinitely and we must stop at the point $\mathbf{x}_2$; otherwise we "leave" the feasible region.

3.2 BASIC FEASIBLE SOLUTIONS

We have developed, in the previous section, a necessary and sufficient condition for an unbounded solution. We also showed that if an optimal solution exists, then an optimal extreme point also exists. The notion of an extreme point is a geometric notion, and an algebraic characterization of extreme points is needed before they can be utilized from a computational point of view.

In this section we introduce basic feasible solutions, and show that they correspond to extreme points. Since an algebraic characterization of the former (and hence the latter) exists, we shall be able to move from one basic feasible solution to another until optimality is reached.

Definition (Basic Feasible Solutions)

Consider the system $\mathbf{Ax} = \mathbf{b}$ and $\mathbf{x} \geq \mathbf{0}$, where $\mathbf{A}$ is an $m \times n$ matrix and $\mathbf{b}$ is an m vector. Suppose that rank $(\mathbf{A}, \mathbf{b})$ = rank $(\mathbf{A}) = m$. After possibly rearranging the columns of $\mathbf{A}$, let $\mathbf{A} = [\mathbf{B}, \mathbf{N}]$ where $\mathbf{B}$ is an $m \times m$ invertible matrix and $\mathbf{N}$ is an $m \times (n - m)$ matrix. The point $\mathbf{x} = \begin{bmatrix} \mathbf{x}_B \\ \mathbf{x}_N \end{bmatrix}$ where

$$\mathbf{x}_B = \mathbf{B}^{-1}\mathbf{b}$$
$$\mathbf{x}_N = \mathbf{0}$$

is called a *basic solution* of the system. If $\mathbf{x}_B \geq \mathbf{0}$, then $\mathbf{x}$ is called a *basic feasible solution* of the system. Here **B** is called the *basic matrix* (or simply the basis) and **N** is called the *nonbasic matrix*. The components of $\mathbf{x}_B$ are called *basic variables*, and the components of $\mathbf{x}_N$ are called *nonbasic variables*. If $\mathbf{x}_B > \mathbf{0}$, then $\mathbf{x}$ is called a *nondegenerate basic feasible solution*, and if at least one component of $\mathbf{x}_B$ is zero, then $\mathbf{x}$ is called a *degenerate basic feasible solution*.

The notion of a basic feasible solution is illustrated by the following two examples.

Example 3.2 (Basic Feasible Solutions)

Consider the polyhedral set defined by the following inequalities (and illustrated in Figure 3.2):

$$\begin{aligned} x_1 + x_2 &\leqslant 6 \\ x_2 &\leqslant 3 \\ x_1, \quad x_2 &\geqslant 0 \end{aligned}$$

By introducing the slack variables x_3 and x_4, the problem is put in the following standard format:

$$\begin{aligned} x_1 + x_2 + x_3 \qquad &= 6 \\ x_2 \qquad + x_4 &= 3 \\ x_1, \ x_2, \ x_3, \ x_4 &\geqslant 0 \end{aligned}$$

Note that the constraint matrix $\mathbf{A} = [\mathbf{a}_1, \mathbf{a}_2, \mathbf{a}_3, \mathbf{a}_4] = \begin{bmatrix} 1 & 1 & 1 & 0 \\ 0 & 1 & 0 & 1 \end{bmatrix}$. From the

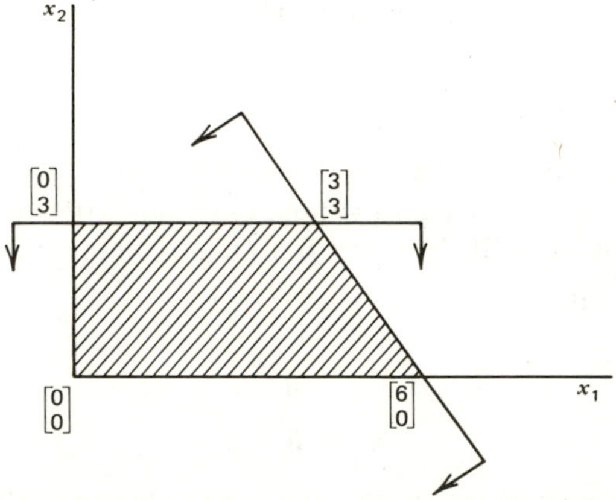

Figure 3.2. **Basic feasible solutions.**

foregoing definition, basic feasible solutions correspond to finding a 2×2 basis $\mathbf{B}$ with nonnegative $\mathbf{B}^{-1}\mathbf{b}$. The following are the possible ways of extracting $\mathbf{B}$ out of $\mathbf{A}$.

1. $\mathbf{B} = [\mathbf{a}_1, \mathbf{a}_2] = \begin{bmatrix} 1 & 1 \\ 0 & 1 \end{bmatrix}$

$$\mathbf{x}_B = \begin{bmatrix} x_1 \\ x_2 \end{bmatrix} = \mathbf{B}^{-1}\mathbf{b} = \begin{bmatrix} 1 & -1 \\ 0 & 1 \end{bmatrix} \begin{bmatrix} 6 \\ 3 \end{bmatrix} = \begin{bmatrix} 3 \\ 3 \end{bmatrix}, \qquad \mathbf{x}_N = \begin{bmatrix} x_3 \\ x_4 \end{bmatrix} = \begin{bmatrix} 0 \\ 0 \end{bmatrix}$$

2. $\mathbf{B} = [\mathbf{a}_1, \mathbf{a}_4] = \begin{bmatrix} 1 & 0 \\ 0 & 1 \end{bmatrix}$

$$\mathbf{x}_B = \begin{bmatrix} x_1 \\ x_4 \end{bmatrix} = \mathbf{B}^{-1}\mathbf{b} = \begin{bmatrix} 1 & 0 \\ 0 & 1 \end{bmatrix} \begin{bmatrix} 6 \\ 3 \end{bmatrix} = \begin{bmatrix} 6 \\ 3 \end{bmatrix}, \qquad \mathbf{x}_N = \begin{bmatrix} x_2 \\ x_3 \end{bmatrix} = \begin{bmatrix} 0 \\ 0 \end{bmatrix}$$

3. $\mathbf{B} = [\mathbf{a}_2, \mathbf{a}_3] = \begin{bmatrix} 1 & 1 \\ 1 & 0 \end{bmatrix}$

$$\mathbf{x}_B = \begin{bmatrix} x_2 \\ x_3 \end{bmatrix} = \mathbf{B}^{-1}\mathbf{b} = \begin{bmatrix} 0 & 1 \\ 1 & -1 \end{bmatrix} \begin{bmatrix} 6 \\ 3 \end{bmatrix} = \begin{bmatrix} 3 \\ 3 \end{bmatrix}, \qquad \mathbf{x}_N = \begin{bmatrix} x_1 \\ x_4 \end{bmatrix} = \begin{bmatrix} 0 \\ 0 \end{bmatrix}$$

4. $\mathbf{B} = [\mathbf{a}_2, \mathbf{a}_4] = \begin{bmatrix} 1 & 0 \\ 1 & 1 \end{bmatrix}$

$$\mathbf{x}_B = \begin{bmatrix} x_2 \\ x_4 \end{bmatrix} = \mathbf{B}^{-1}\mathbf{b} = \begin{bmatrix} 1 & 0 \\ -1 & 1 \end{bmatrix} \begin{bmatrix} 6 \\ 3 \end{bmatrix} = \begin{bmatrix} 6 \\ -3 \end{bmatrix}, \qquad \mathbf{x}_N = \begin{bmatrix} x_1 \\ x_3 \end{bmatrix} = \begin{bmatrix} 0 \\ 0 \end{bmatrix}$$

5. $\mathbf{B} = [\mathbf{a}_3, \mathbf{a}_4] = \begin{bmatrix} 1 & 0 \\ 0 & 1 \end{bmatrix}$

$$\mathbf{x}_B = \begin{bmatrix} x_3 \\ x_4 \end{bmatrix} = \mathbf{B}^{-1}\mathbf{b} = \begin{bmatrix} 1 & 0 \\ 0 & 1 \end{bmatrix} \begin{bmatrix} 6 \\ 3 \end{bmatrix} = \begin{bmatrix} 6 \\ 3 \end{bmatrix}, \qquad \mathbf{x}_N = \begin{bmatrix} x_1 \\ x_2 \end{bmatrix} = \begin{bmatrix} 0 \\ 0 \end{bmatrix}$$

Note that the points corresponding to 1, 2, 3, and 5 above are basic feasible solutions. The point obtained in 4 is a basic solution, but is not feasible because it violates the nonnegativity restrictions. In other words, we have four basic feasible solutions, namely

$$\mathbf{x}_1 = \begin{bmatrix} 3 \\ 3 \\ 0 \\ 0 \end{bmatrix}, \qquad \mathbf{x}_2 = \begin{bmatrix} 6 \\ 0 \\ 0 \\ 3 \end{bmatrix}, \qquad \mathbf{x}_3 = \begin{bmatrix} 0 \\ 3 \\ 3 \\ 0 \end{bmatrix}, \qquad \mathbf{x}_4 = \begin{bmatrix} 0 \\ 0 \\ 6 \\ 3 \end{bmatrix}$$

These points belong to E^4 since after introducing the slack variables we have four variables. These basic feasible solutions, projected in E^2—that is, in the (x_1, x_2) space—give rise to the following four points:

$$\begin{bmatrix} 3 \\ 3 \end{bmatrix}, \quad \begin{bmatrix} 6 \\ 0 \end{bmatrix}, \quad \begin{bmatrix} 0 \\ 3 \end{bmatrix}, \quad \begin{bmatrix} 0 \\ 0 \end{bmatrix}$$

These four points are illustrated in Figure 3.2. Note that these points are precisely the extreme points of the feasible region.

In this example, the possible number of basic feasible solutions is bounded by the number of ways of extracting two columns out of four columns to form the basis. Therefore the number of basic feasible solutions is less or equal to

$$\binom{4}{2} = \frac{4!}{2!2!} = 6.$$

Out of these six possibilities, one point violates the nonnegativity of $\mathbf{B}^{-1}\mathbf{b}$. Furthermore, $\mathbf{a}_1$ and $\mathbf{a}_3$ could not have been used to form a basis since $\mathbf{a}_1 = \mathbf{a}_3 = \begin{bmatrix} 1 \\ 0 \end{bmatrix}$ are linearly dependent, and hence the matrix $\begin{bmatrix} 1 & 1 \\ 0 & 0 \end{bmatrix}$ does not qualify as a basis. This leaves four basic feasible solutions. In general, the number of basic feasible solutions is less than or equal to

$$\binom{n}{m} = \frac{n!}{m!\,(n - m)!}$$

There is another intuitive way of viewing basic solutions and basic feasible solutions. Each constraint, including the nonnegativity constraints, can be associated uniquely with a certain variable. Thus $x_1 \geqslant 0$ can be associated with the variable x_1, and the line $x_1 = 0$ is the boundary of the halfspace corresponding to $x_1 \geqslant 0$. Also, $x_1 + x_2 \leqslant 6$ can be associated with the variable x_3, and $x_3 = 0$ is the boundary of the halfspace corresponding to $x_1 + x_2 \leqslant 6$. Graphically portraying the boundary of the various constraints, we get the graph of Figure 3.3. Now, basic solutions correspond to the intersection of two lines in this graph. The lines correspond to the nonbasic variables. In the graph there are five intersections corresponding to five basic solutions. Note that there is no intersection of the lines $x_2 = 0$ and $x_4 = 0$ and thus no basic solution corre-

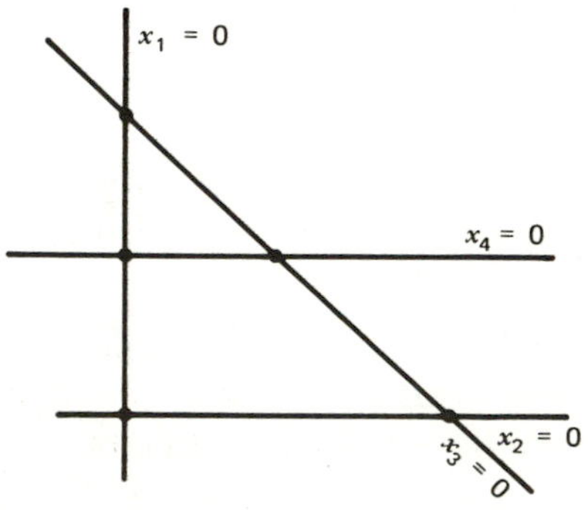

Figure 3.3. Associating basic solutions with nonbasic variables.

sponding to these two variables being nonbasic. As soon as the feasible region is identified, we can distinguish the basic solutions from those that are also basic feasible solutions.

Example 3.3 (Degenerate Basic Feasible Solutions)

Consider the following system of inequalities:

$$\begin{aligned} x_1 + x_2 &\leqslant 6 \\ x_2 &\leqslant 3 \\ x_1 + 2x_2 &\leqslant 9 \\ x_1, \quad x_2 &\geqslant 0 \end{aligned}$$

This system is illustrated in Figure 3.4. Note that the feasible region is precisely the region of Example 3.2 above, since the third restriction $x_1 + 2x_2 \leqslant 9$ is "redundant." After adding the slack variables x_3, x_4, and x_5, we get

$$\begin{aligned} x_1 + x_2 + x_3 &= 6 \\ x_2 + x_4 &= 3 \\ x_1 + 2x_2 + x_5 &= 9 \\ x_1, \quad x_2, \quad x_3, x_4, \quad x_5 &\geqslant 0 \end{aligned}$$

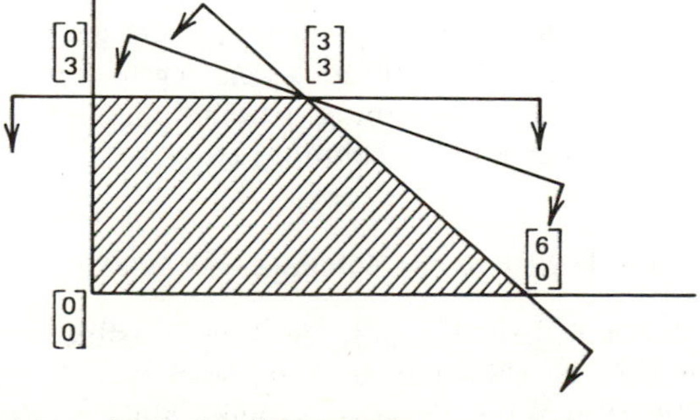

Figure 3.4. **Degenerate basic feasible solutions.**

Note that

$$\mathbf{A} = [\mathbf{a}_1, \mathbf{a}_2, \mathbf{a}_3, \mathbf{a}_4, \mathbf{a}_5] = \begin{bmatrix} 1 & 1 & 1 & 0 & 0 \\ 0 & 1 & 0 & 1 & 0 \\ 1 & 2 & 0 & 0 & 1 \end{bmatrix}.$$

Let us consider the basic feasible solution with $\mathbf{B} = [\mathbf{a}_1, \mathbf{a}_2, \mathbf{a}_3]$.

$$\mathbf{x}_B = \begin{bmatrix} x_1 \\ x_2 \\ x_3 \end{bmatrix} = \begin{bmatrix} 1 & 1 & 1 \\ 0 & 1 & 0 \\ 1 & 2 & 0 \end{bmatrix}^{-1} \begin{bmatrix} 6 \\ 3 \\ 9 \end{bmatrix} = \begin{bmatrix} 0 & -2 & 1 \\ 0 & 1 & 0 \\ 1 & 1 & -1 \end{bmatrix} \begin{bmatrix} 6 \\ 3 \\ 9 \end{bmatrix} = \begin{bmatrix} 3 \\ 3 \\ 0 \end{bmatrix}$$

$$\mathbf{x}_N = \begin{bmatrix} x_4 \\ x_5 \end{bmatrix} = \begin{bmatrix} 0 \\ 0 \end{bmatrix}$$

Note that this basic feasible solution is degenerate since the basic variable $x_3 = 0$. Now consider the basic feasible solution with $\mathbf{B} = [\mathbf{a}_1, \mathbf{a}_2, \mathbf{a}_4]$.

$$\mathbf{x}_B = \begin{bmatrix} x_1 \\ x_2 \\ x_4 \end{bmatrix} = \begin{bmatrix} 1 & 1 & 0 \\ 0 & 1 & 1 \\ 1 & 2 & 0 \end{bmatrix}^{-1} \begin{bmatrix} 6 \\ 3 \\ 9 \end{bmatrix} = \begin{bmatrix} 2 & 0 & -1 \\ -1 & 0 & 1 \\ 1 & 1 & -1 \end{bmatrix} \begin{bmatrix} 6 \\ 3 \\ 9 \end{bmatrix} = \begin{bmatrix} 3 \\ 3 \\ 0 \end{bmatrix}$$

$$\mathbf{x}_N = \begin{bmatrix} x_3 \\ x_5 \end{bmatrix} = \begin{bmatrix} 0 \\ 0 \end{bmatrix}$$

Note that this basic feasible solution gives rise to the same point obtained by $\mathbf{B} = [\mathbf{a}_1, \mathbf{a}_2, \mathbf{a}_3]$. It can be also checked that the basic feasible solution with basis $\mathbf{B} = [\mathbf{a}_1, \mathbf{a}_2, \mathbf{a}_5]$ is given by

$$\mathbf{x}_B = \begin{bmatrix} x_1 \\ x_2 \\ x_5 \end{bmatrix} = \begin{bmatrix} 3 \\ 3 \\ 0 \end{bmatrix} \qquad \mathbf{x}_N = \begin{bmatrix} x_3 \\ x_4 \end{bmatrix} = \begin{bmatrix} 0 \\ 0 \end{bmatrix}$$

Note that all three of the foregoing basic feasible solutions with different *bases* are represented by the single extreme point $(x_1, x_2, x_3, x_4, x_5) = (3, 3, 0, 0, 0)$. Each of the three basic feasible solutions is degenerate since each contains a basic variable at level zero. The remaining extreme points of Figure 3.4 correspond to nondegenerate basic feasible solutions (why?).

Correspondence Between Basic Feasible Solutions and Extreme Points

We shall now show that the collection of basic feasible solutions and the collection of extreme points are equivalent. In other words, a point is a basic feasible solution if and only if it is an extreme point. Since a linear programming problem, with a finite optimal value, has an optimal solution at an extreme point, an optimal basic feasible solution can always be found.

Consider the following problem.

$$\begin{aligned}\text{Minimize} \quad & \mathbf{cx} \\ \text{Subject to} \quad & \mathbf{Ax}=\mathbf{b} \\ & \mathbf{x} \geqslant \mathbf{0}\end{aligned}$$

where $\mathbf{A}$ is an $m \times n$ matrix with rank m. Let $\mathbf{x}$ be an extreme point of the feasible region. We shall show that $\mathbf{x}$ is also a basic feasible solution of the system $\mathbf{Ax} = \mathbf{b}$, $\mathbf{x} \geqslant \mathbf{0}$. Possibly after rearranging the components of $\mathbf{x}$ and the columns of $\mathbf{A}$, let $x_1, x_2, \ldots, x_p$ be positive and $x_{p+1}, \ldots, x_n$ be zero. We first show that $\mathbf{a}_1, \mathbf{a}_2, \ldots, \mathbf{a}_p$ are linearly independent. By contradiction, if these vectors were not linearly independent, then there must exist scalars $\gamma_1, \gamma_2, \ldots, \gamma_p$ not all zero, such that $\sum_{j=1}^{p} \gamma_j \mathbf{a}_j = \mathbf{0}$. We now exhibit two distinct feasible solutions $\mathbf{x}'$ and $\mathbf{x}''$ such that $\mathbf{x} = \frac{1}{2}\mathbf{x}' + \frac{1}{2}\mathbf{x}''$, which violates the assumption that $\mathbf{x}$ is an extreme point. Let $\mathbf{x}'$ and $\mathbf{x}''$ be the following vectors:

$$x_j' = \begin{cases} x_j + \lambda\gamma_j & j = 1, 2, \ldots, p \\ 0 & j = p+1, \ldots, n \end{cases}$$

$$x_j'' = \begin{cases} x_j - \lambda\gamma_j & j = 1, 2, \ldots, p \\ 0 & j = p+1, \ldots, n \end{cases}$$

Since $x_j > 0$ for $j = 1, 2, \ldots, p$, then regardless of the values of $\gamma_1, \gamma_2, \ldots, \gamma_p$, we can choose $\lambda > 0$ such that x_j' and $x_j'' > 0$ for $j = 1, 2, \ldots, p$. Noting that the γ_j's are not identically zero, then $\mathbf{x}' \neq \mathbf{x}''$. Furthermore

$$\begin{aligned}\mathbf{Ax}' &= \sum_{j=1}^{p} \mathbf{a}_j x_j' = \sum_{j=1}^{p} \mathbf{a}_j(x_j + \lambda\gamma_j) = \sum_{j=1}^{p} \mathbf{a}_j x_j + \lambda \sum_{j=1}^{p} \mathbf{a}_j \gamma_j \\ &= \mathbf{b} + \mathbf{0} = \mathbf{b}\end{aligned}$$

Similarly, $\mathbf{Ax}'' = \mathbf{b}$. Therefore, $\mathbf{x}'$ and $\mathbf{x}''$ are distinct feasible solutions, and $\mathbf{x} = \frac{1}{2}\mathbf{x}' + \frac{1}{2}\mathbf{x}''$, which violates the fact that $\mathbf{x}$ is an extreme point. Therefore, $\mathbf{a}_1, \mathbf{a}_2, \ldots, \mathbf{a}_p$ are linearly independent. Since $\mathbf{A}$ has full rank, we can extract $m - p$ vectors from $\mathbf{a}_{p+1}, \ldots, \mathbf{a}_n$ which together with $\mathbf{a}_1, \ldots, \mathbf{a}_p$ form a linearly independent set. Possibly after rearranging the columns, let these vectors be $\mathbf{a}_{p+1}, \ldots, \mathbf{a}_m$. Let $\mathbf{B} = [\mathbf{a}_1, \mathbf{a}_2, \ldots, \mathbf{a}_p, \mathbf{a}_{p+1}, \ldots, \mathbf{a}_m]$. Note that the columns of $\mathbf{B}$ are linearly independent and hence $\mathbf{B}$ is a basis. Further, $\mathbf{x}$ can be decomposed into $\mathbf{x}_B$ and $\mathbf{x}_N$ where $\mathbf{x}_N = \mathbf{0}$ and $\mathbf{x}_B = (x_1, x_2, \ldots, x_p, 0, 0, \ldots, 0)^t$. Finally $\mathbf{Ax} = \mathbf{b}$, and hence $\mathbf{x}$ is indeed a basic feasible solution.

Conversely, suppose that $\mathbf{x}$ is a basic feasible solution of the system $\mathbf{Ax} = \mathbf{b}$, $\mathbf{x} \geqslant \mathbf{0}$. We want to show that $\mathbf{x}$ is an extreme point. Let $\mathbf{B}$ be the basis corresponding to $\mathbf{x}$ and accordingly let $\mathbf{x} = \begin{bmatrix} \mathbf{x}_B \\ \mathbf{0} \end{bmatrix}$. Suppose that $\mathbf{x} = \lambda\mathbf{x}' + (1 - \lambda)\,\mathbf{x}''$ where $0 < \lambda < 1$ and $\mathbf{x}'$ and $\mathbf{x}''$ are feasible. To show that $\mathbf{x}$ is an extreme point, it suffices to show that $\mathbf{x} = \mathbf{x}' = \mathbf{x}''$. Let $\mathbf{x}' = \begin{bmatrix} \mathbf{x}'_B \\ \mathbf{x}'_N \end{bmatrix}$ and $\mathbf{x}'' = \begin{bmatrix} \mathbf{x}''_B \\ \mathbf{x}''_N \end{bmatrix}$. Note that $\mathbf{x}'_N \geqslant \mathbf{0}$ and $\mathbf{x}''_N \geqslant \mathbf{0}$. But since

$$\begin{bmatrix} \mathbf{x}_B \\ \mathbf{0} \end{bmatrix} = \lambda \begin{bmatrix} \mathbf{x}'_B \\ \mathbf{x}'_N \end{bmatrix} + (1 - \lambda) \begin{bmatrix} \mathbf{x}''_B \\ \mathbf{x}''_N \end{bmatrix}, \quad 0 < \lambda < 1$$

and $\mathbf{x}'_N, \mathbf{x}''_N \geqslant \mathbf{0}$, then $\mathbf{x}'_N = \mathbf{x}''_N = \mathbf{0}$. Now $\mathbf{b} = \mathbf{Ax}' = \mathbf{Bx}'_B + \mathbf{Nx}'_N = \mathbf{Bx}'_B$ and hence $\mathbf{x}'_B = \mathbf{B}^{-1}\mathbf{b}$. In other words, $\mathbf{x}'_B = \mathbf{x}_B$, and since $\mathbf{x}'_N = \mathbf{x}_N = \mathbf{0}$, then $\mathbf{x}' = \mathbf{x}$. Similarly $\mathbf{x}'' = \mathbf{x}$. Therefore, $\mathbf{x}$ is an extreme point. This shows that every basic feasible solution is an extreme point and conversely.

Note that every basic feasible solution is equivalent to an extreme point. But there may exist more than one basic feasible solution corresponding to the same extreme point. This case will occur in the presence of degeneracy (as illustrated in Example 3.3). In reference to the preceding proof, this case corresponds to that of an extreme point where the number of positive variables is $p < m$. In this case we can extract $m - p$ vectors to complete the basis. Each possible choice represents a basic feasible solution.

Existence of Extreme Points (Basic Feasible Solutions)

We shall show that every nonempty polyhedral set of the form $X = \{\mathbf{x} : \mathbf{Ax} = \mathbf{b}, \mathbf{x} \geqslant \mathbf{0}\}$ has at least one basic feasible solution. Without loss of generality, suppose that rank $(\mathbf{A}) = m$ and let $\mathbf{x}$ be a feasible solution. Further suppose that $x_1, \ldots, x_p > 0$ and that $x_{p+1} = \cdots = x_n = 0$. If $\mathbf{a}_1, \ldots, \mathbf{a}_p$ are linearly independent, then $\mathbf{x}$ is a basic feasible solution (why?). Otherwise there exist scalars $\gamma_1, \ldots, \gamma_p$ with at least one positive γ_j such that $\sum_{j=1}^{p} \gamma_j \mathbf{a}_j = \mathbf{0}$. Consider the following point $\mathbf{x}'$:

$$x'_j = \begin{cases} x_j - \lambda\gamma_j & j = 1, \ldots, p \\ 0 & j = p + 1, \ldots, n \end{cases}$$

where

$$\lambda = \text{Minimum}\left\{ \frac{x_j}{\gamma_j} : \gamma_j > 0 \right\} = \frac{x_k}{\gamma_k} > 0$$

Let $j \in \{1, \ldots, p\}$. If $\gamma_j \leqslant 0$, then $x'_j > 0$ since both x_j and λ are positive. If

$\gamma_j > 0$, then by the definition of λ we have $x_j/\gamma_j \geqslant \lambda$ and hence $x'_j = x_j - \lambda\gamma_j \geqslant 0$. Thus $\mathbf{x}' \geqslant \mathbf{0}$. Furthermore, $x'_k = 0$ (why?) and hence $\mathbf{x}'$ has at most $p - 1$ positive components. Also $\mathbf{Ax}' = \Sigma_{j=1}^{p} x'_j \mathbf{a}_j = \Sigma_{j=1}^{p}(x_j - \lambda\gamma_j)\mathbf{a}_j = \Sigma_{j=1}^{p} x_j \mathbf{a}_j - \lambda\Sigma_{j=1}^{p}\gamma_j \mathbf{a}_j = \mathbf{b} - \mathbf{0} = \mathbf{b}$. To summarize, we have constructed a feasible point $\mathbf{x}'$ (since $\mathbf{Ax}' = \mathbf{b}$ and $\mathbf{x}' \geqslant \mathbf{0}$) with at most $p - 1$ positive components. If the columns corresponding to these positive components are linearly independent, then $\mathbf{x}'$ is a basic feasible solution. Otherwise the process is repeated. Eventually a basic feasible solution will be obtained.

Let us summarize some of the important facts about the following linear programming problem, where $\mathbf{A}$ is an $m \times n$ matrix with rank m.

$$\begin{aligned} \text{Minimize} \quad & \mathbf{cx} \\ \text{Subject to} \quad & \mathbf{Ax} = \mathbf{b} \\ & \mathbf{x} \geqslant \mathbf{0} \end{aligned}$$

***Theorem* 1**

The collection of extreme points corresponds to the collection of basic feasible solutions, and both are nonempty, provided that the feasible region is not empty.

***Theorem* 2**

Assuming that the feasible region is nonempty, a finite optimal solution exists if and only if $\mathbf{cd}_j \geqslant 0$ for $j = 1, 2, \ldots, l$, where $\mathbf{d}_1, \ldots, \mathbf{d}_l$ are the extreme directions of the feasible region. Otherwise, the optimal solution is unbounded.

***Theorem* 3**

If an optimal solution exists, then an optimal extreme point (or equivalently an optimal basic feasible solution) exists.

Since the number of basic feasible solutions is bounded by $\binom{n}{m}$, one may think of simply listing all basic feasible solutions, and picking the one with the minimal objective value. This is not satisfactory, however, for a number of reasons. Firstly, the number of basic feasible solutions is bounded by $\binom{n}{m}$, which is large, even for moderate values of m and n. Secondly, this simple approach does not tell us if the problem has an unbounded solution that may occur if the feasible region is unbounded. Lastly, if the feasible region is empty, and if we apply the foregoing "simple-minded procedure," we shall discover that the feasible region is empty, only after all possible ways of extracting m columns out of n columns of the matrix $\mathbf{A}$ fail to produce a basic feasible solution, on the grounds that $\mathbf{B}$ does not have an inverse, or else $\mathbf{B}^{-1}\mathbf{b} \ngeqslant \mathbf{0}$.

The simplex method is a clever procedure that moves from an extreme point to another extreme point, with a better (at least not worse) objective. It also discovers whether the feasible region is empty and whether the optimal solution is unbounded. In practice, the method only enumerates a small portion of the extreme points of the feasible region.

3.3 IMPROVING A BASIC FEASIBLE SOLUTION

Given a basic feasible solution, we shall describe a method for obtaining a new basic feasible solution with a better objective value. This is the foundation of the simplex method.

Consider the following linear programming problem.

$$\begin{aligned} \text{Minimize} \quad & \mathbf{cx} \\ \text{Subject to} \quad & \mathbf{Ax} = \mathbf{b} \\ & \mathbf{x} \geqslant \mathbf{0} \end{aligned}$$

where $\mathbf{A}$ is an $m \times n$ matrix with rank m. Suppose that we have a basic feasible solution $\begin{pmatrix} \mathbf{B}^{-1}\mathbf{b} \\ \mathbf{0} \end{pmatrix}$ whose objective value z_0 is given by

$$z_0 = \mathbf{c}\begin{pmatrix} \mathbf{B}^{-1}\mathbf{b} \\ \mathbf{0} \end{pmatrix} = (\mathbf{c}_B, \mathbf{c}_N)\begin{pmatrix} \mathbf{B}^{-1}\mathbf{b} \\ \mathbf{0} \end{pmatrix} = \mathbf{c}_B\mathbf{B}^{-1}\mathbf{b} \tag{3.1}$$

Now let $\mathbf{x} = \begin{pmatrix} \mathbf{x}_B \\ \mathbf{x}_N \end{pmatrix}$ be an arbitrary feasible solution. Then $\mathbf{x}_B \geqslant \mathbf{0}$, $\mathbf{x}_N \geqslant \mathbf{0}$, and $\mathbf{b} = \mathbf{Ax} = \mathbf{Bx}_B + \mathbf{Nx}_N$. Multiplying by $\mathbf{B}^{-1}$ and rearranging the terms, we get

$$\begin{aligned} \mathbf{x}_B &= \mathbf{B}^{-1}\mathbf{b} - \mathbf{B}^{-1}\mathbf{N}\mathbf{x}_N \\ &= \mathbf{B}^{-1}\mathbf{b} - \sum_{j \in R} \mathbf{B}^{-1}\mathbf{a}_j x_j \end{aligned} \tag{3.2}$$

where R is the current set of the indices of the nonbasic variables. Noting Equations (3.2) and (3.1), and letting z denote the objective function at $\mathbf{x}$, we get

$$\begin{aligned} z &= \mathbf{cx} \\ &= \mathbf{c}_B\mathbf{x}_B + \mathbf{c}_N\mathbf{x}_N \\ &= \mathbf{c}_B\left(\mathbf{B}^{-1}\mathbf{b} - \sum_{j \in R} \mathbf{B}^{-1}\mathbf{a}_j x_j\right) + \sum_{j \in R} c_j x_j \\ &= z_0 - \sum_{j \in R} (z_j - c_j)x_j \end{aligned} \tag{3.3}$$

where $z_j = \mathbf{c}_B\mathbf{B}^{-1}\mathbf{a}_j$ for each nonbasic variable.

Equation (3.3) can guide us in the process of improving the current basic feasible solution. Since we are to minimize z, it would be to our advantage to increase x_j (from its current level of zero) whenever $z_j - c_j > 0$. The following rule will be adopted. Fix each nonbasic variable x_j at zero, except for one nonbasic variable x_k with a positive $z_k - c_k(z_k - c_k$ is the most positive $z_j - c_j$, say). From Equation (3.3), the new objective value z is given by

$$z = z_0 - (z_k - c_k)x_k \tag{3.4}$$

Since $z_k - c_k > 0$, it would be to our benefit to increase x_k as much as possible. As x_k is increased, the current basic variables must be modified according to Equation (3.2), and hence $\mathbf{x}_B = \mathbf{B}^{-1}\mathbf{b} - \mathbf{B}^{-1}\mathbf{a}_k x_k = \bar{\mathbf{b}} - \mathbf{y}_k x_k$, where $\mathbf{y}_k = \mathbf{B}^{-1}\mathbf{a}_k$ and $\bar{\mathbf{b}} = \mathbf{B}^{-1}\mathbf{b}$. Denoting the components of $\mathbf{x}_B$ and $\bar{\mathbf{b}}$ by x_{B_1}, $x_{B_2}, \ldots, x_{B_m}$ and $\bar{b}_1, \bar{b}_2, \ldots, \bar{b}_m$, the preceding vector equation reads as follows:

$$\begin{bmatrix} x_{B_1} \\ x_{B_2} \\ \vdots \\ x_{B_r} \\ \vdots \\ x_{B_m} \end{bmatrix} = \begin{bmatrix} \bar{b}_1 \\ \bar{b}_2 \\ \vdots \\ \bar{b}_r \\ \vdots \\ \bar{b}_m \end{bmatrix} - \begin{bmatrix} y_{1k} \\ y_{2k} \\ \vdots \\ y_{rk} \\ \vdots \\ y_{mk} \end{bmatrix} x_k \tag{3.5}$$

If $y_{ik} \leqslant 0$, then x_{B_i} increases as x_k increases and so x_{B_i} continues to be nonnegative. If $y_{ik} > 0$, then x_{B_i} will decrease as x_k increases. In order to satisfy nonnegativity, x_k is increased until the first point at which a basic variable x_{B_r} drops to zero. Examining Equation (3.5), it is then clear that the first basic variable dropping to zero corresponds to the minimum of $\bar{b}_i/y_{ik}$ for positive y_{ik}. More precisely,

$$\frac{\bar{b}_r}{y_{rk}} = \underset{1 \leqslant i \leqslant m}{\text{Minimum}} \left\{ \frac{\bar{b}_i}{y_{ik}} : y_{ik} > 0 \right\} = x_k \tag{3.6}$$

In the absence of degeneracy $\bar{b}_r > 0$, and hence $x_k = \bar{b}_r/y_{rk} > 0$. From Equation (3.4) and the fact that $z_k - c_k > 0$, it then follows that $z < z_0$, and the objective function strictly improves. As x_k increases from level 0 to $\bar{b}_r/y_{rk}$, a new feasible solution is obtained. Substituting $x_k = \bar{b}_r/y_{rk}$ in Equation (3.5) gives the following point:

$$x_{B_i} = \bar{b}_i - \frac{y_{ik}}{y_{rk}}\bar{b}_r \qquad i = 1, 2, \ldots, m$$

$$x_k = \frac{\bar{b}_r}{y_{rk}} \tag{3.7}$$

all other x_j's are zero

From Equation (3.7) above, $x_{B_r} = 0$ and hence at most m variables are positive. The corresponding columns are $\mathbf{a}_{B_1}, \mathbf{a}_{B_2}, \ldots, \mathbf{a}_{B_{r-1}}, \mathbf{a}_k, \mathbf{a}_{B_{r+1}}, \ldots, \mathbf{a}_{B_m}$. Note that these columns are linearly independent since $y_{rk} \neq 0$. (Recall that if $\mathbf{a}_{B_1}, \ldots, \mathbf{a}_{B_r}, \ldots, \mathbf{a}_{B_m}$ are linearly independent, and if $\mathbf{a}_k$ replaces $\mathbf{a}_{B_r}$, then the new columns are linearly independent if and only if $y_{rk} \neq 0$; see Section 2.1). Therefore the point given by Equation (3.7) is a basic feasible solution.

To summarize, we have described a procedure that moves from a basic feasible solution to another basic feasible solution. This is done by increasing the value of a nonbasic variable x_k with positive $z_k - c_k$ and adjusting the current basic variables. In the process, the variable x_{B_r} drops to zero. The variable x_k is said to enter the basis and x_{B_r} is said to leave the basis. In the absence of degeneracy the objective function value strictly decreases and hence the generated points are distinct. Since there is only a finite number of basic feasible solutions, the procedure would terminate in a finite number of steps.

Example 3.4

$$\begin{aligned} \text{Minimize} \quad & x_1 + x_2 \\ \text{Subject to} \quad & x_1 + 2x_2 \leqslant 4 \\ & x_2 \leqslant 1 \\ & x_1, \quad x_2 \geqslant 0 \end{aligned}$$

Introduce the slack variables x_3 and x_4 to put the problem in a standard form. This leads to the following constraint matrix $\mathbf{A}$:

$$\mathbf{A} = [\mathbf{a}_1, \mathbf{a}_2, \mathbf{a}_3, \mathbf{a}_4] = \begin{bmatrix} 1 & 2 & 1 & 0 \\ 0 & 1 & 0 & 1 \end{bmatrix}$$

Consider the basic feasible solution corresponding to $\mathbf{B} = [\mathbf{a}_1, \mathbf{a}_2]$. In other words, x_1 and x_2 are the basic variables while x_3 and x_4 are the nonbasic variables.

$$\mathbf{x}_B = \begin{bmatrix} x_1 \\ x_2 \end{bmatrix} = \mathbf{B}^{-1}\mathbf{b} = \begin{bmatrix} 1 & 2 \\ 0 & 1 \end{bmatrix}^{-1} \begin{bmatrix} 4 \\ 1 \end{bmatrix} = \begin{bmatrix} 1 & -2 \\ 0 & 1 \end{bmatrix} \begin{bmatrix} 4 \\ 1 \end{bmatrix} = \begin{bmatrix} 2 \\ 1 \end{bmatrix}$$

$$\mathbf{x}_N = \begin{bmatrix} x_3 \\ x_4 \end{bmatrix} = \begin{bmatrix} 0 \\ 0 \end{bmatrix}$$

This point is shown in Figure 3.5. In order to improve this basic feasible solution, calculate $z_j - c_j$ for the nonbasic variables.

$$\begin{aligned} z_3 - c_3 &= \mathbf{c}_B\mathbf{B}^{-1}\mathbf{a}_3 - c_3 \\ &= (1, 1)\begin{bmatrix} 1 & -2 \\ 0 & 1 \end{bmatrix}\begin{pmatrix} 1 \\ 0 \end{pmatrix} - 0 \\ &= (1, 1)\begin{pmatrix} 1 \\ 0 \end{pmatrix} - 0 \\ &= 1 \end{aligned}$$

$$\begin{aligned} z_4 - c_4 &= \mathbf{c}_B\mathbf{B}^{-1}\mathbf{a}_4 - c_4 \\ &= (1, 1)\begin{bmatrix} 1 & -2 \\ 0 & 1 \end{bmatrix}\begin{pmatrix} 0 \\ 1 \end{pmatrix} - 0 \\ &= (1, 1)\begin{pmatrix} -2 \\ 1 \end{pmatrix} - 0 \\ &= -1 \end{aligned}$$

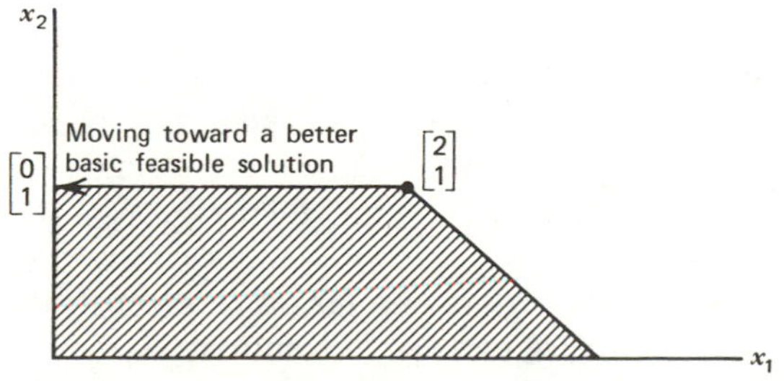

Figure 3.5. Improving a basic feasible solution.

Since $z_3 - c_3 > 0$, then the objective improves by increasing x_3. The modified solution is given by

$$\mathbf{x}_B = \mathbf{B}^{-1}\mathbf{b} - \mathbf{B}^{-1}\mathbf{a}_3 x_3$$

$$\begin{bmatrix} x_1 \\ x_2 \end{bmatrix} = \begin{bmatrix} 2 \\ 1 \end{bmatrix} - \begin{bmatrix} 1 \\ 0 \end{bmatrix} x_3$$

The maximum value of x_3 is 2 (any larger value of x_3 will force x_1 to be negative). Therefore the new basic feasible solution is

$$(x_1, x_2, x_3, x_4) = (0, 1, 2, 0)$$

Here x_3 enters the basis and x_1 leaves the basis. Note that the new point has an objective value equal to 1, which is an improvement over the previous objective value of 3. The improvement is precisely $(z_3 - c_3)x_3 = 2$. The reader is asked to continue the improvement process starting from the new point.

Interpretation of Entering and Leaving the Basis

We now look more closely at the process of entering and leaving the basis, and their interpretation.

INTERPRETATION OF $z_k - c_k$

The criterion $z_k - c_k > 0$ for a nonbasic variable x_k to enter the basis, will be used over and over again throughout the text. It will be helpful at this stage to review the definition of z_k, and make a few comments on the meaning of the entry criterion $z_k - c_k > 0$. Recall that $z = \mathbf{c}_B\bar{\mathbf{b}} - (z_k - c_k)x_k$, where

$$z_k = \mathbf{c}_B\mathbf{B}^{-1}\mathbf{a}_k = \mathbf{c}_B\mathbf{y}_k = \sum_{i=1}^{m} c_{B_i} y_{ik} \tag{3.8}$$

and c_{B_i} is the cost of the ith basic variable. Note that if x_k is raised from zero level, while the other nonbasic variables are kept at zero level, then the basic variables $x_{B_1}, x_{B_2}, \ldots, x_{B_m}$ must be modified according to Equation (3.5). In other words, if x_k is increased by 1 unit, then $x_{B_1}, x_{B_2}, \ldots,$ and x_{B_m} will be decreased respectively by $y_{1k}, y_{2k}, \ldots, y_{mk}$ units (if $y_{ik} < 0$, then x_{B_i} will be increased). The saving (a negative saving means more cost) that results from the modification of the basic variables, as a result of increasing x_k by 1 unit, is therefore, $\sum_{i=1}^{m} c_{B_i} y_{ik}$, which is z_k (see Equation 3.8). However, the cost of increasing x_k itself by 1 unit is c_k. Hence $z_k - c_k$ is the saving minus the cost of increasing x_k by 1 unit. Naturally, if $z_k - c_k$ is positive, it will be to our advantage to increase x_k. For each unit of x_k, the cost will be reduced by an amount $z_k - c_k$ and hence it will be to our advantage to increase x_k as much as possible. On the other hand, if $z_k - c_k < 0$, then by increasing x_k, the net saving is negative, and this action will result in a larger cost. So this action is prohibited. Finally if $z_k - c_k = 0$, then increasing x_k will lead to a different solution, with the same cost. So whether x_k is kept at zero level, or increased, no change in cost takes place.

Now suppose that x_k is a basic variable. In particular, suppose that x_k is the tth basic variable, that is, $x_k = x_{B_t}$, $c_k = c_{B_t}$, and $\mathbf{a}_k = \mathbf{a}_{B_t}$. Recall that $z_k = \mathbf{c}_B\mathbf{B}^{-1}\mathbf{a}_k = \mathbf{c}_B\mathbf{B}^{-1}\mathbf{a}_{B_t}$. But $\mathbf{B}^{-1}\mathbf{a}_{B_t}$ is a vector of zeros except for one at the tth position (see Exercise 2.12). Therefore, $z_k = c_{B_t}$, and hence $z_k - c_k = c_{B_t} - c_{B_t} = 0$.

Leaving the Basis and the Blocking Variable

Suppose that we decided to increase a nonbasic variable x_k with a positive $z_k - c_k$. From Equation (3.4), the larger the value of x_k, the smaller is the objective z. As x_k is increased, the basic variables are modified according to Equation (3.5). If the vector $\mathbf{y}_k$ has any positive component(s), then the corresponding basic variable(s) is decreased as x_k is increased. Therefore the nonbasic variable x_k cannot be indefinitely increased, because otherwise the nonnegativity of the basic variables will be violated. The first basic variable x_{B_r} that drops to zero is called the *blocking variable* because it blocked further increase of x_k. Thus x_k enters the basis and x_{B_r} leaves the basis.

Example 3.5

$$\begin{aligned} \text{Minimize} \quad & 2x_1 - x_2 \\ \text{Subject to} \quad & -x_1 + x_2 \leqslant 2 \\ & 2x_1 + x_2 \leqslant 6 \\ & x_1,\ x_2 \geqslant 0 \end{aligned}$$

Introduce the slack variables x_3 and x_4. This leads to the following constraints:

$$\begin{aligned} -x_1 + x_2 + x_3 \qquad &= 2 \\ 2x_1 + x_2 \qquad + x_4 &= 6 \\ x_1, \quad x_2, \quad x_3, \quad x_4 &\geqslant 0 \end{aligned}$$

Consider the basic feasible solution with basis $\mathbf{B} = [\mathbf{a}_1, \mathbf{a}_2] = \begin{bmatrix} -1 & 1 \\ 2 & 1 \end{bmatrix}$ and

$$\mathbf{B}^{-1} = \begin{bmatrix} -\frac{1}{3} & \frac{1}{3} \\ \frac{2}{3} & \frac{1}{3} \end{bmatrix}.$$

$$\mathbf{x}_B = \mathbf{B}^{-1}\mathbf{b} - \mathbf{B}^{-1}\mathbf{N}\mathbf{x}_N$$

$$\begin{bmatrix} x_{B_1} \\ x_{B_2} \end{bmatrix} = \begin{bmatrix} x_1 \\ x_2 \end{bmatrix} = \begin{bmatrix} -\frac{1}{3} & \frac{1}{3} \\ \frac{2}{3} & \frac{1}{3} \end{bmatrix}\begin{bmatrix} 2 \\ 6 \end{bmatrix} - \begin{bmatrix} -\frac{1}{3} & \frac{1}{3} \\ \frac{2}{3} & \frac{1}{3} \end{bmatrix}\begin{bmatrix} 1 & 0 \\ 0 & 1 \end{bmatrix}\begin{bmatrix} x_3 \\ x_4 \end{bmatrix}$$

$$= \begin{bmatrix} \frac{4}{3} \\ \frac{10}{3} \end{bmatrix} - \begin{bmatrix} -\frac{1}{3} \\ \frac{2}{3} \end{bmatrix} x_3 - \begin{bmatrix} \frac{1}{3} \\ \frac{1}{3} \end{bmatrix} x_4 \tag{3.9}$$

Currently $x_3 = x_4 = 0$, $x_1 = \frac{4}{3}$ and $x_2 = \frac{10}{3}$. Note that

$$z_4 - c_4 = \mathbf{c}_B\mathbf{B}^{-1}\mathbf{a}_4 - c_4 = (2, -1)\begin{bmatrix} -\frac{1}{3} & \frac{1}{3} \\ \frac{2}{3} & \frac{1}{3} \end{bmatrix}\begin{bmatrix} 0 \\ 1 \end{bmatrix} - 0 = \tfrac{1}{3} > 0$$

Hence the objective improves by introducing x_4 in the basis. Then x_3 is kept at zero level, x_4 is increased, and x_1 and x_2 are modified according to Equation (3.9). We see that x_4 can be increased to 4, at which instant x_1 drops to zero. Any further increase of x_4 results in violating the nonnegativity of x_1, and so x_1 is called the *blocking variable*. With $x_4 = 4$ and $x_3 = 0$, the modified values of x_1 and x_2 are 0 and 2 respectively. The new basic feasible solution is

$$(x_1, x_2, x_3, x_4) = (0, 2, 0, 4)$$

Note that $\mathbf{a}_4$ replaces $\mathbf{a}_1$; that is, x_1 drops from the basis and x_4 enters the basis. The new set of basic and nonbasic variables are given below:

$$\mathbf{x}_B = \begin{bmatrix} x_{B_1} \\ x_{B_2} \end{bmatrix} = \begin{bmatrix} x_4 \\ x_2 \end{bmatrix} = \begin{bmatrix} 4 \\ 2 \end{bmatrix}, \qquad \mathbf{x}_N = \begin{bmatrix} x_3 \\ x_1 \end{bmatrix} = \begin{bmatrix} 0 \\ 0 \end{bmatrix}$$

Moving from the old to the new basic feasible solution is illustrated in Figure 3.6. Note that as x_4 increases by 1 unit, x_1 decreases by $\frac{1}{3}$ unit and x_2 decreases by $\frac{1}{3}$ unit; that is, we move in the direction $(-\frac{1}{3}, -\frac{1}{3})$ in the (x_1, x_2) space. This continues until we are blocked by the nonnegativity restriction $x_1 \geqslant 0$. At this point x_1 drops to zero and leaves the basis.

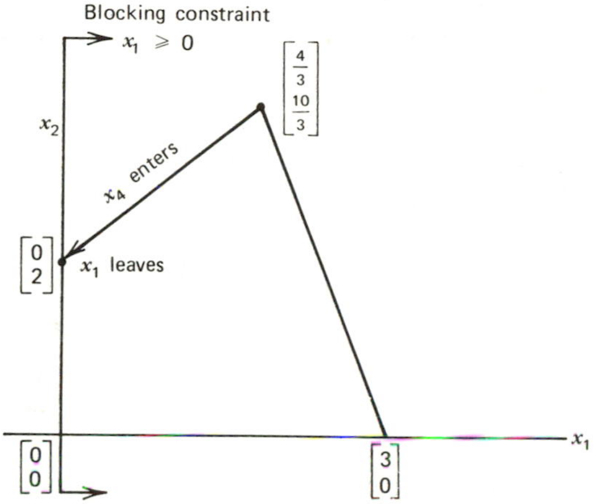

Figure 3.6. **Blocking variable (Constraint).**

3.4 TERMINATION: OPTIMALITY AND UNBOUNDEDNESS

We have discussed a procedure that moves from one basic feasible solution to an improved basic feasible solution, by introducing one variable into the basis, and removing another variable from the basis. The criteria for entering and leaving are summarized below.

1. *Entering*: x_k may enter if $z_k - c_k > 0$
2. *Leaving*: x_{B_r} may leave if

$$\frac{\bar{b}_r}{y_{rk}} = \underset{1 \leqslant i \leqslant m}{\text{Minimum}} \left\{ \frac{\bar{b}_i}{y_{ik}} : y_{ik} > 0 \right\}$$

Two logical questions immediately arise. What happens if each nonbasic variable x_j has $z_j - c_j \leqslant 0$? In this case no nonbasic variable is eligible for entering the basis. Second, suppose that $z_k - c_k > 0$, and hence x_k is eligible to enter the basis, but we cannot find any positive component y_{ik}, that is, $\mathbf{y}_k \leqslant \mathbf{0}$. As the reader may have suspected, the first case says that we have already reached the optimal solution, and the second case says that the optimal solution is unbounded. These cases will be discussed in more detail in this section.

Termination with an Optimal Solution

Consider the following problem, where $\mathbf{A}$ is an $m \times n$ matrix with rank m.

$$\begin{aligned} \text{Minimize} \quad & \mathbf{cx} \\ \text{Subject to} \quad & \mathbf{Ax} = \mathbf{b} \\ & \mathbf{x} \geqslant \mathbf{0} \end{aligned}$$

Suppose that $\mathbf{x}^*$ is a basic feasible solution with basis $\mathbf{B}$; that is, $\mathbf{x}^* = \begin{pmatrix} \mathbf{B}^{-1}\mathbf{b} \\ \mathbf{0} \end{pmatrix}$. Let z^* denote the objective of $\mathbf{x}^*$, that is, $z^* = \mathbf{c}_B\mathbf{B}^{-1}\mathbf{b}$. Suppose further that $z_j - c_j \leqslant 0$ for all nonbasic variables, and hence there are no nonbasic variables that are eligible to enter the basis. Let $\mathbf{x}$ be any feasible solution with objective value z. Then from Equation (3.3) we have

$$z^* - z = \sum_{j \in R} (z_j - c_j)x_j \tag{3.10}$$

Since $z_j - c_j \leqslant 0$ and $x_j \geqslant 0$ for all variables, then $z^* \leqslant z$. This holds for every feasible vector $\mathbf{x}$ and therefore $\mathbf{x}^*$ is an optimal basic feasible solution.

Unique and Alternative Optimal Solutions

We can get more information from Equation (3.10). If $z_j - c_j < 0$ for all nonbasic components, then the current optimal solution is unique. To show this, let $\mathbf{x}$ be any feasible solution that is distinct from $\mathbf{x}^*$. Then there is at least one nonbasic component x_j that is positive, because if all nonbasic components are zero, $\mathbf{x}$ would not be distinct from $\mathbf{x}^*$. From Equation (3.10) it follows that $z > z^*$, and hence $\mathbf{x}^*$ is the unique optimal solution.

Now consider the case where $z_j - c_j \leqslant 0$ for all nonbasic components, but $z_k - c_k = 0$ for at least one nonbasic variable x_k. As x_k is increased, we get (in the absence of degeneracy) points that are distinct from $\mathbf{x}^*$ but have the same objective value (why?). If x_k is increased until it is blocked by a basic variable, we get an alternative optimal basic feasible solution. The process of increasing x_k from level zero until it is blocked generates an infinite number of alternative optimal solutions.

Example 3.6

$$\begin{aligned}
\text{Minimize} \quad & -3x_1 + x_2 \\
\text{Subject to} \quad & x_1 + 2x_2 + x_3 = 4 \\
& -x_1 + x_2 + x_4 = 1 \\
& x_1, \quad x_2, \; x_3, \; x_4 \geqslant 0
\end{aligned}$$

Consider the basic feasible solution with basis $\mathbf{B} = [\mathbf{a}_1, \mathbf{a}_4] = \begin{bmatrix} 1 & 0 \\ -1 & 1 \end{bmatrix}$ and $\mathbf{B}^{-1} = \begin{bmatrix} 1 & 0 \\ 1 & 1 \end{bmatrix}$. The corresponding point is given by

$$\mathbf{x}_B = \begin{bmatrix} x_1 \\ x_4 \end{bmatrix} = \mathbf{B}^{-1}\mathbf{b} = \begin{bmatrix} 1 & 0 \\ 1 & 1 \end{bmatrix}\begin{bmatrix} 4 \\ 1 \end{bmatrix} = \begin{bmatrix} 4 \\ 5 \end{bmatrix}, \qquad \mathbf{x}_N = \begin{bmatrix} x_2 \\ x_3 \end{bmatrix} = \begin{bmatrix} 0 \\ 0 \end{bmatrix}$$

and the objective value is -12. To see if we can improve this solution, calculate $z_2 - c_2$ and $z_3 - c_3$:

$$\begin{aligned} z_2 - c_2 &= \mathbf{c}_B\mathbf{B}^{-1}\mathbf{a}_2 - c_2 \\ &= (-3, 0)\begin{bmatrix} 1 & 0 \\ 1 & 1 \end{bmatrix}\begin{bmatrix} 2 \\ 1 \end{bmatrix} - 1 \\ &= (-3, 0)\begin{bmatrix} 2 \\ 3 \end{bmatrix} - 1 \\ &= -7 \end{aligned}$$

$$\begin{aligned} z_3 - c_3 &= \mathbf{c}_B\mathbf{B}^{-1}\mathbf{a}_3 - c_3 \\ &= (-3, 0)\begin{bmatrix} 1 & 0 \\ 1 & 1 \end{bmatrix}\begin{bmatrix} 1 \\ 0 \end{bmatrix} - 0 \\ &= (-3, 0)\begin{bmatrix} 1 \\ 1 \end{bmatrix} - 0 \\ &= -3 \end{aligned}$$

Since both $z_2 - c_2 < 0$ and $z_3 - c_3 < 0$, then the basic feasible solution $(x_1, x_2, x_3, x_4) = (4, 0, 0, 5)$ is the unique optimal point. This unique optimal solution is illustrated in Figure 3.7*a*. Now consider a new problem where the objective function $-2x_1 - 4x_2$ is to be minimized over the same region. Again, consider the same point (4, 0, 0, 5). The objective value is -8. Calculate $z_2 - c_2$

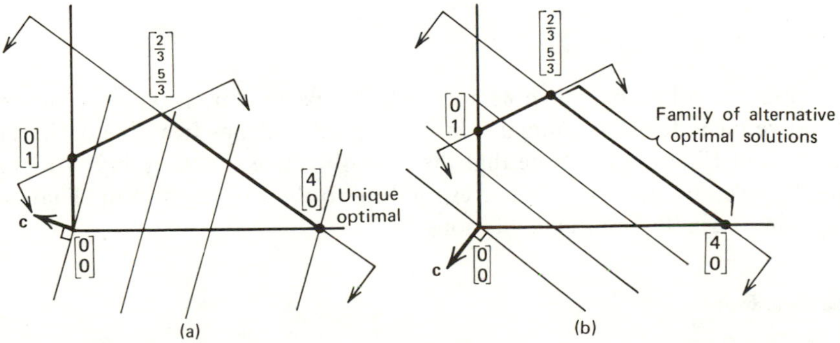

Figure 3.7. Termination criterion: (*a*) Unique optimal. (*b*) Alternative optima.

and $z_3 - c_3$ as follows:

$$\begin{aligned} z_2 - c_2 &= \mathbf{c}_B\mathbf{B}^{-1}\mathbf{a}_2 - c_2 \\ &= (-2, 0)\begin{bmatrix} 2 \\ 3 \end{bmatrix} + 4 \\ &= 0 \end{aligned}$$

$$\begin{aligned} z_3 - c_3 &= \mathbf{c}_B\mathbf{B}^{-1}\mathbf{a}_3 - c_3 \\ &= (-2, 0)\begin{bmatrix} 1 \\ 1 \end{bmatrix} - 0 \\ &= -2 \end{aligned}$$

In this case, the given basic feasible solution is optimal, but it is no longer a unique optimal solution. We see that by increasing x_2, a class of optimal solutions is obtained. Actually, if we increase x_2, keep $x_3 = 0$, and modify x_1 and x_4, we get

$$\begin{aligned} \begin{bmatrix} x_1 \\ x_4 \end{bmatrix} &= \mathbf{B}^{-1}\mathbf{b} - \mathbf{B}^{-1}\mathbf{a}_2 x_2 \\ &= \begin{bmatrix} 4 \\ 5 \end{bmatrix} - \begin{bmatrix} 2 \\ 3 \end{bmatrix} x_2 \end{aligned}$$

For any $x_2 \leqslant \frac{5}{3}$, the solution

$$\begin{bmatrix} x_1 \\ x_2 \\ x_3 \\ x_4 \end{bmatrix} = \begin{bmatrix} 4 - 2x_2 \\ x_2 \\ 0 \\ 5 - 3x_2 \end{bmatrix}$$

is an optimal solution with objective -8. In particular, if $x_2 = \frac{5}{3}$, we get an alternative basic feasible optimal solution, where x_4 drops from the basis. This is illustrated in Figure 3.7*b*. Note that the new objective function lines are parallel to the hyperplane $x_1 + x_2 = 4$ corresponding to the first constraint. That is why we obtain alternative optimal solutions.

Unboundedness

Suppose that we have a basic feasible solution of the system $\mathbf{Ax} = \mathbf{b}$, $\mathbf{x} \geqslant \mathbf{0}$, with objective value z_0. Now let us consider the case when we find a nonbasic variable x_k with $z_k - c_k > 0$ and $\mathbf{y}_k \leqslant \mathbf{0}$. This variable is eligible to enter the basis since increasing it will improve the objective function. From Equation (3.3) we have

$$z = z_0 - (z_k - c_k)x_k$$

Since we are minimizing the objective z, and since $z_k - c_k > 0$, then it is to our benefit to increase x_k indefinitely, which will make z go to $-\infty$. The reason that we were not able to do this before was that the increase in the value of x_k was blocked by a basic variable. This puts a "ceiling" on x_k beyond which a basic variable will be negative. But if blocking is not encountered, there is no reason why we should stop increasing x_k. This is precisely the case when $\mathbf{y}_k \leqslant \mathbf{0}$. Recall that from Equation (3.5) we have

$$\mathbf{x}_B = \mathbf{B}^{-1}\mathbf{b} - \mathbf{y}_k x_k$$

and so if $\mathbf{y}_k \leqslant \mathbf{0}$, then x_k can be increased indefinitely without any of the basic variables becoming negative. Therefore the solution $\mathbf{x}$ (where $\mathbf{x}_B = \mathbf{B}^{-1}\mathbf{b} - \mathbf{y}_k x_k$, x_k is arbitrarily large and other nonbasic components are zero) is feasible and its objective value $z = z_0 - (z_k - c_k)x_k$, which approaches $-\infty$ as x_k approaches $+\infty$.

To summarize, if we have a basic feasible solution with $z_k - c_k > 0$ for some nonbasic variable x_k, and meanwhile $\mathbf{y}_k \leqslant \mathbf{0}$, then the optimal is unbounded with objective $-\infty$. This is obtained by increasing x_k indefinitely and adjusting the values of the current basic variables, and is equivalent to moving along the ray:

$$\left\{ \begin{bmatrix} \mathbf{B}^{-1}\mathbf{b} \\ \hline 0 \\ \vdots \\ 0 \\ \vdots \\ 0 \end{bmatrix} + x_k \begin{bmatrix} -\mathbf{y}_k \\ \hline 0 \\ \vdots \\ 1 \\ \vdots \\ 0 \end{bmatrix} : x_k \geqslant 0 \right\}$$

Note that the vertex of the ray is the current basic feasible solution $\begin{pmatrix} \mathbf{B}^{-1}\mathbf{b} \\ \mathbf{0} \end{pmatrix}$ and the direction of the ray is

$$\mathbf{d} = \begin{bmatrix} -\mathbf{y}_k \\ \hline 0 \\ \vdots \\ 1 \\ \vdots \\ 0 \end{bmatrix}$$

where the 1 appears in the kth position. It may be noted that

$$\mathbf{cd} = (\mathbf{c}_B, \mathbf{c}_N)\mathbf{d} = -\mathbf{c}_B\mathbf{y}_k + c_k = -z_k + c_k$$

But since $c_k - z_k < 0$ (because x_k was eligible to enter the basis), then $\mathbf{cd} < 0$, which is the necessary and sufficient condition for unboundedness. In Exercise 3.28 we ask the reader to verify that $\mathbf{d}$ given above is indeed an (extreme) direction of the feasible region.

Example 3.7 (Unboundedness)

Minimize $-x_1 - 3x_2$

$$\begin{aligned} \text{Subject to} \quad x_1 - 2x_2 &\leq 4 \\ -x_1 + x_2 &\leq 3 \\ x_1, \quad x_2 &\geq 0 \end{aligned}$$

The problem, illustrated in Figure 3.8, clearly has an unbounded optimal solution. After introducing the slack variables x_3 and x_4, we get the constraint matrix $\mathbf{A} = \begin{bmatrix} 1 & -2 & 1 & 0 \\ -1 & 1 & 0 & 1 \end{bmatrix}$. Now consider the basic feasible solution whose basis $\mathbf{B}$ is $[\mathbf{a}_3, \mathbf{a}_4] = \begin{bmatrix} 1 & 0 \\ 0 & 1 \end{bmatrix}$.

$$\mathbf{x}_B = \begin{bmatrix} x_3 \\ x_4 \end{bmatrix} = \begin{bmatrix} 1 & 0 \\ 0 & 1 \end{bmatrix} \begin{bmatrix} 4 \\ 3 \end{bmatrix} = \begin{bmatrix} 4 \\ 3 \end{bmatrix}, \qquad \mathbf{x}_N = \begin{bmatrix} x_1 \\ x_2 \end{bmatrix} = \begin{bmatrix} 0 \\ 0 \end{bmatrix}.$$

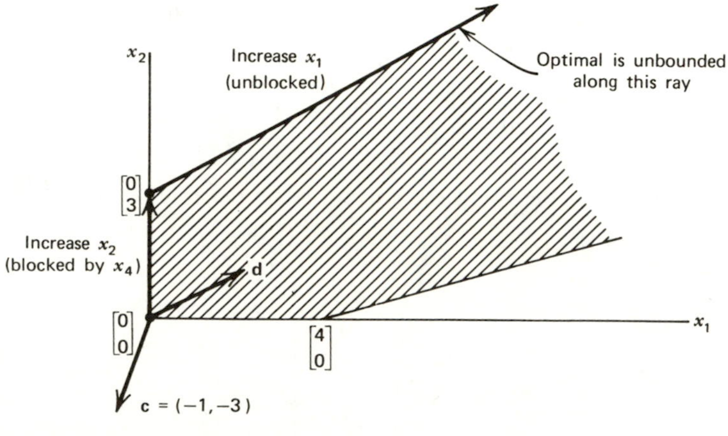

Figure 3.8. **Unbounded optimal.**

Calculate $z_1 - c_1$ and $z_2 - c_2$ as follows:

$$z_1 - c_1 = \mathbf{c}_B\mathbf{B}^{-1}\mathbf{a}_1 - c_1$$

$$= (0, 0)\begin{bmatrix} 1 \\ -1 \end{bmatrix} + 1$$

$$= 1$$

$$z_2 - c_2 = \mathbf{c}_B\mathbf{B}^{-1}\mathbf{a}_2 - c_2$$

$$= (0, 0)\begin{bmatrix} -2 \\ 1 \end{bmatrix} + 3$$

$$= 3$$

So we increase x_2 with the most positive $z_j - c_j$. Note that $\mathbf{x}_B = \mathbf{B}^{-1}\mathbf{b} - \mathbf{B}^{-1}\mathbf{a}_2 x_2$, and hence

$$\begin{bmatrix} x_3 \\ x_4 \end{bmatrix} = \begin{bmatrix} 4 \\ 3 \end{bmatrix} - \begin{bmatrix} -2 \\ 1 \end{bmatrix} x_2$$

The maximum value of x_2 is 3, at which instant x_4 drops to zero. Therefore the new basic feasible solution is $(x_1, x_2, x_3, x_4) = (0, 3, 10, 0)$. The new basis $\mathbf{B}$ is $[\mathbf{a}_3, \mathbf{a}_2] = \begin{bmatrix} 1 & -2 \\ 0 & 1 \end{bmatrix}$ with inverse $\begin{bmatrix} 1 & 2 \\ 0 & 1 \end{bmatrix}$. Calculate $z_1 - c_1$ and $z_4 - c_4$ as follows:

$$z_1 - c_1 = \mathbf{c}_B\mathbf{B}^{-1}\mathbf{a}_1 - c_1$$

$$= (0, -3)\begin{bmatrix} 1 & 2 \\ 0 & 1 \end{bmatrix}\begin{bmatrix} 1 \\ -1 \end{bmatrix} + 1$$

$$= (0, -3)\begin{bmatrix} -1 \\ -1 \end{bmatrix} + 1$$

$$= 4$$

$$z_4 - c_4 = \mathbf{c}_B\mathbf{B}^{-1}\mathbf{a}_4 - c_4$$

$$= (0, -3)\begin{bmatrix} 1 & 2 \\ 0 & 1 \end{bmatrix}\begin{bmatrix} 0 \\ 1 \end{bmatrix} - 0$$

$$= (0, -3)\begin{bmatrix} 2 \\ 1 \end{bmatrix}$$

$$= -3$$

Note that $z_1 - c_1 > 0$ and $\mathbf{y}_1 = \mathbf{B}^{-1}\mathbf{a}_1 = \begin{bmatrix} -1 \\ -1 \end{bmatrix} \leqslant \begin{bmatrix} 0 \\ 0 \end{bmatrix}$. Therefore the optimal solution is unbounded. In this case, if x_1 is increased and x_4 is kept zero, we get the following solution:

$$\mathbf{x}_B = \mathbf{B}^{-1}\mathbf{b} - \mathbf{B}^{-1}\mathbf{a}_1 x_1$$

$$\begin{bmatrix} x_3 \\ x_2 \end{bmatrix} = \begin{bmatrix} 10 \\ 3 \end{bmatrix} - \begin{bmatrix} -1 \\ -1 \end{bmatrix} x_1 = \begin{bmatrix} 10 + x_1 \\ 3 + x_1 \end{bmatrix}$$

$$x_4 = 0$$

Note that this solution is feasible for all $x_1 \geqslant 0$. In particular,

$$x_1 - 2x_2 + x_3 = x_1 - 2(3 + x_1) + (10 + x_1) = 4,$$

and

$$-x_1 + x_2 + x_4 = -x_1 + (3 + x_1) + 0 = 3$$

Furthermore, $z = -9 - 4x_1$, which approaches $-\infty$ as x_1 approaches ∞. Therefore the optimal solution is unbounded by moving along the ray

$$\{(0, 3, 10, 0) + x_1(1, 1, 1, 0) : x_1 \geqslant 0\}$$

Again note that the necessary and sufficient condition for unboundedness holds, namely

$$\mathbf{cd} = (-1, -3, 0, 0)\begin{bmatrix} 1 \\ 1 \\ 1 \\ 0 \end{bmatrix} = -4 < 0$$

3.5 THE SIMPLEX METHOD

All the machinery that is needed to describe the simplex algorithm, and to prove its convergence in a finite number of iterations (in the absence of degeneracy), has been generated. Given a basic feasible solution, we can either improve it if $z_k - c_k > 0$ for some nonbasic variable x_k, or stop with an optimal point if $z_j - c_j \leqslant 0$ for all nonbasic variables. If $z_k - c_k > 0$, and the vector $\mathbf{y}_k$ contains at least one positive component, then the increase in x_k will be blocked by one of the current basic variables, which drops to zero and leaves the basis. On the other hand, if $z_k - c_k > 0$ and $\mathbf{y}_k \leqslant \mathbf{0}$, then x_k can be increased indefinitely, and the optimal solution is unbounded and has value $-\infty$. This discussion is exactly what the simplex method does.

We now give a summary of the simplex method for solving the following linear programming problem.

$$\begin{aligned} \text{Minimize} \quad & \mathbf{cx} \\ \text{Subject to} \quad & \mathbf{Ax} = \mathbf{b} \\ & \mathbf{x} \geqslant \mathbf{0} \end{aligned}$$

where $\mathbf{A}$ is an $m \times n$ matrix with rank m (the requirement that rank $(\mathbf{A}) = m$ will be relaxed in Chapter 4).

The Simplex Algorithm (Minimization Problem)

INITIALIZATION STEP

Choose a starting basic feasible solution with basis $\mathbf{B}$. (Several procedures for finding an initial basis will be described in Chapter 4.)

MAIN STEP

1. Solve the system $\mathbf{Bx}_B = \mathbf{b}$ (with unique solution $\mathbf{x}_B = \mathbf{B}^{-1}\mathbf{b} = \bar{\mathbf{b}}$). Let $\mathbf{x}_B = \bar{\mathbf{b}}$, $\mathbf{x}_N = \mathbf{0}$, and $z = \mathbf{c}_B\mathbf{x}_B$.
2. Solve the system $\mathbf{wB} = \mathbf{c}_B$ (with unique solution $\mathbf{w} = \mathbf{c}_B\mathbf{B}^{-1}$). Calculate $z_j - c_j = \mathbf{wa}_j - c_j$ for all nonbasic variables. Let

$$z_k - c_k = \underset{j \in R}{\text{Maximum}}\ z_j - c_j$$

where R is the current set of indices associated with the nonbasic variables. If $z_k - c_k \leqslant 0$, then stop with the current basic feasible solution as an optimal solution. Otherwise go to step 3.
3. Solve the system $\mathbf{By}_k = \mathbf{a}_k$ (with unique solution $\mathbf{y}_k = \mathbf{B}^{-1}\mathbf{a}_k$). If $\mathbf{y}_k \leqslant \mathbf{0}$, then stop with the conclusion that the optimal solution is unbounded along the ray

$$\left\{ \begin{bmatrix} \bar{\mathbf{b}} \\ \mathbf{0} \end{bmatrix} + x_k \begin{bmatrix} -\mathbf{y}_k \\ \mathbf{e}_k \end{bmatrix} : x_k \geqslant 0 \right\}$$

where $\mathbf{e}_k$ is an $n - m$ vector of zeros except for a 1 at the kth position. If $\mathbf{y}_k \nleqslant \mathbf{0}$, go to step 4.

4. Here x_k enters the basis and the blocking variable x_{B_r} leaves the basis, where the index r is determined by the following minimum ratio test:

$$\frac{\bar{b}_r}{y_{rk}} = \underset{1 \leqslant i \leqslant m}{\text{Minimum}} \left\{ \frac{\bar{b}_i}{y_{ik}} : y_{ik} > 0 \right\}$$

Update the basis $\mathbf{B}$ where $\mathbf{a}_k$ replaces $\mathbf{a}_{B_r}$, the index set R and repeat step 1.

Modification for a Maximization Problem

A maximization problem can be transformed into a minimization problem by multiplying the objective coefficients by -1. A maximization problem can also be handled directly as follows. Let $z_k - c_k$ instead be the minimum $z_j - c_j$ for j nonbasic; the stopping criterion is that $z_k - c_k \geqslant 0$. Otherwise, the steps are as above.

Finite Convergence of the Simplex Method in the Absence of Degeneracy

Note that at each iteration (one pass through the main step) one of the following three actions is executed. We may stop with an optimal extreme point if $z_k - c_k \leqslant 0$; we may stop with an unbounded solution if $z_k - c_k > 0$ and $\mathbf{y}_k \leqslant \mathbf{0}$; or else we generate a new basic feasible solution if $z_k - c_k > 0$ and $\mathbf{y}_k \nleqslant \mathbf{0}$. In the absence of degeneracy, $\bar{b}_r > 0$ and hence $x_k = \bar{b}_r / y_{rk} > 0$. Therefore the difference between the objective values at the previous iteration and the current iteration is $x_k\ (z_k - c_k) > 0$. In other words, the objective function strictly decreases at each iteration and hence the basic feasible solutions generated by the simplex method are distinct. Since there is only a finite number of basic feasible solutions, the method would stop in a finite number of steps with a finite optimal solution or with an unbounded optimal solution. From this discussion the following theorem is obvious.

Theorem 4 (Finite Convergence)

In the absence of degeneracy, the simplex method stops in a finite number of iterations, either with an optimal basic feasible solution or with the conclusion that the optimal is unbounded.

In the presence of degeneracy, it is possible that $\bar{b}_r = 0$, and hence the maximum increase in the entering variable x_k is 0. In this case the objective value remains the same as that of the previous iteration. It is therefore possible, though highly unlikely in practice, that during the simplex procedure, we move indefinitely through a sequence of bases, all corresponding to the same extreme point and having the same objective value. This is called *cycling* and will be discussed in more detail in Chapter 4.

Example 3.8

$$\begin{aligned} \text{Minimize} \quad & -x_1 - 3x_2 \\ \text{Subject to} \quad & 2x_1 + 3x_2 \leqslant 6 \\ & -x_1 + x_2 \leqslant 1 \\ & x_1, \quad x_2 \geqslant 0 \end{aligned}$$

The problem is illustrated in Figure 3.9. After introducing the nonnegative slack variables x_3 and x_4, we get the following constraints:

$$\begin{aligned} 2x_1 + 3x_2 + x_3 \quad &= 6 \\ -x_1 + x_2 + \quad x_4 &= 1 \\ x_1, \quad x_2, \quad x_3, x_4 &\geqslant 0 \end{aligned}$$

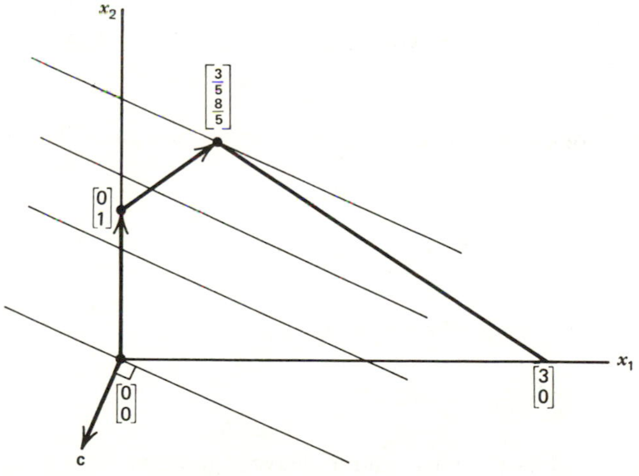

Figure 3.9. **Example of the simplex method.**

Iteration 1

Let $\mathbf{B} = [\mathbf{a}_3, \mathbf{a}_4] = \begin{bmatrix} 1 & 0 \\ 0 & 1 \end{bmatrix}$ and $\mathbf{N} = [\mathbf{a}_1, \mathbf{a}_2] = \begin{bmatrix} 2 & 3 \\ -1 & 1 \end{bmatrix}$. Solving the system $\mathbf{B}\mathbf{x}_B = \mathbf{b}$ leads to $x_{B_1} = x_3 = 6$ and $x_{B_2} = x_4 = 1$. The nonbasic variables are x_1 and x_2 and the objective $z = \mathbf{c}_B\mathbf{x}_B = (0, 0)\begin{bmatrix} 6 \\ 1 \end{bmatrix} = 0$. In order to determine which variable enters the basis, calculate $z_j - c_j = \mathbf{c}_B\mathbf{B}^{-1}\mathbf{a}_j - c_j = \mathbf{w}\mathbf{a}_j - c_j$. First we find $\mathbf{w}$ by solving the system $\mathbf{wB} = \mathbf{c}_B$:

$$(w_1, w_2)\begin{bmatrix} 1 & 0 \\ 0 & 1 \end{bmatrix} = (0, 0) \Rightarrow w_1 = w_2 = 0$$

$$z_1 - c_1 = \mathbf{wa}_1 - c_1 = 1$$

$$z_2 - c_2 = \mathbf{wa}_2 - c_2 = 3$$

Therefore x_2 is increased. In order to determine x_2 we need to calculate $\mathbf{y}_2$ by solving the system $\mathbf{By}_2 = \mathbf{a}_2$:

$$\begin{bmatrix} 1 & 0 \\ 0 & 1 \end{bmatrix}\begin{bmatrix} y_{12} \\ y_{22} \end{bmatrix} = \begin{bmatrix} 3 \\ 1 \end{bmatrix} \Rightarrow y_{12} = 3 \quad \text{and} \quad y_{22} = 1$$

The variable x_{B_r} leaving the basis is determined by the following minimum ratio test:

$$\text{Minimum}\left\{ \frac{\bar{b}_1}{y_{12}}, \frac{\bar{b}_2}{y_{22}} \right\} = \text{Minimum}\left\{ \tfrac{6}{3}, \tfrac{1}{1} \right\} = 1$$

Therefore the index $r = 2$; that is, $x_{B_2} = x_4$ leaves the basis. This is also obvious by noting that

$$\begin{bmatrix} x_{B_1} \\ x_{B_2} \end{bmatrix} = \begin{bmatrix} x_3 \\ x_4 \end{bmatrix} = \begin{bmatrix} 6 \\ 1 \end{bmatrix} - \begin{bmatrix} 3 \\ 1 \end{bmatrix} x_2 \tag{3.11}$$

and x_4 first drops to zero when $x_2 = 1$.

Iteration 2

The variable x_2 enters the basis and x_4 leaves the basis:

$$\mathbf{B} = [\mathbf{a}_3, \mathbf{a}_2] = \begin{bmatrix} 1 & 3 \\ 0 & 1 \end{bmatrix} \quad \text{and} \quad \mathbf{N} = [\mathbf{a}_1, \mathbf{a}_4] = \begin{bmatrix} 2 & 0 \\ -1 & 1 \end{bmatrix}$$

Now $\mathbf{x}_B$ can be determined by solving $\mathbf{Bx}_B = \mathbf{b}$ or simply by noting that $x_2 = 1$ in Equation (3.11).

$$\begin{bmatrix} x_{B_1} \\ x_{B_2} \end{bmatrix} = \begin{bmatrix} x_3 \\ x_2 \end{bmatrix} = \begin{bmatrix} 3 \\ 1 \end{bmatrix} \qquad \begin{bmatrix} x_1 \\ x_4 \end{bmatrix} = \begin{bmatrix} 0 \\ 0 \end{bmatrix}$$

The objective value is $z = -3$. Calculate $\mathbf{w}$ by $\mathbf{wB} = \mathbf{c}_B$:

$$(w_1, w_2)\begin{bmatrix} 1 & 3 \\ 0 & 1 \end{bmatrix} = (0, -3) \Rightarrow w_1 = 0 \quad \text{and} \quad w_2 = -3$$

$$\begin{aligned} z_1 - c_1 &= \mathbf{wa}_1 - c_1 \\ &= (0, -3)\begin{bmatrix} 2 \\ -1 \end{bmatrix} + 1 = 4 \end{aligned}$$

The variable x_4 left the basis in the previous iteration and cannot enter the basis in this iteration since $z_4 - c_4 < 0$ (see Exercise 3.33). Therefore x_1 is increased. Solve the system $\mathbf{By}_1 = \mathbf{a}_1$:

$$\begin{bmatrix} 1 & 3 \\ 0 & 1 \end{bmatrix}\begin{bmatrix} y_{11} \\ y_{21} \end{bmatrix} = \begin{bmatrix} 2 \\ -1 \end{bmatrix} \Rightarrow y_{11} = 5 \quad \text{and} \quad y_{21} = -1$$

Since $y_{21} < 0$, then $x_{B_1} = x_3$ leaves the basis as x_1 is increased. This is also clear by noting that

$$\begin{bmatrix} x_{B_1} \\ x_{B_2} \end{bmatrix} = \begin{bmatrix} x_3 \\ x_2 \end{bmatrix} = \begin{bmatrix} 3 \\ 1 \end{bmatrix} - \begin{bmatrix} 5 \\ -1 \end{bmatrix} x_1$$

and x_3 drops to zero when $x_1 = \frac{3}{5}$.

Iteration 3

Here x_1 enters the basis and x_3 leaves the basis:

$$\mathbf{B} = [\mathbf{a}_1, \mathbf{a}_2] = \begin{bmatrix} 2 & 3 \\ -1 & 1 \end{bmatrix} \quad \text{and} \quad \mathbf{N} = [\mathbf{a}_3, \mathbf{a}_4] = \begin{bmatrix} 1 & 0 \\ 0 & 1 \end{bmatrix}$$

$$\begin{bmatrix} x_{B_1} \\ x_{B_2} \end{bmatrix} = \begin{bmatrix} x_1 \\ x_2 \end{bmatrix} = \begin{bmatrix} \frac{3}{5} \\ \frac{8}{5} \end{bmatrix} \quad \mathbf{x}_N = \begin{bmatrix} x_3 \\ x_4 \end{bmatrix} = \begin{bmatrix} 0 \\ 0 \end{bmatrix}$$

The objective value $z = \frac{-27}{5}$. Calculate $\mathbf{w}$ by $\mathbf{wB} = \mathbf{c}_B$:

$$(w_1, w_2)\begin{bmatrix} 2 & 3 \\ -1 & 1 \end{bmatrix} = (-1, -3) \Rightarrow w_1 = \frac{-4}{5} \quad \text{and} \quad w_2 = \frac{-3}{5}$$

The variable x_3 left the basis in the last iteration and cannot enter the basis in this iteration (because $z_3 - c_3 < 0$).

$$\begin{aligned} z_4 - c_4 &= \mathbf{wa}_4 - c_4 \\ &= \left(\frac{-4}{5}, \frac{-3}{5}\right)\begin{bmatrix} 0 \\ 1 \end{bmatrix} - 0 = \frac{-3}{5} \end{aligned}$$

Therefore $z_j - c_j < 0$ for all nonbasic variables, and the current point is optimal. The optimal solution is therefore

$$(x_1, x_2, x_3, x_4) = \left(\tfrac{3}{5}, \tfrac{8}{5}, 0, 0\right)$$

with objective value $\frac{-27}{5}$. Figure 3.9 displays the sequence of steps that the simplex method took to reach the optimal point.

3.6 THE SIMPLEX METHOD IN TABLEAU FORMAT

At each iteration the following linear systems of equations need be solved: $\mathbf{B}\mathbf{x}_B = \mathbf{b}$, $\mathbf{w}\mathbf{B} = \mathbf{c}_B$, and $\mathbf{B}\mathbf{y}_k = \mathbf{a}_k$. Various procedures for solving and updating these systems will lead to different algorithms that all lie under the general framework of the simplex method described above. In this section we describe the simplex method in tableau format. In subsequent chapters we shall describe several procedures for solving the preceding systems for general linear programs as well as for problems with special structure such as network flow problems.

Suppose that we have a starting basic feasible solution $\mathbf{x}$ with basis $\mathbf{B}$. The linear programming problem can be represented as follows.

$$\begin{aligned} &\text{Minimize} && z \\ &\text{Subject to} && z - \mathbf{c}_B\mathbf{x}_B - \mathbf{c}_N\mathbf{x}_N = 0 && (3.12) \\ & && \mathbf{B}\mathbf{x}_B + \mathbf{N}\mathbf{x}_N = \mathbf{b} && (3.13) \\ & && \mathbf{x}_B, \quad \mathbf{x}_N \geqslant \mathbf{0} \end{aligned}$$

From Equation (3.13) we have

$$\mathbf{x}_B + \mathbf{B}^{-1}\mathbf{N}\mathbf{x}_N = \mathbf{B}^{-1}\mathbf{b} \tag{3.14}$$

Multiplying (3.14) by $\mathbf{c}_B$ and adding to Equation (3.12), we get

$$z + \mathbf{0}\mathbf{x}_B + (\mathbf{c}_B\mathbf{B}^{-1}\mathbf{N} - \mathbf{c}_N)\mathbf{x}_N = \mathbf{c}_B\mathbf{B}^{-1}\mathbf{b} \tag{3.15}$$

Currently $\mathbf{x}_N = \mathbf{0}$, and from Equations (3.14) and (3.15) we get $\mathbf{x}_B = \mathbf{B}^{-1}\mathbf{b}$ and $z = \mathbf{c}_B\mathbf{B}^{-1}\mathbf{b}$. Also, from (3.14) and (3.15) we can conveniently represent the current basic feasible solution in the following tableau. Here we think of z as a (basic) variable to be minimized. The objective row will be referred to as row 0 and the remaining rows are rows 1 through m. The right-hand-side column

(RHS) will denote the values of the basic variables (including the objective function). The basic variables are identified on the far left column.

	z	$\mathbf{x}_B$	$\mathbf{x}_N$	RHS	
z	1	$\mathbf{0}$	$\mathbf{c}_B\mathbf{B}^{-1}\mathbf{N} - \mathbf{c}_N$	$\mathbf{c}_B\mathbf{B}^{-1}\mathbf{b}$	Row 0
$\mathbf{x}_B$	$\mathbf{0}$	$\mathbf{I}$	$\mathbf{B}^{-1}\mathbf{N}$	$\mathbf{B}^{-1}\mathbf{b}$	Rows 1 through m

Not only does this tableau give us the value of the objective function $\mathbf{c}_B\mathbf{B}^{-1}\mathbf{b}$ and the basic variables $\mathbf{B}^{-1}\mathbf{b}$ on the right-hand side, but it also gives us all the information we need to proceed with the simplex method. Actually the cost row gives us $\mathbf{c}_B\mathbf{B}^{-1}\mathbf{N} - \mathbf{c}_N$, which consists of the $z_j - c_j$'s for the nonbasic variables. So row zero will tell us if we are at the optimal solution (if each $z_j - c_j \leqslant 0$), and which nonbasic variable to increase otherwise. If x_k is increased, then the vector $\mathbf{y}_k = \mathbf{B}^{-1}\mathbf{a}_k$, which is stored in the tableau in rows 1 through m under the variable x_k, will help us determine by how much x_k can be increased. If $\mathbf{y}_k \leqslant \mathbf{0}$, then x_k can be increased indefinitely without being blocked, and the optimal objective is unbounded. On the other hand, if $\mathbf{y}_k \nleqslant \mathbf{0}$, that is, if $\mathbf{y}_k$ has at least one positive component, then the increase in x_k will be blocked by one of the current basic variables, which drops to zero. The minimum ratio test (which can be performed since $\mathbf{B}^{-1}\mathbf{b} = \bar{\mathbf{b}}$ and $\mathbf{y}_k$ are both available in the tableau) determines the blocking variable. We would like to have a scheme that will do the following.

1. Update the basic variables and their values.
2. Update the $z_j - c_j$ values of the new nonbasic variables.
3. Update the $\mathbf{y}_j$ columns.

Pivoting

All of the foregoing tasks can be simultaneously accomplished by a simple pivoting operation. If x_k enters the basis and x_{B_r} leaves the basis, then pivoting on y_{rk} can be stated as follows.

1. Divide row r by y_{rk}.
2. For $i = 1, 2, \ldots, m$ and $i \neq r$, update the ith row by adding to it $-y_{ik}$ times the new rth row.
3. Update row zero by adding to it $c_k - z_k$ times the new rth row. The two tableaux of Tables 3.1 and 3.2 represent the situation immediately before and after pivoting.

Table 3.1 Before Pivoting

	z	x_{B_1}		x_{B_r}		x_{B_m}		x_j		x_k		RHS
z	1	0	$\cdots$	0	$\cdots$	0	$\cdots$	$z_j - c_j$	$\cdots$	$z_k - c_k$	$\cdots$	$\mathbf{c}_B\bar{\mathbf{b}}$
x_{B_1}	0	1	$\cdots$	0	$\cdots$	0	$\cdots$	y_{1j}	$\cdots$	y_{1k}	$\cdots$	$\bar{b}_1$
$\vdots$	$\vdots$	$\vdots$		$\vdots$		$\vdots$		$\vdots$		$\vdots$		$\vdots$
x_{B_r}	0	0	$\cdots$	1	$\cdots$	0	$\cdots$	y_{rj}	$\cdots$	$\boxed{y_{rk}}$	$\cdots$	$\bar{b}_r$
$\vdots$	$\vdots$	$\vdots$		$\vdots$		$\vdots$		$\vdots$		$\vdots$		$\vdots$
x_{B_m}	0	0	$\cdots$	0	$\cdots$	1	$\cdots$	y_{mj}	$\cdots$	y_{mk}	$\cdots$	$\bar{b}_m$

Table 3.2 After Pivoting

	z	x_{B_1}		x_{B_r}		x_{B_m}				x_k		RHS
z	1	0	$\cdots$	$\dfrac{c_k - z_k}{y_{rk}}$	$\cdots$	0	$\cdots$	$(z_j - c_j) - \dfrac{y_{rj}}{y_{rk}}(z_k - c_k)$	$\cdots$	0	$\cdots$	$\mathbf{c}_B\bar{\mathbf{b}} - (z_k - c_k)\dfrac{\bar{b}_r}{y_{rk}}$
x_{B_1}	0	1	$\cdots$	$\dfrac{-y_{1k}}{y_{rk}}$	$\cdots$	0	$\cdots$	$y_{1j} - \dfrac{y_{rj}}{y_{rk}} y_{1k}$	$\cdots$	0	$\cdots$	$\bar{b}_1 - \dfrac{y_{1k}}{y_{rk}} \bar{b}_r$
$\vdots$	$\vdots$	$\vdots$		$\vdots$		$\vdots$		$\vdots$		$\vdots$		$\vdots$
x_k	0	0	$\cdots$	$\dfrac{1}{y_{rk}}$	$\cdots$	0	$\cdots$	$\dfrac{y_{rj}}{y_{rk}}$	$\cdots$	1	$\cdots$	$\dfrac{\bar{b}_r}{y_{rk}}$
$\vdots$	$\vdots$	$\vdots$		$\vdots$		$\vdots$		$\vdots$		$\vdots$		$\vdots$
x_{B_m}	0	0	$\cdots$	$\dfrac{-y_{mk}}{y_{rk}}$	$\cdots$	1	$\cdots$	$y_{mj} - \dfrac{y_{rj}}{y_{rk}} y_{mk}$	$\cdots$	0	$\cdots$	$\bar{b}_m - \dfrac{y_{mk}}{y_{1k}} \bar{b}_r$

Let us examine the implications of the pivoting operation.

1. The variable x_k entered the basis and x_{B_r} left the basis. This is illustrated on the left-hand side of the tableau by replacing x_{B_r} with x_k. For the purpose of the following iteration, the new x_{B_r} is now x_k.
2. The right-hand side of the tableau gives the current values of the basic variables (review Equation 3.7). The nonbasic variables are kept zero.
3. Suppose that the original columns of the new basic and nonbasic variables are $\hat{\mathbf{B}}$ and $\hat{\mathbf{N}}$ respectively. Through a sequence of elementary row operations (characterized by pivoting at the intermediate iterations), the original tableau reduces to the current tableau with $\hat{\mathbf{B}}$ replaced by $\mathbf{I}$. From Chapter 2 we know that this is equivalent to premultiplication by $\hat{\mathbf{B}}^{-1}$. Thus, pivoting results in a new tableau that gives the new $\hat{\mathbf{B}}^{-1}\hat{\mathbf{N}}$ under the nonbasic variables, an updated set of $z_j - c_j$'s for the new nonbasic variables, and the values of the new basic variables and objective function.

The Simplex Method in Tableau Format (Minimization Problem)

INITIALIZATION STEP

Find an initial basic feasible solution with basis $\mathbf{B}$. Form the following initial tableau.

	z	$\mathbf{x}_B$	$\mathbf{x}_N$	RHS
z	1	$\mathbf{0}$	$\mathbf{c}_B\mathbf{B}^{-1}\mathbf{N} - \mathbf{c}_N$	$\mathbf{c}_B\bar{\mathbf{b}}$
$\mathbf{x}_B$	$\mathbf{0}$	$\mathbf{I}$	$\mathbf{B}^{-1}\mathbf{N}$	$\bar{\mathbf{b}}$

MAIN STEP

Let $z_k - c_k = \text{Maximum}\{z_j - c_j : j \in R\}$. If $z_k - c_k \leqslant 0$, then stop; the current solution is optimal. Otherwise examine $\mathbf{y}_k$. If $\mathbf{y}_k \leqslant \mathbf{0}$, then stop; the optimal solution is unbounded along the ray

$$\left\{\begin{bmatrix} \mathbf{B}^{-1}\mathbf{b} \\ \mathbf{0} \end{bmatrix} + x_k \begin{bmatrix} -\mathbf{y}_k \\ \mathbf{e}_k \end{bmatrix} : x_k \geqslant 0\right\}$$

where $\mathbf{e}_k$ is a vector of zeros except for a 1 at the kth position. If $\mathbf{y}_k \nleqslant \mathbf{0}$, determine the index r as follows:

$$\frac{\bar{b}_r}{y_{rk}} = \underset{1 \leqslant i \leqslant m}{\text{Minimum}} \left\{ \frac{\bar{b}_i}{y_{ik}} : y_{ik} > 0 \right\}$$

Update the tableau by pivoting at y_{rk}. Update the basic and nonbasic variables where x_k enters the basis and x_{B_r} leaves the basis, and repeat the main step.

Example 3.9

$$\begin{aligned} \text{Minimize} \quad & x_1+x_2-4x_3 \\ \text{Subject to} \quad & x_1+x_2+2x_3 \leqslant 9 \\ & x_1+x_2-\ \ x_3 \leqslant 2 \\ & -x_1+x_2+\ \ x_3 \leqslant 4 \\ & x_1,\ x_2,\ \ x_3 \geqslant 0 \end{aligned}$$

Introduce the nonnegative slack variables x_4, x_5, and x_6. The problem becomes the following.

$$\begin{aligned} \text{Minimize} \quad & x_1+x_2-4x_3+0x_4+0x_5+0x_6 \\ \text{Subject to} \quad & x_1+x_2+2x_3+x_4 = 9 \\ & x_1+x_2-x_3+x_5 = 2 \\ & -x_1+x_2+x_3+x_6 = 4 \\ & x_1,\ x_2,\ x_3,\ x_4,\ x_5,\ x_6 \geqslant 0 \end{aligned}$$

Since $\mathbf{b} \geqslant \mathbf{0}$, then we can choose our initial basis as $\mathbf{B} = [\mathbf{a}_4, \mathbf{a}_5, \mathbf{a}_6] = \mathbf{I}_3$, and we indeed have $\mathbf{B}^{-1}\mathbf{b} = \bar{\mathbf{b}} \geqslant \mathbf{0}$. This gives the following initial tableau.

Iteration 1

	z	x_1	x_2	x_3	x_4	x_5	x_6	RHS
z	1	−1	−1	4	0	0	0	0
x_4	0	1	1	2	1	0	0	9
x_5	0	1	1	−1	0	1	0	2
x_6	0	−1	1	(1)	0	0	1	4

Iteration 2

	z	x_1	x_2	x_3	x_4	x_5	x_6	RHS
z	1	3	−5	0	0	0	−4	−16
x_4	0	(3)	−1	0	1	0	−2	1
x_5	0	0	2	0	0	1	1	6
x_3	0	−1	1	1	0	0	1	4

Iteration 3

	z	x_1	x_2	x_3	x_4	x_5	x_6	RHS
z	1	0	-4	0	-1	0	-2	-17
x_1	0	1	$-\frac{1}{3}$	0	$\frac{1}{3}$	0	$-\frac{2}{3}$	$\frac{1}{3}$
x_5	0	0	2	0	0	1	1	6
x_3	0	0	$\frac{2}{3}$	1	$\frac{1}{3}$	0	$\frac{1}{3}$	$\frac{13}{3}$

This is the optimal tableau since $z_j - c_j \leqslant 0$ for all nonbasic variables. The optimal solution is given by

$$x_1 = \tfrac{1}{3},\ x_2 = 0,\ x_3 = \tfrac{13}{3}$$

$$z = -17$$

Note that the current optimal basis consists of the columns $\mathbf{a}_1$, $\mathbf{a}_5$, and $\mathbf{a}_3$, namely

$$\mathbf{B} = [\mathbf{a}_1, \mathbf{a}_5, \mathbf{a}_3] = \begin{bmatrix} 1 & 0 & 2 \\ 1 & 1 & -1 \\ -1 & 0 & 1 \end{bmatrix}$$

Interpretation of Entries in the Simplex Tableau

Consider the following typical simplex tableau and assume nondegeneracy.

	z	$\mathbf{x}_B$	$\mathbf{x}_N$	RHS
z	1	$\mathbf{0}$	$\mathbf{c}_B\mathbf{B}^{-1}\mathbf{N} - \mathbf{c}_N$	$\mathbf{c}_B\mathbf{B}^{-1}\mathbf{b}$
$\mathbf{x}_B$	$\mathbf{0}$	$\mathbf{I}$	$\mathbf{B}^{-1}\mathbf{N}$	$\mathbf{B}^{-1}\mathbf{b}$

The tableau may be thought of as a representation of both the basic variables $\mathbf{x}_B$ and the cost variable z in terms of the nonbasic variables $\mathbf{x}_N$. The nonbasic variables can therefore be thought of as independent variables, whereas $\mathbf{x}_B$ and z are dependent variables. From row zero we have

$$\begin{aligned} z &= \mathbf{c}_B\mathbf{B}^{-1}\mathbf{b} - (\mathbf{c}_B\mathbf{B}^{-1}\mathbf{N} - \mathbf{c}_N)\mathbf{x}_N \\ &= \mathbf{c}_B\mathbf{B}^{-1}\mathbf{b} + \sum_{j \in R} (c_j - z_j)x_j \end{aligned}$$

and hence the rate of change of z as a function of a typical nonbasic variable x_j, namely $\partial z/\partial x_j$, is simply $c_j - z_j$. In order to minimize z, we should increase x_j if $\partial z/\partial x_j < 0$, that is, if $z_j - c_j > 0$.

Also, the basic variables can be represented in terms of the nonbasic variables as follows:

$$\begin{aligned}\mathbf{x}_B &= \mathbf{B}^{-1}\mathbf{b} - \mathbf{B}^{-1}\mathbf{N}\mathbf{x}_N \\ &= \mathbf{B}^{-1}\mathbf{b} - \sum_{j \in R} \mathbf{B}^{-1}\mathbf{a}_j x_j \\ &= \mathbf{B}^{-1}\mathbf{b} - \sum_{j \in R} \mathbf{y}_j x_j\end{aligned}$$

Therefore $\partial \mathbf{x}_B / \partial x_j = -\mathbf{y}_j$; that is, $-\mathbf{y}_j$ is the rate of change of the basic variables as a function of the nonbasic variable x_j. In other words, if x_j increases by 1 unit, then the ith basic variable x_{B_i} decreases by an amount y_{ij}, or simply $\partial x_{B_i} / \partial x_j = -y_{ij}$. A column $\mathbf{y}_j$ can be alternatively interpreted as follows. Recall that $\mathbf{B}\mathbf{y}_j = \mathbf{a}_j$, and hence $\mathbf{y}_j$ represents the linear combination of the basic columns that are needed to represent $\mathbf{a}_j$. More specifically,

$$\mathbf{a}_j = \sum_{i=1}^{m} \mathbf{a}_{B_i} y_{ij}$$

The simplex tableau also gives us a convenient way of predicting the rate of change of the objective function and the value of the basic variables as a function of the right-hand-side vector $\mathbf{b}$. Since the right-hand-side vector usually represents scarce resources, we can predict the rate of change of the objective function as the availability of the resources is varied. In particular,

$$z = \mathbf{c}_B \mathbf{B}^{-1}\mathbf{b} - \sum_{j \in R} (z_j - c_j) x_j$$

and hence $\partial z / \partial \mathbf{b} = \mathbf{c}_B \mathbf{B}^{-1}$. If the original identity consists of slack variables with zero costs, then the elements of row zero at the final tableau under the slacks give $\mathbf{c}_B \mathbf{B}^{-1}\mathbf{I} - \mathbf{0} = \mathbf{c}_B \mathbf{B}^{-1}$, which is $\partial z / \partial \mathbf{b}$. More specifically, if we let $\mathbf{w} = \mathbf{c}_B \mathbf{B}^{-1}$, then $\partial z / \partial b_i = w_i$.

Similarly, the rate of change of the basic variables as a function of the right-hand-side vector $\mathbf{b}$ is given by

$$\frac{\partial \mathbf{x}_B}{\partial \mathbf{b}} = \mathbf{B}^{-1}$$

In particular, $\partial x_{B_i} / \partial \mathbf{b}$ is the ith row of $\mathbf{B}^{-1}$, $\partial \mathbf{x}_B / \partial b_j$ is the jth column of $\mathbf{B}^{-1}$, and $\partial x_{B_i} / \partial b_j$ is the (i, j) entry of $\mathbf{B}^{-1}$. If the tableau corresponds to a degenerate basic feasible solution, then as a nonbasic variable x_j increases, at least one of the basic variables may become immediately negative (see Equation 3.5) destroying feasibility. In this case a change of basis is necessary to restore feasibility, leading to nondifferentiability of the objective value as a function of x_j. The advanced reader will note that, in this case, the quantities given in this section correspond to one-sided directional derivatives.

Identifying $\mathbf{B}^{-1}$ from the Simplex Tableau

The basis inverse matrix can be identified from the simplex tableau as follows. Assume that the original tableau has an identity matrix. The process of reducing the basis matrix $\mathbf{B}$ of the original tableau to an identity matrix in the current tableau, is equivalent to premultiplying rows 1 through m of the original tableau by $\mathbf{B}^{-1}$ to produce the current tableau (why?). This also converts the identity matrix of the original tableau to $\mathbf{B}^{-1}$. Therefore, $\mathbf{B}^{-1}$ can be extracted from the current tableau as the submatrix in rows 1 through m under the original identity columns.

Example 3.10

To illustrate the interpretations of the simplex tableau, consider Example 3.9 at iteration 2. Then

$$\frac{\partial z}{\partial x_1} = -3, \quad \frac{\partial z}{\partial x_2} = 5, \quad \frac{\partial z}{\partial x_6} = 4$$

$$\frac{\partial x_4}{\partial x_1} = -3, \quad \frac{\partial x_5}{\partial x_1} = 0, \quad \frac{\partial x_3}{\partial x_6} = -1$$

$$\frac{\partial \mathbf{x}_B}{\partial x_2} = \begin{bmatrix} 1 \\ -2 \\ -1 \end{bmatrix}$$

$$\frac{\partial z}{\partial b_1} = \frac{\partial z}{\partial b_2} = 0, \quad \frac{\partial z}{\partial b_3} = -4$$

$$\frac{\partial x_5}{\partial b_2} = 1, \quad \frac{\partial x_4}{\partial b_3} = -2$$

$$\mathbf{B}^{-1} = \begin{bmatrix} 1 & 0 & -2 \\ 0 & 1 & 1 \\ 0 & 0 & 1 \end{bmatrix}$$

The vector $\mathbf{a}_2$ can be represented as a linear combination of the basic columns as follows: $\mathbf{a}_2 = -1\mathbf{a}_4 + 2\mathbf{a}_5 + \mathbf{a}_3$. At iteration 3 we have

$$\frac{\partial z}{\partial x_2} = 4, \quad \frac{\partial z}{\partial x_6} = 2$$

$$\frac{\partial x_5}{\partial x_2} = -2, \quad \frac{\partial x_5}{\partial x_6} = -1, \quad \frac{\partial x_3}{\partial x_2} = -\frac{2}{3}$$

$$\frac{\partial \mathbf{x}_B}{\partial x_2} = \begin{bmatrix} \frac{1}{3} \\ -2 \\ \frac{-2}{3} \end{bmatrix}$$

$$\frac{\partial z}{\partial b_1} = -1, \quad \frac{\partial z}{\partial b_2} = 0, \quad \frac{\partial z}{\partial b_3} = -2$$

$$\frac{\partial \mathbf{x}_B}{\partial b_3} = \begin{bmatrix} \frac{-2}{3} \\ 1 \\ \frac{1}{3} \end{bmatrix}$$

$$\frac{\partial x_1}{\partial b_1} = \frac{1}{3}, \quad \frac{\partial x_3}{\partial b_1} = \frac{1}{3}$$

$$\mathbf{B}^{-1} = \begin{bmatrix} \frac{1}{3} & 0 & \frac{-2}{3} \\ 0 & 1 & 1 \\ \frac{1}{3} & 0 & \frac{1}{3} \end{bmatrix}$$

The vector $\mathbf{a}_2$ can be represented as a linear combination of the basic columns as follows: $\mathbf{a}_2 = -\frac{1}{3}\mathbf{a}_1 + 2\mathbf{a}_5 + \frac{2}{3}\mathbf{a}_3$.

3.7 BLOCK PIVOTING

Throughout this chapter we have considered the possibility of entering a single nonbasic variable into the basis at each iteration. Recall that whenever a nonbasic variable enters the basis we must ensure that the new set of variables, the current basic set minus the exit variable plus the entering variable, (1) also forms a basis, (2) remains feasible, that is, $x_{B_i} \geqslant 0$ for all i, and (3) the value of the objective function either remains constant or decreases (for a minimization problem). It is possible to enter sets of nonbasic variables so long as the same three conditions above are satisfied. The process of introducing several nonbasic variables simultaneously is called *block pivoting*. However, in the case of multiple entries, the foregoing conditions become harder to ensure.

Suppose that we enter several nonbasic variables into the basis in such a way that condition 2 above is maintained. Note that

$$z = \mathbf{c}_B\mathbf{B}^{-1}\mathbf{b} - \sum_{j\in R}(z_j - c_j)x_j$$

If we, for example, use the entry criterion that $z_j - c_j > 0$ for all entering variables, we shall ensure that the value of z will either remain constant or else decrease.

With respect to the question whether the new set still forms a basis, we must extend the rule that the pivot element be nonzero. Consider the basic equations before pivoting:

$$\mathbf{x}_B = \mathbf{B}^{-1}\mathbf{b} - \mathbf{B}^{-1}\mathbf{N}\mathbf{x}_N$$

Let $\bar{\mathbf{b}} = \mathbf{B}^{-1}\mathbf{b}$, $\mathbf{Y}_N = \mathbf{B}^{-1}\mathbf{N}$, $\mathbf{x}_B = \begin{pmatrix}\mathbf{x}_{B_1}\\ \mathbf{x}_{B_2}\end{pmatrix}$, $\mathbf{x}_N = \begin{pmatrix}\mathbf{x}_{N_1}\\ \mathbf{x}_{N_2}\end{pmatrix}$ where the vector $\mathbf{x}_{N_1}$ enters and the vector $\mathbf{x}_{B_2}$ leaves the basis. Here $\mathbf{x}_{B_2}$ and $\mathbf{x}_{N_1}$ each contain the same number of variables (why?). On partitioning the basic system, we get

$$\begin{bmatrix}\mathbf{I}_1 & \mathbf{0}\\ \mathbf{0} & \mathbf{I}_2\end{bmatrix}\begin{pmatrix}\mathbf{x}_{B_1}\\ \mathbf{x}_{B_2}\end{pmatrix} = \begin{pmatrix}\bar{\mathbf{b}}_1\\ \bar{\mathbf{b}}_2\end{pmatrix} - \begin{bmatrix}\mathbf{Y}_{N_{11}} & \mathbf{Y}_{N_{12}}\\ \mathbf{Y}_{N_{21}} & \mathbf{Y}_{N_{22}}\end{bmatrix}\begin{pmatrix}\mathbf{x}_{N_1}\\ \mathbf{x}_{N_2}\end{pmatrix}$$

On rearranging, we get

$$\begin{bmatrix}\mathbf{I}_1 & \mathbf{Y}_{N_{11}}\\ \mathbf{0} & \mathbf{Y}_{N_{21}}\end{bmatrix}\begin{pmatrix}\mathbf{x}_{B_1}\\ \mathbf{x}_{N_1}\end{pmatrix} = \begin{pmatrix}\bar{\mathbf{b}}_1\\ \bar{\mathbf{b}}_2\end{pmatrix} - \begin{bmatrix}\mathbf{0} & \mathbf{Y}_{N_{12}}\\ \mathbf{I}_2 & \mathbf{Y}_{N_{22}}\end{bmatrix}\begin{pmatrix}\mathbf{x}_{B_2}\\ \mathbf{x}_{N_2}\end{pmatrix}$$

Now, the new set of variables, $\mathbf{x}_{B_1}$ and $\mathbf{x}_{N_1}$, will form a basis if and only if the matrix $\begin{bmatrix}\mathbf{I}_1 & \mathbf{Y}_{N_{11}}\\ \mathbf{0} & \mathbf{Y}_{N_{12}}\end{bmatrix}$ can be converted into the identity matrix; that is, if this matrix is invertible. From Chapter 2 we know that this matrix is invertible if and only if the determinant of the square matrix $\mathbf{Y}_{N_{12}}$ is nonzero. This is a natural extension of the rule for one variable entry. The new rule for maintaining a basis is as follows. Consider the square submatrix formed by the elements in the leaving rows and entering columns of the current tableau. If the determinant of this submatrix is nonzero, the new set will form a basis.

Rules for checking feasibility of the new basic set are more involved. Except in special circumstances, such as network flows, the rules for feasibility are difficult to check. This is primarily the reason why block pivoting is generally avoided in practice.

EXERCISES

3.1 Consider the linear program: Minimize $\mathbf{cx}$ subject to $\mathbf{Ax} \leqslant \mathbf{b}$, $\mathbf{x} \geqslant \mathbf{0}$, where $\mathbf{c}$ is a nonzero vector. Suppose that the point $\mathbf{x}_0$ is such that $\mathbf{Ax}_0 < \mathbf{b}$ and $\mathbf{x}_0 > \mathbf{0}$. Show that $\mathbf{x}_0$ cannot be an optimal solution.

3.2 Consider the following linear programming problem.

$$\begin{array}{ll} \text{Maximize} & x_1 + 3x_2 \\ \text{Subject to} & -x_1 + x_2 \leqslant 4 \\ & -x_1 + 2x_2 \leqslant 12 \\ & x_1 + x_2 \leqslant 10 \\ & x_1, \quad x_2 \geqslant 0 \end{array}$$

a. Sketch the feasible region in the (x_1, x_2) space and identify the optimal solution.
b. Identify all the extreme points and reformulate the problem in terms of the convex combinations of the extreme points. Solve the resulting problem.
c. Suppose that the third constraint is dropped. Identify the extreme points and directions and reformulate the problem in terms of convex combinations of the extreme points and linear combinations of the extreme directions. Solve the resulting problem, identify the optimal solution of the original problem, and interpret the solution.
d. Is the procedure described in (b) and (c) above practical for solving larger problems? Discuss.

3.3 Consider the region defined by the constraints $\mathbf{Ax} \geqslant \mathbf{b}$, where $\mathbf{A}$ is an $m \times n$ matrix where $m > n$. Further suppose that rank $(\mathbf{A}) = n$. Show that $\mathbf{x}_0$ is an extreme point of the region if and only if the following decomposition of $\mathbf{A}$ and $\mathbf{b}$ is possible.

$$\mathbf{A} = \begin{bmatrix} \mathbf{A}_1 \\ \mathbf{A}_2 \end{bmatrix} \begin{array}{l} n \text{ rows} \\ m - n \text{ rows} \end{array} \qquad \mathbf{b} = \begin{bmatrix} \mathbf{b}_1 \\ \mathbf{b}_2 \end{bmatrix} \begin{array}{l} n \text{ rows} \\ m - n \text{ rows} \end{array}$$

$$\mathbf{A}_1\mathbf{x}_0 = \mathbf{b}_1, \qquad \mathbf{A}_2\mathbf{x}_0 \geqslant \mathbf{b}_2$$

$$\text{rank}\,(\mathbf{A}_1) = n$$

3.4 Consider the following linear programming problem.

$$\text{Maximize} \quad 2x_1 + x_2 - x_3$$

$$\begin{aligned} \text{Subject to} \quad & x_1 + x_2 + 2x_3 \leqslant 6 \\ & x_1 + 4x_2 - x_3 \leqslant 4 \\ & x_1, \quad x_2, \quad x_3 \geqslant 0 \end{aligned}$$

Find the optimal solution by evaluating the objective function at all extreme points of the constraint set. Show that this approach is valid in this problem. Now suppose that the first constraint is replaced by $x_1 + x_2 - 2x_3 \leqslant 6$. Can the same approach for finding the optimal point be used? Explain why.

3.5 Consider the following linear programming problem.

$$\text{Maximize} \quad x_1 + 2x_2 + 4x_3 \qquad + 5x_5 + x_6$$

$$\begin{aligned} \text{Subject to} \quad & 2x_1 + 6x_2 + 3x_3 + 2x_4 + 3x_5 + 4x_6 \leqslant 600 \\ & x_1, \quad x_2, \quad x_3, \quad x_4, \quad x_5, \quad x_6 \geqslant 0 \end{aligned}$$

This problem has one constraint in addition to the nonnegativity constraints, and is called a *knapsack problem*. Find all basic feasible solutions of the problem, and find the optimal by comparing these basic feasible solutions.

3.6 Consider the following constraints.

$$\begin{aligned} x_1 + 2x_2 &\leqslant 6 \\ x_1 - x_2 &\leqslant 4 \\ x_2 &\leqslant 2 \\ x_1, \quad x_2 &\geqslant 0 \end{aligned}$$

a. Draw the feasible region.
b. Identify the extreme points, and at each extreme point identify the basic and nonbasic variables.
c. Suppose that a move is made from the extreme point $\begin{bmatrix} 4 \\ 0 \end{bmatrix}$ to the extreme point $\begin{bmatrix} \frac{14}{3} \\ \frac{2}{3} \end{bmatrix}$ in the (x_1, x_2) space. Specify which variable entered the basis and which variable left the basis.

3.7 Consider the polyhedral set consisting of points (x_1, x_2) such that

$$x_1 + x_2 \leqslant 1$$
$$x_1, \quad x_2 \quad \text{unrestricted}$$

Verify geometrically, and algebraically that this set has no extreme points. Formulate an equivalent set in a higher dimension where all variables are restricted to be nonnegative. Show that extreme points of the new set indeed exist.

3.8 Show that, in the absence of degeneracy, there is one-to-one correspondence between feasible bases and extreme points. Develop similar results in the presence of degeneracy. Give an example.

3.9 Consider the following system.

$$\begin{aligned} x_1 + x_2 &\leqslant 2 \\ -x_1 + 2x_2 &\leqslant 3 \\ x_1, \quad x_2 &\geqslant 0 \end{aligned}$$

The point $(\frac{1}{2}, \frac{1}{2})$ is feasible. Verify whether it is basic. If not, use the method described in the text for reducing it to a basic feasible solution.

3.10 Solve the following problem.

$$\begin{aligned} \text{Maximize} \quad & 5x_1 + 4x_2 \\ \text{Subject to} \quad & x_1 + 2x_2 \leqslant 6 \\ & 2x_1 - x_2 \leqslant 4 \\ & 5x_1 + 3x_2 \leqslant 15 \\ & x_1, \quad x_2 \geqslant 0 \end{aligned}$$

a. Graphically.
b. By the simplex method.

3.11 Solve the following linear programming problem by the simplex method, at each iteration identifying $\mathbf{B}$ and $\mathbf{B}^{-1}$.

$$\begin{aligned} \text{Maximize} \quad & 3x_1 + 2x_2 \\ \text{Subject to} \quad & 2x_1 - 3x_2 \leqslant 3 \\ & -x_1 + x_2 \leqslant 5 \\ & x_1, \quad x_2 \geqslant 0 \end{aligned}$$

3.12 Solve the following problem by the simplex method starting with the basic feasible solution $(x_1, x_2) = (4, 0)$.

$$\begin{array}{ll} \text{Maximize} & -x_1+2x_2 \\ \text{Subject to} & 3x_1+4x_2=12 \\ & 2x_1-\ \ x_2\leqslant 12 \\ & x_1,\quad x_2\geqslant 0 \end{array}$$

(*Hint*. Identify the initial basis and find its inverse.)

3.13 Consider the following problem.

$$\begin{array}{ll} \text{Maximize} & -3x_1-2x_2 \\ \text{Subject to} & -x_1+\ \ x_2\leqslant\ 1 \\ & 6x_1+4x_2\leqslant 24 \\ & x_1 \qquad\quad \geqslant 0 \\ & \qquad\quad x_2\geqslant 2 \end{array}$$

a. Solve the problem graphically.
b. Set up the problem in tableau form for the simplex method, and obtain an initial basic feasible solution.
c. Perform one pivot. After one pivot
   i. indicate the basic vectors
   ii. indicate the values of all variables
   iii. is the solution optimal?
d. Draw the requirements space representation.
   i. give the possible bases
   ii. give the possible feasible bases
e. Relate each basis in *d* (i) to a point in *a*.

3.14 Consider the constraints $\mathbf{Ax} = \mathbf{b}$, $\mathbf{x} \geqslant \mathbf{0}$ and assume that they form a bounded region. Consider the following two problems, where x_n is the nth component of $\mathbf{x}$.

$$\begin{array}{ll} \text{Minimize} & x_n \\ \text{Subject to} & \mathbf{Ax} = \mathbf{b} \\ & \ \ \mathbf{x} \geqslant \mathbf{0} \end{array}$$

$$\begin{array}{ll} \text{Maximixe} & x_n \\ \text{Subject to} & \mathbf{Ax} = \mathbf{b} \\ & \ \ \mathbf{x} \geqslant \mathbf{0} \end{array}$$

Let the optimal objectives of both problems be x_n' and x_n''. Let x_n be any number in the interval $[x_n', x_n'']$. Show that there exists a feasible point whose nth component is equal to x_n.

3.15 Suppose we have a basic feasible solution of the system $\mathbf{Ax} = \mathbf{b}$, $\mathbf{x} \geqslant \mathbf{0}$ with basis $\mathbf{B}$. Suppose that $z_k - c_k > 0$ and x_k is introduced into the basis and x_{B_r} is removed from the basis. Denote the new basis by $\hat{\mathbf{B}}$. Show algebraically that after pivoting:
a. The column under x_j is $(\hat{\mathbf{B}})^{-1}\mathbf{a}_j$.
b. The right-hand side is $(\hat{\mathbf{B}})^{-1}\mathbf{b}$.
c. The new cost row is composed of $(\mathbf{c}_{\hat{B}})(\hat{\mathbf{B}})^{-1}\mathbf{a}_j - c_j$.
(*Hint*. Suppose

$$\mathbf{B} = (\mathbf{a}_1, \mathbf{a}_2, \ldots, \mathbf{a}_r, \ldots, \mathbf{a}_m)$$

$$\hat{\mathbf{B}} = (\mathbf{a}_1, \mathbf{a}_2, \ldots, \mathbf{a}_k, \ldots, \mathbf{a}_m)$$

First show that $\hat{\mathbf{B}} = \mathbf{BE}$, and $(\hat{\mathbf{B}})^{-1} = \mathbf{E}^{-1}\mathbf{B}^{-1}$, where

$$\mathbf{E} = \begin{bmatrix} 1 & 0 \ldots & \overset{\overset{r\text{th column}}{\downarrow}}{y_{1k}} & \ldots & 0 \\ 0 & 1 \ldots & y_{2k} & \ldots & 0 \\ \vdots & \vdots & \vdots & & \vdots \\ 0 & 0 \ldots & y_{rk} & \ldots & 0 \\ \vdots & \vdots & \vdots & & \vdots \\ 0 & 0 \ldots & y_{mk} & \ldots & 1 \end{bmatrix} \begin{array}{l} \\ \\ \\ \leftarrow r\text{th row} \\ \\ \\ \end{array}$$

This form is usually called the *product form of the inverse* and is discussed in more detail in Section 5.1).

3.16 Suppose that we have a basic feasible solution that is nondegenerate. Further, suppose that an improving nonbasic variable enters the basis. Show that if the minimum ratio criterion for exiting from the basis occurs at a unique index, then the resulting basic feasible solution is also nondegenerate.

3.17 Solve the following problem by the simplex method.

$$\begin{array}{lrcl} \text{Maximize} & x_1 - 2x_2 + x_3 & & \\ \text{Subject to} & x_1 + x_2 + x_3 & \leqslant & 12 \\ & 2x_1 + x_2 - x_3 & \leqslant & 6 \\ & -x_1 + 3x_2 \phantom{{}+x_3} & \leqslant & 9 \\ & x_1, \quad x_2, \quad x_3 & \geqslant & 0 \end{array}$$

3.18 Consider the problem

$$\begin{array}{ll} \text{Maximize} & 2x_1 + x_2 - 3x_3 + 5x_4 \\ \text{Subject to} & x_1 + 2x_2 + 4x_3 - x_4 \leqslant 6 \\ & 2x_1 + 3x_2 - x_3 + x_4 \leqslant 12 \\ & x_1 + x_3 + x_4 \leqslant 4 \\ & x_1, \quad x_2, \quad x_3, \quad x_4 \geqslant 0 \end{array}$$

Find a basic feasible solution with the basic variables as x_1, x_2, and x_4. Is this solution optimal? If not, then starting with this solution find the optimal solution.

3.19 Use the simplex method to solve the following problem. Note that the variables are unrestricted in sign.

$$\begin{array}{ll} \text{Minimize} & 3x_1 - x_2 \\ \text{Subject to} & x_1 - 3x_2 \geqslant -3 \\ & 2x_1 + 3x_2 \geqslant -6 \\ & 2x_1 + x_2 \leqslant 8 \\ & 4x_1 - x_2 \leqslant 16 \end{array}$$

3.20 An agricultural mill produces feed for cattle. The cattle feed consists of three main ingredients: corn, lime, and fish meal. These ingredients contain three nutrients: protein, calcium, and vitamins. The following table gives the nutrient contents per pound of each ingredient.

	Ingredient		
NUTRIENT	CORN	LIME	FISH MEAL
Protein	25	15	25
Calcium	15	30	20
Vitamins	5	12	8

The protein, calcium, and vitamins content per pound of the cattle feed must be in the following intervals respectively: [18, 22], [20, ∞), and [6, 12]. If the selling prices per pound of corn, lime, and fish meal are respectively \$0.10, \$0.08, and \$0.12, find the least expensive mix.
(*Hint*. First find a basis $\mathbf{B}$ such that $\mathbf{B}^{-1}\mathbf{b} \geqslant \mathbf{0}$.)

3.21 A firm makes three products 1, 2, and 3. Each product requires production time in three departments as shown below.

PRODUCT	DEPARTMENT 1	DEPARTMENT 2	DEPARTMENT 3
1	3 hr/unit	2 hr/unit	1 hr/unit
2	4 hr/unit	1 hr/unit	3 hr/unit
3	2 hr/unit	2 hr/unit	3 hr/unit

There are 600, 400, 300 hours of production time available in the three departments, respectively. If each unit of products 1, 2, and 3 contribute \$2, \$4, and \$2.5 to profit respectively, find the optimal product mix.

3.22 Solve Exercise 1.9 as a linear model by the simplex method. Find the objective $36/x_2x_3$ corresponding to the optimal point obtained from the simplex method. By trial and error see if you can find a feasible point whose objective $36/x_2x_3$ is better than the objective obtained above.

3.23 Solve Exercise 1.10 to find the number of barrels of each crude oil that satisfy the demand and minimize the total cost.
(*Hint*. First find a basis $\mathbf{B}$ with $\mathbf{B}^{-1}\mathbf{b} \geqslant \mathbf{0}$.)

3.24 A nut packager has on hand 150 pounds of peanuts, 100 pounds of cashews, and 50 pounds of almonds. The packager can sell three kinds of mixtures of these nuts: a cheap mix consisting of 80% peanuts and 20% cashews; a party mix with 50% peanuts, 30% cashews, and 20% almonds; and a deluxe mix with 20% peanuts, 50% cashews, and 30% almonds. If the 12-ounce can of the cheap mix, the party mix, and the deluxe mix can be sold for \$0.90, \$1.10, and \$1.30 respectively, how many cans of each type would the packager produce in order to maximize his return?

3.25 Suppose, a priori, we know that a solution cannot be optimal unless it involved a variable at a positive value. Show that this variable can be eliminated and that the reduced system with one less equation and variable can be solved in its place. Illustrate by an example.

3.26 Consider the linear program: Minimize $\mathbf{cx}$ subject to $\mathbf{Ax} \geqslant \mathbf{b}$, $\mathbf{x} \geqslant \mathbf{0}$. Converting the inequality constraints to equality constraints, suppose that the optimal basis is $\mathbf{B}$. Show that $\mathbf{w} = \mathbf{c}_B\mathbf{B}^{-1} \geqslant \mathbf{0}$.

3.27 Suppose that some tableau for a linear programming problem exhibits an indication of unboundedness. Considering only the basic vectors and the nonbasic vector corresponding to the column giving the unboundedness indication, demonstrate which quantities, if any, satisfy the definition of an extreme point. Also demonstrate which quantities, if any, satisfy the definition of an extreme ray. Give an example.

3.28 A necessary and sufficient condition for unboundedness of the objective value of a minimization problem is that there exists a direction of the feasible region such that $\mathbf{cd} < 0$. A condition for unboundedness in the simplex method is the existence of an index j such that $z_j - c_j > 0$ and

$\mathbf{y}_j \leqslant \mathbf{0}$. Discuss in detail the correspondence between the two conditions. (*Hint*. Let

$$\mathbf{d} = \begin{bmatrix} -\mathbf{y}_j \\ \hline 0 \\ \vdots \\ 1 \\ \vdots \\ 0 \end{bmatrix}$$

where the 1 appears at the jth position. Show that $\mathbf{d}$ is a direction of the set and that $\mathbf{cd} = c_j - z_j$. Can you show that $\mathbf{d}$ is an extreme direction?)

3.29 Consider the following problem.

$$\begin{aligned} \text{Maximize} \quad & 3x_1 + 2x_2 - x_3 + x_4 \\ \text{Subject to} \quad & 2x_1 - 4x_2 - x_3 + x_4 \leqslant 8 \\ & x_1 + x_2 + 2x_3 - 3x_4 \leqslant 10 \\ & x_1 - x_2 - 4x_3 + x_4 \leqslant 3 \\ & x_1, \quad x_2, \quad x_3, \quad x_4 \geqslant 0 \end{aligned}$$

Use the simplex method to verify that the optimal solution is unbounded. Make use of the final simplex tableau to construct a feasible solution with an objective of at least 3000. Make use of the final tableau to construct a direction $\mathbf{d}$ such that $\mathbf{cd} > 0$.

3.30 Prove or give a counterexample. In order to have a basic variable in a particular constraint, that variable must have a nonzero coefficient in its original column and the particular row.

3.31 We showed in the text that $z_j - c_j = 0$ for a basic variable. Interpret this result.

3.32 Consider the linear programming problem: Maximize $\mathbf{cx}$ subject to $\mathbf{Ax} = \mathbf{b}$, $\mathbf{x} \geqslant \mathbf{0}$ where $\mathbf{A}$ is an $m \times n$ matrix of rank m. Suppose an optimal solution with basis $\mathbf{B}$ is at hand. Further, suppose that $\mathbf{b}$ is replaced by $\mathbf{b} + \lambda\mathbf{d}$ where λ is a scalar and $\mathbf{d}$ is a fixed nonzero vector of dimension m. Give a condition such that the basis $\mathbf{B}$ will be optimal for all $\lambda \geqslant 0$.

3.33 Show that in the simplex method if a variable x_j leaves the basis, it cannot enter the basis in the next iteration.

3.34 Can a vector that is inserted at one iteration in the simplex method be removed immediately at the next iteration? When can this occur and when is it impossible?

3.35 Find a nonbasic feasible optimal solution of the following problem.

$$
\begin{array}{ll}
\text{Maximize} & 11x_1 + 2x_2 - x_3 + 3x_4 + 4x_5 + x_6 \\
\text{Subject to} & 5x_1 + x_2 - x_3 + 2x_4 + x_5 = 12 \\
& -14x_1 - 3x_2 + 3x_3 - 5x_4 + x_6 = 2 \\
& 2x_1 + \frac{1}{2}x_2 - \frac{1}{2}x_3 + \frac{1}{2}x_4 \leqslant \frac{5}{2} \\
& 3x_1 + \frac{1}{2}x_2 + \frac{1}{2}x_3 + \frac{3}{2}x_4 \leqslant 3 \\
& x_1, \quad x_2, \quad x_3, \quad x_4, \quad x_5, \; x_6 \geqslant 0
\end{array}
$$

(*Hint*. Let the initial basis consist of x_5, x_6, and the slack variables of the last two constraints. Then find alternative optimal solutions of the problem.)

3.36 The following mathematical formulation describes a problem of allocating three resources to the annual production of three commodities by a manufacturing firm. The amounts of the three products to be produced are denoted by x_1, x_2, and x_3. The objective function reflects the dollar contribution to profit of these products.

$$
\begin{array}{ll}
\text{Maximize} & 10x_1 + 15x_2 + 5x_3 \\
\text{Subject to} & 2x_1 + x_2 \leqslant 6000 \\
& 3x_1 + 3x_2 + x_3 \leqslant 9000 \\
& x_1 + 2x_2 + 2x_3 \leqslant 4000 \\
& x_1, \quad x_2, \quad x_3 \geqslant 0
\end{array}
$$

a. Without using the simplex method, verify that the optimal basis consists of the slack variable of the first constraint, x_1, and x_2.
b. Make use of the information in (a) to write the optimal tableau.
c. The Research and Development Department proposes a new product whose production coefficients are represented by $[2, 4, 1]^t$. If the contribution to profit is \$12 per unit of this new product, should this product be produced? If so, what is the new optimal program?
d. What is the minimal contribution to profit that should be expected before production of this new product would actually increase the value of the objective function?

3.37 Write a precise argument showing that in the absence of degeneracy, and assuming feasibility, the simplex method will provide an optimal solution, or show unboundedness, of a linear program in a finite number of steps.

3.38 Consider the following linear program.

$$\begin{array}{ll} \text{Minimize} & -x_1 - 2x_2 + x_3 \\ \text{Subject to} & 2x_1 + x_2 + x_3 \leqslant 6 \\ & 2x_2 - x_3 \leqslant 3 \\ & x_1, \quad x_2, \quad x_3 \geqslant 0 \end{array}$$

a. Find the optimal solution by the simplex method. At each iteration identify $\mathbf{B}$, $\mathbf{N}$, $\mathbf{B}^{-1}$, $\mathbf{B}^{-1}\mathbf{N}$, $\mathbf{c}_B\mathbf{B}^{-1}$, and the $z_j - c_j$'s.
b. At optimality, find $\partial x_1/\partial x_3$, $\partial x_2/\partial x_4$, $\partial z/\partial x_5$, $\partial \mathbf{x}_B/\partial \mathbf{x}_N$, where x_4 and x_5 are the slack variables. Interpret these values.
c. Suppose that c_1 is changed from -1 to $-1 + \Delta_1$, and c_2 is changed from -2 to $-2 + \Delta_2$. Find the region in the (Δ_1, Δ_2) space that will maintain optimality of the vector you obtained in (a).
d. Suppose a new activity x_6 is considered. Here $c_6 = -3$, $a_{16} = 3$, and $a_{26} = 3$. Is it worthwhile to produce the activity? If your answer is yes, find the new optimal solution.
e. Suppose that b_1 is changed from 6 to $6 + \Delta$. Find the range of Δ that will maintain optimality of the basis found in part (a).
f. Make use of the final tableau to represent the column $\mathbf{a}_3$ as a linear combination of $\mathbf{a}_1$ and $\mathbf{a}_2$.

3.39 Consider the following linear programming problem.

$$\begin{array}{ll} \text{Maximize} & 2x_1 + 12x_2 + 7x_3 \\ \text{Subject to} & x_1 + 3x_2 + 2x_3 \leqslant 10{,}000 \\ & 2x_1 + 2x_2 + x_3 \leqslant 4{,}000 \\ & x_1, \quad x_2, \quad x_3 \geqslant 0 \end{array}$$

The optimal solution is shown below, where z is the objective function, and x_4 and x_5 are the slack variables.

	z	x_1	x_2	x_3	x_4	x_5	RHS
z	1	12	2	0	0	7	28,000
x_4	0	−3	−1	0	1	−2	2,000
x_3	0	2	2	1	0	1	4,000

a. What are the rates of increase of the objective as a function of the right-hand side of the first and second constraints respectively?
b. Suppose that the right-hand side of the second constraint is changed to $4000 + \Delta$. What is the range of Δ that will keep the basis of the foregoing tableau optimal?
c. Find explicitly the optimal value z as a function of Δ.
d. Suppose that increasing the right-hand side of the second constraint involved expanding a manufacturing department. This will involve a fixed cost as well as a variable cost that is a function of Δ. In particular the cost as a function of Δ is

$$h(\Delta) = \begin{cases} 0 & \text{if } \Delta = 0 \\ 3000 + 3\Delta & \text{if } \Delta > 0 \end{cases}$$

What is the break-even point, where the cost and the added profit will balance? What do you recommend for the optimal value of Δ?

3.40 Solve Exercise 1.11 by the simplex method.

3.41 Solve Exercise 1.12 by the simplex method. Suppose that extra man-hours can be obtained by allowing overtime at the average of \$12 per hour. Would it be profitable to increase the man-hours? If so, by how much? How would this increase the profit?

3.42 The starting and current tableaux of a given problem are shown below. Find the values of the unknowns a through l.

Starting Tableau

z	x_1	x_2	x_3	x_4	x_5	RHS
1	a	1	-2	0	0	0
0	b	c	d	1	0	6
0	-1	3	e	0	1	1

Current Tableau

z	x_1	x_2	x_3	x_4	x_5	RHS
1	0	7	j	k	l	9
0	g	2	-1	$\frac{1}{2}$	0	f
0	h	i	1	$\frac{1}{2}$	1	4

3.43 The following is the current simplex tableau of a given maximization problem. The objective is to maximize $5x_1 + 3x_2$, and the slack variables are x_3 and x_4. The constraints are of the $\leqslant$ type.

	z	x_1	x_2	x_3	x_4	RHS
z	1	b	1	f	g	10
x_3	0	c	0	1	$\frac{1}{5}$	2
x_1	0	d	e	0	1	a

a. Find the unknowns a through g.
b. Find $\mathbf{B}^{-1}$.
c. Find $\partial x_3/\partial x_2$, $\partial z/\partial b_1$, $\partial z/\partial x_4$, $\partial x_1/\partial b_2$.
d. Is the tableau optimal?

3.44 The following is the current simplex tableau of a linear programming problem. The objective is to minimize $-28x_4 - x_5 - 2x_6$, and x_1, x_2, and x_3 are the slack variables.

	z	x_1	x_2	x_3	x_4	x_5	x_6	RHS
z	1	b	c	0	0	-1	g	-14
x_6	0	3	0	$-\frac{14}{3}$	0	1	1	a
x_2	0	6	d	2	0	$\frac{5}{2}$	0	5
x_4	0	0	e	f	1	0	0	0

a. Find the values of the unknowns a through g in the tableau.
b. Find $\mathbf{B}^{-1}$.
c. Find $\partial x_2/\partial x_1$, $\partial z/\partial x_5$, $\partial x_6/\partial b_3$.
d. Without explicitly finding the basic vectors $\mathbf{a}_6$, $\mathbf{a}_2$, $\mathbf{a}_4$, give the representation of the vector $\mathbf{a}_5$ in terms of these basic vectors.

3.45 Consider the problem: Maximize $\mathbf{cx}$ subject to $\mathbf{Ax} = \mathbf{b}$, $\mathbf{x}$ unrestricted in sign. Under what conditions does this problem have a bounded optimal solution?

3.46 Consider a linear programming problem in which some of the variables are unrestricted in sign. What are the conditions for a bounded optimal solution? Without introducing additional variables, show how the entry and exit criteria of the simplex method can be modified such that the unrestricted variables are handled directly. How does the simplex method recognize reaching an optimal solution in this case? Illustrate by solving the following problem.

$$
\begin{aligned}
\text{Minimize} \quad & -2x_1 + x_2 \\
\text{Subject to} \quad & x_1 + x_2 \leqslant 4 \\
& x_1 - x_2 \leqslant 6 \\
& x_1 \geqslant 0 \\
& x_2 \quad \text{unrestricted}
\end{aligned}
$$

3.47 Consider the following simplex tableau for a minimization problem (the constraints are of the $\leqslant$ type and x_3, x_4, and x_5 are the slacks).

z	x_1	x_2	x_3	x_4	x_5	RHS
1	0	a	0	b	0	f
0	1	-2	0	4	0	c
0	0	-1	1	5	0	d
0	0	0	0	7	1	e

Suppose that $a < 0$, $b \leqslant 0$, and $c, d, e \geqslant 0$

a. Find $\mathbf{B}^{-1}$.
b. Find $\mathbf{B}$.
c. Is the tableau optimal?
d. Give the original tableau.
e. From the tableau identify $\mathbf{c}_B\mathbf{B}^{-1}$ and give its interpretation.

Now suppose that $a > 0$, $b \leqslant 0$; and $c, d, e \geqslant 0$.

f. Is the new tableau optimal?
g. Give an extreme direction.
h. Let $a = 5$ and $f = -10$. Give a feasible solution with $z = -200$.

3.48 Construct a detailed flow diagram of the simplex method. What are the number of operations (additions, subtractions, multiplications, divisions) that are needed at each simplex iteration? Using FORTRAN (or another language), convert the flow diagram into a simplex code.

NOTES AND REFERENCES

1. This chapter describes the simplex algorithm of Dantzig (developed in 1947 and published at a later date in 1949 [86]). The material of this chapter is standard and can be found in most linear programming books.
2. In Section 3.1 we proved optimality at an extreme point via the representation theorem. The reader may note that the simplex algorithm itself gives a constructive proof of this fact.
3. The material on block pivoting is due to Tucker [440]. For further reading on block pivoting, see Dantzig [97] and Lasdon [305].

FOUR: STARTING SOLUTION AND CONVERGENCE

In the previous chapter, we developed the simplex method with the assumption that an initial basic feasible solution is at hand. In many cases, such a solution is not readily available, and some work may be needed to get the simplex method started. In this chapter we describe two procedures (the two-phase method and the big-M method), both involving *artificial variables* to obtain an initial basic feasible solution to a slightly modified set of constraints. The simplex method is used to eliminate the artificial variables and to solve the original problem. We also discuss in more detail the difficulties associated with degeneracy. In particular we show that the simplex method converges in a finite number of steps, even in the presence of degeneracy, provided that a special rule for exiting from the basis is adopted.

4.1 THE INITIAL BASIC FEASIBLE SOLUTION

Recall that the simplex method starts with a basic feasible solution and moves to an improved basic feasible solution, until the optimal point is reached or else unboundedness of the objective function is verified. However, in order to initialize the simplex method, a basis $\mathbf{B}$ with $\bar{\mathbf{b}} = \mathbf{B}^{-1}\mathbf{b} \geqslant \mathbf{0}$ must be available. We shall show that the simplex method can always be initiated with a very simple basis, namely the identity.

Easy Case

Suppose that the constraints are of the form $\mathbf{Ax} \leqslant \mathbf{b}$, $\mathbf{x} \geqslant \mathbf{0}$ where $\mathbf{A}$ is an $m \times n$ matrix and $\mathbf{b}$ is an m nonnegative vector. By adding the slack vector $\mathbf{x}_s$, the constraints are put in the following standard form: $\mathbf{Ax} + \mathbf{x}_s = \mathbf{b}$, $\mathbf{x} \geqslant \mathbf{0}$, $\mathbf{x}_s \geqslant \mathbf{0}$. Note that the new $m \times (m + n)$ constraint matrix $(\mathbf{A}, \mathbf{I})$ has rank m, and a basic feasible solution of this system is at hand, by letting $\mathbf{x}_s = \mathbf{b}$ be the basic vector, and $\mathbf{x} = \mathbf{0}$ be the nonbasic vector. With this starting basic feasible solution, the simplex method can be applied.

Some Bad Cases

In many occasions, finding a starting basic feasible solution is not as straightforward as the case described above. To illustrate, suppose that the constraints are of the form $\mathbf{Ax} \leqslant \mathbf{b}$, $\mathbf{x} \geqslant \mathbf{0}$ but the vector $\mathbf{b}$ is not nonnegative. In this case, after introducing the slack vector $\mathbf{x}_s$, we cannot let $\mathbf{x} = \mathbf{0}$, because $\mathbf{x}_s = \mathbf{b}$ violates the nonnegativity requirement.

Another situation occurs when the constraints are of the form $\mathbf{Ax} \geqslant \mathbf{b}$, $\mathbf{x} \geqslant \mathbf{0}$, where $\mathbf{b} \geqslant \mathbf{0}$. After subtracting the slack vector $\mathbf{x}_s$, we get $\mathbf{Ax} - \mathbf{x}_s = \mathbf{b}$, $\mathbf{x} \geqslant \mathbf{0}$, and $\mathbf{x}_s \geqslant \mathbf{0}$. Again, there is no obvious way of picking a basis $\mathbf{B}$ from the matrix $(\mathbf{A}, -\mathbf{I})$ with $\bar{\mathbf{b}} = \mathbf{B}^{-1}\mathbf{b} \geqslant \mathbf{0}$.

In general, any linear programming problem can be transformed into a problem of the following form.

$$\begin{aligned} \text{Minimizes} \quad & \mathbf{cx} \\ \text{Subject to} \quad & \mathbf{Ax} = \mathbf{b} \\ & \mathbf{x} \geqslant \mathbf{0} \end{aligned}$$

where $\mathbf{b} \geqslant \mathbf{0}$ (if $b_i < 0$, the ith row can be multiplied by -1). This can be accomplished by introducing slack variables and by simple manipulation of the constraints and variables. If $\mathbf{A}$ contains an identity matrix, then an immediate basic feasible solution is at hand, by simply letting $\mathbf{B} = \mathbf{I}$, and since $\mathbf{b} \geqslant \mathbf{0}$, then $\mathbf{B}^{-1}\mathbf{b} = \bar{\mathbf{b}} \geqslant \mathbf{0}$. Otherwise, something else must be done.

Example 4.1

a. Consider the following constraints:

$$\begin{aligned} x_1 + 2x_2 &\leqslant 4 \\ -x_1 + x_2 &\leqslant 1 \\ x_1, \quad x_2 &\geqslant 0 \end{aligned}$$

After adding the slack variables x_3 and x_4, we get

$$\begin{aligned} x_1 + 2x_2 + x_3 \qquad &= 4 \\ -x_1 + x_2 \qquad + x_4 &= 1 \\ x_1, \quad x_2, \quad x_3, \quad x_4 &\geqslant 0 \end{aligned}$$

An obvious starting basic feasible solution is given by $\mathbf{x}_B = \begin{bmatrix} x_3 \\ x_4 \end{bmatrix} = \begin{bmatrix} 4 \\ 1 \end{bmatrix}$ and $\mathbf{x}_N = \begin{bmatrix} x_1 \\ x_2 \end{bmatrix} = \begin{bmatrix} 0 \\ 0 \end{bmatrix}$.

b. Consider the following constraints:

$$\begin{aligned} x_1 + x_2 + x_3 &\leqslant 6 \\ -2x_1 + 3x_2 + 2x_3 &\geqslant 3 \\ x_2, \quad x_3 &\geqslant 0 \end{aligned}$$

Note that x_1 is unrestricted. So the change of variable $x_1 = x_1^+ - x_1^-$ is made. Also the slack variables x_4 and x_5 are introduced. This leads to the following constraints in standard form:

$$\begin{aligned} x_1^+ - x_1^- + x_2 + x_3 + x_4 \qquad &= 6 \\ -2x_1^+ + 2x_1^- + 3x_2 + 2x_3 \qquad - x_5 &= 3 \\ x_1^+, \quad x_1^-, \quad x_2, \quad x_3, \quad x_4, \quad x_5 &\geqslant 0 \end{aligned}$$

Note that the constraint matrix does not contain an identity and no obvious feasible basis $\mathbf{B}$ can be extracted.

c. Consider the following constraints:

$$
\begin{aligned}
x_1 + x_2 - 2x_3 &\leqslant -3 \\
-2x_1 + x_2 + 3x_3 &\leqslant 7 \\
x_1, \quad x_2, \quad x_3 &\geqslant 0
\end{aligned}
$$

Since the right-hand side of the first constraint is negative, the first inequality is multiplied by -1. Introducing the slack variables x_4 and x_5 leads to the following system:

$$
\begin{aligned}
-x_1 - x_2 + 2x_3 - x_4 \quad &= 3 \\
-2x_1 + x_2 + 3x_3 \quad + x_5 &= 7 \\
x_1, \quad x_2, \quad x_3, \quad x_4, x_5 &\geqslant 0
\end{aligned}
$$

Note again that this constraint matrix contains no identity.

Artificial Variables

After manipulating the constraints and introducing slack variables, suppose that the constraints are put in the format $\mathbf{Ax} = \mathbf{b}$, $\mathbf{x} \geqslant \mathbf{0}$ where $\mathbf{A}$ is an $m \times n$ matrix and $\mathbf{b} \geqslant \mathbf{0}$ is an m vector. Further suppose that $\mathbf{A}$ has no identity submatrix (if $\mathbf{A}$ has an identity submatrix then we have an obvious starting basic feasible solution). In this case we shall resort to artificial variables to get a starting basic feasible solution, and then use the simplex method itself to get rid of these artificial variables.

To illustrate, suppose that we change the restrictions by adding an artificial vector $\mathbf{x}_a$ leading to the system $\mathbf{Ax} + \mathbf{x}_a = \mathbf{b}$, $\mathbf{x} \geqslant \mathbf{0}$, $\mathbf{x}_a \geqslant \mathbf{0}$. Note that by construction, we forced an identity matrix corresponding to the artificial vector. This gives an immediate basic feasible solution of the new system, namely $\mathbf{x}_a = \mathbf{b}$ and $\mathbf{x} = \mathbf{0}$. Even though we now have a starting basic feasible solution, and the simplex method can be applied, we have in effect changed the problem. In order to get back to our original problem, we must force these artificial variables to zero, because $\mathbf{Ax} = \mathbf{b}$ if and only if $\mathbf{Ax} + \mathbf{x}_a = \mathbf{b}$ with $\mathbf{x}_a = \mathbf{0}$. In other words, *artificial variables* are only a tool to get the simplex method started, but we must guarantee that these variables will eventually drop to zero.

At this stage, it is worthwhile to note the difference between slack and artificial variables. A slack variable is introduced to put the problem in equality form, and the slack variable can very well be positive, which means that the inequality holds as a strict inequality. *Artificial variables*, however, are not legitimate variables, and they may be introduced to facilitate the initiation of the simplex method. These artificial variables, however, must eventually drop to zero in order to attain feasibility in the original problem.

Example 4.2

Consider the following constraints:

$$\begin{aligned} x_1 + 2x_2 &\geq 4 \\ -3x_1 + 4x_2 &\geq 5 \\ 2x_1 + x_2 &\leq 6 \\ x_1, x_2 &\geq 0 \end{aligned}$$

Introducing the slack variables x_3, x_4, and x_5, we get

$$\begin{aligned} x_1 + 2x_2 - x_3 \qquad\qquad\qquad &= 4 \\ -3x_1 + 4x_2 \qquad - x_4 \qquad &= 5 \\ 2x_1 + x_2 \qquad\qquad\qquad + x_5 &= 6 \\ x_1, \quad x_2, \quad x_3, \quad x_4, \quad x_5 &\geq 0 \end{aligned}$$

This constraint matrix has no identity submatrix. We can introduce three artificial variables to obtain a starting basic feasible solution. Note, however, that x_5 appears in the last row only, and it has coefficient 1. So we only need to introduce two artificial variables x_6 and x_7, which leads to the following system.

$$\begin{array}{lccr} \overbrace{\qquad\textit{Legitimate variables}\qquad} & \overbrace{\quad\textit{Artificial variables}\quad} & & \\ x_1 + 2x_2 - x_3 & +x_6 \qquad\quad & = 4 \\ -3x_1 + 4x_2 \quad - x_4 & \qquad\quad +x_7 & = 5 \\ 2x_1 + x_2 \qquad\qquad + x_5 & & = 6 \\ x_1, \; x_2, \; x_3, \; x_4, \; x_5, & x_6, \quad x_7 & \geq 0 \end{array}$$

Now we have an immediate starting basic feasible solution of the new system, namely $x_5 = 6$, $x_6 = 4$, and $x_7 = 5$. The rest of the variables are nonbasic and have value zero. Needless to say, we eventually would like for the artificial variables x_6 and x_7 to drop to zero.

4.2 THE TWO-PHASE METHOD

There are various methods that can be used to eliminate the artificial variables. One of these methods is to minimize the sum of the artificial variables, subject to the constraints $\mathbf{Ax} + \mathbf{x}_a = \mathbf{b}$, $\mathbf{x} \geqslant \mathbf{0}$ and $\mathbf{x}_a \geqslant \mathbf{0}$. If the original problem has a feasible solution, then the optimal value of this problem is zero, where all the artificial variables drop to zero. More importantly, as the artificial variables drop to zero, they leave the basis, and legitimate variables enter instead. Eventually all artificial variables leave the basis (this is not always the case, because we may have an artificial variable in the basis at level zero; this will be discussed later in greater detail). The basis then consists of legitimate variables. In other words, we get a basic feasible solution of the original system $\mathbf{Ax} = \mathbf{b}$, $\mathbf{x} \geqslant \mathbf{0}$, and the simplex method can be started with the original objective $\mathbf{cx}$. If, on the other hand, after solving this problem we have a positive artificial variable, then the original problem has no feasible solutions (why?). This procedure is called the *two-phase method.* In the first phase we reduce artificial variables to value zero, or conclude that the original problem has no feasible solutions. In the former case, the second phase minimizes the original objective function starting with the basic feasible solution obtained at the end of the phase I. The two-phase method is outlined below.

Phase I

Solve the following linear program starting with the basic feasible solution $\mathbf{x} = \mathbf{0}$ and $\mathbf{x}_a = \mathbf{b}$.

$$\begin{aligned} \text{Minimize} \quad & \mathbf{1x}_a \\ \text{Subject to} \quad & \mathbf{Ax}+\mathbf{x}_a=\mathbf{b} \\ & \mathbf{x},\ \mathbf{x}_a \geqslant \mathbf{0} \end{aligned}$$

If at optimality $\mathbf{x}_a \neq \mathbf{0}$, then stop; the original problem has no feasible solutions. Otherwise let the basic and nonbasic legitimate variables be $\mathbf{x}_B$ and $\mathbf{x}_N$. (We are assuming that all artificial variables left the basis. The case where some artificial variables remain in the basis at zero level will be discussed later.) Go to phase II.

Phase II

Solve the following linear program starting with the basic feasible solution $\mathbf{x}_B = \mathbf{B}^{-1}\mathbf{b}$ and $\mathbf{x}_N = \mathbf{0}$.

$$\begin{aligned} \text{Minimize} \quad & \mathbf{c}_B\mathbf{x}_B + \mathbf{c}_N\mathbf{x}_N \\ \text{Subject to} \quad & \mathbf{x}_B + \mathbf{B}^{-1}\mathbf{N}\mathbf{x}_N = \mathbf{B}^{-1}\mathbf{b} \\ & \mathbf{x}_B,\ \mathbf{x}_N \geqslant \mathbf{0} \end{aligned}$$

The optimal solution of the original problem is given by the optimal solution of the foregoing problem.

Example 4.3

$$\begin{aligned} \text{Minimize} \quad & x_1 - 2x_2 \\ \text{Subject to} \quad & x_1 + x_2 \geq 2 \\ & -x_1 + x_2 \geq 1 \\ & x_2 \leq 3 \\ & x_1, \quad x_2 \geq 0 \end{aligned}$$

The feasible region and the path taken by phase I and phase II to reach the optimal point are shown in Figure 4.1. After introducing the slack variables x_3, x_4, x_5, the following problem is obtained.

$$\begin{aligned} \text{Minimize} \quad & x_1 - 2x_2 \\ \text{Subject to} \quad & x_1 + x_2 - x_3 \qquad\qquad = 2 \\ & -x_1 + x_2 \qquad - x_4 \qquad = 1 \\ & \qquad\; x_2 \qquad\qquad + x_5 = 3 \\ & x_1, \quad x_2, \quad x_3, \quad x_4, \quad x_5 \geq 0 \end{aligned}$$

An initial identity is not available. So introduce the artificial variables x_6 and x_7 (note that the last constraint does not need an artificial variable). Phase I starts by minimizing $x_0 = x_6 + x_7$.

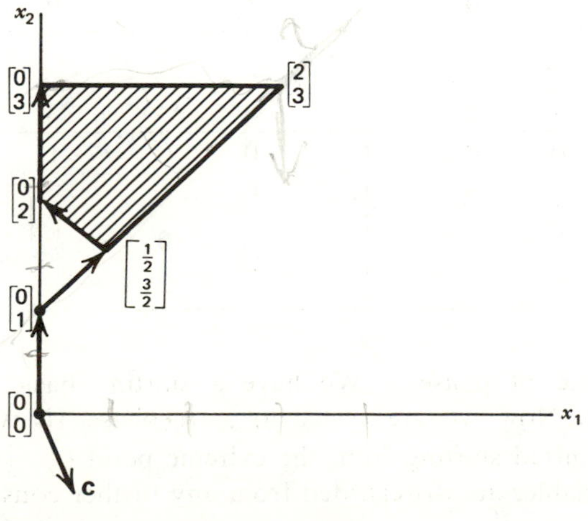

Figure 4.1. **Example of the two-phase method.**

PHASE I

x_0	x_1	x_2	x_3	x_4	x_5	x_6 (Artificials)	x_7 (Artificials)	RHS
1	0	0	0	0	0	−1	−1	0
0	1	1	−1	0	0	1	0	2
0	−1	1	0	−1	0	0	1	1
0	0	1	0	0	1	0	0	3

Add rows 1 and 2 to row 0 so that $z_6 - c_6 = z_7 - c_7 = 0$ will be displayed.

	x_0	x_1	x_2	x_3	x_4	x_5	x_6	x_7	RHS
x_0	1	0	2	−1	−1	0	0	0	3
x_6	0	1	1	−1	0	0	1	0	2
x_7	0	−1	(1)	0	−1	0	0	1	1
x_5	0	0	1	0	0	1	0	0	3

	x_0	x_1	x_2	x_3	x_4	x_5	x_6	x_7	RHS
x_0	1	2	0	−1	1	0	0	−2	1
x_6	0	(2)	0	−1	1	0	1	−1	1
x_2	0	−1	1	0	−1	0	0	1	1
x_5	0	1	0	0	1	1	0	−1	2

	x_0	x_1	x_2	x_3	x_4	x_5	x_6	x_7	RHS
x_0	1	0	0	0	0	0	−1	−1	0
x_1	0	1	0	$-\frac{1}{2}$	$\frac{1}{2}$	0	$\frac{1}{2}$	$-\frac{1}{2}$	$\frac{1}{2}$
x_2	0	0	1	$-\frac{1}{2}$	$-\frac{1}{2}$	0	$\frac{1}{2}$	$\frac{1}{2}$	$\frac{3}{2}$
x_5	0	0	0	$\frac{1}{2}$	$\frac{1}{2}$	1	$-\frac{1}{2}$	$-\frac{1}{2}$	$\frac{3}{2}$

This is the end of phase I. We have a starting basic feasible solution, $(x_1, x_2) = (\frac{1}{2}, \frac{3}{2})$. Now we are ready to start phase II, where the original objective is minimized starting from the extreme point $(\frac{1}{2}, \frac{3}{2})$ (see Figure 4.1). The artificial variables are disregarded from any further consideration.

PHASE II

z	x_1	x_2	x_3	x_4	x_5	RHS
1	-1	2	0	0	0	0
0	1	0	$-\frac{1}{2}$	$\frac{1}{2}$	0	$\frac{1}{2}$
0	0	1	$-\frac{1}{2}$	$-\frac{1}{2}$	0	$\frac{3}{2}$
0	0	0	$\frac{1}{2}$	$\frac{1}{2}$	1	$\frac{3}{2}$

Multiply rows 1 and 2 by 1 and -2 respectively and add to row 0 producing $z_1 - c_1 = z_2 - c_2 = 0$.

	z	x_1	x_2	x_3	x_4	x_5	RHS
z	1	0	0	$\frac{1}{2}$	$\frac{3}{2}$	0	$-\frac{5}{2}$
x_1	0	1	0	$-\frac{1}{2}$	$\textcircled{\frac{1}{2}}$	0	$\frac{1}{2}$
x_2	0	0	1	$-\frac{1}{2}$	$-\frac{1}{2}$	0	$\frac{3}{2}$
x_5	0	0	0	$\frac{1}{2}$	$\frac{1}{2}$	1	$\frac{3}{2}$

	z	x_1	x_2	x_3	x_4	x_5	RHS
z	1	-3	0	2	0	0	-4
x_4	0	2	0	-1	1	0	1
x_2	0	1	1	-1	0	0	2
x_5	0	-1	0	$\textcircled{1}$	0	1	1

	z	x_1	x_2	x_3	x_4	x_5	RHS
z	1	-1	0	0	0	-2	-6
x_4	0	1	0	0	1	1	2
x_2	0	0	1	0	0	1	3
x_3	0	-1	0	1	0	1	1

Since $z_j - c_j \leqslant 0$ for all nonbasic variables, the optimal point (0, 3) with objective -6 is reached. Note that phase I moved from the infeasible point (0, 0), to the infeasible point (0, 1), and finally to the feasible point $(\frac{1}{2}, \frac{3}{2})$. From this extreme point, phase II moved to the feasible point (0, 2), and finally to the optimal point (0, 3). This is illustrated in Figure 4.1. The purpose of phase I is to get us to an extreme point of the feasible region, while phase II takes us from this feasible point to the optimal point.

Analysis of the Two-Phase Method

At the end of phase I either $\mathbf{x}_a \neq \mathbf{0}$ or else $\mathbf{x}_a = \mathbf{0}$. These two cases are discussed in detail below.

Case A: $\mathbf{x}_a \neq \mathbf{0}$

If $\mathbf{x}_a \neq \mathbf{0}$, then the original problem has no feasible solutions, because if there is an $\mathbf{x} \geq \mathbf{0}$ with $\mathbf{A}\mathbf{x} = \mathbf{b}$, then $\begin{pmatrix}\mathbf{x}\\ \mathbf{0}\end{pmatrix}$ is a feasible solution of the phase I problem and $\mathbf{0}(\mathbf{x}) + \mathbf{1}(\mathbf{0}) = 0 < \mathbf{1}\mathbf{x}_a$, violating optimality of $\mathbf{x}_a$.

Example 4.4 (Empty Feasible Region)

Minimize $-3x_1+4x_2$

$$\begin{aligned} \text{Subject to} \quad x_1 + x_2 &\leq 4 \\ 2x_1 + 3x_2 &\geq 18 \\ x_1, \quad x_2 &\geq 0 \end{aligned}$$

The constraints admit no feasible points, as shown in Figure 4.2. This will be detected by phase I. Introducing the slack variables x_3 and x_4, we get the following constraints in standard form:

$$\begin{aligned} x_1 + x_2 + x_3 \qquad &= 4 \\ 2x_1 + 3x_2 \qquad - x_4 &= 18 \\ x_1, \quad x_2, \; x_3, \; x_4 &\geq 0 \end{aligned}$$

Since no convenient basis exists, introduce the artificial variable x_5 into the second constraint. Phase I is used to get rid of the artificial.

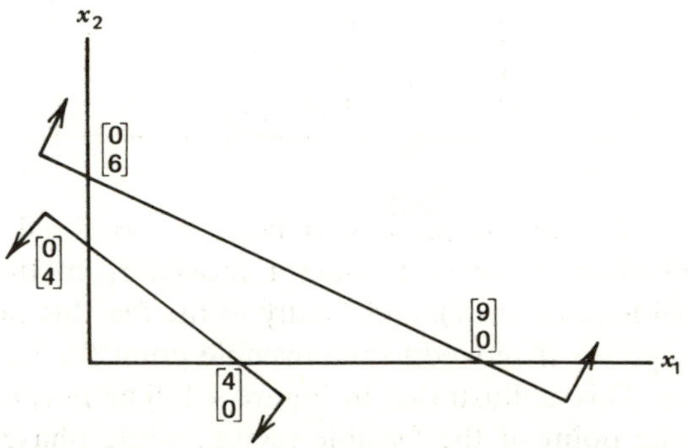

Figure 4.2. Empty feasible region.

PHASE I

x_0	x_1	x_2	x_3	x_4	x_5	RHS
1	0	0	0	0	−1	0
0	1	1	1	0	0	4
0	2	3	0	−1	1	18

Add row 2 to row 0 so that $z_5 - c_5 = 0$ is displayed.

	x_0	x_1	x_2	x_3	x_4	x_5	RHS
x_0	1	2	3	0	−1	0	18
x_3	0	1	(1)	1	0	0	4
x_5	0	2	3	0	−1	1	18

	x_0	x_1	x_2	x_3	x_4	x_5	RHS
x_0	1	−1	0	−3	−1	0	6
x_2	0	1	1	1	0	0	4
x_5	0	−1	0	−3	−1	1	6

The optimality criterion of the simplex method, namely $z_j - c_j \leqslant 0$, holds for all variables; but the artificial $x_5 > 0$. We conclude that the original problem has no feasible solutions.

Case B: $\mathbf{x}_a = \mathbf{0}$

This case is further divided into two subcases. In subcase B1 all the artificial variables are out of the basis at the end of phase I. The subcase B2 corresponds to the presence of at least one artificial in the basis at zero level. These cases are discussed below.

Subcase B1 (All Artificials Are Out of the Basis)

Since at the end of phase I we have a basic feasible solution, and since $\mathbf{x}_a$ is out of the basis, then the basis consists entirely of legitimate variables. If the legitimate vector $\mathbf{x}$ is decomposed accordingly into $\mathbf{x}_B$ and $\mathbf{x}_N$, then at the end of phase I we have the following tableau.

	x_0	$\mathbf{x}_B$	$\mathbf{x}_N$	$\mathbf{x}_a$	RHS
x_0	1	$\mathbf{0}$	$\mathbf{0}$	$-\mathbf{1}$	$\mathbf{0}$
$\mathbf{x}_B$	$\mathbf{0}$	$\mathbf{I}$	$\mathbf{B}^{-1}\mathbf{N}$	$\mathbf{B}^{-1}$	$\mathbf{B}^{-1}\mathbf{b}$

Now phase II can be started, with the original objective, after discarding the columns corresponding to $\mathbf{x}_a$. (These columns may be kept for the purpose of bookkeeping since they would present $\mathbf{B}^{-1}$ at each iteration. Note, however, that an artificial variable should never be permitted to enter the basis again.) The $z_j - c_j$'s for the nonbasic variables are given by the vector $\mathbf{c}_B\mathbf{B}^{-1}\mathbf{N} - \mathbf{c}_N$, which can be easily calculated from the matrix $\mathbf{B}^{-1}\mathbf{N}$ stored in the final tableau of phase I. The following initial tableau of phase II is constructed. Starting with this tableau, the simplex method is used to find the optimal solution.

	z	$\mathbf{x}_B$	$\mathbf{x}_N$	RHS
z	1	$\mathbf{0}$	$\mathbf{c}_B\mathbf{B}^{-1}\mathbf{N} - \mathbf{c}_N$	$\mathbf{c}_B\mathbf{B}^{-1}\mathbf{b}$
$\mathbf{x}_B$	$\mathbf{0}$	$\mathbf{I}$	$\mathbf{B}^{-1}\mathbf{N}$	$\mathbf{B}^{-1}\mathbf{b}$

Subcase B2 (Some Artificials Are in the Basis at Zero Level)

In this case we may proceed directly to phase II, or else eliminate the artificial basic variables, and then proceed to phase II. These two actions are discussed in further detail below.

PROCEED DIRECTLY TO PHASE II

First eliminate the columns corresponding to the nonbasic artificial variables of phase I. The starting tableau of phase II consists of some legitimate variables and some artificial variables at zero level. The cost row consisting of $z_j - c_j$'s is constructed for the original objective function so that all legitimate variables that are basic have $z_j - c_j = 0$. The cost coefficients of the artificial variables are given value zero (justify!). While solving the phase II problem by the simplex method, we must be careful that artificial variables never reach a positive level (since this would destroy feasibility). To illustrate, consider the following tableau, where for simplicity we assume that the basis consists of the legitimate variables $x_1, x_2, \ldots, x_k$ and the artificial variables $x_{n+k+1}, \ldots, x_{n+m}$ (the artificial variables $x_{n+1}, \ldots, x_{n+k}$ left the basis during phase I).

	z	x_1	$\cdots$	x_k	$x_{k+1} \cdots x_j \cdots x_n$	x_{n+k+1}	$\cdots$	x_{n+m}	RHS
z	1	0	$\cdots$	0	$z_j - c_j$	0	$\cdots$	0	$\mathbf{c}_B\bar{\mathbf{b}}$
x_1		1			y_{1j}	0	$\cdots$	0	$\bar{b}_1$
x_2			1		y_{2j}	0	$\cdots$	0	$\bar{b}_2$
$\vdots$			$\ddots$		$\vdots$	$\vdots$		$\vdots$	$\vdots$
x_k				1	y_{kj}	0	$\cdots$	0	$\bar{b}_k$
x_{n+k+1}		0	$\cdots$	0	$y_{k+1,j}$	1			0
$\vdots$		$\vdots$		$\vdots$	$\vdots$		$\ddots$		$\vdots$
x_{n+r}		0	$\cdots$	0	y_{rj}		1		0
$\vdots$		$\vdots$		$\vdots$	$\vdots$		$\ddots$		$\vdots$
x_{n+m}		0	$\cdots$	0	y_{mj}			1	0

Suppose that $z_j - c_j > 0$, and so x_j is eligible to enter the basis. If $y_{ij} \geqslant 0$ for $i = k + 1, \ldots, m$, then the artificial variable x_{n+i} will remain zero as x_j enters the basis and the usual minimum ratio test is performed. If, on the other hand, at least one component $y_{rj} < 0$ $(r = k + 1, \ldots,$ or $m)$, then the artificial variable x_{n+r} becomes positive as x_j is increased. This action must be prohibited since it would destroy feasibility. This can be done by pivoting at y_{rj} rather than using the usual minimum ratio test.† Even though $y_{rj} < 0$, pivoting at y_{rj} would maintain feasibility since the right-hand side at the corresponding row is zero. In this case x_j enters the basis and the artificial x_{n+r} leaves the basis, and the objective value remains constant. With this slight modification the simplex method is used to solve phase II.

FIRST ELIMINATE THE BASIC ARTIFICIAL VARIABLES AT THE END OF PHASE I

Rather than adopting the preceding rule, which guarantees that artificial variables will always be zero during phase II, we can eliminate the artificial variables altogether before proceeding to phase II. The following is a typical tableau (possibly after rearranging) at the end of phase I. The objective row and column do not play any role in the subsequent analysis and are hence omitted.

	BASIC LEGITIMATE VARIABLES				NONBASIC LEGITIMATE VARIABLES	NONBASIC ARTIFICIAL VARIABLES	BASIC ARTIFICIAL VARIABLES				
	x_1	x_2	$\cdots$	x_k	$x_{k+1} \cdots x_n$	$x_{n+1} \cdots x_{n+k}$	x_{n+k+1}		$\cdots$	x_{n+m}	RHS
x_1	1						0	0	$\cdots$	0	$\bar{b}_1$
x_2		1			$\mathbf{R}_1$	$\mathbf{R}_3$	0	0	$\cdots$	0	$\bar{b}_2$
$\vdots$			$\ddots$				$\vdots$	$\vdots$		$\vdots$	$\vdots$
x_k				1			0	0	$\cdots$	0	b_k
x_{n+k+1}	0	0	$\cdots$	0			1				0
$\vdots$	$\vdots$	$\vdots$		$\vdots$	$\mathbf{R}_2$	$\mathbf{R}_4$		1	$\ddots$		0 $\vdots$
x_{n+m}	0	0	$\cdots$	0						1	0

We now attempt to drive the artificial variables $x_{n+k+1}, \ldots, x_{n+m}$ out of the basis by placing $m - k$ of the nonbasic legitimate variables $x_{k+1}, \ldots, x_n$ into the basis. For instance, x_{n+k+1} can be driven out of the basis by pivoting at any nonzero element of the first row of $\mathbf{R}_2$. The corresponding nonbasic legitimate variable enters and x_{n+k+1} leaves the basis, and the tableau is updated. This process is continued. If all artificial variables drop from the basis, then the basis

†Actually, if any one of the y_{ij}'s is positive for $r = k + 1, \ldots, m$, then we may use the usual minimum ratio test as x_j will enter the basis at zero level.

would consist entirely of legitimate columns, and we proceed to phase II as described in subcase B1. If, on the other hand, we reach a tableau where the matrix $\mathbf{R}_2 = \mathbf{0}$, none of the artificial variables can leave the basis by introducing $x_{k+1}, x_{k+2}, \ldots,$ or x_n. Denoting $(x_1, x_2, \ldots, x_k)^t$ and $(x_{k+1}, \ldots, x_n)^t$ by $\mathbf{x}_1$ and $\mathbf{x}_2$ respectively, and decomposing $\mathbf{A}$ and $\mathbf{b}$ accordingly into $\left[\begin{array}{c|c} \mathbf{A}_{11} & \mathbf{A}_{12} \\ \hline \mathbf{A}_{21} & \mathbf{A}_{22} \end{array}\right]$ and $\begin{bmatrix} \mathbf{b}_1 \\ \mathbf{b}_2 \end{bmatrix}$, it is clear that the system

$$\begin{array}{r} \\ k \\ m-k \end{array} \begin{array}{c} \quad k \qquad n-k \\ \left[\begin{array}{c|c} \mathbf{A}_{11} & \mathbf{A}_{12} \\ \hline \mathbf{A}_{21} & \mathbf{A}_{22} \end{array}\right] \end{array} \begin{array}{c} \\ \begin{bmatrix} \mathbf{x}_1 \\ \mathbf{x}_2 \end{bmatrix} = \begin{bmatrix} \mathbf{b}_1 \\ \mathbf{b}_2 \end{bmatrix} \end{array}$$

is transformed into

$$\begin{array}{r} \\ k \\ m-k \end{array} \begin{array}{c} k \quad n-k \\ \left[\begin{array}{c|c} \mathbf{I} & \mathbf{R}_1 \\ \hline \mathbf{0} & \mathbf{0} \end{array}\right] \end{array} \begin{array}{c} \\ \begin{bmatrix} \mathbf{x}_1 \\ \mathbf{x}_2 \end{bmatrix} = \begin{bmatrix} \bar{\mathbf{b}}_1 \\ \mathbf{0} \end{bmatrix} \end{array}$$

through a sequence of elementary matrix operations. This shows that rank $(\mathbf{A}, \mathbf{b}) = k < m$; that is, the last $m - k$ equations are *mathematically redundant* and $\mathbf{R}_1 = \mathbf{A}_{11}^{-1}\mathbf{A}_{12}$ and $\bar{\mathbf{b}}_1 = \mathbf{A}_{11}^{-1}\mathbf{b}_1$. The last $m - k$ rows of the last tableau can be thrown away, and we can proceed to phase II without the artificial variables. The basic variables are $x_1, x_2, \ldots, x_k$ and the nonbasic variables are $x_{k+1}, \ldots, x_n$. The starting tableau of phase II is depicted below, where $\mathbf{c}_B = (c_1, c_2, \ldots, c_k)$.

z	$x_1 \cdots x_k$	$x_{k+1} \cdots x_n$	RHS
1	$\mathbf{0}$	$\mathbf{c_B A}_{11}^{-1}\mathbf{A}_{12} - \mathbf{c}_N$	$\mathbf{c}_B\bar{\mathbf{b}}_1$
0	$\mathbf{I}$	$\mathbf{A}_{11}^{-1}\mathbf{A}_{12} = \mathbf{R}_1$	$\bar{\mathbf{b}}_1$

Example 4.5 (Redundancy)

Minimize $\quad -x_1+2x_2-3x_3$

Subject to
$$\begin{aligned} x_1 + x_2 + x_3 &= 6 \\ -x_1 + x_2 + 2x_3 &= 4 \\ 2x_2 + 3x_3 &= 10 \\ x_3 &\leqslant 2 \\ x_1, \quad x_2, \quad x_3 &\geqslant 0 \end{aligned}$$

We need to introduce a slack variable x_4. The constraint matrix **A** is given below:

$$\mathbf{A} = \begin{bmatrix} 1 & 1 & 1 & 0 \\ -1 & 1 & 2 & 0 \\ 0 & 2 & 3 & 0 \\ 0 & 0 & 1 & 1 \end{bmatrix}$$

Note that the matrix is not of full rank. Actually if we add the first two row of **A**, we get the third row; that is, any one of the first three constraints is redundant and can be thrown away. We shall proceed as if this fact were not known, however, and introduce the artificial variables x_5, x_6, and x_7. The phase I objective is: Minimize $x_0 = x_5 + x_6 + x_7$. Phase I proceeds as follows.

PHASE I

x_0	x_1	x_2	x_3	x_4	x_5	x_6	x_7	RHS
1	0	0	0	0	−1	−1	−1	0
1	1	1	1	0	1	0	0	6
0	−1	1	2	0	0	1	0	4
0	0	2	3	0	0	0	1	10
0	0	0	1	1	0	0	0	2

Add rows 1, 2, and 3 to row 0, to display $z_5 - c_5 = z_6 - c_6 = z_7 - c_7 = 0$.

	x_0	x_1	x_2	x_3	x_4	x_5	x_6	x_7	RHS
x_0	1	0	4	6	0	0	0	0	20
x_5	0	1	1	1	0	1	0	0	6
x_6	0	−1	1	2	0	0	1	0	4
x_7	0	0	2	3	0	0	0	1	10
x_4	0	0	0	(1)	1	0	0	0	2

	x_0	x_1	x_2	x_3	x_4	x_5	x_6	x_7	RHS
x_0	1	0	4	0	−6	0	0	0	8
x_5	0	1	1	0	−1	1	0	0	4
x_6	0	−1	(1)	0	−2	0	1	0	0
x_7	0	0	2	0	−3	0	0	1	4
x_3	0	0	0	1	1	0	0	0	2

	x_0	x_1	x_2	x_3	x_4	x_5	x_6	x_7	RHS
x_0	1	4	0	0	2	0	-4	0	8
x_5	0	⓶	0	0	1	1	-1	0	4
x_2	0	-1	1	0	-2	0	1	0	0
x_7	0	2	0	0	1	0	-2	1	4
x_3	0	0	0	1	1	0	0	0	2

	x_0	x_1	x_2	x_3	x_4	x_5	x_6	x_7	RHS
x_0	1	0	0	0	0	-2	-2	0	0
x_1	0	1	0	0	$\frac{1}{2}$	$\frac{1}{2}$	$-\frac{1}{2}$	0	2
x_2	0	0	1	0	$-\frac{3}{2}$	$\frac{1}{2}$	$\frac{1}{2}$	0	2
x_7	0	0	0	0	0	-1	-1	1	0
x_3	0	0	0	1	1	0	0	0	2

Since all the artificial variables are at level zero, we may proceed to phase II with a basic feasible solution of the original problem. We can either proceed directly with the artificial x_7 into the basis at zero level, or attempt to eliminate x_7 from the basis. The only legitimate nonbasic variable is x_4, and it has zero coefficient in row 3 corresponding to x_7. This shows that the third row (constraint of the original problem) is redundant and can be thrown away. This will be done as we move to phase II.

PHASE II

Obviously $z_1 - c_1 = z_2 - c_2 = z_3 - c_3 = 0$. Thus x_5 and x_6 are nonbasic artificial variables and will not be introduced in the phase II problem. In order to complete row 0 we need to calculate $z_4 - c_4$:

$$z_4 - c_4 = \mathbf{c}_B\mathbf{B}^{-1}\mathbf{a}_4 - c_4$$

$$= (-1, 2, -3)\begin{pmatrix} \frac{1}{2} \\ -\frac{3}{2} \\ 1 \end{pmatrix} - 0$$

$$= -\frac{13}{2}$$

Since we are minimizing and $z_4 - c_4 \leqslant 0$ for the only nonbasic variable, then we stop; the solution obtained from phase I is optimal. The tableau below displays the optimal solution.

	z	x_1	x_2	x_3	x_4	RHS
z	1	0	0	0	$-\frac{13}{2}$	-4
x_1	0	1	0	0	$\frac{1}{2}$	2
x_2	0	0	1	0	$-\frac{3}{2}$	2
x_3	0	0	0	1	1	2

Organization of the Tableau in the Two-Phase Method

To eliminate the need for recomputing the $z_j - c_j$'s when the phase I objective function is replaced by the phase II (original) objective function, an extra row could be added to the tableau representing the original cost coefficients. The following represents the setup of the initial tableau (not yet in basic form).

	z	x_0	$x_1 \cdots x_n$	ARTIFICIALS $x_{n+1} \cdots x_{n+m}$	RHS	
z	1	0	$-c_1 \cdots -c_n$	$0 \cdots 0$	0	← Phase II Objective
x_0	0	1	$0 \cdots 0$	$-1 \cdots -1$	0	← Phase I Objective
x_{n+1}	0	0	$a_{11} \cdots a_{1n}$	$1 \cdots 0$	b_1	
$\vdots$	$\vdots$	$\vdots$	$\vdots \quad \vdots$	$\vdots \quad \vdots$	$\vdots$	
x_{n+m}	0	0	$a_{m1} \cdots a_{mn}$	$0 \cdots 1$	b_m	

To convert this tableau to basic form (that is, unit vectors for all basic variables) we must perform preliminary pivoting to obtain zeros for x_{n+1} through x_{n+m} in the x_0 (phase I objective) row. This is done by successively adding every row (except the z row) to the x_0 row.

Once the initial tableau has been converted to basic form, the simplex method is applied to the resulting tableau using the x_0 row as the (phase I) objective function row until optimality is achieved. During phase I the z row is transformed just as any other row of the tableau would be, to maintain unit vectors for the basic variables. Note, however, that during phase I the basis entry is solely determined by the entries in row x_0.

At the end of phase I, if x_0 is positive the problem is found to be infeasible and the process is terminated. Otherwise, the x_0 row and x_0 column are deleted and phase II is initiated (after possibly eliminating artificial variables), with the values in the z row being the correct values of $z_j - c_j$ (why?) for the phase II objective.

4.3 THE BIG-M METHOD

Recall that artificial variables constitute a tool that can be used to initiate the simplex method. However, the presence of artificial variables at a positive level, by construction, means that the current point is an infeasible solution of the original system. The two-phase method is one way to get rid of the artificials. However, during phase I of the two-phase method the original cost coefficients are essentially ignored. Phase I of the two-phase method seeks any basic feasible solution, not necessarily a good one. Another possibility for eliminating the artificial variables is to assign coefficients for these variables in the original objective function in such a way as to make their presence in the basis at a positive level, very unattractive from the objective function point of view. To illustrate, suppose that we want to solve the following linear programming problem, where $\mathbf{b} \geqslant \mathbf{0}$.

$$\begin{aligned} \text{Minimize} \quad & \mathbf{cx} \\ \text{Subject to} \quad & \mathbf{Ax}=\mathbf{b} \\ & \mathbf{x} \geqslant \mathbf{0} \end{aligned}$$

If no convenient basis is known, we introduce the artificial vector $\mathbf{x}_a$, which leads to the following system:

$$\begin{aligned} \mathbf{Ax} + \mathbf{x}_a &= \mathbf{b} \\ \mathbf{x}, \mathbf{x}_a &\geqslant \mathbf{0} \end{aligned}$$

The starting basic feasible solution is given by $\mathbf{x}_a = \mathbf{b}$ and $\mathbf{x} = \mathbf{0}$. In order to reflect the undesirability of a nonzero artificial vector, the objective function is modified such that a large penalty is paid for any such solution. More specifically consider the following problem.

$$\begin{aligned} \text{Minimize} \quad & \mathbf{cx} + M\mathbf{1x}_a \\ \text{Subject to} \quad & \mathbf{Ax} + \mathbf{x}_a = \mathbf{b} \\ & \mathbf{x}, \quad \mathbf{x}_a \geqslant \mathbf{0} \end{aligned}$$

where M is a very large positive number. The term $M\mathbf{1x}_a$ can be interpreted as a penalty to be paid by any solution with $\mathbf{x}_a \neq \mathbf{0}$. Even though the starting solution $\mathbf{x} = \mathbf{0}$, $\mathbf{x}_a = \mathbf{b}$ is feasible to the new constraints, it has a very unattractive objective value, namely $M\mathbf{1b}$. Therefore the simplex method itself will try to get the artificial variables out of the basis, and then continue to find the optimal solution of the original problem.

The big-M method is illustrated by the following numerical example. Validation of the method and possible complications will be discussed later.

Example 4.6

$$\text{Minimize} \quad x_1 - 2x_2$$

$$\begin{aligned} \text{Subject to} \quad x_1 + x_2 &\geq 2 \\ -x_1 + x_2 &\geq 1 \\ x_2 &\leq 3 \\ x_1, \quad x_2 &\geq 0 \end{aligned}$$

This example was solved earlier by the two-phase method (Example 4.3). The slack variables x_3, x_4, x_5 are introduced and the artificial variables x_6 and x_7 are incorporated in the first two constraints. The modified objective function is $z = x_1 - 2x_2 + Mx_6 + Mx_7$, where M is a large positive number. This leads to the following sequence of tableaux.

z	x_1	x_2	x_3	x_4	x_5	x_6	x_7	RHS
1	−1	2	0	0	0	$-M$	$-M$	0
0	1	1	−1	0	0	1	0	2
0	−1	1	0	−1	0	0	1	1
0	0	1	0	0	1	0	0	3

Multiply rows 1 and 2 by M and add to row 0.

	z	x_1	x_2	x_3	x_4	x_5	x_6	x_7	RHS
z	1	−1	$2+2M$	$-M$	$-M$	0	0	0	$3M$
x_6	0	1	1	−1	0	0	1	0	2
x_7	0	−1	(1)	0	−1	0	0	1	1
x_5	0	0	1	0	0	1	0	0	3

	z	x_1	x_2	x_3	x_4	x_5	x_6	x_7	RHS
z	1	$1+2M$	0	$-M$	$2+M$	0	0	$-2-2M$	$-2+M$
x_6	0	(2)	0	−1	1	0	1	−1	1
x_2	0	−1	1	0	−1	0	0	1	1
x_5	0	1	0	0	1	1	0	−1	2

	z	x_1	x_2	x_3	x_4	x_5	x_6	x_7	RHS
z	1	0	0	$\frac{1}{2}$	$\frac{3}{2}$	0	$-\frac{1}{2}-M$	$-\frac{3}{2}-M$	$-\frac{5}{2}$
x_1	0	1	0	$-\frac{1}{2}$	$\frac{1}{2}$ (circled)	0	$\frac{1}{2}$	$-\frac{1}{2}$	$\frac{1}{2}$
x_2	0	0	1	$-\frac{1}{2}$	$-\frac{1}{2}$	0	$\frac{1}{2}$	$\frac{1}{2}$	$\frac{3}{2}$
x_5	0	0	0	$\frac{1}{2}$	1	1	$-\frac{1}{2}$	$\frac{3}{2}$	$\frac{3}{2}$

	z	x_1	x_2	x_3	x_4	x_5	x_6	x_7	RHS
z	1	-3	0	2	0	0	$-2-M$	$-M$	-4
x_4	0	2	0	-1	1	0	1	-1	1
x_2	0	1	1	-1	0	0	1	0	2
x_5	0	-1	0	1 (circled)	0	1	-1	0	1

	z	x_1	x_2	x_3	x_4	x_5	x_6	x_7	RHS
z	1	-1	0	0	0	-2	$-M$	$-M$	-6
x_4	0	1	0	0	1	1	0	-1	2
x_2	0	0	1	0	0	1	0	0	3
x_3	0	-1	0	1	0	1	-1	0	1

Since $z_j - c_j \leqslant 0$ for each nonbasic variable, the last tableau gives the optimal solution. The sequence of points generated in the (x_1x_2) space is illustrated in Figure 4.3.

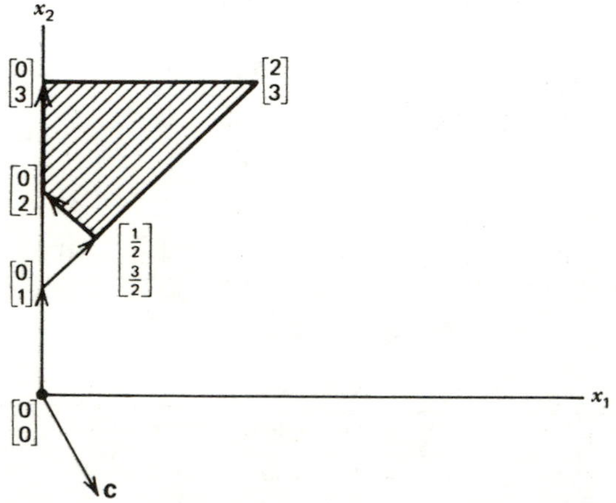

Figure 4.3. **Example of the big-M method.**

Analysis of the Big-M Method

We now discuss in more detail the possible cases that may arise while solving the big-M problem. The original problem P and the big-M problem $P(M)$ are stated below, where the vector $\mathbf{b}$ is nonnegative.

$$\begin{array}{lll} \textit{Problem P}: & \text{Minimize} & \mathbf{c}\,\mathbf{x} \\ & \text{Subject to} & \mathbf{A}\mathbf{x} = \mathbf{b} \\ & & \mathbf{x} \geq \mathbf{0} \end{array}$$

$$\begin{array}{lll} \textit{Problem P(M)}: & \text{Minimize} & \mathbf{c}\,\mathbf{x} + M\mathbf{1}\mathbf{x}_a \\ & \text{Subject to} & \mathbf{A}\mathbf{x} + \mathbf{x}_a = \mathbf{b} \\ & & \mathbf{x},\ \ \mathbf{x}_a \geq \mathbf{0} \end{array}$$

Since problem $P(M)$ has a feasible solution (say $\mathbf{x} = \mathbf{0}$ and $\mathbf{x}_a = \mathbf{b}$), then while solving it by the simplex method one of the following two cases may arise.

1. We shall arrive at an optimal solution of $P(M)$.
2. Conclude that $P(M)$ has an unbounded optimal solution, that is, $z \to -\infty$.

Of course, we are interested in conclusions about problem P and not $P(M)$. The following analysis will help us to draw such conclusions.

Case A: Finite Optimal Solution of $P(M)$.

Under this case, we have two possibilities. First, the optimal solution to $P(M)$ has all artificials at value zero, and second, not all artificials are zero. These cases are discussed below.

Subcase A1: $(\mathbf{x}^*, \mathbf{0})$ is an optimal solution of $P(M)$.

In this case $\mathbf{x}^*$ is an optimal solution to problem P. This can be easily verified as follows. Let $\mathbf{x}$ be any feasible solution to problem P, and note that $(\mathbf{x}, \mathbf{0})$ is a feasible solution to problem $P(M)$. Since $(\mathbf{x}^*, \mathbf{0})$ is an optimal solution of problem $P(M)$, then $\mathbf{c}\mathbf{x}^* + 0 \leq \mathbf{c}\mathbf{x} + 0$, that is, $\mathbf{c}\mathbf{x}^* \leq \mathbf{c}\mathbf{x}$. Since $\mathbf{x}^*$ is a feasible solution of problem P, then $\mathbf{x}^*$ is indeed an optimal solution of P. This case was illustrated by Example 4.6 above.

Subcase A2: $(\mathbf{x}^*, \mathbf{x}_a^*)$ is an optimal solution of $P(M)$ and $\mathbf{x}_a^* \neq \mathbf{0}$.

If M is a very large positive number, then we conclude that there exists no feasible solution of P. To illustrate this case, suppose on the contrary that $\mathbf{x}$ was a feasible solution of P. Then $(\mathbf{x}, \mathbf{0})$ would be a feasible solution of $P(M)$ and by optimality of $(\mathbf{x}^*, \mathbf{x}_a^*)$ we have

$$\mathbf{cx}^* + M\mathbf{1x}_a^* \leqslant \mathbf{cx} + 0 = \mathbf{cx}$$

Since M is very large, $\mathbf{x}_a^* \geqslant \mathbf{0}$ and $\mathbf{x}_a^* \neq \mathbf{0}$, and since $\mathbf{x}_a^*$ corresponds to one of the finite number of basic feasible solutions, the preceding inequality is impossible (why?). Therefore $\mathbf{x}$ could not have been a feasible solution of P. A formal proof is left as Exercise 4.13.

Example 4.7 (No Feasible Solutions)

$$\begin{aligned}
\text{Minimize} \quad & -x_1 - 3x_2 + x_3 \\
\text{Subject to} \quad & x_1 + x_2 + 2x_3 \leqslant 4 \\
& -x_1 + x_3 \geqslant 4 \\
& x_3 \geqslant 3 \\
& x_1, \quad x_2, \quad x_3 \geqslant 0
\end{aligned}$$

Introduce the slack variables x_4, x_5, and x_6. Also, introduce the two artificial variables x_7 and x_8 for the last two constraints with cost coefficients equal to M. This leads to the following sequence of tableaux.

z	x_1	x_2	x_3	x_4	x_5	x_6	x_7	x_8	RHS
1	1	3	-1	0	0	0	$-M$	$-M$	0
0	1	1	2	1	0	0	0	0	4
0	-1	0	1	0	-1	0	1	0	4
0	0	0	1	0	0	-1	0	1	3

Multiply rows 2 and 3 by M and add to row 0.

	z	x_1	x_2	x_3	x_4	x_5	x_6	x_7	x_8	RHS
z	1	$1-M$	3	$-1+2M$	0	$-M$	$-M$	0	0	$7M$
x_4	0	1	1	$\textcircled{2}$	1	0	0	0	0	4
x_7	0	-1	0	1	0	-1	0	1	0	4
x_8	0	0	0	1	0	0	-1	0	1	3

	z	x_1	x_2	x_3	x_4	x_5	x_6	x_7	x_8	RHS
z	1	$\frac{3}{2} - 2M$	$\frac{7}{2} - M$	0	$\frac{1}{2} - M$	$-M$	$-M$	0	0	$2+3M$
x_3	0	$\frac{1}{2}$	$\frac{1}{2}$	1	$\frac{1}{2}$	0	0	0	0	2
x_7	0	$-\frac{3}{2}$	$-\frac{1}{2}$	0	$-\frac{1}{2}$	-1	0	1	0	2
x_8	0	$-\frac{1}{2}$	$-\frac{1}{2}$	0	$-\frac{1}{2}$	0	-1	0	1	1

Since $M > 0$ is very large, then $z_j - c_j \leqslant 0$ for all nonbasic variables; that is, the simplex optimality criteria hold. Since the artificials x_7 and x_8 still remain in the basis at a positive level, then we conclude that the original problem has no feasible solutions.

Case B: $P(M)$ has an unbounded optimal solution, that is, $z \to -\infty$.

Suppose that during the solution of the big-M problem, the updated column $\mathbf{y}_k$ is $\leqslant \mathbf{0}$, where the index k is that of the *most positive* $z_j - c_j$. Then problem $P(M)$ has an unbounded optimal solution. In the meantime, if all artificials are equal to zero, then the original problem has an unbounded optimal solution. Otherwise, if at least one artificial variable is positive, then the original problem is infeasible. These two subcases are discussed in more detail below.

Subcase B1 : $z_k - c_k =$ Maximum $(z_i - c_i) > 0$, $\mathbf{y}_k \leqslant \mathbf{0}$, and all artificials are equal to zero.

In this case we have a feasible solution of the original problem. Furthermore, since problem $P(M)$ is unbounded, then there is a direction, $(\mathbf{d}_1, \mathbf{d}_2) \geqslant (\mathbf{0}, \mathbf{0})$ of the set $\{(\mathbf{x}, \mathbf{x}_a) : \mathbf{A}\mathbf{x} + \mathbf{x}_a = \mathbf{b}, \mathbf{x} \geqslant \mathbf{0}, \mathbf{x}_a \geqslant \mathbf{0}\}$ such that $\mathbf{c}\mathbf{d}_1 + M\mathbf{1}\mathbf{d}_2 < 0$ (recall that this is the necessary and sufficient condition for unboundedness of $P(M)$). Since M is very large and $\mathbf{d}_2 \geqslant \mathbf{0}$, then $\mathbf{c}\mathbf{d}_1 + M\mathbf{1}\mathbf{d}_2 < 0$ implies that $\mathbf{d}_2 = \mathbf{0}$ and hence $\mathbf{c}\mathbf{d}_1 < 0$. Thus we have found a direction $\mathbf{d}_1$ of the set $\{\mathbf{x} : \mathbf{A}\mathbf{x} = \mathbf{b}, \mathbf{x} \geqslant \mathbf{0}\}$ such that $\mathbf{c}\mathbf{d}_1 < 0$ (why?). This implies that problem P has an unbounded optimal solution.

Example 4.8

$$
\begin{array}{lrl}
\text{Minimize} & -x_1 - x_2 & \\
\text{Subject to} & x_1 - x_2 - x_3 & = 1 \\
& -x_1 + x_2 + 2x_3 - x_4 & = 1 \\
& x_1, \; x_2, \; x_3, \; x_4 & \geqslant 0
\end{array}
$$

Introduce the artificial variables x_5 and x_6 with cost coefficient M. This leads to the following sequence of tableaux.

z	x_1	x_2	x_3	x_4	x_5	x_6	RHS
1	1	1	0	0	$-M$	$-M$	0
0	1	-1	-1	0	1	0	1
0	-1	1	2	-1	0	1	1

Multiply rows 1 and 2 by M and add to row 0.

	z	x_1	x_2	x_3	x_4	x_5	x_6	RHS
z	1	1	1	M	$-M$	0	0	$2M$
x_5	0	1	-1	-1	0	1	0	1
x_6	0	-1	1	(2)	-1	0	1	1

	z	x_1	x_2	x_3	x_4	x_5	x_6	RHS
z	1	$1+\frac{1}{2}M$	$1-\frac{1}{2}M$	0	$-\frac{1}{2}M$	0	$-\frac{1}{2}M$	$\frac{3}{2}M$
x_5	0	($\frac{1}{2}$)	$-\frac{1}{2}$	0	$-\frac{1}{2}$	1	$\frac{1}{2}$	$\frac{3}{2}$
x_3	0	$-\frac{1}{2}$	$\frac{1}{2}$	1	$-\frac{1}{2}$	0	$\frac{1}{2}$	$\frac{1}{2}$

	z	x_1	x_2	x_3	x_4	x_5	x_6	RHS
z	1	0	2	0	1	$-M-2$	$-M-1$	-3
x_1	0	1	-1	0	-1	2	1	3
x_3	0	0	0	1	-1	1	1	2

The most positive $z_j - c_j$ corresponds to x_2 and $\mathbf{y}_2 \leqslant \mathbf{0}$. Therefore the big-$M$ problem is unbounded. Furthermore, the artificial variables x_5 and x_6 are equal to zero, and hence the original program has an unbounded optimal solution along the ray $\{(3, 0, 2, 0) + \lambda(1, 1, 0, 0) : \lambda \geqslant 0\}$.

Subcase B2: $z_k - c_k = \text{Maximum } (z_j - c_j) > 0$, $y_k \leqslant 0$, and not all artificials are equal to zero.

In this case we show that there could be no feasible solution of the original problem. To illustrate, suppose that the basis consists of the first m columns, where columns 1 through p are formed by real variables and columns $p + 1$ through m are formed by artificial variables. The corresponding tableau is depicted below.

	$x_1 \cdots x_p$	$x_{p+1} \cdots x_m$	$x_{m+1} \cdots$	x_j	$\cdots x_n$	RHS
x_1	$1 \cdots 0$	$0 \cdots 0$	$\cdots$	y_{1j}	$\cdots$	$\bar{b}_1$
	$0 \cdots 0$	$0 \cdots 0$	$\cdots$	y_{2j}	$\cdots$	
$\vdots$				$\vdots$		$\vdots$
x_p	$0 \cdots 1$	$0 \cdots 0$	$\cdots$	y_{pj}	$\cdots$	$\bar{b}_p$
x_{p+1}	$0 \cdots 0$	$1 \cdots 0$	$\cdots$	$y_{p+1,j}$	$\cdots$	$\bar{b}_{p+1}$
$\vdots$	$0 \cdots 0$	$0 \cdots 0$		$\vdots$		$\vdots$
x_m	$0 \cdots 0$	$0 \cdots 1$	$\cdots$	y_{mj}	$\cdots$	$\bar{b}_m$

Noting that $c_i = M$ for $i = p + 1, \ldots, m$, then for $j = m + 1, \ldots, n$ we get

$$z_j - c_j = \sum_{i=1}^{p} c_i y_{ij} + M\left(\sum_{i=p+1}^{m} y_{ij}\right) - c_j \tag{4.1}$$

First note that $\sum_{i=p+1}^{m} y_{ij} \leqslant 0$ for all $j = m + 1, \ldots, n$. To show this, first recall that $\mathbf{y}_k \leqslant \mathbf{0}$ and hence $\sum_{i=p+1}^{m} y_{ik} \leqslant 0$ holds trivially. By contradiction, suppose that $\sum_{i=p+1}^{m} y_{ij} > 0$ for some nonbasic variable x_j. From Equation (4.1) and since M is very large, then $z_j - c_j$ is a large positive number, violating the definition of $z_k - c_k$ (recall that $z_k - c_k = \text{Maximum } (z_l - c_l)$ and $\mathbf{y}_k \leqslant \mathbf{0}$). Therefore $\sum_{i=p+1}^{m} y_{ij} \leqslant 0$ for all $j = m + 1, \ldots, n$. Summing the last $m - p$ equations of the above tableau, we get

$$\sum_{i=p+1}^{m} x_i + \sum_{j=m+1}^{n} x_j\left(\sum_{i=p+1}^{m} y_{ij}\right) = \sum_{i=p+1}^{m} \bar{b}_i \tag{4.2}$$

By contradiction, suppose that problem P has a feasible solution. Then $x_i = 0$ for all artificial variables and hence $x_i = 0$ for $i = p + 1, \ldots, m$. Also $x_j \geqslant 0$ and $\sum_{i=p+1}^{m} y_{ij} \leqslant 0$ for $j = m + 1, \ldots, n$ as shown above. Therefore the left-hand side of Equation (4.2) is $\leqslant 0$. But the right-hand side is positive since not all artificials are equal to zero. This contradiction shows that there could be no feasible solution of the original problem.

Note that we cannot conclude that the original problem is inconsistent if some $z_j - c_j > 0$, $\mathbf{y}_j \leqslant \mathbf{0}$ and not all artificials are zero. It is imperative that we use the *most positive* $z_j - c_j$. See Exercise 4.21.

Example 4.9

$$\begin{aligned} \text{Minimize} \quad & -x_1 - x_2 \\ \text{Subject to} \quad & x_1 - x_2 \geqslant 1 \\ & -x_1 + x_2 \geqslant 1 \\ & x_1, \ x_2 \geqslant 0 \end{aligned}$$

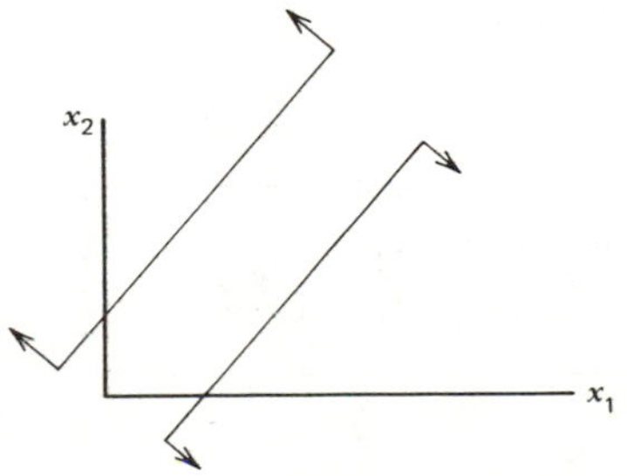

Figure 4.4. Empty feasible region.

This problem has no feasible solutions, as shown in Figure 4.4. We shall show that the optimal solution of the big-M problem is unbounded. Introduce the slack variables x_3 and x_4 and the artificial variables x_5 and x_6.

z	x_1	x_2	x_3	x_4	x_5	x_6	RHS
1	1	1	0	0	$-M$	$-M$	0
0	1	-1	-1	0	1	0	1
0	-1	1	0	-1	0	1	1

Multiply rows 1 and 2 by M and add to row 0.

	z	x_1	x_2	x_3	x_4	x_5	x_6	RHS
z	1	1	1	$-M$	$-M$	0	0	$2M$
x_5	0	1	-1	-1	0	1	0	1
x_6	0	-1	1	0	-1	0	1	1

	z	x_1	x_2	x_3	x_4	x_5	x_6	RHS
z	1	0	2	$1-M$	$-M$	-1	0	$2M-1$
x_1	0	1	-1	-1	0	1	0	1
x_6	0	0	0	-1	-1	1	1	2

The last tableau indicates unboundedness since $z_2 - c_2 > 0$ and $\mathbf{y}_2 \leqslant \mathbf{0}$. Note, however, that x_6 is positive, so we conclude that the original problem has no

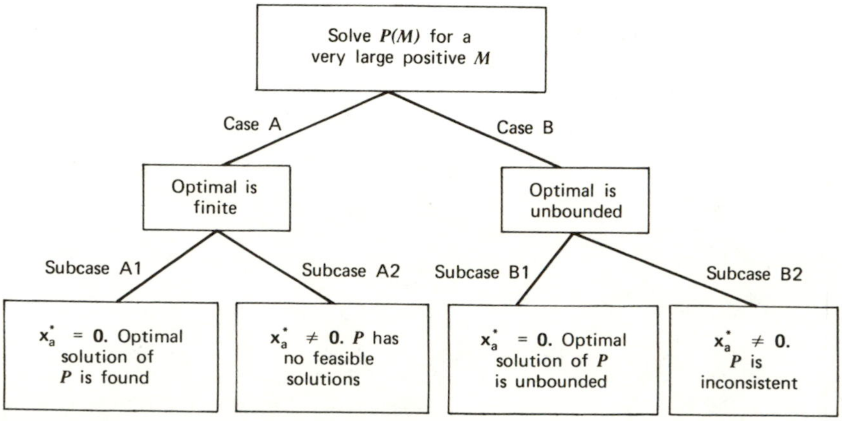

Figure 4.5. Analysis of the big-M method.

feasible solutions. This is also clear by examining the second row, which reads

$$x_5 + x_6 = 2 + x_3 + x_4$$

Since x_3, $x_4 \geqslant 0$, then $x_5 + x_6$ will always be positive, indicating that the original system is inconsistent.

The possible cases (A1, A2, B1, and B2) that may arise during the course of solving problem $P(M)$ are summarized in Figure 4.5.

Comparison of the Two-Phase and Big-M Methods

The big-M method has two important disadvantages in comparison to the two-phase method. First, in order to conduct the big-M method we must select a value for M. Without solving the linear program it is difficult to determine just how large M should be in order to ensure that the artificial variables are driven out of the basis. In Exercise 4.18 we ask the reader to consider this question.

The second major difficulty with the big-M method is that a large value of M will completely dominate the other cost coefficients and may result in serious round-off error problems in a computer.

4.4 THE SINGLE ARTIFICIAL VARIABLE TECHNIQUE

Thus far we have described two methods to initiate the simplex algorithm by the use of artificial variables. In this section we discuss a procedure that requires only a single artificial variable to get started. Consider the following problem.

$$\begin{aligned} \text{Minimize} \quad & \mathbf{cx} \\ \text{Subject to} \quad & \mathbf{Ax} = \mathbf{b} \\ & \mathbf{x} \geqslant \mathbf{0} \end{aligned}$$

Suppose that we can partition the constraint matrix $\mathbf{A}$ into $\mathbf{A} = [\mathbf{B}, \mathbf{N}]$, where $\mathbf{B}$ is a basis matrix, not necessarily feasible. This can certainly be done if the constraints were originally inequalities (that is, $\mathbf{Ax} \geqslant \mathbf{b}$ or $\mathbf{Ax} \leqslant \mathbf{b}$) by utilizing the slacks as basic variables. Index the basic variables from 1 to m.

Multiplying the constraints by $\mathbf{B}^{-1}$ (which would be $\pm\mathbf{I}$ for slack variables), we get

$$\mathbf{I}\mathbf{x}_B + \mathbf{B}^{-1}\mathbf{N}\mathbf{x}_N = \bar{\mathbf{b}}$$

where $\bar{\mathbf{b}} = \mathbf{B}^{-1}\mathbf{b}$ and $\bar{\mathbf{b}} \ngeqslant \mathbf{0}$ (if $\bar{\mathbf{b}} \geqslant \mathbf{0}$, we have a starting basic feasible solution).

To this system let us add a single artificial variable, x_a, with a coefficient of -1 in each constraint:

$$\mathbf{I}\mathbf{x}_B + \mathbf{B}^{-1}\mathbf{N}\mathbf{x}_N - \mathbf{1}x_a = \bar{\mathbf{b}}$$

Now, introduce x_a into the basis by selecting the pivot row r as follows:

$$\bar{b}_r = \underset{1 \leqslant i \leqslant m}{\text{Minimum}} \{ \bar{b}_i \}$$

Note that $\bar{b}_r < 0$. On pivoting in row r (that is, inserting x_a and removing x_r), we get the new right-hand-side values

$$\bar{b}'_r = -\bar{b}_r (\geqslant 0)$$

$$\bar{b}'_i = \bar{b}_i - \bar{b}_r (\geqslant 0)$$

Thus by entering x_a and exiting x_r we have constructed a basic feasible solution to the enlarged system (the one including the artificial variable). Starting with this solution, the simplex method can be used, employing either the two-phase or big-M method.

Example 4.10

$$\begin{array}{lrcl} \text{Minimize} & 2x_1 + 3x_2 & & \\ \text{Subject to} & x_1 + x_2 & \geqslant & 3 \\ & -2x_1 + x_2 & \geqslant & 2 \\ & x_1, \quad x_2 & \geqslant & 0 \end{array}$$

Subtracting the slack variables x_3 and x_4 and multiplying by -1, we get the basic system

$$\begin{aligned} -x_1 - x_2 + x_3 &= -3 \\ 2x_1 - x_2 + x_4 &= -2 \end{aligned}$$

Appending a single artificial variable x_5 with activity vector $\begin{pmatrix} -1 \\ -1 \end{pmatrix}$ to the initial tableau of the phase I problem, we get the following tableau.

	x_0	x_1	x_2	x_3	x_4	x_5	RHS
x_0	1	0	0	0	0	−1	0
x_3	0	−1	−1	1	0	(−1)	−3
x_4	0	2	−1	0	1	−1	−2

Pivoting in the x_3 row and x_5 column, we get the following tableau.

	x_0	x_1	x_2	x_3	x_4	x_5	RHS
x_0	1	1	1	−1	0	0	3
x_5	0	1	1	−1	0	1	3
x_4	0	3	0	−1	1	0	1

The tableau above is ready for application of the two-phase method. Subsequent tableaux are not shown.

An analysis of the two-phase method and the big-M method for the single artificial variable technique discussed in this section can be made similar to the analysis of Sections 4.2 and 4.3. The details are left to the reader in Exercise 4.28.

4.5 DEGENERACY AND CYCLING

In Chapter 3 we developed the simplex method with the assumption that an initial basic feasible solution is known. We then described in this chapter how to obtain such a starting basic feasible solution. In Chapter 3 we also showed that the simplex method converges in a finite number of steps provided that the basic feasible solutions visited were nondegenerate. In the remainder of this chapter we examine the problems of degeneracy and cycling more closely and give a rule that prevents cycling, and hence guarantees finite convergence of the simplex algorithm.

We have seen that the simplex algorithm moves among basic feasible solutions until optimality is reached, or else unboundedness is verified. Since the number of basic feasible solutions is finite, the simplex method would converge in a finite number of steps, provided that none of the bases are repeated. Now suppose that we have a basic feasible solution with basis **B**. Further suppose that we have a nonbasic variable x_k with $z_k - c_k > 0$ (for a minimization problem). Therefore x_k enters the basis and x_{B_r} leaves the basis, where the index r is determined as follows:

$$\frac{\bar{b}_r}{y_{rk}} = \underset{1 \leq i \leq m}{\text{Minimum}} \left\{ \frac{\bar{b}_i}{y_{ik}} : y_{ik} > 0 \right\}$$

where $\bar{\mathbf{b}} = \mathbf{B}^{-1}\mathbf{b}$ and $\mathbf{y}_k = \mathbf{B}^{-1}\mathbf{a}_k$. Column $\mathbf{a}_k$ enters the basis and $\mathbf{a}_{B_r}$ leaves the basis. The basic feasible solutions before and after pivoting are given by the following.

Before pivoting	*After pivoting*	
$\mathbf{x}_B = \bar{\mathbf{b}}$	$\mathbf{x}_B = \bar{\mathbf{b}} - \dfrac{\bar{b}_r}{y_{rk}}\mathbf{y}_k$	
$x_k = 0$	$x_k = \dfrac{\bar{b}_r}{y_{rk}}$	(4.3)
all other x_j's $= 0$	all other x_j's $= 0$	

Furthermore, the difference in the objective function before and after pivoting is given by $(\bar{b}_r/y_{rk})(z_k - c_k)$. In the absence of degeneracy we have $\bar{\mathbf{b}} = \mathbf{B}^{-1}\mathbf{b} > \mathbf{0}$. In particular $\bar{b}_r > 0$. Since $y_{rk} > 0$ and $z_k - c_k > 0$, then the objective function strictly decreases at each iteration, which in turn implies that the basic feasible solutions generated are distinct (why?). Now suppose that $\bar{b}_r = 0$ (which can occur only in the presence of degeneracy). In this case the objective function remains constant. Furthermore, examining the values of all the variables in Equation (4.3), it is evident that we have the same extreme point before and after pivoting, represented by different bases, however (since $\mathbf{a}_k$ entered and $\mathbf{a}_{B_r}$ left the basis). As the process is repeated, it is conceivable that another degenerate pivot is taken, resulting in the same extreme point with a different basis representation. It is therefore possible, though highly unlikely, that we may stay at a nonoptimal extreme point, and pivot through a sequence of bases $\mathbf{B}_1$, $\mathbf{B}_2, \ldots, \mathbf{B}_t$ where $\mathbf{B}_t = \mathbf{B}_1$. If the same sequence of pivots is used over and over again, we shall *cycle* forever among the bases $\mathbf{B}_1, \mathbf{B}_2, \ldots, \mathbf{B}_t = \mathbf{B}_1$ without reaching the optimal solution. Example 4.11 below illustrates this problem of cycling.

Example 4.11 (Cycling)

Consider the following example given by Beale.

$$\begin{array}{lllllllll}
\text{Minimize} & & & & -\frac{3}{4}x_4 & +20x_5 & -\frac{1}{2}x_6 & +6x_7 & \\
\text{Subject to} & x_1 & & & +\frac{1}{4}x_4 & -8x_5 & -x_6 & +9x_7 & =0 \\
& & x_2 & & +\frac{1}{2}x_4 & -12x_5 & -\frac{1}{2}x_6 & +3x_7 & =0 \\
& & & x_3 & & & +x_6 & & =1 \\
& x_1, & x_2, & x_3, & x_4, & x_5, & x_6, & x_7 & \geq 0
\end{array}$$

The optimal solution is given by $x_1 = \frac{3}{4}$, $x_4 = x_6 = 1$ and all other variables equal to zero. The optimal objective value is $-\frac{5}{4}$. The following rules are adopted. The entering variable is that with the most positive $z_j - c_j$, and the leaving variable is determined by the minimum ratio test, where ties are broken arbitrarily.

	z	x_1	x_2	x_3	x_4	x_5	x_6	x_7	RHS
z	1	0	0	0	$\frac{3}{4}$	-20	$\frac{1}{2}$	-6	0
x_1	0	1	0	0	($\frac{1}{4}$)	-8	-1	9	0
x_2	0	0	1	0	$\frac{1}{2}$	-12	$-\frac{1}{2}$	3	0
x_3	0	0	0	1	0	0	1	0	1

	z	x_1	x_2	x_3	x_4	x_5	x_6	x_7	RHS
z	1	-3	0	0	0	4	$\frac{7}{2}$	-33	0
x_4	0	4	0	0	1	-32	-4	36	0
x_2	0	-2	1	0	0	(4)	$\frac{3}{2}$	-15	0
x_3	0	0	0	1	0	0	1	0	1

	z	x_1	x_2	x_3	x_4	x_5	x_6	x_7	RHS
z	1	-1	-1	0	0	0	2	-18	0
x_4	0	-12	8	0	1	0	(8)	-84	0
x_5	0	$-\frac{1}{2}$	$\frac{1}{4}$	0	0	1	$\frac{3}{8}$	$-\frac{15}{4}$	0
x_3	0	0	0	1	0	0	1	0	1

	z	x_1	x_2	x_3	x_4	x_5	x_6	x_7	RHS
z	1	2	-3	0	$-\frac{1}{4}$	0	0	3	0
x_6	0	$-\frac{3}{2}$	1	0	$\frac{1}{8}$	0	1	$-\frac{21}{2}$	0
x_5	0	$\frac{1}{16}$	$-\frac{1}{8}$	0	$-\frac{3}{64}$	1	0	($\frac{3}{16}$)	0
x_3	0	$\frac{3}{2}$	-1	1	$-\frac{1}{8}$	0	0	$\frac{21}{2}$	1

	z	x_1	x_2	x_3	x_4	x_5	x_6	x_7	RHS
z	1	1	-1	0	$\frac{1}{2}$	-16	0	0	0
x_6	0	(2)	-6	0	$-\frac{5}{2}$	56	1	0	0
x_7	0	$\frac{1}{3}$	$-\frac{2}{3}$	0	$-\frac{1}{4}$	$\frac{16}{3}$	0	1	0
x_3	0	-2	6	1	$\frac{5}{2}$	-56	0	0	1

	z	x_1	x_2	x_3	x_4	x_5	x_6	x_7	RHS
z	1	0	2	0	$\frac{7}{4}$	-44	$-\frac{1}{2}$	0	0
x_1	0	1	-3	0	$-\frac{5}{4}$	28	$\frac{1}{2}$	0	0
x_7	0	0	$\textcircled{\frac{1}{3}}$	0	$\frac{1}{6}$	-4	$-\frac{1}{6}$	1	0
x_3	0	0	0	1	0	0	1	0	1

	z	x_1	x_2	x_3	x_4	x_5	x_6	x_7	RHS
z	1	0	0	0	$\frac{3}{4}$	-20	$\frac{1}{2}$	-6	0
x_1	0	1	0	0	$\frac{1}{4}$	-8	-1	9	0
x_2	0	0	1	0	$\frac{1}{2}$	-12	$-\frac{1}{2}$	3	0
x_3	0	0	0	1	0	0	1	0	1

We see that the last tableau above is identical to the first tableau. All the tableaux correspond to the extreme point (0, 0, 1, 0, 0, 0, 0) with different bases. The foregoing sequence of pivots generated the bases $\mathbf{B}_1$, $\mathbf{B}_2$, $\mathbf{B}_3$, $\mathbf{B}_4$, $\mathbf{B}_5$, $\mathbf{B}_6$, and $\mathbf{B}_7$, where $\mathbf{B}_7 = \mathbf{B}_1 = [\mathbf{a}_1, \mathbf{a}_2, \mathbf{a}_3]$. If the same sequence of pivots are used over and over again, the simplex algorithm will cycle forever among these bases without reaching the optimal point.

A Rule that Prevents Cycling

Even though cycling is very unlikely, and actually it is not an easy task to formulate a problem that cycles, it is of theoretical interest to develop a rule that prevents cycling. We give such a rule here, and illustrate it by the problem of Example 4.11. Validation of the rule is postponed until the next section. Consider the following linear programming problem.

$$\begin{aligned} \text{Minimize} \quad & \mathbf{cx} \\ \text{Subject to} \quad & \mathbf{Ax} = \mathbf{b} \\ & \mathbf{x} \geqslant \mathbf{0} \end{aligned}$$

where $\mathbf{A}$ is an $m \times n$ matrix of rank m. Since the simplex method is usually started with the initial basis as the identity matrix (corresponding to slack and/or artificial variables), we shall assume that the first m columns of $\mathbf{A}$ form the identity. The following rule, which specifies the variable leaving the basis if the simplex minimum ratio test produces several candidates, will guarantee noncycling.

Exiting Rule

Given a basic feasible solution with basis **B**, suppose that the nonbasic variable x_k is chosen to enter the basis (say $0 < z_k - c_k = \text{Maximum } z_j - c_j$). The index r of the variable x_{B_r} leaving the basis is determined as follows. Let

$$I_0 = \left\{ r : \frac{\bar{b}_r}{y_{rk}} = \underset{1 \leqslant i \leqslant m}{\text{Minimum}} \left\{ \frac{\bar{b}_i}{y_{ik}} : y_{ik} > 0 \right\} \right\}$$

If I_0 is a singleton, namely $I_0 = \{r\}$, then x_{B_r} leaves the basis. Otherwise form I_1 as follows:

$$I_1 = \left\{ r : \frac{y_{r1}}{y_{rk}} = \underset{i \in I_0}{\text{Minimum}} \left\{ \frac{y_{i1}}{y_{ik}} \right\} \right\}$$

If I_1 is singleton, namely $I_1 = \{r\}$, then x_{B_r} leaves the basis. Otherwise form I_2. In general I_j is formed from I_{j-1} as follows:

$$I_j = \left\{ r : \frac{y_{rj}}{y_{rk}} = \underset{i \in I_{j-1}}{\text{Minimum}} \left\{ \frac{y_{ij}}{y_{ik}} \right\} \right\}$$

Eventually, for some $j \leqslant m$, I_j will be a singleton (why?). If $I_j = \{r\}$, then x_{B_r} leaves the basis.

Before we illustrate the preceding rule, let us briefly discuss its implications. First we use the usual minimum ratio test as an exiting criterion. If this test gives a unique index, then the corresponding variable leaves the basis. In case of a tie we try to break it by replacing the right-hand side in the minimum ratio calculation by the first column $\mathbf{y}_1$, and by only using the rows corresponding to the tie. If the tie is still not broken, the second column is used, and so forth. When or before column m is reached, the tie must be broken, for if this were not the case, we have two rows of the matrix $\mathbf{B}^{-1} = (\mathbf{y}_1, \mathbf{y}_2, \ldots, \mathbf{y}_m)$, which are proportional (why?). This is impossible, however, in view of linear independence of the rows of $\mathbf{B}^{-1}$.

Example 4.12

We now solve the problem of Example 4.11 using the additional rule for exiting from the basis.

	z	x_1	x_2	x_3	x_4	x_5	x_6	x_7	RHS
z	1	0	0	0	$\frac{3}{4}$	-20	$\frac{1}{2}$	-6	0
x_1	0	1	0	0	$\frac{1}{4}$	-8	-1	9	0
x_2	0	0	1	0	$\textcircled{\frac{1}{2}}$	-12	$-\frac{1}{2}$	3	0
x_3	0	0	0	1	0	0	1	0	1

Here $I_0 = \{1, 2\}$, $I_1 = \{2\}$, and therefore $x_{B_2} = x_2$ leaves the basis. Note that in Example 4.11, x_1 left the basis during the first iteration.

	z	x_1	x_2	x_3	x_4	x_5	x_6	x_7	RHS
z	1	0	$-\frac{3}{2}$	0	0	-2	$\frac{5}{4}$	$-\frac{21}{2}$	0
x_1	0	1	$-\frac{1}{2}$	0	0	-2	$-\frac{3}{4}$	$\frac{15}{2}$	0
x_4	0	0	2	0	1	-24	-1	6	0
x_3	0	0	0	1	0	0	$\textcircled{1}$	0	1

Here $I_0 = \{3\}$. Therefore $x_{B_3} = x_3$ leaves.

	z	x_1	x_2	x_3	x_4	x_5	x_6	x_7	RHS
z	1	0	$-\frac{3}{2}$	$-\frac{5}{4}$	0	-2	0	$-\frac{21}{2}$	$-\frac{5}{4}$
x_1	0	1	$-\frac{1}{2}$	$\frac{3}{4}$	0	-2	0	$\frac{15}{2}$	$\frac{3}{4}$
x_4	0	0	2	1	1	-24	0	6	1
x_6	0	0	0	1	0	0	1	0	1

The foregoing tableau gives the optimal solution, since $z_j - c_j \leqslant 0$ for all nonbasic variables.

4.6 LEXICOGRAPHIC VALIDATION OF CYCLING PREVENTION

In this section we show that the rule adopted in the previous section indeed prevents cycling. We do this by showing that none of the previous bases visited by the simplex method are repeated. In view of the finite number of bases, this automatically guarantees stopping in finite number of iterations.

Lexicographically Positive Vectors

In order to facilitate the proof of finite convergence, it will be convenient to introduce the notion of a lexicographically positive vector. A vector $\mathbf{x}$ is called *lexicographically positive* (denoted by $\mathbf{x} \succ \mathbf{0}$) if the following two requirements hold:

1. $\mathbf{x}$ is not identically zero.
2. The first nonzero component of $\mathbf{x}$ is positive.

For example, $(0, 2, -1, 3)$, $(2, 1, -3, 1)$, and $(0, 0, 1, -1)$ are lexicographically positive vectors whereas $(-1, 1, 2, 3)$, $(0, 0, 0, 0)$, and $(0, 0, -2, 1)$ are not. A *lexicographically nonnegative* vector, denoted by $\succcurlyeq \mathbf{0}$, is either the zero vector or else a lexicographically positive vector. In order to prove that none of the bases generated by the simplex method is repeated, we first show that each row of the $m \times (m + 1)$ matrix $(\bar{\mathbf{b}}, \mathbf{B}^{-1})$ is lexicographically positive at each iteration, where $\bar{\mathbf{b}} = \mathbf{B}^{-1}\mathbf{b}$. Indeed, in the absence of degeneracy $\bar{\mathbf{b}} > \mathbf{0}$, and therefore each row of $(\bar{\mathbf{b}}, \mathbf{B}^{-1})$ is clearly lexicographically positive.

First recall that the original basis is $\mathbf{I}$, and since $\mathbf{b} \geqslant \mathbf{0}$, then each row of the matrix $(\bar{\mathbf{b}}, \mathbf{B}^{-1}) = (\mathbf{b}, \mathbf{I})$ is lexicographically positive. (If a feasible basis that is different from the identity is available, we still have a starting solution that satisfies the lexicographic positive condition. See Exercise 4.48.) In view of this, the preceding result will be proved, if we can show the following: if each row of $(\bar{\mathbf{b}}, \mathbf{B}^{-1})$ is $\succ \mathbf{0}$, then each row of $(\hat{\mathbf{b}}, \hat{\mathbf{B}}^{-1})$ is $\succ \mathbf{0}$ where $\hat{\mathbf{B}}$ is the new basis obtained after pivoting and $\hat{\mathbf{b}} = \hat{\mathbf{B}}^{-1}\mathbf{b}$. Consider the following two tableaux before and after introducing x_k, and recall that the first m columns (ignoring the z column) in these tableaux represent $\mathbf{B}^{-1}$ and $\hat{\mathbf{B}}^{-1}$ respectively (since the first m columns of the original problem form the identity). Here $\bar{z}$ denotes the objective value $\mathbf{c}_B\bar{\mathbf{b}}$ before pivoting.

Before Pivoting

	z	x_1	$\cdots$	x_j	$\cdots$	x_m	$x_{m+1}\ \cdots$	x_k	$\cdots\ x_n$	RHS
z	1	$z_1 - c_1$	$\cdots$	$z_j - c_j$	$\cdots$	$z_m - c_m$	$\cdots$	$z_k - c_k$	$\cdots$	$\bar{z}$
x_{B_1}	0	y_{11}	$\cdots$	y_{1j}	$\cdots$	y_{1m}	$\cdots$	y_{1k}	$\cdots$	$\bar{b}_1$
$\vdots$	$\vdots$	$\vdots$		$\vdots$				$\vdots$		$\vdots$
x_{B_i}	0	y_{i1}	$\cdots$	y_{ij}	$\cdots$	y_{im}	$\cdots$	y_{ik}	$\cdots$	$\bar{b}_i$
$\vdots$	$\vdots$	$\vdots$		$\vdots$		$\vdots$		$\vdots$		$\vdots$
x_{B_r}	0	y_{r1}	$\cdots$	y_{rj}	$\cdots$	y_{rm}	$\cdots$	(y_{rk})	$\cdots$	$\bar{b}_r$
$\vdots$	$\vdots$	$\vdots$		$\vdots$		$\vdots$		$\vdots$		$\vdots$
x_{B_m}	0	y_{m1}	$\cdots$	y_{mj}	$\cdots$	y_{mm}	$\cdots$	y_{mk}	$\cdots$	$\bar{b}_m$

After Pivoting

	z	x_1	$\cdots$	x_j	$\cdots$	x_m	$x_{m+1} \cdots x_k \cdots x_n$	RHS
z	1	$(z_1 - c_1) - \frac{y_{r1}}{y_{rk}}(z_k - c_k)$	$\cdots$	$(z_j - c_j) - \frac{y_{rj}}{y_{rk}}(z_k - c_k)$	$\cdots$	$(z_m - c_m) - \frac{y_{rm}}{y_{rk}}(z_k - c_k)$	$\cdots 0 \cdots$	$\bar{z} - \frac{\bar{b}_r}{y_{rk}}(z_k - c_k)$
x_{B_1}	0	$y_{11} - \frac{y_{r1}}{y_{rk}} y_{1k}$	$\cdots$	$y_{1j} - \frac{y_{rj}}{y_{rk}} y_{1k}$	$\cdots$	$y_{1m} - \frac{y_{rm}}{y_{rk}} y_{1k}$	$\cdots 0 \cdots$	$\bar{b}_1 - \frac{\bar{b}_r}{y_{rk}} y_{1k}$
$\vdots$	$\vdots$	$\vdots$		$\vdots$		$\vdots$	$\vdots$	$\vdots$
x_{B_i}	0	$y_{i1} - \frac{y_{r1}}{y_{rk}} y_{ik}$	$\cdots$	$y_{ij} - \frac{y_{rj}}{y_{rk}} y_{ik}$	$\cdots$	$y_{im} - \frac{y_{rm}}{y_{rk}} y_{ik}$	$\cdots 0 \cdots$	$\bar{b}_i - \frac{\bar{b}_r}{y_{rk}} y_{ik}$
$\vdots$	$\vdots$	$\vdots$		$\vdots$		$\vdots$	$\vdots$	$\vdots$
x_k	0	$\frac{y_{r1}}{y_{rk}}$	$\cdots$	$\frac{y_{rj}}{y_{rk}}$	$\cdots$	$\frac{y_{rm}}{y_{rk}}$	$\cdots 1 \cdots$	$\frac{\bar{b}_r}{y_{rk}}$
$\vdots$	$\vdots$	$\vdots$		$\vdots$		$\vdots$	$\vdots$	$\vdots$
x_{B_m}	0	$y_{m1} - \frac{y_{r1}}{y_{rk}} y_{mk}$	$\cdots$	$y_{mj} - \frac{y_{rj}}{y_{rk}} y_{mk}$	$\cdots$	$y_{mm} - \frac{y_{rm}}{y_{rk}} y_{mk}$	$\cdots 0 \cdots$	$\bar{b}_m - \frac{\bar{b}_r}{y_{rk}} y_{mk}$

Consider a typical row i of $(\hat{\mathbf{b}}, \hat{\mathbf{B}}^{-1})$. From the foregoing tableau this row is given by

$$\left(\bar{b}_i - \frac{\bar{b}_r}{y_{rk}} y_{ik}, y_{i1} - \frac{y_{r1}}{y_{rk}} y_{ik}, \ldots, y_{im} - \frac{y_{rm}}{y_{rk}} y_{ik}\right) \quad i \neq r \tag{4.4}$$

$$\left(\frac{\bar{b}_r}{y_{rk}}, \frac{y_{r1}}{y_{rk}}, \ldots, \frac{y_{rm}}{y_{rk}}\right) \quad i = r \tag{4.5}$$

Since $y_{rk} > 0$ and the rth row is $\succ \mathbf{0}$ before pivoting, then from Equation (4.5) the rth row after pivoting is also $\succ \mathbf{0}$. Now consider $i \neq r$. There are two mutually exclusive cases: either $i \notin I_0$ or else $i \in I_0$. First suppose that $i \notin I_0$. If $y_{ik} \leqslant 0$, then from Equation (4.4) we see that the ith row after pivoting is given by

$$(\bar{b}_i, y_{i1}, \ldots, y_{im}) - \frac{y_{ik}}{y_{rk}} (\bar{b}_r, y_{r1}, \ldots, y_{rm})$$

which is the sum of two vectors that are lexicographically positive and lexicographically nonnegative (why?), and is hence $\succ \mathbf{0}$. Now suppose that $y_{ik} > 0$. By definition of I_0 and since $i \notin I_0$, then $\bar{b}_r/y_{rk} < \bar{b}_i/y_{ik}$ and hence $\bar{b}_i - (\bar{b}_r/y_{rk})y_{ik} > 0$. From Equation (4.4) the ith row is $\succ \mathbf{0}$. Now consider the case $i \in I_0$. Then $y_{ik} > 0$ and $\bar{b}_i - (\bar{b}_r/y_{rk})y_{ik} = 0$. There are two mutually exhaustive cases: either $i \notin I_1$, or else $i \in I_1$. In the former case, by definition of I_1, $y_{i1} - (y_{r1}/y_{rk})y_{ik} > 0$ and from Equation (4.4) the ith row is $\succ \mathbf{0}$. If, on the other hand, $i \in I_1$, then $y_{i1} - (y_{r1}/y_{rk})y_{ik} = 0$ and we examine whether $i \in I_2$ or not. This process is continued until termination in at most $m + 1$ steps, with the conclusion that each row of $(\hat{\mathbf{b}}, \hat{\mathbf{B}}^{-1})$ is $\succ \mathbf{0}$.

The foregoing analysis shows that each row of $(\mathbf{B}^{-1}\mathbf{b}, \mathbf{B}^{-1})$ is lexicographically positive at any given iteration. This fact will be used shortly to prove finite convergence. First note, by examining row 0 before and after pivoting, that

$$(\mathbf{c}_B\mathbf{B}^{-1}\mathbf{b}, \mathbf{c}_B\mathbf{B}^{-1}) - (\mathbf{c}_{\hat{B}}\hat{\mathbf{B}}^{-1}\mathbf{b}, \mathbf{c}_{\hat{B}}\hat{\mathbf{B}}^{-1}) = \frac{z_k - c_k}{y_{rk}} (\bar{b}_r, y_{r1}, y_{r2}, \ldots, y_{rm})$$

Note that $(\bar{b}_r, y_{r1}, \ldots, y_{rm})$ is the rth row of $(\bar{\mathbf{b}}, \mathbf{B}^{-1})$ and is therefore $\succ \mathbf{0}$. Since $z_k - c_k > 0$ and $y_{rk} > 0$, it is therefore evident that

$$(\mathbf{c}_B\mathbf{B}^{-1}\mathbf{b}, \mathbf{c}_B\mathbf{B}^{-1}) - (\mathbf{c}_{\hat{B}}\hat{\mathbf{B}}^{-1}\mathbf{b}, \mathbf{c}_{\hat{B}}\hat{\mathbf{B}}^{-1}) \succ \mathbf{0}$$

Now we are ready to show that the rule of Section 4.5 will indeed prevent cycling. We do this by showing that the bases developed by the simplex method are distinct. Suppose by contradiction that a sequence of bases $\mathbf{B}_1, \mathbf{B}_2, \ldots, \mathbf{B}_t$ is generated where $\mathbf{B}_1 = \mathbf{B}_t$. From the preceding analysis we have

$$\left(\mathbf{c}_{B_j}\mathbf{B}_j^{-1}\mathbf{b}, \mathbf{c}_{B_j}\mathbf{B}_j^{-1}\right) - \left(\mathbf{c}_{B_{j+1}}\mathbf{B}_{j+1}^{-1}\mathbf{b}, \mathbf{c}_{B_{j+1}}\mathbf{B}_{j+1}^{-1}\right) \succ \mathbf{0} \quad \text{for } j = 1, 2, \ldots, t-1$$

Adding over $j = 1, 2, \ldots, t-1$ and noting that $\mathbf{B}_1$ is assumed equal to $\mathbf{B}_t$, we get $\mathbf{0} \succ \mathbf{0}$, which is impossible. This contradiction asserts that the bases visited by the simplex algorithm are distinct. Since there is but a finite number of bases, convergence in a finite number of steps is established.

EXERCISES

4.1 Solve the following problem by the two-phase simplex method.

$$\begin{aligned} \text{Maximize} \quad & 2x_1 - x_2 + x_3 \\ \text{Subject to} \quad & x_1 + x_2 - 2x_3 \leqslant 8 \\ & 4x_1 - x_2 + x_3 \geqslant 2 \\ & 2x_1 + 3x_2 - x_3 \geqslant 4 \\ & x_1, \quad x_2, \quad x_3 \geqslant 0 \end{aligned}$$

4.2 Consider the following linear programming problem.

$$\begin{aligned} \text{Maximize} \quad & x_1 + 2x_2 \\ \text{Subject to} \quad & x_1 + x_2 \geqslant 1 \\ & -x_1 + x_2 \leqslant 3 \\ & x_2 \leqslant 5 \\ & x_1, \quad x_2 \geqslant 0 \end{aligned}$$

a. Solve the problem geometrically.
b. Solve the problem by the two-phase simplex method. Show that the points generated by phase I correspond to basic solutions of the original system.

4.3 Phase I of the two-phase method can be made use of to check redundancy. Suppose that we have the following three constraints:

$$\begin{aligned} x_1 - x_2 &\geqslant 2 \\ 2x_1 + 3x_2 &\geqslant 4 \\ 3x_1 + 2x_2 &\geqslant 6 \end{aligned}$$

Note that the third constraint can be obtained by adding the first two constraints. Would phase I detect this kind of redundancy? If not, what kind of redundancy will it detect? Is the type of redundancy of the foregoing problem equivalent to degeneracy? Discuss.

4.4 Solve the following problem by the two-phase simplex method.

$$\begin{aligned} \text{Minimize} \quad & x_1+3x_2-x_3 \\ \text{Subject to} \quad & x_1+x_2+x_3 \geq 3 \\ & -x_1+2x_2 \geq 2 \\ & -x_1+5x_2+x_3 \leq 4 \\ & x_1, \quad x_2, \quad x_3 \geq 0 \end{aligned}$$

4.5 Show how phase I of the simplex method can be used to solve n simultaneous linear equations in n unknowns. Show how the following cases can be detected:

a. Inconsistency of the system.
b. Redundancy of the equations.
c. Unique solution.

Also show how the inverse matrix corresponding to the system of equations can be found in (c). Illustrate by solving the following system.

$$\begin{aligned} x_1+2x_2+x_3 &= 4 \\ -x_1-x_2+2x_3 &= 3 \\ x_1-x_2+x_3 &= 2 \end{aligned}$$

4.6 Solve the following problem by the two-phase method.

$$\begin{aligned} \text{Maximize} \quad & -x_1-2x_2 \\ \text{Subject to} \quad & 3x_1+4x_2 \leq 20 \\ & 2x_1-x_2 \geq 2 \\ & x_1, \quad x_2 \geq 0 \end{aligned}$$

4.7 Solve the following problem by the two-phase method.

$$\begin{aligned} \text{Maximize} \quad & 5x_1-2x_2+x_3 \\ \text{Subject to} \quad & x_1+4x_2+x_3 \leq 6 \\ & 2x_1+x_2+3x_3 \geq 2 \\ & x_1, \quad x_2 \geq 0 \\ & x_3 \quad \text{unrestricted} \end{aligned}$$

4.8 Solve the following problem by the two-phase method.

$$\begin{aligned} \text{Maximize} \quad & 4x_1+5x_2-3x_3 \\ \text{Subject to} \quad & x_1+x_2+x_3=10 \\ & x_1-x_2 \geqslant 1 \\ & 2x_1+3x_2+x_3 \leqslant 20 \\ & x_1, \quad x_2, \quad x_3 \geqslant 0 \end{aligned}$$

4.9 Use the big-M method to solve the following problem.

$$\begin{aligned} \text{Minimize} \quad & -2x_1+2x_2+x_3+x_4 \\ \text{Subject to} \quad & x_1+2x_2+x_3+x_4 \leqslant 2 \\ & x_1-x_2+x_3+5x_4 \geqslant 4 \\ & 2x_1-x_2+x_3 \geqslant 2 \\ & x_1, \quad x_2, \quad x_3, \quad x_4 \geqslant 0 \end{aligned}$$

4.10 Use the big-M method to solve the following problem.

$$\begin{aligned} \text{Maximize} \quad & x_1-x_2+x_3 \\ \text{Subject to} \quad & x_1+x_2+2x_3 \geqslant 4 \\ & x_1-2x_2+x_3 \leqslant 2 \\ & x_1, \quad x_2, \quad x_3 \geqslant 0 \end{aligned}$$

4.11 Is it possible that the optimal solution of the big-M problem be unbounded and in the meantime the optimal solution of the original problem be bounded? Discuss in detail.

4.12 Consider the problem: Maximize $\mathbf{cx}$ subject to $\mathbf{Ax} = \mathbf{b}$, $\mathbf{x} \geqslant \mathbf{0}$, where $\mathbf{A}$ is an $m \times n$ matrix and $\mathbf{b} \geqslant \mathbf{0}$. Add the artificial vector $\mathbf{x}_a$ and consider the following big-M problem: Maximize $\mathbf{cx} - M\mathbf{1x}_a$ subject to $\mathbf{Ax} + \mathbf{x}_a = \mathbf{b}$, $\mathbf{x} \geqslant \mathbf{0}$, $\mathbf{x}_a \geqslant \mathbf{0}$. Show that there exists an $\overline{M} > 0$ such that for all $M \geqslant \overline{M}$ either the basic feasible solution $(\mathbf{x}^*, \mathbf{x}_a^*)$ solves the big-M problem or else the solution of the big-M problem is unbounded along the ray $\{(\mathbf{x}^*, \mathbf{x}_a^*) + \lambda(\mathbf{d}^*, \mathbf{d}_a^*) : \lambda \geqslant 0\}$ where $(\mathbf{d}^*, \mathbf{d}_a^*)$ is an extreme direction. Interpret both cases. (This establishes the fact that for M larger than some number, regardless of whether the original problem has a feasible region or not, the solution of the big-M problem is achieved at some fixed extreme point or along some fixed ray.)

4.13 Suppose that a linear programming problem admits feasible points. Utilize the result of Exercise 4.12 to show that if M is large enough, then a finite optimal solution of the big-M problem must have all artificials equal to zero. Give all details.

4.14 Is it possible that the region in E^n given by

$$\mathbf{Ax} = \mathbf{b}$$
$$\mathbf{x} \geqslant \mathbf{0}$$

is bounded, whereas the region in E^{n+m} given by

$$\begin{aligned} \mathbf{Ax} + \mathbf{x}_a &= \mathbf{b} \\ \mathbf{x},\ \mathbf{x}_a &\geqslant \mathbf{0} \end{aligned}$$

is unbounded? What are the implications of your answer on using the big-M method as a solution procedure?

4.15 Solve the following problem by the big-M method.

$$\begin{aligned} \text{Minimize} \quad & 2x_1 + 4x_2 - x_4 \\ \text{Subject to} \quad & x_1 + 2x_2 - x_3 + x_4 \leqslant 2 \\ & 2x_1 + x_2 + 2x_3 + 3x_4 = 4 \\ & x_1 - x_3 + x_4 \geqslant 3 \\ & x_1,\ x_2,\ x_4 \geqslant 0 \\ & x_3 \quad \text{unrestricted} \end{aligned}$$

4.16 Solve the following problem by the big-M method.

$$\begin{aligned} \text{Maximize} \quad & 2x_1 + 4x_2 + 4x_3 - 3x_4 \\ \text{Subject to} \quad & x_1 + x_2 + x_3 = 4 \\ & x_1 + 4x_2 + 4x_4 = 8 \\ & x_1,\ x_2,\ x_3,\ x_4 \geqslant 0 \end{aligned}$$

4.17 Indicate whether the following statement is true or false. If M is chosen extremely large, then the two-phase method and the big-M method will generate the same sequence of bases. Discuss in detail.

4.18 Compare the two-phase and the big-M methods. What are the advantages and disadvantages of each? How large should M be chosen?

4.19 Suppose that either phase I is completed or the bounded optimal solution of the big-M problem is found. Further suppose that there exists at least one artificial at a positive level indicating that the original system $\mathbf{Ax} = \mathbf{b}$ and $\mathbf{x} \geqslant \mathbf{0}$ has no solution. Can you differentiate between the following two cases?

a. The system $\mathbf{Ax} = \mathbf{b}$ is inconsistent.

b. The system $\mathbf{Ax} = \mathbf{b}$ is consistent but $\mathbf{Ax} = \mathbf{b}$ implies that $\mathbf{x} \ngeqslant \mathbf{0}$.

4.20 Suppose that the big-M method is used to solve a linear programming problem. Further suppose that $z_k - c_k = \text{Maximum } (z_j - c_j) > 0$. Show that the original problem is infeasible if not all artificials are equal to zero and $y_{ik} \leqslant 0$ for each i such that x_{B_i} is an artificial variable. (Note that this gives a more general result than Subcase B2 of Section 4.3.)

4.21 Show that problem P could have an unbounded optimal solution even if during the solution of the big-M problem a nonbasic variable x_j is found such that $z_j - c_j > 0$, $\mathbf{y}_j \leqslant \mathbf{0}$ and not all artificials are zero. (This shows that it is important to examine the most positive $z_j - c_j$). (*Hint*. Consider Example 4.8 and introduce x_1 rather than x_3 in the first iteration.)

4.22 *Geometric redundancy* occurs when deletion of a constraint does not alter the feasible set. How can geometric redundancy be detected? (*Hint*. Consider the objective of minimizing x_s where x_s is a particular slack variable.)

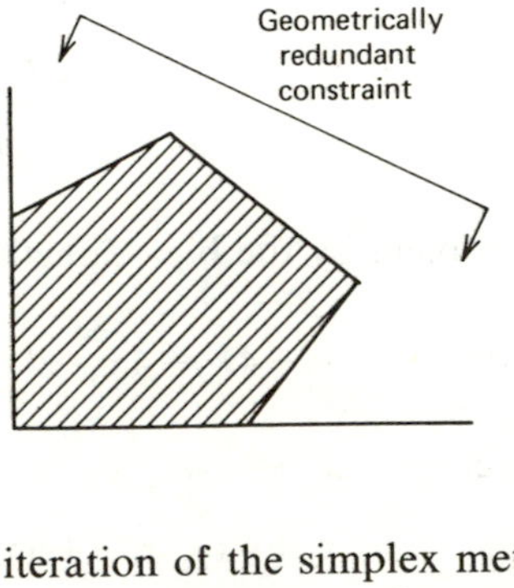

4.23 Suppose that at some iteration of the simplex method the slack variable x_s is basic in the ith row. Show that if $y_{ij} \leqslant 0$, $j = 1, \ldots, n$, $j \neq s$, then the constraint associated with x_s is geometrically redundant. Also show that the ith constraint and variable x_s may be deleted from the current tableau without changing the feasible set.

4.24 Solve the following linear program by both the two-phase method and the big-M method.

$$\begin{array}{lrcl} \text{Minimize} & 3x_1 - 2x_2 + 5x_3 & & \\ \text{Subject to} & x_1 + 2x_2 + x_3 & \geq & 5 \\ & -3x_1 + x_2 - x_3 & \leq & 4 \\ & x_1, \quad x_2, \quad x_3 & \geq & 0 \end{array}$$

4.25 Solve the following problem by the big-M method.

$$\begin{array}{lrcl} \text{Minimize} & 3x_1 + 2x_2 + 4x_3 + 8x_4 & & \\ \text{Subject to} & x_1 + 2x_2 + 5x_3 + 6x_4 & \geq & 8 \\ & -2x_1 + 5x_2 + 3x_3 - 5x_4 & \leq & 3 \\ & x_1, \quad x_2, \quad x_3, \quad x_4 & \geq & 0 \end{array}$$

4.26 Solve the following problem by the big-M method.

$$\begin{array}{lrcl} \text{Maximize} & 2x_1 - x_2 & & \\ \text{Subject to} & x_1 + x_2 & \leq & 3 \\ & -x_1 + x_2 & \geq & 1 \\ & x_1, \quad x_2 & \geq & 0 \end{array}$$

4.27 Solve the following problem by the big-M method.

$$\begin{array}{lrcl} \text{Maximize} & 5x_1 - 2x_2 + x_3 & & \\ \text{Subject to} & x_1 + 4x_2 + x_3 & \leq & 6 \\ & 2x_1 + x_2 + 3x_3 & \geq & 2 \\ & x_1, \qquad\quad x_3 & \geq & 0 \\ & x_2 & & \text{unrestricted} \end{array}$$

4.28 Discuss in detail all the possible cases that may arise when using the single artificial variable technique with both the two-phase method and the big-M method.

4.29 Use the single artificial variable technique to solve the following problem.

$$\begin{aligned}
\text{Maximize}\quad & 4x_1+5x_2+7x_3-x_4\\
\text{Subject to}\quad & x_1+x_2+2x_3-x_4\geq 1\\
& 2x_1-6x_2+3x_3+x_4\leq -3\\
& x_1+4x_2+3x_3+2x_4=-5\\
& x_1,\ x_2,\ x_4\geq 0\\
& x_3 \quad \text{unrestricted}
\end{aligned}$$

4.30 Use the single artificial variable technique to solve the following linear programming problem.

$$\begin{aligned}
\text{Minimize}\quad & -x_1-2x_2+x_3\\
\text{Subject to}\quad & x_1+x_2+x_3\geq 4\\
& 2x_1 \qquad -x_3\geq 3\\
& x_2+x_3\leq 2\\
& x_1,\ x_2,\ x_3\geq 0
\end{aligned}$$

4.31 Discuss the advantages and disadvantages of using a single artificial variable compared with a method using several artificial variables.

4.32 Suppose it is possible to get the constraints of a linear program to the form $\mathbf{I}\mathbf{x}_B+\mathbf{B}^{-1}\mathbf{N}\mathbf{x}_N=\bar{\mathbf{b}}$, where $\bar{\mathbf{b}}=\mathbf{B}^{-1}\mathbf{b}\ngeq\mathbf{0}$. Show that a single artificial variable x_a with activity vector $\hat{\mathbf{b}}$ (where $\hat{\mathbf{b}}\leq\bar{\mathbf{b}}$) can be added and a basic feasible solution would be easily obtained.

4.33 Construct detailed flow diagrams of the two-phase method and the big-M method. Using FORTRAN (or another language), code either of the two methods.

4.34 A manufacturer wishes to find the optimal weekly production of items A, B, and C that maximizes his profit. The unit profit and the minimal weekly production of these items are respectively \$2.00, \$2.00, and \$4.00; and 100 units, 60 units, and 60 units. Products A, B, and C are processed on three machines. The hours required per item per machine are summarized below.

MACHINE	ITEM A	B	C
1	0	1	2
2	1	1	1
3	2	1	1

The number of hours of machines 1, 2, and 3 available per week are 240, 400, and 360 respectively. Find the optimal production schedule.

4.35 A farmer has 200 acres and 18,000 man-hours available. He wishes to determine the acreage allocated to the following products: corn, wheat, okra, tomatos, and green beans. The farmer must produce at least 250 tons of corn to feed his hogs and cattle, and he must produce at least 80 tons of wheat, which he precontracted. The tonnage and labor in man-hours per acre of the different products are summarized below.

	CORN	WHEAT	OKRA	TOMATOS	BEANS
Tons/acre	10	4	4	8	6
Man-hours/acre	120	150	100	80	120

The corn, wheat, okra, tomatos, and beans can be sold for $120.00, $150.00, $60.00, $80.00, and $55.00 per ton. Find the optimal solution.

4.36 A company manufactures stoves and ovens. The company has three warehouses and two retail stores. Sixty, 80, and 50 stoves and 80, 50, and 50 ovens are available at the three warehouses respectively. One hundred and 90 stoves and 60 and 120 ovens are required at the retail stores respectively. The unit shipping costs, which apply to both the stoves and ovens, from the warehouses to the retail stores are given below.

WAREHOUSE	STORE 1	2
1	3	5
2	2	3
3	6	3

Find the shipping pattern that minimizes the total transportation cost by the simplex method.

4.37 A manufacturer wishes to plan the production of two items A and B for the months of March, April, May, and June. The demands that must be met are given below.

	March	April	May	June
Item A	400	500	600	400
Item B	600	600	700	600

Suppose that the inventory of A and B at the end of February is 100 and 150 respectively. Further suppose that at least 150 units of item B must be available at the end of June. The inventory holding costs of items A and B during any month are given by \$1.00 and \$0.80 times the inventory of the item at the end of the month. Furthermore, because of space limitation, the sum of items A and B in stock cannot exceed 250 during any month. Finally, the maximum number of items A and B that can be produced during any given month is 500 and 650 respectively.

a. Formulate the production problem as a linear program. The objective is to minimize the total inventory cost (the production cost is assumed constant).
b. Find the optimal production/inventory pattern.
c. Management is considering installing a new manufacturing system for item B at the end of April. This would raise the maximum items that can be produced per month from 650 to 700 and meanwhile would reduce the unit manufacturing cost from \$8.00 to \$6.50. Assess the benefits of this system in reducing the total manufacturing plus inventory costs. If you were a member of the management team, discuss how you would assess whether the new system is cost effective.
d. Suppose that management decided to introduce the new system. A market research indicated that item B can be backlogged without serious dissatisfaction of customers. It was the management's assessment that each unit of unsatisfied demand during any month must be charged an additional \$1.00. Formulate the production/inventory problem and find the optimal solution by the simplex method.

4.38 A trucking company owns three types of trucks, type I, type II, and type III. These trucks are equipped to haul three different types of machines per load according to the following chart.

	Truck Type		
	I	II	III
Machine A	1	1	1
Machine B	0	1	2
Machine C	2	1	1

Trucks of type I, II, and III cost \$400, \$600, and \$900 per trip, respectively. We are interested to find how many trucks of each type should be sent to haul 12 machines of type A, 10 machines of type B, and 16 machines of type C. Formulate the problem and then solve it by the simplex method.

(This is an integer programming problem; you may ignore the integer requirements.)

4.39 A company produces refrigerators, stoves, and dishwashers. During the coming year, sales are expected to be the following:

	QUARTER			
PRODUCT	1	2	3	4
Refrigerators	1500	1000	2000	1200
Stoves	1500	1500	1200	1500
Dishwashers	1000	2000	1500	2500

The company wants a production schedule that meets the demand requirements. Management also has decided that the inventory level for each product must be at least 150 units at the end of each quarter. There is no inventory of any product at the start of the first quarter.

During a quarter only 18,000 hours of production time are available. A refrigerator requires 2 hours, a stove 4 hours, and a dishwahser 3 hours of production time. Refrigerators cannot be manufactured in the fourth quarter because the company plans to modify tooling for a new product line.

Assume that each item left in inventory at the end of a quarter incurs a holding cost of \$5. The company wants to plan its production schedule over the year, in such a way that meets the quarterly demands and minimizes the total inventory cost. Formulate the problem and then solve it by the simplex method.

4.40 A manufacturer of metal sheets received an order for producing 2000 sheets of size $2' \times 4'$ and 1000 sheets of size $4' \times 7'$. Two standard sheets are available of sizes $10' \times 3000'$ and $11' \times 2000'$. The engineering staff decided that the following four cutting patterns are suitable for this order. Formulate the problem of meeting the order and minimizing the waste as a linear program and solve it by the simplex method.

Pattern 1

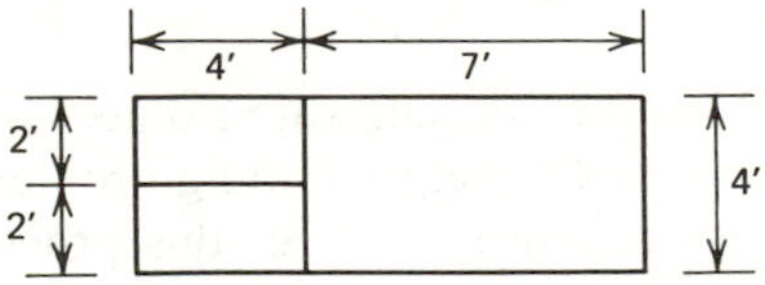

Pattern 2

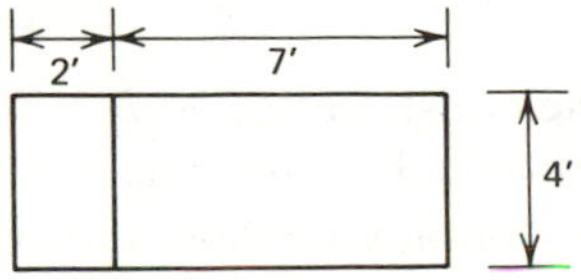

Pattern 3

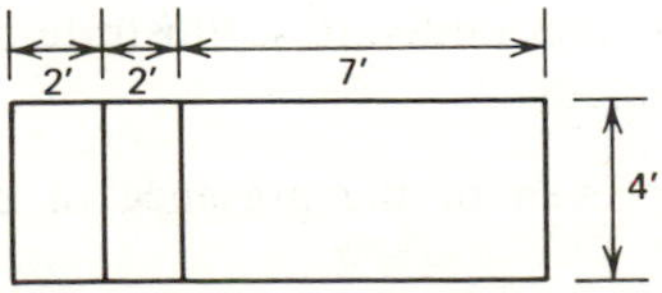

Pattern 4

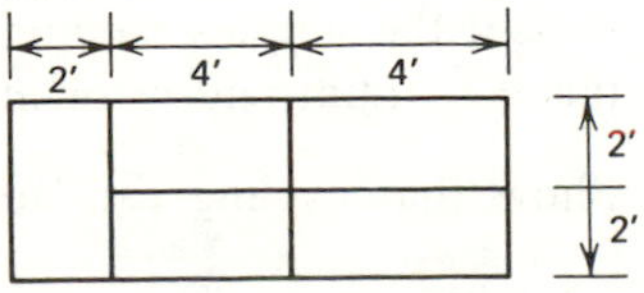

4.41 Solve the following problem, using the additional exiting rule for noncycling.

$$\begin{array}{lrl} \text{Maximize} & x_1+2x_2+\ x_3 & \\ \text{Subject to} & x_1+4x_2+6x_3 & \leqslant 4 \\ & -x_1+\ x_2+4x_3 & \leqslant 1 \\ & x_1+3x_2+\ x_3 & \leqslant 6 \\ & x_1,\quad x_2,\quad x_3 & \geqslant 0 \end{array}$$

4.42 Consider the following region.

$$\begin{aligned} x_1 - x_2 + x_3 &\leqslant 4 \\ -x_1 + x_2 - 2x_3 &\leqslant 3 \\ 4x_1 + x_2 - 5x_3 &\leqslant 2 \\ x_1,\quad x_2,\quad x_3 &\geqslant 0 \end{aligned}$$

Recall that $\mathbf{d}$ is a direction of the region if $\mathbf{Ad} \leqslant \mathbf{0}$, $\mathbf{d} \geqslant \mathbf{0}$, and $\mathbf{d}$ is nonzero. Thus, in order to find directions of the region, we may solve the following problem.

$$\begin{array}{lrl} \text{Maximize} & d_1+d_2+\ d_3 & \\ \text{Subject to} & d_1-d_2+\ d_3 & \leqslant 0 \\ & -d_1+d_2-2d_3 & \leqslant 0 \\ & 4d_1+d_2-5d_3 & \leqslant 0 \\ & d_1+d_2+\ d_3 & \leqslant 1 \\ & d_1,\ d_2,\quad d_3 & \geqslant 0 \end{array}$$

The constraint $d_1 + d_2 + d_3 \leqslant 1$ is added for normalization; otherwise the optimal objective may reach $+\infty$. Solve this direction-finding problem by the simplex method with the additional exiting rule. Does this procedure generate extreme directions? Why or why not? Can the normalization constraint $d_1 + d_2 + d_3 \leqslant 1$ be deleted? If so, describe how to find directions if the simplex method indicates unboundedness. Illustrate by deleting this constraint and resolving the problem.

4.43 Show that cycling can never occur even in the presence of degeneracy provided that a unique minimum is obtained in the computation

$$\underset{1 \leqslant i \leqslant m}{\text{Minimum}} \left\{ \frac{\bar{b}_i}{y_{ik}} : y_{ik} > 0 \right\}$$

where $\bar{\mathbf{b}} = \mathbf{B}^{-1}\mathbf{b}$, $\mathbf{y}_k = \mathbf{B}^{-1}\mathbf{a}_k$, and x_k is the entering variable.

4.44 Consider the following problem.

$$\begin{aligned} \text{Maximize} \quad & 3x_1 + 4x_2 \\ \text{Subject to} \quad & 2x_1 + x_2 \leqslant 8 \\ & -x_1 + 2x_2 \leqslant 6 \\ & x_1 + x_2 \leqslant 6 \\ & x_1, \quad x_2 \geqslant 0 \end{aligned}$$

a. Solve the problem geometrically and verify that the optimal point is a degenerate basic feasible solution.
b. Solve the problem by the simplex method.
c. From (a), identify the constraint that causes degeneracy and resolve the problem after throwing this constraint away. Note that degeneracy disappears and the same optimal solution is obtained.
d. Can you show in general that degenerate basic feasible solutions can be made nondegenerate by throwing some constraints away without affecting the feasible region?

4.45 The additional rule for exiting from the basis is designed to specify which variable leaves the basis if the minimum ratio test results in a tie or several ties. Show in detail that the exiting rule would specify a unique variable to leave the basis after at most $m + 1$ columns are examined, namely $\bar{\mathbf{b}}, \mathbf{y}_1, \mathbf{y}_2, \ldots,$ and $\mathbf{y}_m$.

4.46 We showed that the row vector $(\mathbf{c}_B\mathbf{B}^{-1}\mathbf{b}, \mathbf{c}_B\mathbf{B}^{-1})$ is lexicographically decreasing from one iteration to another. Give an economic interpretation of this fact. (*Hint*. Note that $z = \mathbf{c}_B\mathbf{B}^{-1}\mathbf{b}$ and that $\partial z / \partial \mathbf{b} = \mathbf{c}_B\mathbf{B}^{-1}$.)

4.47 Suppose that we have an optimal extreme point of a minimization linear programming problem. In the presence of degeneracy, is it possible that this extreme point corresponds to a basic feasible solution such that $z_j - c_j > 0$ for at least one nonbasic variable? If this were the case, are we guaranteed of another basic feasible solution corresponding to the same extreme point where $z_j - c_j \leqslant 0$ for all nonbasic variables? Why or why not? Illustrate by a numerical example.

4.48 In order to prove finite convergence of the simplex method using the noncycling rule, we assumed that the first m columns of the constraint matrix form the identity. Show that this assumption can be relaxed provided that we have any basic feasible solution. (*Hint*. Let $\mathbf{B}$ be the starting basis and consider the following equivalent problem.)

$$\begin{aligned} \text{Minimize} \quad & \mathbf{0x}_B + (\mathbf{c}_N - \mathbf{c}_B\mathbf{B}^{-1}\mathbf{N})\mathbf{x}_N \\ \text{Subject to} \quad & \mathbf{Ix}_B + \mathbf{B}^{-1}\mathbf{N}\mathbf{x}_N = \mathbf{B}^{-1}\mathbf{b} \\ & \mathbf{x}_B, \quad \mathbf{x}_N \geqslant \mathbf{0} \end{aligned}$$

4.49 Consider the following problem.

$$\begin{aligned} \text{Minimize} \quad & (\mathbf{cx}, \mathbf{cY}) \\ \text{Subject to} \quad & \mathbf{A}(\mathbf{x}, \mathbf{Y}) = (\mathbf{b}, \mathbf{I}) \\ & (\mathbf{x}, \mathbf{Y}) \succcurlyeq \mathbf{0} \end{aligned}$$

where $\mathbf{A}$ is an $m \times n$ matrix, $\mathbf{c}$ is an n vector, and the variables are the n vector $\mathbf{x}$ and the $n \times m$ matrix $\mathbf{Y}$. The objective function is a row vector, and the minimization is taken in the lexicographic sense, that is, $(\mathbf{cx}_2, \mathbf{cY}_2) \prec (\mathbf{cx}_1, \mathbf{cY}_1)$ if and only if $(\mathbf{cx}_1, \mathbf{cY}_1) - (\mathbf{cx}_2, \mathbf{cY}_2) \succ \mathbf{0}$. Each row of the matrix $(\mathbf{x}, \mathbf{Y})$ is restricted to be lexicographically nonnegative, which means that each row is zero or $\succ \mathbf{0}$.

a. Let $\mathbf{x}$ be a basic feasible solution of the system $\mathbf{Ax} = \mathbf{b}$, $\mathbf{x} \geqslant \mathbf{0}$ with basis $\mathbf{B}$. Show that $\mathbf{x} = \begin{pmatrix} \mathbf{B}^{-1}\mathbf{b} \\ \mathbf{0} \end{pmatrix}$ and $\mathbf{Y} = \begin{pmatrix} \mathbf{B}^{-1} \\ \mathbf{0} \end{pmatrix}$ is a feasible solution of the foregoing problem provided that $(\mathbf{B}^{-1}\mathbf{b}, \mathbf{B}^{-1}) \succ \mathbf{0}$.
b. Show that the simplex method with the exiting rule of Section 4.5 generates the sequence $(\mathbf{x}_1, \mathbf{Y}_1)$, $(\mathbf{x}_2, \mathbf{Y}_2) \ldots$ where $(\mathbf{cx}_{j-1}, \mathbf{cY}_{j-1}) - (\mathbf{cx}_j, \mathbf{cY}_j) \succ \mathbf{0}$ for all j. Interpret this fact emphasizing the relationship between the bases generated by the simplex method and the foregoing problem.

4.50 Consider the following problem.

$$\begin{aligned} \text{Minimize} \quad & \mathbf{cx} \\ \text{Subject to} \quad & \mathbf{Ax} = \mathbf{b} \\ & \mathbf{x} \geqslant \mathbf{0} \end{aligned}$$

Assume that the first m columns of $\mathbf{A}$ form the identity and assume that

$\mathbf{b} \geqslant \mathbf{0}$. Given a basis $\mathbf{B}$, the corresponding basic feasible solution is nondegenerate if $\mathbf{B}^{-1}\mathbf{b} > \mathbf{0}$. Consider the following perturbation procedure of Charnes. Replace $\mathbf{b}$ by $\mathbf{b} + \Sigma_{j=1}^{m} \mathbf{a}_j \epsilon^j$ where ϵ is a very small positive number. Now suppose that we have a basis $\mathbf{B}$, where $\mathbf{B}^{-1}(\mathbf{b} + \Sigma_{j=1}^{m} \mathbf{a}_j \epsilon^j) = \bar{\mathbf{b}} + \Sigma_{j=1}^{m} \mathbf{y}_j \epsilon^j > \mathbf{0}$. Suppose that x_k is chosen to enter the basis and the following minimum ratio test is made:

$$\underset{1 \leqslant i \leqslant m}{\text{Minimum}} \left\{ \frac{\bar{b}_i + \sum_{j=1}^{m} y_{ij}\epsilon^j}{y_{ik}} : y_{ik} > 0 \right\}$$

a. Show that the minimum ratio occurs at a unique index r. Show that the method of finding this index is precisely the rule of Section 4.5.
b. Show that the new right-hand side after pivoting is positive and the objective function strictly improves even in the presence of degeneracy.
c. Show that cycling will not occur if the rule in (a) is adopted. Interpret this in terms of the perturbed problem.
d. Show that all the computations can be carried without explicitly replacing the right-hand side with $\mathbf{b} + \Sigma_{j=1}^{m} \mathbf{a}_j \epsilon^j$ and without explicitly assigning a value to ϵ.

NOTES AND REFERENCES

1. The use of artificial variables to obtain a starting basic feasible solution was first published by Dantzig [87] in 1951.
2. The single artificial variable technique of Section 4.4 can be viewed as the dual of a similar technique that adds a new row to obtain a starting basic dual feasible solution. The latter is discussed in Section 6.6.
3. The cycling example of Section 4.5 is due to Beale [27]. The proof of the cycling prevention rule via lexicographic ordering was published by Dantzig, Orden, and Wolfe [112] in 1954. The cycling prevention rule can also be interpreted as a perturbation technique, as briefly described in Exercise 4.50. This technique was independently devised by Charnes [61] and published in 1952.

FIVE: SPECIAL SIMPLEX FORMS AND OPTIMALITY CONDITIONS

In this chapter we describe some special methods for using the simplex procedure, or slight modifications of it. The formats considered here will prove advantageous in later chapters. The revised simplex method, which proceeds through the same steps as the simplex method but keeps all pertinent information in a smaller array, is described in Section 5.1. In Section 5.2 we describe a slight modification of the simplex method for dealing with lower and upper bounds of the variables without introducing slack variables. The remainder of the chapter is devoted to some geometric aspects of the simplex method. In particular the Kuhn-Tucker optimality conditions are discussed.

5.1 THE REVISED SIMPLEX METHOD

The *revised simplex* method is a systematic procedure for implementing the steps of the simplex method in a smaller array, thus saving storage space. Let us begin by reviewing the steps of the simplex method.

Steps of the Simplex Method (Minimization Problem)

Suppose that we are given a basic feasible solution with basis $\mathbf{B}$ (and basis inverse $\mathbf{B}^{-1}$). Then:

1. The basic feasible solution is given by $\mathbf{x}_B = \mathbf{B}^{-1}\mathbf{b} = \bar{\mathbf{b}}$ and $\mathbf{x}_N = \mathbf{0}$. The objective $z = \mathbf{c}_B\mathbf{B}^{-1}\mathbf{b} = \mathbf{c}_B\bar{\mathbf{b}}$.
2. Calculate $\mathbf{w} = \mathbf{c}_B\mathbf{B}^{-1}$. For each nonbasic variable, calculate $z_j - c_j = \mathbf{c}_B\mathbf{B}^{-1}\mathbf{a}_j - c_j = \mathbf{w}\mathbf{a}_j - c_j$. Let $z_k - c_k = \text{Maximum } z_j - c_j$. If $z_k - c_k \leqslant 0$, then stop; the current solution is optimal. Otherwise go to step 3.
3. Calculate $\mathbf{y}_k = \mathbf{B}^{-1}\mathbf{a}_k$. If $\mathbf{y}_k \leqslant \mathbf{0}$, then stop; the optimal solution is unbounded. Otherwise determine the index of the variable x_{B_r} leaving the basis as follows:

$$\frac{\bar{b}_r}{y_{rk}} = \underset{1 \leqslant i \leqslant m}{\text{Minimum}} \left\{ \frac{\bar{b}_i}{y_{ik}} : y_{ik} > 0 \right\}$$

Update $\mathbf{B}$ by replacing $\mathbf{a}_{B_r}$ with $\mathbf{a}_k$ and go to step 1.

Examining the preceding steps, it becomes clear that the simplex method can be executed using a smaller array. Suppose that we have a basic feasible solution with a known $\mathbf{B}^{-1}$. The following array is constructed where $\mathbf{w} = \mathbf{c}_B\mathbf{B}^{-1}$ and $\bar{\mathbf{b}} = \mathbf{B}^{-1}\mathbf{b}$.

BASIS INVERSE	RHS
$\mathbf{w}$	$\mathbf{c}_B\bar{\mathbf{b}}$
$\mathbf{B}^{-1}$	$\bar{\mathbf{b}}$

Note that the right-hand side denotes the values of the objective function and the basic variables. Since $\mathbf{w}$ is known, step 2 above can be performed (outside the tableau) in order to determine whether to stop or to introduce a new variable into the basis. Suppose that $z_k - c_k > 0$. Using $\mathbf{B}^{-1}$ we may compute $\mathbf{y}_k = \mathbf{B}^{-1}\mathbf{a}_k$. If $\mathbf{y}_k \leqslant \mathbf{0}$, we stop with an unbounded optimal solution. Otherwise the column $\left[\frac{z_k - c_k}{\mathbf{y}_k}\right]$ is inserted to the right of the above array, leading to the following tableau.

BASIS INVERSE	RHS		x_k
$\mathbf{w}$	$\mathbf{c}_B\bar{\mathbf{b}}$		$z_k - c_k$
$\mathbf{B}^{-1}$	$\bar{b}_1$ $\bar{b}_2$ $\vdots$ $\bar{b}_r$ $\vdots$ $\bar{b}_m$		y_{1k} y_{2k} $\vdots$ y_{rk} $\vdots$ y_{mk}

The index r of step 3 can now be calculated by the usual minimum ratio test. More importantly, pivoting at y_{rk} gives the new values of $\mathbf{w}$, $\mathbf{B}^{-1}$, $\bar{\mathbf{b}}$, and $\mathbf{c}_B\bar{\mathbf{b}}$, and the process is repeated. We leave it as an exercise to the reader to verify that pivoting indeed updates the $(m + 1) \times (m + 1)$ array.

The revised simplex method converges in a finite number of steps provided that a noncycling rule for determining the exit variable in case of a tie is adopted. This is obvious since the revised simplex method carries exactly the same steps of the simplex method, with the exception that only a part of the tableau is presented and other information is generated only as required. The following is a summary of the revised simplex method.

Summary of the Revised Simplex Method in Tableau Format
(Minimization Problem)

INITIALIZATION STEP

Find an initial basic feasible solution with basis inverse $\mathbf{B}^{-1}$. Calculate $\mathbf{w} = \mathbf{c}_B\mathbf{B}^{-1}$, $\bar{\mathbf{b}} = \mathbf{B}^{-1}\mathbf{b}$, and form the following array.

BASIS INVERSE	RHS
$\mathbf{w}$	$\mathbf{c}_B\bar{\mathbf{b}}$
$\mathbf{B}^{-1}$	$\bar{\mathbf{b}}$

MAIN STEP

For each nonbasic variable, calculate $z_j - c_j = \mathbf{w}\mathbf{a}_j - c_j$. Let $z_k - c_k = \text{Maximum } z_j - c_j$. If $z_k - c_k \leqslant 0$, stop; the current basic feasible solution is optimal. Otherwise calculate $\mathbf{y}_k = \mathbf{B}^{-1}\mathbf{a}_k$. If $\mathbf{y}_k \leqslant \mathbf{0}$, stop; the optimal solution is unbounded. If $\mathbf{y}_k \nleq \mathbf{0}$, insert the column $\begin{bmatrix} z_k - c_k \\ \mathbf{y}_k \end{bmatrix}$ to the right of the tableau leading to the following tableau.

BASIS INVERSE	RHS		x_k
$\mathbf{w}$	$\mathbf{c}_B\bar{\mathbf{b}}$		$z_k - c_k$
$\mathbf{B}^{-1}$	$\bar{\mathbf{b}}$		$\mathbf{y}_k$

Determine the index r as follows:

$$\frac{\bar{b}_r}{y_{rk}} = \underset{1 \leq i \leq m}{\text{Minimum}} \left\{ \frac{\bar{b}_i}{y_{ik}} : y_{ik} > 0 \right\}$$

Pivot at y_{rk}. This updates the tableau. Now the column corresponding to x_k is completely eliminated from the tableau and the main step is repeated.

Example 5.1

$$\text{Minimize } -x_1 - 2x_2 + x_3 - x_4 - 4x_5 + 2x_6$$

$$\begin{aligned} \text{Subject to} \quad x_1 + x_2 + x_3 + x_4 + x_5 + x_6 &\leq 6 \\ 2x_1 - x_2 - 2x_3 + x_4 \qquad\qquad &\leq 4 \\ x_3 + x_4 + 2x_5 + x_6 &\leq 4 \\ x_1, \quad x_2, \quad x_3, \; x_4, \quad x_5, \quad x_6 &\geq 0 \end{aligned}$$

Introduce the slack variables x_7, x_8, x_9. The initial basis is $\mathbf{B} = [\mathbf{a}_7, \mathbf{a}_8, \mathbf{a}_9] = \mathbf{I}_3$. Also, $\mathbf{w} = \mathbf{c}_B\mathbf{B}^{-1} = (0, 0, 0)$ and $\bar{\mathbf{b}} = \mathbf{b}$.

Iteration 1

	BASIS INVERSE			RHS
z	0	0	0	0
x_7	1	0	0	6
x_8	0	1	0	4
x_9	0	0	1	4

Here $\mathbf{w} = (0, 0, 0)$. Noting that $z_j - c_j = \mathbf{w}\mathbf{a}_j - c_j$, we get

$z_1 - c_1 = 1, z_2 - c_2 = 2, z_3 - c_3 = -1,$

$z_4 - c_4 = 1, z_5 - c_5 = 4, z_6 - c_6 = -2$

Therefore $k = 5$ and x_5 enters the basis:

$$\mathbf{y}_5 = \mathbf{B}^{-1}\mathbf{a}_5 = \begin{bmatrix} 1 & 0 & 0 \\ 0 & 1 & 0 \\ 0 & 0 & 1 \end{bmatrix} \begin{bmatrix} 1 \\ 0 \\ 2 \end{bmatrix} = \begin{bmatrix} 1 \\ 0 \\ 2 \end{bmatrix}$$

Insert the vector

$$\left[\frac{z_5 - c_5}{\mathbf{y}_5} \right] = \begin{bmatrix} 4 \\ 1 \\ 0 \\ 2 \end{bmatrix}$$

to the right of the above tableau and pivot at $y_{35} = 2$.

	BASIS INVERSE			RHS	x_5
z	0	0	0	0	4
x_7	1	0	0	6	1
x_8	0	1	0	4	0
x_9	0	0	1	4	(2)

	BASIS INVERSE			RHS
z	0	0	-2	-8
x_7	1	0	$-\frac{1}{2}$	4
x_8	0	1	0	4
x_5	0	0	$\frac{1}{2}$	2

Iteration 2

Here $\mathbf{w} = (0, 0, -2)$. Noting that $z_j - c_j = \mathbf{w}\mathbf{a}_j - c_j$, we get

$z_1 - c_1 = 1, z_2 - c_2 = 2, z_3 - c_3 = -3,$

$z_4 - c_4 = -1, z_6 - c_6 = -4,$

$z_9 - c_9 = -2.$

Therefore $k = 2$ and x_2 enters the basis:

$$\mathbf{y}_2 = \mathbf{B}^{-1}\mathbf{a}_2 = \begin{bmatrix} 1 & 0 & -\frac{1}{2} \\ 0 & 1 & 0 \\ 0 & 0 & \frac{1}{2} \end{bmatrix} \begin{bmatrix} 1 \\ -1 \\ 0 \end{bmatrix} = \begin{bmatrix} 1 \\ -1 \\ 0 \end{bmatrix}$$

Insert the vector

$$\begin{bmatrix} z_2 - c_2 \\ \hline \mathbf{y}_2 \end{bmatrix} = \begin{bmatrix} 2 \\ 1 \\ -1 \\ 0 \end{bmatrix}$$

to the right of the above tableau and pivot at y_{12}.

	BASIS INVERSE			RHS	x_2
z	0	0	-2	-8	2
x_7	1	0	$-\frac{1}{2}$	4	①
x_8	0	1	0	4	-1
x_5	0	0	$\frac{1}{2}$	2	0

	BASIS INVERSE			RHS
z	-2	0	-1	-16
x_2	1	0	$-\frac{1}{2}$	4
x_8	1	1	$-\frac{1}{2}$	8
x_5	0	0	$\frac{1}{2}$	2

Iteration 3

Here $\mathbf{w} = (-2, 0, -1)$. Noting that $z_j - c_j = \mathbf{w}\mathbf{a}_j - c_j$, we get

$z_1 - c_1 = -1, z_3 - c_3 = -4, z_4 - c_4 = -2,$

$z_6 - c_6 = -5, z_9 - c_9 = -1.$

Since $z_j - c_j \leq 0$ for all nonbasic variables (x_7 just left the basis and so $z_7 - c_7 < 0$), we stop; the basic feasible solution of the foregoing tableau is optimal.

Comparison Between the Simplex and the Revised Simplex Methods

It may be helpful to give a brief comparison between the simplex and the revised simplex methods. For the revised method we need an $(m + 1) \times (m + 2)$ array as opposed to an $(m + 1) \times (n + 1)$ array for the simplex method. If n is significantly larger than m, this would result in a substantial saving in computer core storage. The number of multiplications (division is considered a multiplication) and additions (subtraction is considered an addition) per iteration of both procedures are given in Table 5.1 below. In Exercise 5.3 we ask the reader to verify the validity of the entries of the table.

Table 5.1 Comparison of the Simplex and the Revised Simplex Methods

METHOD		OPERATION		
		PIVOTING	$z_j - c_j$'s	TOTAL
Simplex	Multiplications	$(m + 1)(n - m + 1)$		$m(n - m) + n + 1$
	Additions	$m(n - m + 1)$		$m(n - m + 1)$
Revised Simplex	Multiplications	$(m + 1)^2$	$m(n - m)$	$m(n - m) + (m + 1)^2$
	Additions	$m(m + 1)$	$m(n - m)$	$m(n + 1)$

From Table 5.1 we see that the number of operations required during an iteration of the simplex method is slightly less than those required for the revised simplex method. Note, however, that for most practical problems the density d (number of nonzero elements divided by total number of elements) of nonzero elements in the constraint matrix is usually small (in many cases $d \leqslant 0.05$). The revised simplex method can take advantage of this situation while calculating $z_j - c_j$. Note that $z_j = \mathbf{w}\mathbf{a}_j$ and we can skip zero elements of $\mathbf{a}_j$ while performing the calculation $\mathbf{w}\mathbf{a}_j = \Sigma_{i=1}^{m} w_i a_{ij}$. Therefore the number of operations in the revised simplex method for calculating the $z_j - c_j$'s is given by d times the entries of Table 5.1, substantially reducing the total number of operations. While pivoting, for both the simplex and the revised simplex methods, no operations are skipped because the current tableaux usually fill quickly with nonzero entries, even if the original constraint matrix was sparse.

To summarize, if n is significantly larger than m, and if the density d is small, the computational effort of the revised simplex method is significantly smaller than that of the simplex method. Also, in the revised simplex method, the use of the original data for calculating the $z_j - c_j$'s and the updated column $\mathbf{y}_k$ tends to reduce the cumulative round-off error.

Product Form of the Inverse

We now discuss another implementation of the revised simplex method where the inverse of the basis is stored as the product of elementary matrices (an *elementary matrix* is a square matrix that differs from the identity in only one row or one column).

Consider a basis $\mathbf{B}$ composed of the columns $\mathbf{a}_{B_1}, \mathbf{a}_{B_2}, \ldots, \mathbf{a}_{B_m}$ and suppose that $\mathbf{B}^{-1}$ is known. Now suppose that the nonbasic column $\mathbf{a}_k$ replaces $\mathbf{a}_{B_r}$, resulting in the new basis $\hat{\mathbf{B}}$. We wish to find $\hat{\mathbf{B}}^{-1}$ in terms of $\mathbf{B}^{-1}$. Noting that $\mathbf{a}_k = \mathbf{B}\mathbf{y}_k$ and $\mathbf{a}_{B_i} = \mathbf{B}\mathbf{e}_i$ where $\mathbf{e}_i$ is a vector of zeros except for 1 at the ith position, we have

$$\hat{\mathbf{B}} = (\mathbf{a}_{B_1}, \mathbf{a}_{B_2}, \ldots, \mathbf{a}_{B_{r-1}}, \mathbf{a}_k, \mathbf{a}_{B_{r+1}}, \ldots, \mathbf{a}_{B_m})$$

$$= (\mathbf{B}\mathbf{e}_1, \mathbf{B}\mathbf{e}_2, \ldots, \mathbf{B}\mathbf{e}_{r-1}, \mathbf{B}\mathbf{y}_k, \mathbf{B}\mathbf{e}_{r+1}, \ldots, \mathbf{B}\mathbf{e}_m)$$

$$= \mathbf{B}\mathbf{T}$$

where $\mathbf{T}$ is the identity with the rth column replaced by $\mathbf{y}_k$. The inverse of $\mathbf{T}$, which we shall denote by $\mathbf{E}$, is given below.

$$
\begin{array}{c}
r\text{th column} \\
\downarrow \\
\mathbf{E} =
\begin{bmatrix}
1 & 0 & \cdots & 0 & -y_{1k}/y_{rk} & 0 & \cdots & 0 \\
0 & 1 & \cdots & 0 & -y_{2k}/y_{rk} & 0 & \cdots & 0 \\
\vdots & \vdots & & \vdots & \vdots & \vdots & & \vdots \\
0 & 0 & \cdots & 0 & 1/y_{rk} & 0 & \cdots & 0 \\
\vdots & \vdots & & \vdots & \vdots & \vdots & & \vdots \\
0 & 0 & \cdots & 0 & -y_{mk}/y_{rk} & 0 & \cdots & 1
\end{bmatrix}
\begin{array}{l} \\ \\ \\ \leftarrow r\text{th row} \\ \\ \\ \end{array}
\end{array}
$$

Therefore $\hat{\mathbf{B}}^{-1} = \mathbf{T}^{-1}\mathbf{B}^{-1} = \mathbf{E}\mathbf{B}^{-1}$ where the elementary matrix $\mathbf{E}$ is specified above. To summarize, the basis inverse at a new iteration can be obtained by premultiplying the basis inverse at the previous iteration by an elementary matrix $\mathbf{E}$. Needless to say, only the nonidentity column $\mathbf{g}$ and its position r need be stored to specify $\mathbf{E}$.

Let the basis $\mathbf{B}_1$ at the first iteration be the identity $\mathbf{I}$. Then the basis inverse $\mathbf{B}_2^{-1}$ at iteration 2 is $\mathbf{B}_2^{-1} = \mathbf{E}_1\mathbf{B}_1^{-1} = \mathbf{E}_1\mathbf{I} = \mathbf{E}_1$ where $\mathbf{E}_1$ is the elementary matrix corresponding to the first iteration. Similarly $\mathbf{B}_3^{-1} = \mathbf{E}_2\mathbf{B}_2^{-1} = \mathbf{E}_2\mathbf{E}_1$, and in general

$$\mathbf{B}_t^{-1} = \mathbf{E}_{t-1}\mathbf{E}_{t-2} \cdots \mathbf{E}_2\mathbf{E}_1 \tag{5.1}$$

Equation (5.1), which specifies the basis inverse as the product of elementary matrices, is called the *product form of the inverse*. Using this form, all the steps of the simplex method can be performed without pivoting. First, it will be helpful to elaborate on multiplying a vector by an elementary matrix.

POST MULTIPLYING

Let E be an elementary matrix with nonidentity column g appearing at the rth position. Let c be a row vector. Then

$$
\mathbf{c}\mathbf{E} = (c_1, c_2, \ldots, c_m)
\begin{bmatrix}
1 & 0 & \cdots & g_1 & \cdots & 0 \\
0 & 1 & \cdots & g_2 & \cdots & 0 \\
\vdots & \vdots & & \vdots & & \\
& & & & & \\
0 & 0 & \cdots & g_m & \cdots & 1
\end{bmatrix}
\overset{\text{position } r}{\underset{\downarrow}{}}
$$

$$
= \left(c_1, c_2, \ldots, c_{r-1}, \sum_{i=1}^{m} c_i g_i, c_{r+1}, \ldots, c_m \right)
$$

$$
= (c_1, c_2, \ldots, c_{r-1}, \mathbf{c}\mathbf{g}, c_{r+1}, \ldots, c_m) \tag{5.2}
$$

In other words, $\mathbf{c}\mathbf{E}$ is equal to c except that the rth component is replaced by $\mathbf{c}\mathbf{g}$.

PREMULTIPLYING

Let a be an m vector. Then

$$
\mathbf{E}\mathbf{a} =
\begin{bmatrix}
1 & \cdots & g_1 & \cdots & 0 \\
\vdots & & \vdots & & \vdots \\
0 & \cdots & g_r & \cdots & 0 \\
\vdots & & \vdots & & \vdots \\
0 & \cdots & g_m & \cdots & 1
\end{bmatrix}
\begin{bmatrix}
a_1 \\
\vdots \\
a_r \\
\vdots \\
a_m
\end{bmatrix}
$$

$$
=
\begin{bmatrix}
a_1 + g_1 a_r \\
\vdots \\
g_r a_r \\
\vdots \\
a_m + g_m a_r
\end{bmatrix}
=
\begin{bmatrix}
a_1 \\
\vdots \\
0 \\
\vdots \\
a_m
\end{bmatrix}
+ a_r
\begin{bmatrix}
g_1 \\
\vdots \\
g_r \\
\vdots \\
g_m
\end{bmatrix}
$$

In other words,

$$\mathbf{Ea} = \hat{\mathbf{a}} + a_r\mathbf{g} \tag{5.3}$$

where $\hat{\mathbf{a}}$ is equal to $\mathbf{a}$ except that the rth component a_r is replaced by zero.

With the foregoing formulas for post- and premultiplying a vector by an elementary matrix, the revised simplex method can be executed without pivoting. The following discussion elaborates on the simplex calculations.

COMPUTING THE VECTOR $\mathbf{w} = \mathbf{c}_B\mathbf{B}^{-1}$

At iteration t we wish to calculate the vector $\mathbf{w}$. Note that

$$\mathbf{w} = \mathbf{c}_B\mathbf{B}_t^{-1} = \mathbf{c}_B\mathbf{E}_{t-1}\mathbf{E}_{t-2} \cdots \mathbf{E}_2\mathbf{E}_1$$

Computing $\mathbf{w}$ can be iteratively performed as follows. First compute $\mathbf{c}_B\mathbf{E}_{t-1}$ according to Equation (5.2). Then apply (5.2) to calculate $(\mathbf{c}_B\mathbf{E}_{t-1})\mathbf{E}_{t-2}$, and so forth. After $\mathbf{w}$ is computed, we can calculate $z_j - c_j = \mathbf{wa}_j - c_j$ for nonbasic variables. From this we either stop or else decide to introduce a nonbasic variable x_k.

COMPUTING THE UPDATED COLUMN $\mathbf{y}_k$ AND THE RIGHT-HAND-SIDE $\bar{\mathbf{b}}$

If x_k is to enter the basis at iteration t, then $\mathbf{y}_k$ is calculated as follows:

$$\mathbf{y}_k = \mathbf{B}_t^{-1}\mathbf{a}_k = \mathbf{E}_{t-1}\mathbf{E}_{t-2} \cdots \mathbf{E}_2\mathbf{E}_1\mathbf{a}_k$$

This computation can be executed by successively applying Equation (5.3). If $\mathbf{y}_k \leqslant \mathbf{0}$, we stop with the conclusion that the optimal solution is unbounded. Otherwise the usual minimum ratio test determines the index r of the variable x_{B_r} leaving the basis. Thus x_k enters and x_{B_r} leaves the basis. A new elementary matrix $\mathbf{E}_t$ is generated where the nonidentity column $\mathbf{g}$ is given by

$$\begin{bmatrix} \dfrac{-y_{1k}}{y_{rk}} \\ \vdots \\ \dfrac{1}{y_{rk}} \\ \vdots \\ \dfrac{-y_{mk}}{y_{rk}} \end{bmatrix}$$

and appears at position r. The new right-hand side is given by

$$\mathbf{B}_{t+1}^{-1}\mathbf{b} = \mathbf{E}_t\mathbf{B}_t^{-1}\mathbf{b}$$

Since $\mathbf{B}_t^{-1}\mathbf{b}$ is known from the last iteration, then a single application of Equation (5.3) updates the right-hand-side vector $\bar{\mathbf{b}}$.

UPDATING THE BASIS INVERSE

The basis inverse is updated by generating $\mathbf{E}_t$ as discussed above. It is worthwhile noting that the number of elementary matrices required to represent the basis inverse increases by 1 at each iteration. If this number becomes large, it would be necessary to reinvert the basis and represent it as the product of m elementary matrices (see Exercise 5.7). It is emphasized that each elementary matrix is completely described by its nonidentity column and its position. Therefore an elementary matrix $\mathbf{E}$ could be stored as $\begin{bmatrix} \mathbf{g} \\ r \end{bmatrix}$ where $\mathbf{g}$ is the nonidentity column and r is its position.

Example 5.2

$$\begin{aligned}
\text{Minimize} \quad & -x_1 - 2x_2 + x_3 \\
\text{Subject to} \quad & x_1 + x_2 + x_3 \leq 4 \\
& -x_1 + 2x_2 - 2x_3 \leq 6 \\
& 2x_1 + x_2 \leq 5 \\
& x_1, \quad x_2, \quad x_3 \geq 0
\end{aligned}$$

Introduce the slack variables x_4, x_5, and x_6. The original basis consists of x_4, x_5, and x_6.

Iteration 1

$$\bar{\mathbf{b}} = \begin{bmatrix} 4 \\ 6 \\ 5 \end{bmatrix}$$

$$\mathbf{x}_B = \begin{bmatrix} x_{B_1} \\ x_{B_2} \\ x_{B_3} \end{bmatrix} = \begin{bmatrix} x_4 \\ x_5 \\ x_6 \end{bmatrix} = \begin{bmatrix} 4 \\ 6 \\ 5 \end{bmatrix} \qquad \mathbf{x}_N = \begin{bmatrix} x_1 \\ x_2 \\ x_3 \end{bmatrix} = \begin{bmatrix} 0 \\ 0 \\ 0 \end{bmatrix}$$

$$z = 0$$

$$\mathbf{w} = \mathbf{c}_B = (0, 0, 0)$$

Note that $z_j - c_j = \mathbf{wa}_j - c_j$. Therefore

$$z_1 - c_1 = 1,\ z_2 - c_2 = 2,\ z_3 - c_3 = -1$$

Thus, $k = 2$, and x_2 enters the basis.

$$\mathbf{y}_2 = \mathbf{a}_2 = \begin{bmatrix} 1 \\ 2 \\ 1 \end{bmatrix}$$

Here x_{B_r} leaves the basis where r is determined by

$$\text{Minimum}\left\{ \frac{\bar{b}_1}{y_{12}}, \frac{\bar{b}_2}{y_{22}}, \frac{\bar{b}_3}{y_{32}} \right\} = \text{Minimum}\left\{ \frac{4}{1}, \frac{6}{2}, \frac{5}{1} \right\} = 3$$

Therefore $r = 2$; that is, $x_{B_2} = x_5$ leaves the basis and x_2 enters the basis. The nonidentity column of $\mathbf{E}_1$ is given by

$$\mathbf{g} = \begin{bmatrix} -\dfrac{y_{12}}{y_{22}} \\ \dfrac{1}{y_{22}} \\ -\dfrac{y_{32}}{y_{22}} \end{bmatrix} = \begin{bmatrix} -\dfrac{1}{2} \\ \dfrac{1}{2} \\ -\dfrac{1}{2} \end{bmatrix}.$$

and $\mathbf{E}_1$ is represented by $\begin{bmatrix} \mathbf{g} \\ 2 \end{bmatrix}$.

Iteration 2

Update $\bar{\mathbf{b}}$. Noting Equation (5.3), we have

$$\bar{\mathbf{b}} = \mathbf{E}_1 \begin{bmatrix} 4 \\ 6 \\ 5 \end{bmatrix} = \begin{bmatrix} 4 \\ 0 \\ 5 \end{bmatrix} + 6 \begin{bmatrix} -\frac{1}{2} \\ \frac{1}{2} \\ -\frac{1}{2} \end{bmatrix} = \begin{bmatrix} 1 \\ 3 \\ 2 \end{bmatrix}$$

$$\mathbf{x}_B = \begin{bmatrix} x_{B_1} \\ x_{B_2} \\ x_{B_3} \end{bmatrix} = \begin{bmatrix} x_4 \\ x_2 \\ x_6 \end{bmatrix} = \begin{bmatrix} 1 \\ 3 \\ 2 \end{bmatrix} \qquad \mathbf{x}_N = \begin{bmatrix} x_1 \\ x_5 \\ x_3 \end{bmatrix} = \begin{bmatrix} 0 \\ 0 \\ 0 \end{bmatrix}$$

$$z = 0 - \bar{b}_2(z_2 - c_2) = -6$$

$$\mathbf{w} = \mathbf{c}_B \mathbf{E}_1 = (0, -2, 0)\mathbf{E}_1.$$

Noting Equation (5.2), then $\mathbf{w} = (0, -1, 0)$. Note that $z_j - c_j = \mathbf{w}\mathbf{a}_j - c_j$. Therefore

$z_1 - c_1 = 2, z_3 - c_3 = 1$

Thus, $k = 1$ and x_1 enters the basis. Noting Equation (5.3),

$$\mathbf{y}_1 = \mathbf{E}_1\mathbf{a}_1 = \mathbf{E}_1\begin{bmatrix} 1 \\ -1 \\ 2 \end{bmatrix} = \begin{bmatrix} 1 \\ 0 \\ 2 \end{bmatrix} - \begin{bmatrix} -\frac{1}{2} \\ \frac{1}{2} \\ -\frac{1}{2} \end{bmatrix} = \begin{bmatrix} \frac{3}{2} \\ -\frac{1}{2} \\ \frac{5}{2} \end{bmatrix}$$

Then x_{B_r} leaves the basis where r is determined by

$$\text{Minimum}\left\{\frac{\bar{b}_1}{y_{11}}, \frac{\bar{b}_3}{y_{31}}\right\} = \text{Minimum}\left\{\frac{1}{\frac{3}{2}}, \frac{2}{\frac{5}{2}}\right\} = \frac{2}{3}$$

Therefore $r = 1$; that is, $x_{B_1} = x_4$ leaves and x_1 enters the basis. The nonidentity column of $\mathbf{E}_2$ is given by

$$\mathbf{g} = \begin{bmatrix} \dfrac{1}{y_{11}} \\ -\dfrac{y_{21}}{y_{11}} \\ -\dfrac{y_{31}}{y_{11}} \end{bmatrix} = \begin{bmatrix} \dfrac{2}{3} \\ \dfrac{1}{3} \\ -\dfrac{5}{3} \end{bmatrix}$$

Also, $\mathbf{E}_2$ is represented by $\begin{bmatrix} \mathbf{g} \\ 1 \end{bmatrix}$.

Iteration 3

Update $\bar{\mathbf{b}}$. Noting Equation (5.3), we have

$$\bar{\mathbf{b}} = \mathbf{E}_2\begin{bmatrix} 1 \\ 3 \\ 2 \end{bmatrix} = \begin{bmatrix} 0 \\ 3 \\ 2 \end{bmatrix} + 1\begin{bmatrix} \frac{2}{3} \\ \frac{1}{3} \\ -\frac{5}{3} \end{bmatrix} = \begin{bmatrix} \frac{2}{3} \\ \frac{10}{3} \\ \frac{1}{3} \end{bmatrix}$$

$$\mathbf{x}_B = \begin{bmatrix} x_{B_1} \\ x_{B_2} \\ x_{B_3} \end{bmatrix} = \begin{bmatrix} x_1 \\ x_2 \\ x_6 \end{bmatrix} = \begin{bmatrix} \frac{2}{3} \\ \frac{10}{3} \\ \frac{1}{3} \end{bmatrix} \qquad \mathbf{x}_N = \begin{bmatrix} x_4 \\ x_5 \\ x_3 \end{bmatrix} = \begin{bmatrix} 0 \\ 0 \\ 0 \end{bmatrix}$$

$$z = -6 - \bar{b}_1(z_1 - c_1) = -\tfrac{22}{3}$$

$$\mathbf{w} = \mathbf{c}_B\mathbf{E}_2\mathbf{E}_1 = (-1, -2, 0)\mathbf{E}_2\mathbf{E}_1.$$

Applying Equation (5.2) twice, we get

$$\mathbf{c}_B\mathbf{E}_2 = (-\tfrac{4}{3}, -2, 0)$$

$$\mathbf{w} = (\mathbf{c}_B\mathbf{E}_2)\mathbf{E}_1 = (-\tfrac{4}{3}, -\tfrac{1}{3}, 0)$$

Note that $z_j - c_j = \mathbf{w}\mathbf{a}_j - c_j$. Therefore

$$z_3 - c_3 = -\tfrac{5}{3}, z_5 - c_5 = -\tfrac{1}{3}$$

Since $z_j - c_j \leqslant 0$ for all nonbasic variables, then the optimal solution is at hand. The objective value is $-\frac{22}{3}$ and

$$(x_1, x_2, x_3, x_4, x_5, x_6) = (\tfrac{2}{3}, \tfrac{10}{3}, 0, 0, 0, \tfrac{1}{3})$$

5.2 THE SIMPLEX METHOD FOR BOUNDED VARIABLES

In most practical problems the variables are usually bounded. A typical variable x_j is bounded from below by l_j and from above by u_j. If we denote the lower and upper bound vectors by $\mathbf{l}$ and $\mathbf{u}$ respectively, we get the following linear program with bounded variables.

$$\begin{aligned} \text{Minimize} \quad & \mathbf{cx} \\ \text{Subject to} \quad & \mathbf{Ax} = \mathbf{b} \\ & \mathbf{l} \leqslant \mathbf{x} \leqslant \mathbf{u} \end{aligned}$$

If $\mathbf{l} = \mathbf{0}$, the usual nonnegativity restrictions are obtained. In fact, any lower bound vector can be transformed into the zero vector by using the change of variables $\mathbf{x}' = \mathbf{x} - \mathbf{l}$. The most straightforward (and the least efficient) method of handling the constraints $\mathbf{l} \leqslant \mathbf{x} \leqslant \mathbf{u}$ is to introduce the slack vectors $\mathbf{x}_1$ and $\mathbf{x}_2$, leading to the constraints $\mathbf{x} + \mathbf{x}_1 = \mathbf{u}$ and $\mathbf{x} - \mathbf{x}_2 = \mathbf{l}$. This increases the number of equality constraints from m to $m + 2n$ and the number of variables from n to $3n$. Even if $\mathbf{l} = \mathbf{0}$ or is transformed into $\mathbf{0}$ as discussed above, the slack vector $\mathbf{x}_1$ is needed, which increases both the constraints and variables by n.

From the foregoing discussion it is clear that the problem size (and hence the computational effort) would increase significantly if the constraints $\mathbf{l} \leqslant \mathbf{x} \leqslant \mathbf{u}$ are treated in the usual manner by introducing slack vectors. The simplex method with bounded variables handles these constraints implicitly in a fashion similar to that used by the simplex method to handle the constraints $\mathbf{x} \geqslant \mathbf{0}$. The algorithm of this section moves from a basic feasible solution to an improved basic feasible solution of the system $\mathbf{Ax} = \mathbf{b}$, $\mathbf{l} \leqslant \mathbf{x} \leqslant \mathbf{u}$, until optimality is reached or unboundness is verified.

Definition (Basic Feasible Solutions)

Consider the system $\mathbf{Ax} = \mathbf{b}$ and $\mathbf{l} \leqslant \mathbf{x} \leqslant \mathbf{u}$, where $\mathbf{A}$ is an $m \times n$ matrix of rank m. The vector $\mathbf{x}$ is a *basic feasible solution* of the system if $\mathbf{A}$ can be decomposed into $[\mathbf{B}, \mathbf{N}_1, \mathbf{N}_2]$, where the matrix $\mathbf{B}$ has rank m, and $\mathbf{l}_B \leqslant \mathbf{B}^{-1}\mathbf{b} - \mathbf{B}^{-1}\mathbf{N}_1\mathbf{l}_{N_1} - \mathbf{B}^{-1}\mathbf{N}_2\mathbf{u}_{N_2} = \mathbf{x}_B \leqslant \mathbf{u}_B$, $\mathbf{x}_{N_1} = \mathbf{l}_{N_1}$, and $\mathbf{x}_{N_2} = \mathbf{u}_{N_2}$. The matrix $\mathbf{B}$ is called the *basis*, $\mathbf{x}_B$ are the *basic variables*, and $\mathbf{x}_{N_1}$ and $\mathbf{x}_{N_2}$ are the nonbasic variables at their lower and upper limits respectively. If, in addition, $\mathbf{l}_B < \mathbf{x}_B < \mathbf{u}_B$, then $\mathbf{x}$ is called a *nondegenerate basic feasible solution*; otherwise it is called a *degenerate basic feasible solution*.

Note that a basic feasible solution is obtained by assigning $n - m$ of the variables at their lower and/or upper bounds, and then solving uniquely for $\mathbf{x}_B$, such that $\mathbf{x}_B$ lies between its lower and upper limits. Therefore a nonbasic variable x_j is equal either to its lower bound or to its upper bound.

Example 5.3

Consider the region given by

$$\begin{aligned} x_1 + x_2 &\leqslant 5 \\ -x_1 + 2x_2 &\leqslant 4 \\ 0 \leqslant x_1 &\leqslant 4 \\ -1 \leqslant x_2 &\leqslant 4 \end{aligned}$$

First introduce the slack variables x_3 and x_4. This gives the following system (note that close examination of the system shows that u_3 and u_4 can be replaced by 6 and 10 respectively):

$$\begin{aligned} x_1 + x_2 + x_3 \qquad &= 5 \\ -x_1 + 2x_2 \qquad + x_4 &= 4 \\ 0 \leqslant x_1 &\leqslant 4 \\ -1 \leqslant x_2 &\leqslant 4 \\ 0 \leqslant x_3 &\leqslant \infty \\ 0 \leqslant x_4 &\leqslant \infty \end{aligned}$$

We would like to find all the basic feasible solutions of this system. This can be accomplished by extracting a basis of the first two constraints, solving the basic variables in terms of the nonbasic variables, and then assigning the nonbasic variables at their lower or upper bounds. To illustrate the method select, say, $\mathbf{a}_2$

and $\mathbf{a}_4$ as the basic vectors. Then

$$\mathbf{B} = [\mathbf{a}_2, \mathbf{a}_4] = \begin{bmatrix} 1 & 0 \\ 2 & 1 \end{bmatrix} \qquad \mathbf{B}^{-1} = \begin{bmatrix} 1 & 0 \\ -2 & 1 \end{bmatrix}$$

Multiplying the first two constraints by $\mathbf{B}^{-1}$ and transferring x_1 and x_3 to the right-hand side, we get

$$x_2 = \quad 5 - x_1 - x_3$$

$$x_4 = -6 + 3x_1 + 2x_3$$

Now assign x_1 at its lower and upper bounds and x_3 at its lower bound and solve for x_2 and x_4.

1. $x_1 = 0$, $x_3 = 0 \Rightarrow x_2 = 5$ and $x_4 = -6$. Since $x_4 < 0$, this is not a basic feasible solution.
2. $x_1 = 4$, $x_3 = 0 \Rightarrow x_2 = 1$ and $x_4 = 6$. Therefore $(x_1, x_2, x_3, x_4) = (4, 1, 0, 6)$ is a basic feasible solution.

The other basic solutions can be obtained in a similar manner. If all of the possible bases were enumerated, we would see that the basic feasible solutions are (2, 3, 0, 0), (0, 2, 3, 0), (4, 1, 0, 6), (0, − 1, 6, 6), and (4, − 1, 2, 10). Projecting these points in the (x_1, x_2) space, we get the extreme points shown in Figure 5.1. In other words, in this example the basic feasible solutions and the extreme points of the system $\mathbf{Ax} = \mathbf{b}$, $\mathbf{l} \leqslant \mathbf{x} \leqslant \mathbf{u}$ are equivalent. This result is true in general, and the proof is very similar to the case $\mathbf{Ax} = \mathbf{b}$ and $\mathbf{x} \geqslant \mathbf{0}$, and is hence left as an exercise for the reader (see Exercise 5.11).

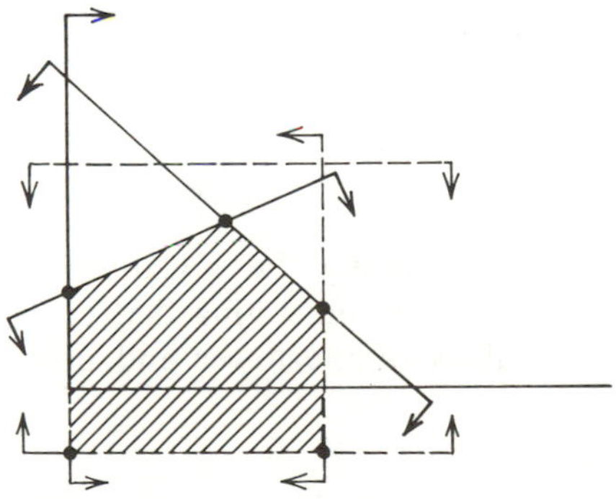

Figure 5.1. **Basic feasible solutions.**

Improving a Basic Feasible Solution

We now know how to characterize a basic feasible solution, and we also know that an optimal basic feasible solution exists provided that the feasible region is not empty and the optimal is finite (why?). Note, however, that the number of basic feasible solutions is large (The number of basic feasible solutions is bounded above by $\binom{n}{m} 2^{n-m}$. For each possible way of extracting a basis there are 2^{n-m} ways of fixing the nonbasic variables at their lower and/or upper bounds). Therefore a systematic way of moving among the basic feasible solutions is needed.

Now suppose that we have a basis $\mathbf{B}$ and suppose that the nonbasic matrix is decomposed into $\mathbf{N}_1$ and $\mathbf{N}_2$, that is, $\mathbf{A} = [\mathbf{B}, \mathbf{N}_1, \mathbf{N}_2]$. Accordingly, the vector $\mathbf{x}$ is decomposed into $[\mathbf{x}_B, \mathbf{x}_{N_1}, \mathbf{x}_{N_2}]$ and $\mathbf{c}$ is decomposed into $[\mathbf{c}_B, \mathbf{c}_{N_1}, \mathbf{c}_{N_2}]$. Both the basic variables and the objective function can be represented in terms of the independent (that is, nonbasic) vectors $\mathbf{x}_{N_1}$ and $\mathbf{x}_{N_2}$ as follows:

$$\mathbf{x}_B = \mathbf{B}^{-1}\mathbf{b} - \mathbf{B}^{-1}\mathbf{N}_1\mathbf{x}_{N_1} - \mathbf{B}^{-1}\mathbf{N}_2\mathbf{x}_{N_2} \tag{5.4}$$

$$\begin{aligned} z &= \mathbf{c}_B\mathbf{x}_B + \mathbf{c}_{N_1}\mathbf{x}_{N_1} + \mathbf{c}_{N_2}\mathbf{x}_{N_2} \\ &= \mathbf{c}_B\left(\mathbf{B}^{-1}\mathbf{b} - \mathbf{B}^{-1}\mathbf{N}_1\mathbf{x}_{N_1} - \mathbf{B}^{-1}\mathbf{N}_2\mathbf{x}_{N_2}\right) + \mathbf{c}_{N_1}\mathbf{x}_{N_1} + \mathbf{c}_{N_2}\mathbf{x}_{N_2} \\ &= \mathbf{c}_B\mathbf{B}^{-1}\mathbf{b} + \left(\mathbf{c}_{N_1} - \mathbf{c}_B\mathbf{B}^{-1}\mathbf{N}_1\right)\mathbf{x}_{N_1} + \left(\mathbf{c}_{N_2} - \mathbf{c}_B\mathbf{B}^{-1}\mathbf{N}_2\right)\mathbf{x}_{N_2} \end{aligned} \tag{5.5}$$

Suppose that we have a current basic feasible solution where $\mathbf{x}_{N_1} = \mathbf{l}_{N_1}$, $\mathbf{x}_{N_2} = \mathbf{u}_{N_2}$, and $\mathbf{l}_B \leqslant \mathbf{x}_B \leqslant \mathbf{u}_B$. This solution is represented by the following tableau. The right-hand-side column gives the true values of z and $\mathbf{x}_B$ (denoted by $\hat{z}$ and $\hat{\mathbf{b}}$) where $\mathbf{x}_{N_1} = \mathbf{l}_{N_1}$ and $\mathbf{x}_{N_2} = \mathbf{u}_{N_2}$ are substituted in Equations (5.4) and (5.5). We emphasize that this column does not give $\mathbf{c}_B\mathbf{B}^{-1}\mathbf{b}$ and $\mathbf{B}^{-1}\mathbf{b}$.

	z	$\mathbf{x}_B$	$\mathbf{x}_{N_1}$	$\mathbf{x}_{N_2}$	RHS
z	1	$\mathbf{0}$	$\mathbf{c}_B\mathbf{B}^{-1}\mathbf{N}_1 - \mathbf{c}_{N_1}$	$\mathbf{c}_B\mathbf{B}^{-1}\mathbf{N}_2 - \mathbf{c}_{N_2}$	$\hat{z}$
$\mathbf{x}_B$	$\mathbf{0}$	$\mathbf{I}$	$\mathbf{B}^{-1}\mathbf{N}_1$	$\mathbf{B}^{-1}\mathbf{N}_2$	$\hat{\mathbf{b}}$

Now we try to improve the objective by investigating the possibility of modifying the nonbasic variables. From Equation (5.5) and noting that $\mathbf{c}_{N_1} - \mathbf{c}_B\mathbf{B}^{-1}\mathbf{N}_1$ and $\mathbf{c}_{N_2} - \mathbf{c}_B\mathbf{B}^{-1}\mathbf{N}_2$ give the $c_j - z_j$ values of the lower and upper bounded nonbasic variables respectively, we get

$$z = \mathbf{c}_B\mathbf{B}^{-1}\mathbf{b} - \sum_{j \in R_1} (z_j - c_j)x_j - \sum_{j \in R_2} (z_j - c_j)x_j \tag{5.6}$$

where R_1 is the set of indices of nonbasic variables at their lower bounds and R_2 is the set of indices of nonbasic variables at their upper bounds. For $j \in R_1$ and $z_j - c_j > 0$ it would be to our benefit to increase x_j from its current value of l_j.

Similarly, for $j \in R_2$ and $z_j - c_j < 0$, it would be to our benefit to decrease x_j from its current value of u_j. As in the simplex method, we shall modify the value of only one nonbasic variable while all other nonbasic variables are fixed. The index k of this nonbasic variable is determined as follows. First examine

$$\text{Maximum}\left(\underset{j \in R_1}{\text{Maximum}}\, z_j - c_j \quad , \quad \underset{j \in R_2}{\text{Maximum}}\, c_j - z_j \right)$$

If this maximum is positive, then k is the index where the maximum is achieved. If it corresponds to R_1, then x_k is increased from its current level of l_k, and if it corresponds to R_2, then x_k is decreased from its current level of u_k. If the maximum is $\leqslant 0$, then $z_j - c_j \leqslant 0$ for all $j \in R_1$ and $z_j - c_j \geqslant 0$ for all $j \in R_2$. Examining Equation (5.6), this indicates that the current solution is optimal.

To summarize, given a basic feasible solution, if $z_j - c_j \leqslant 0$ for all nonbasic variables at their lower bounds, and if $z_j - c_j \geqslant 0$ for all nonbasic variables at their upper bounds, then we stop with the conclusion that the current solution is optimal. Otherwise we choose a nonbasic variable x_k according to the foregoing rule. If x_k is at its lower bound, then it is increased; otherwise it is decreased. These two cases are discussed in detail below.

Increasing x_k from its Current level l_k

Let $x_k = l_k + \Delta_k$ where Δ_k is the increase in x_k (currently $\Delta_k = 0$). Noting that all other nonbasic variables are fixed and that the current value of $\mathbf{x}_B$ and z are respectively $\hat{\mathbf{b}}$ and $\hat{z}$, substituting $x_k = l_k + \Delta_k$ in Equations (5.4) and (5.6), we get

$$\begin{aligned} \mathbf{x}_B &= \mathbf{B}^{-1}\mathbf{b} - \mathbf{B}^{-1}\mathbf{N}_1\mathbf{l}_{N_1} - \mathbf{B}^{-1}\mathbf{N}_2\mathbf{u}_{N_2} - \mathbf{B}^{-1}\mathbf{a}_k\Delta_k \\ &= \hat{\mathbf{b}} - \mathbf{y}_k\Delta_k \end{aligned} \tag{5.7}$$

$$\begin{aligned} z &= \mathbf{c}_B\mathbf{B}^{-1}\mathbf{b} - \sum_{j \in R_1} (z_j - c_j)l_j - \sum_{j \in R_2} (z_j - c_j)u_j - (z_k - c_k)\Delta_k \\ &= \hat{z} - (z_k - c_k)\Delta_k \end{aligned} \tag{5.8}$$

Since $z_k - c_k > 0$ (why?), then from Equation (5.8) it is to our benefit to increase Δ_k as much as possible. However, as Δ_k increases, the basic variables are modified according to Equation (5.7). The increase in Δ_k may be blocked as follows.

1. A BASIC VARIABLE DROPS TO ITS LOWER BOUND

Denote the value of Δ_k at which a basic variable drops to its lower bound by γ_1. From Equation (5.7) we have $\mathbf{l}_B \leqslant \mathbf{x}_B = \hat{\mathbf{b}} - \mathbf{y}_k\Delta_k$. Therefore $\mathbf{y}_k\Delta_k \leqslant \hat{\mathbf{b}} - \mathbf{l}_B$. If $\mathbf{y}_k \leqslant \mathbf{0}$, then Δ_k can be made arbitrarily large without violating this inequality and so $\gamma_1 = \infty$ (that is, no basic variable drops to its lower bound). Otherwise γ_1 is given by the minimum ratio shown below. Therefore

$$\gamma_1 = \begin{cases} \underset{1 \leqslant i \leqslant m}{\text{Minimum}} \left\{ \dfrac{\hat{b}_i - l_{B_i}}{y_{ik}} : y_{ik} > 0 \right\} = \dfrac{\hat{b}_r - l_{B_r}}{y_{rk}} & \text{if } \mathbf{y}_k \nleqslant \mathbf{0} \\ \infty \cdots\cdots\cdots\cdots\cdots & \text{if } \mathbf{y}_k \leqslant 0 \end{cases} \tag{5.9}$$

The basic variable that reaches its lower bound is a candidate for x_{B_r}.

2. A BASIC VARIABLE REACHES ITS UPPER BOUND

Denote the value of Δ_k at which a basic variable reaches its upperbound by γ_2. From Equation (5.7) $\hat{\mathbf{b}} - \mathbf{y}_k\Delta_k = \mathbf{x}_B \leqslant \mathbf{u}_B$ and hence $-\mathbf{y}_k\Delta_k \leqslant \mathbf{u}_B - \hat{\mathbf{b}}$. If $\mathbf{y}_k \geqslant \mathbf{0}$, then Δ_k can be made arbitrarily large without violating this inequality and so $\gamma_2 = \infty$ (that is, no basic variable reaches its upper bound). Otherwise, γ_2 is given by the minimum ratio shown below. Therefore

$$\gamma_2 = \begin{cases} \underset{1 \leqslant i \leqslant m}{\text{Minimum}} \left\{ \dfrac{u_{B_i} - \hat{b}_i}{-y_{ik}} : y_{ik} < 0 \right\} = \dfrac{u_{B_r} - \hat{b}_r}{-y_{rk}} & \text{if } \mathbf{y}_k \ngeqslant \mathbf{0} \\ \infty \cdots\cdots\cdots\cdots\cdots & \text{if } \mathbf{y}_k \geqslant \mathbf{0} \end{cases} \tag{5.10}$$

The basic variable that reaches its upper bound is a candidate for x_{B_r}.

3. x_k ITSELF REACHES ITS UPPER BOUND

The value of Δ_k at which x_k reaches its upper bound u_k is obviously $u_k - l_k$.

These three cases give the maximum increase in Δ_k before being blocked by a variable or by x_k itself. Obviously Δ_k is given by

$$\Delta_k = \text{Minimum}(\gamma_1, \gamma_2, u_k - l_k) \tag{5.11}$$

If $\Delta_k = \infty$, then the increase in x_k is *unblocked* and by Equation (5.8) the optimal solution is unbounded. If, on the other hand, $\Delta_k < \infty$, a new basic feasible solution is obtained where $x_k = l_k + \Delta_k$ and the basic variables are modified according to Equation (5.7).

Updating the Tableau When the Nonbasic Variable Increases

The current tableau must be updated to reflect the new basic feasible solution. If $\Delta_k = u_k - l_k$, then no change of basis is made and x_k is still nonbasic, except this time it is at its upper bound. Only the RHS column is changed to reflect the new value of the objective function and the new values of the basic variables. According to Equations (5.8) and (5.7), $\hat{z}$ is replaced by $\hat{z} - (z_k - c_k)\Delta_k$ and $\hat{\mathbf{b}}$ is replaced by $\hat{\mathbf{b}} - \mathbf{y}_k\Delta_k$. On the other hand, if Δ_k is given by γ_1 or γ_2, then x_k enters the basis and x_{B_r} leaves the basis, where the index r is determined according to Equation (5.9) if $\Delta_k = \gamma_1$ or according to (5.10) if $\Delta_k = \gamma_2$. The tableau except the RHS column is updated by pivoting at y_{rk}. Note that y_{rk} may be either positive or negative. Since the right-hand side is computed separately, this should cause no alarm. The right-hand-side column is updated according to Equations (5.8) and (5.7) except that the rth component of the new vector $\hat{\mathbf{b}}$ is replaced by $l_k + \Delta_k$ to reflect the value of x_k, which has just entered the basis.

Alternately, the right-hand-side vector can be updated directly with the rest of the tableau. This, however, requires three distinct operations (which may be performed in any order). First, we multiply the nonbasic entering column by its current value (either l_k or u_k) and add the result to the RHS vector. Next, we multiply the basic leaving column by the value it will assume (either l_{B_r} or u_{B_r}) and subtract the result from the RHS. Finally, we perform a normal pivot operation on the adjusted RHS vector.

Decreasing x_k from its Current Level u_k

This case is very similar to that of increasing x_k and is only discussed briefly. In this case $z_k - c_k < 0$ and $x_k = u_k - \Delta_k$, where $\Delta_k \geqslant 0$ denotes the decrease in x_k. Noting Equations (5.4) and (5.6), we get

$$\mathbf{x}_B = \hat{\mathbf{b}} + \mathbf{y}_k\Delta_k \tag{5.12}$$

$$z = \hat{z} + (z_k - c_k)\Delta_k \tag{5.13}$$

The maximum increase in Δ_k is given by Equation (5.11) where γ_1 and γ_2 are specified below:

$$\gamma_1 = \begin{cases} \underset{1 \leqslant i \leqslant m}{\text{Minimum}} \left\{ \dfrac{\hat{b}_i - l_{B_i}}{-y_{ik}} : y_{ik} < 0 \right\} = \dfrac{\hat{b}_r - l_{B_r}}{-y_{rk}} & \text{if } \mathbf{y}_k \not\geqslant \mathbf{0} \\ \infty \cdots\cdots\cdots & \text{if } \mathbf{y}_k \geqslant \mathbf{0} \end{cases} \tag{5.14}$$

$$\gamma_2 = \begin{cases} \underset{1 \leqslant i \leqslant m}{\text{Minimum}} \left\{ \dfrac{u_{B_i} - \hat{b}_i}{y_{ik}} : y_{ik} > 0 \right\} = \dfrac{u_{B_r} - \hat{b}_r}{y_{rk}} & \text{if } \mathbf{y}_k \not\leqslant \mathbf{0} \\ \infty \cdots\cdots\cdots & \text{if } \mathbf{y}_k \leqslant \mathbf{0} \end{cases} \tag{5.15}$$

If $\Delta_k = \infty$, then the decrease of x_k is unblocked and by Equation (5.13) the optimal solution is unbounded. If $\Delta_k < \infty$, then a new basic feasible solution is obtained where $x_k = u_k - \Delta_k$ and the basic variables are modified according to Equation (5.12).

Updating the Tableau When the Nonbasic Variable Decreases

If $\Delta_k = u_k - l_k$, then x_k is still nonbasic but at its lower bound. The tableau is unchanged except for the RHS column, which is updated according to Equations (5.13) and (5.12). If Δ_k is given by γ_1 or γ_2, then x_k enters the basis and x_{B_r} leaves the basis where r is determined by Equation (5.14) if $\Delta_k = \gamma_1$ and by (5.15) if $\Delta_k = \gamma_2$. The tableau except for the RHS is updated by pivoting at y_{rk}. Again y_{rk} could be either positive or negative. The RHS column is updated according to Equations (5.13) and (5.12) except that the rth component of the new vector $\hat{\mathbf{b}}$ is replaced by $u_k - \Delta_k$ to reflect the value of x_k which has just entered the basis. We may also again utilize the alternative method, described before, to update the RHS vector.

Getting Started

If no basic feasible solution is conveniently available, we may start the lower-upper bound simplex method with artificial variables. This is accomplished by (1) setting all of the original variables to one of their bounds, (2) adjusting the RHS values accordingly, (3) multiplying rows, as necessary, by -1 to get $\hat{b}_i \geq 0$, and (4) adding artificial columns. The two-phase or the big-M method may be employed to drive the artificial variables out of the basis.

We now have all the ingredients of the simplex method with bounded variables. In the absence of degeneracy, note that the procedure described above moves from one basic feasible solution to an improved basic feasible solution and therefore must stop in a finite number of iterations. Verification of this fact and the handling of the degenerate case are left as an exercise for the reader (see Exercise 5.15). We give below a summary of the algorithm.

Summary of the Simplex Method for Bounded Variables (Minimization Problem)

INITIALIZATION STEP

Find a starting basic feasible solution (use artificials if necessary). Let $\mathbf{x}_B$ be the basic variables and let $\mathbf{x}_{N_1}$ and $\mathbf{x}_{N_2}$ be the nonbasic variables at their lower and upper bounds respectively. Form the following tableau where $\hat{z} = \mathbf{c}_B\mathbf{B}^{-1}\mathbf{b} + (\mathbf{c}_{N_1} - \mathbf{c}_B\mathbf{B}^{-1}\mathbf{N}_1)\mathbf{l}_{N_1} + (\mathbf{c}_{N_2} - \mathbf{c}_B\mathbf{B}^{-1}\mathbf{N}_2)\mathbf{u}_{N_2}$ and $\hat{\mathbf{b}} = \mathbf{B}^{-1}\mathbf{b} - \mathbf{B}^{-1}\mathbf{N}_1\mathbf{l}_{N_1} - \mathbf{B}^{-1}\mathbf{N}_2\mathbf{u}_{N_2}$.

	z	$\mathbf{x}_B$	$\mathbf{x}_{N_1}$	$\mathbf{x}_{N_2}$	RHS
z	1	$\mathbf{0}$	$\mathbf{c}_B\mathbf{B}^{-1}\mathbf{N}_1 - \mathbf{c}_{N_1}$	$\mathbf{c}_B\mathbf{B}^{-1}\mathbf{N}_2 - \mathbf{c}_{N_2}$	$\hat{z}$
$\mathbf{x}_B$	0	$\mathbf{I}$	$\mathbf{B}^{-1}\mathbf{N}_1$	$\mathbf{B}^{-1}\mathbf{N}_2$	$\hat{\mathbf{b}}$

MAIN STEP

1. If $z_j - c_j \leqslant 0$ for nonbasic variables at their lower bound and $z_j - c_j \geqslant 0$ for nonbasic variables at their upper bound, than the current solution is optimal. Otherwise if one of these conditions is violated for the index k, then go to step 2 if x_k is at its lower bound and step 3 if x_k is at its upper bound.
2. The variable x_k is increased from its current value of l_k to $l_k + \Delta_k$. The value of Δ_k is given by Equation by (5.11) where γ_1 and γ_2 are given by Equations (5.9) and (5.10). If $\Delta_k = \infty$, stop; the optimal solution is unbounded. Otherwise the tableau is updated, as described previously. Repeat step 1.
3. The variable x_k is decreased from its current value of u_k to $u_k - \Delta_k$. The value of Δ_k is given by Equation (5.11) where γ_1 and γ_2 are given by Equations (5.14) and (5.15). If $\Delta_k = \infty$, stop; the optimal solution is unbounded. Otherwise, the tableau is updated as described previously. Repeat step 1.

It will be helpful to distinguish between nonbasic variables at their lower and upper bounds during the simplex iterations. This is done by flagging the corresponding columns by l and u respectively.

Example 5.4

Minimize $-2x_1 - 4x_2 - x_3$

$$\begin{aligned} \text{Subject to} \quad 2x_1 + x_2 + x_3 &\leqslant 10 \\ x_1 + x_2 - x_3 &\leqslant 4 \\ 0 \leqslant x_1 &\leqslant 4 \\ 0 \leqslant x_2 &\leqslant 6 \\ 1 \leqslant x_3 &\leqslant 4 \end{aligned}$$

Introduce the slack variables x_4 and x_5. These are bounded below by 0 and bounded above by ∞. Initially the basic variables are x_4 and x_5, and the nonbasic variables at their lower bound are $x_1 = x_2 = 0$ and $x_3 = 1$. Note that the objective is -1 and the values of the basic variables x_4 and x_5 are given by $10 - 1 = 9$ and $4 + 1 = 5$.

Iteration 1

	z	l x_1	l x_2	l x_3	x_4	x_5	RHS
z	1	2	4	1	0	0	−1
x_4	0	2	1	1	1	0	9
x_5	0	1	(1)	−1	0	1	5

The maximum value of $z_j - c_j$ for lower bounded nonbasic variables is 4 corresponding to x_2. Therefore $x_k = x_2$ is increased. Then $\mathbf{y}_2 = \begin{bmatrix} 1 \\ 1 \end{bmatrix}$ and Δ_2 is given by Minimum $(\gamma_1, \gamma_2, u_2 - l_2)$ = Minimum $(\gamma_1, \gamma_2, 6)$. Also γ_1 and γ_2 are given according to Equations (5.9) and (5.10) as follows. First,

$$\gamma_1 = \text{Minimum}\left\{ \frac{9-0}{1}, \frac{5-0}{1} \right\} = 5$$

corresponding to $x_{B_2} = x_5$, that is, $r = 2$. This means that Δ_2 can be increased to value 5 before a basic variable drops to its lower bound. Second, $\gamma_2 = \infty$, which means that Δ_2 can be increased indefinitely without any basic variable reaching its upper bound.

Therefore Δ_2 = Minimum $(5, \infty, 6) = 5$. The objective is replaced by $-1 - (z_2 - c_2)\Delta_2 = -1 - 4 \times 5 = -21$ and

$$\begin{bmatrix} x_4 \\ x_5 \end{bmatrix} = \begin{bmatrix} 9 \\ 5 \end{bmatrix} - \mathbf{y}_2\Delta_2 = \begin{bmatrix} 9 \\ 5 \end{bmatrix} - \begin{bmatrix} 1 \\ 1 \end{bmatrix} 5 = \begin{bmatrix} 4 \\ 0 \end{bmatrix}.$$

The value of x_2 is given by $\Delta_2 = 5$. Then x_2 enters and x_5 leaves. The tableau is updated by pivoting at y_{22}.

Iteration 2

	z	l x_1	x_2	l x_3	x_4	l x_5	RHS
z	1	−2	0	5	0	−4	−21
x_4	0	1	0	2	1	−1	4
x_2	0	1	1	(−1)	0	1	5

All nonbasic variables are at their lower bounds and the maximum value of $z_j - c_j$ is 5, corresponding to x_3. Therefore x_3 enters, $\mathbf{y}_3 = \begin{bmatrix} 2 \\ -1 \end{bmatrix}$ and Δ_3 is given by

$\Delta_3 = \text{Minimum}(\gamma_1, \gamma_2, u_3 - l_3) = \text{Minimum}(\gamma_1, \gamma_2, 3)$

The values of γ_1 and γ_2 are obtained from Equations (5.9) and (5.10) as follows. First,

$$\gamma_1 = \frac{4 - 0}{2} = 2$$

corresponding to $x_{B_r} = x_{B_1} = x_4$, that is, $r = 1$. This means that x_4 drops to its lower limit as x_3 is increased. Second,

$$\gamma_2 = \frac{6 - 5}{1} = 1$$

corresponding to $x_{B_r} = x_{B_2} = x_2$, that is, $r = 2$. This means that x_2 reaches its upper bound as x_3 is increased. Therefore $\Delta_3 = \text{Minimum } (2, 1, 3) = 1 = \gamma_2$. Now $x_3 = 1 + \Delta_3 = 2$.

The objective is replaced by $-21 - (z_3 - c_3)\Delta_3 = -21 - 5 \times 1 = -26$.

$$\begin{bmatrix} x_4 \\ x_2 \end{bmatrix} = \begin{bmatrix} 4 \\ 5 \end{bmatrix} - \mathbf{y}_3\Delta_3 = \begin{bmatrix} 4 \\ 5 \end{bmatrix} - \begin{bmatrix} 2 \\ -1 \end{bmatrix} 1 = \begin{bmatrix} 2 \\ 6 \end{bmatrix}$$

Here x_3 enters the basis and x_2 reaches its upper bound and leaves the basis. The tableau (except the RHS, which was updated separately) is updated by pivoting at $y_{r3} = y_{23} = -1$. If we had chosen to use the alternative method for updting the RHS column, we would first replace the RHS column of iteration 2 by

$$\begin{bmatrix} -21 \\ 4 \\ 5 \end{bmatrix} + 1 \begin{bmatrix} 5 \\ 2 \\ -1 \end{bmatrix} - 6 \begin{bmatrix} 0 \\ 0 \\ 1 \end{bmatrix} = \begin{bmatrix} -16 \\ 6 \\ -2 \end{bmatrix}$$

Upon pivoting at y_{23} $(= -1)$, we would obtain the right-hand-side column below.

Iteration 3

		l	u			l	
	z	x_1	x_2	x_3	x_4	x_5	RHS
z	1	3	5	0	0	1	−26
x_4	0	(3)	2	0	1	1	2
x_3	0	−1	−1	1	0	−1	2

The maximum value of $z_j - c_j$ for nonbasic variables at their lower bound is 3, corresponding to x_1. Therefore x_1 is increased. Here $\mathbf{y}_1 = \begin{bmatrix} 3 \\ -1 \end{bmatrix}$ and Δ_1 is given by

$$\Delta_1 = \text{Minimum}(\gamma_1, \gamma_2, u_1 - l_1) = \text{Minimum}(\gamma_1, \gamma_2, 4)$$

The values of γ_1 and γ_2 are given by Equations (5.9) and (5.10) as follows. First,

$$\gamma_1 = \frac{2-0}{3} = \frac{2}{3}$$

corresponding to $x_{B_r} = x_{B_1} = x_4$, that is, $r = 1$. This means that as Δ_1 is increased to $\frac{2}{3}$, x_4 drops to its lower limit and drops from the basis. Second,

$$\gamma_2 = \frac{4-2}{1} = 2$$

corresponding to $x_{B_r} = x_{B_2} = x_3$, that is, $r = 3$. This means that as Δ_1 is increased to 2, x_3 reaches its upper limit and drops from the basis.

Therefore $\Delta_1 = \text{Minimum}\ (\frac{2}{3}, 2, 4) = \frac{2}{3}$. So $x_1 = \frac{2}{3}$. The objective is replaced by $-26 - (z_1 - c_1)\Delta_1 = -26 - 3 \times \frac{2}{3} = -28$.

$$\begin{bmatrix} x_4 \\ x_3 \end{bmatrix} = \begin{bmatrix} 2 \\ 2 \end{bmatrix} - \mathbf{y}_1\Delta_1 = \begin{bmatrix} 2 \\ 2 \end{bmatrix} - \begin{bmatrix} 3 \\ -1 \end{bmatrix}\frac{2}{3} = \begin{bmatrix} 0 \\ \frac{8}{3} \end{bmatrix}$$

Here x_1 enters the basis and x_4 leaves the basis.

The tableau is updated by pivoting at $y_{11} = 3$, and the RHS is updated separately where $z = -28$, $x_1 = \frac{2}{3}$, and $x_3 = \frac{8}{3}$.

Iteration 4

			u		l	l	
	z	x_1	x_2	x_3	x_4	x_5	RHS
z	1	0	3	0	-1	0	-28
x_1	0	1	$\frac{2}{3}$	0	$\frac{1}{3}$	$\frac{1}{3}$	$\frac{2}{3}$
x_3	0	0	$-\frac{1}{3}$	1	$\frac{1}{3}$	$-\frac{2}{3}$	$\frac{8}{3}$

Since $z_j - c_j \geqslant 0$ for nonbasic variables at their upper bound and $z_j - c_j \leqslant 0$ for nonbasic variables at their lower bound, then the foregoing tableau gives an optimal solution (is it unique?). The variables are given by $(x_1, x_2, x_3, x_4, x_5) = (\frac{2}{3}, 6, \frac{8}{3}, 0, 0)$ and the objective is -28.

5.3 THE KUHN-TUCKER CONDITIONS AND THE SIMPLEX METHOD

In this section we develop the necessary and sufficient Kuhn-Tucker optimality conditions for a linear programming problem. These conditions will be used as a general framework for many algorithms during the remainder of the book.

The Kuhn-Tucker Conditions for Inequality Constraints

Consider the following linear programming problem.

$$\begin{aligned} \text{Minimize} \quad & \mathbf{cx} \\ \text{Subject to} \quad & \mathbf{Ax} \geqslant \mathbf{b} \\ & \mathbf{x} \geqslant \mathbf{0} \end{aligned}$$

where $\mathbf{c}$ is an n vector, $\mathbf{b}$ is an m vector, and $\mathbf{A}$ is an $m \times n$ matrix. The Kuhn-Tucker conditions can be stated as follows. The vector $\mathbf{x}$ is an optimal solution of the foregoing problem if there exist an n vector $\mathbf{v}$ and an m vector $\mathbf{w}$ such that the following three conditions hold. Conversely, if the following three conditions hold, then $\mathbf{x}$ is an optimal solution of the foregoing problem.

$$\mathbf{Ax} \geqslant \mathbf{b}, \qquad \mathbf{x} \geqslant \mathbf{0} \tag{5.16}$$

$$\mathbf{c} - \mathbf{wA} - \mathbf{v} = \mathbf{0}, \qquad \mathbf{w} \geqslant \mathbf{0}, \mathbf{v} \geqslant \mathbf{0} \tag{5.17}$$

$$\mathbf{w}(\mathbf{Ax} - \mathbf{b}) = 0, \qquad \mathbf{vx} = 0 \tag{5.18}$$

Before proceeding any further, let us briefly discuss these three optimality conditions. The first condition (5.16) merely states that the candidate point must be feasible; that is, it must satisfy the constraints of the problem. This is usually referred to as *primal feasibility*. The second condition (5.17) is usually referred to as *dual feasibility*, since it corresponds to feasibility of a problem closely related to the original problem. This problem is called the *dual* problem and will be discussed in detail in Chapter 6. Here $\mathbf{w}$ and $\mathbf{v}$ are called the *Lagrangian multipliers* (or *dual variables*) corresponding to the constraints $\mathbf{Ax} \geqslant \mathbf{b}$ and $\mathbf{x} \geqslant \mathbf{0}$ respectively. Finally, the third condition (5.18) is usually referred to as *complementary slackness*. Since $\mathbf{w} \geqslant \mathbf{0}$ and $\mathbf{Ax} \geqslant \mathbf{b}$, then $\mathbf{w}(\mathbf{Ax} - \mathbf{b}) = 0$ if and only if either w_i is 0 or else the ith slack variable is 0. Similarly $\mathbf{vx} = 0$ if and only if either x_j is 0 or else v_j is 0.

The following example illustrates the Kuhn-Tucker conditions.

Example 5.5

$$\begin{aligned} \text{Minimize} \quad & -x_1 - 3x_2 \\ \text{Subject to} \quad & x_1 - 2x_2 \geqslant -4 \\ & -x_1 - x_2 \geqslant -4 \\ & x_1, \quad x_2 \geqslant 0 \end{aligned}$$

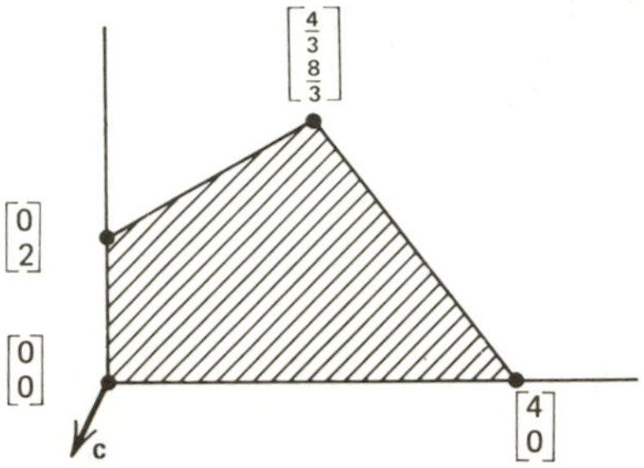

Figure 5.2. **Verification of the KT conditions.**

Equations (5.16), (5.17), and (5.18) represent a useful tool in verifying whether a certain point is optimal. To illustrate, suppose we were told that the optimal solution of the foregoing problem is the point (0, 0). We see geometrically in Figure 5.2, or by using the simplex method, that (0, 0) is not the optimal point. First, inequality (5.16) holds since (0, 0) is indeed a feasible point. Since none of the first two constraints is binding (that is, $+x_1 - 2x_2 > -4$ and $-x_1 - x_2 > -4$), then $w_1 = w_2 = 0$, in order to satisfy Equation (5.18). Since $\mathbf{w} = \mathbf{0}$, then from Equation (5.17) we have to have $\mathbf{c} = \mathbf{v}$, that is, $\mathbf{v} = (-1, -3)$. This violates nonnegativity of $\mathbf{v}$, however. Therefore, (0, 0) could not be an optimal solution of this problem.

Now suppose that we were told that the optimal point is $(\frac{4}{3}, \frac{8}{3})$. In order to check whether this is a true statement, we can use the Kuhn-Tucker conditions. Since $x_1, x_2 > 0$, then $v_1 = v_2 = 0$ in order to satisfy complementary slackness. From Equation (5.17), $\mathbf{w}$ must satisfy the equations $\mathbf{c} - \mathbf{wA} = \mathbf{0}$, that is,

$$
\begin{aligned}
w_1 - w_2 &= -1 \\
-2w_1 - w_2 &= -3
\end{aligned}
$$

and hence $w_1 = \frac{2}{3}$ and $w_2 = \frac{5}{3}$. Note that $\mathbf{w} \geqslant \mathbf{0}$ and $\mathbf{Ax} = \mathbf{b}$ and hence $\mathbf{w}(\mathbf{Ax} - \mathbf{b}) = 0$. Therefore conditions (5.16), (5.17), and (5.18) hold and $(\frac{4}{3}, \frac{8}{3})$ is indeed an optimal point.

Geometric Intepretation of the Optimality Conditions

Before proceeding to prove the Kuhn-Tucker conditions, let us examine their geometric interpretation. As we mentioned earlier, condition (5.16) merely states that the point $\mathbf{x}$ must be feasible. Now let us examine (5.17) and (5.18) more carefully. Given a feasible point $\mathbf{x}$, we can immediately determine the *binding* (also referred to as *active*) and the *nonbinding* (also referred to as *inactive*) constraints, that is, the constraints that hold as equalities and those that hold as strict inequalities. If a constraint holds as a strict inequality, such as $\mathbf{a}^i\mathbf{x} > b_i$ (where $\mathbf{a}^i$ is the ith row of $\mathbf{A}$), then condition (5.18) requires that $w_i = 0$, and similarly if $x_j > 0$ then $v_j = 0$. Since this is the case, then conditions (5.17) and (5.18) reduce to

$$\mathbf{c} = \sum_{i \in I} w_i \mathbf{a}^i + \sum_{j \in J} v_j \mathbf{e}_j$$

$$w_i \geqslant 0 \qquad i \in I$$

$$v_j \geqslant 0 \qquad j \in J$$

where

$I = \{i : \mathbf{a}^i \mathbf{x} = b_i\}$, the set of binding constraints

$J = \{j : x_j = 0\}$, the set of binding nonnegativity constraints

$\mathbf{e}_j$ is a vector of zeros except for a 1 at the jth position.

From this discussion, it is clear that conditions (5.17) and (5.18) reduce to the simple criterion that $\mathbf{c}$, the gradient of the objective function, can be represented as a nonnegative combination of the gradients of the binding constraints, where $\mathbf{a}^i$ is the gradient of the constraint $\mathbf{a}^i\mathbf{x} \geqslant b_i$ and $\mathbf{e}_j$ is the gradient of the constraint $x_j \geqslant 0$. In other words, the Kuhn-Tucker conditions hold if $\mathbf{x}$ is feasible, and the gradient of the objective function $\mathbf{c}$ lies in the cone generated by the gradients of the binding constraints. Since the Kuhn-Tucker conditions are both necessary and sufficient, then a point is optimal if and only if $\mathbf{c}$ lies in the prescribed cone.

Example 5.6

Minimize $-x_1 - 3x_2$

$$\begin{aligned} \text{Subject to} \quad x_1 - 2x_2 &\geqslant -4 \\ -x_1 - x_2 &\geqslant -4 \\ x_1, \quad x_2 &\geqslant 0 \end{aligned}$$

The gradients of the objective function and the constraints are given below:

$\mathbf{c} = (-1, -3)$

$\mathbf{a}^1 = (1, -2)$

$\mathbf{a}^2 = (-1, -1)$

$\mathbf{e}_1 = (1, 0)$

$\mathbf{e}_2 = (0, 1)$

Let us consider the four extreme points of Figure 5.3.

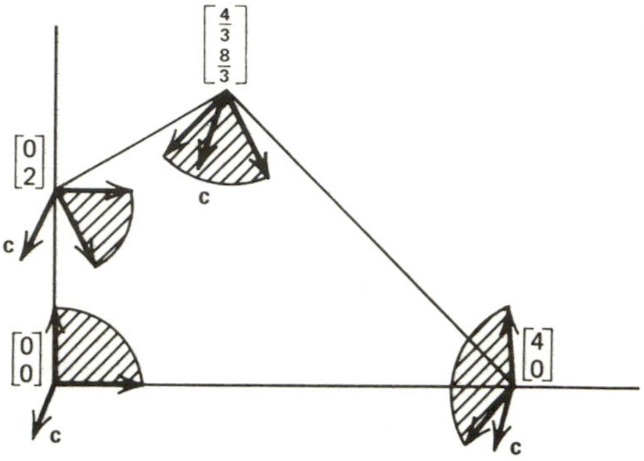

Figure 5.3. **Geometry of the KT conditions.**

1. The extreme point $\mathbf{x} = (0, 0)$. The binding constraints are the nonnegativity constraints. We see from Figure 5.3 that $\mathbf{c}$ does not belong to the cone of the gradients $\mathbf{e}_1$ and $\mathbf{e}_2$ of the binding constraints. Therefore (0, 0) is not optimal.
2. The extreme point $\mathbf{x} = (0, 2)$. The binding constraints are $x_1 - 2x_2 \geqslant -4$ and $x_1 \geqslant 0$. Here $\mathbf{c}$ does not belong to the cone generated by the gradients $\mathbf{a}^1$ and $\mathbf{e}_1$ of the binding constraints. So (0, 2) is not optimal.
3. The extreme point $(\frac{4}{3}, \frac{8}{3})$. The binding constraints are $x_1 - 2x_2 \geqslant -4$ and $-x_1 - x_2 \geqslant -4$. Here $\mathbf{c}$ belongs to the cone of the gradients $\mathbf{a}^1$ and $\mathbf{a}^2$ of the binding constraints. Therefore $(\frac{4}{3}, \frac{8}{3})$ is an optimal point.
4. The extreme point (4, 0). The binding constraints are $-x_1 - x_2 \geqslant -4$ and $x_2 \geqslant 0$. Here $\mathbf{c}$ does not belong to the cone generated by the gradients $\mathbf{a}^2$ and $\mathbf{e}_2$ of the binding constraints and hence is not optimal.

Proof of the Kuhn-Tucker Conditions

First let us show that the Kuhn-Tucker conditions are sufficient for optimality. Suppose that $\mathbf{x}$ is a feasible solution of the problem and suppose that there exist vectors $\mathbf{w}$ and $\mathbf{v}$ such that conditions (5.16), (5.17), and (5.18) hold. We shall show that $\mathbf{x}$ is indeed an optimal solution. Let $\mathbf{x}'$ be any feasible point satisfying $\mathbf{Ax}' \geqslant \mathbf{b}$ and $\mathbf{x}' \geqslant \mathbf{0}$. From condition (5.17), $\mathbf{c} - \mathbf{wA} - \mathbf{v} = \mathbf{0}$ and we get

$$0 = (\mathbf{c} - \mathbf{wA} - \mathbf{v})(\mathbf{x} - \mathbf{x}') = (\mathbf{cx} - \mathbf{cx}') - \mathbf{wAx} - \mathbf{vx} + \mathbf{wAx}' + \mathbf{vx}'$$

From condition (5.18), $\mathbf{wAx} = \mathbf{wb}$ and $\mathbf{vx} = 0$. Therefore

$$0 = (\mathbf{cx} - \mathbf{cx}') + \mathbf{w}(\mathbf{Ax}' - \mathbf{b}) + \mathbf{vx}'$$

Since $\mathbf{w} \geqslant \mathbf{0}$, $\mathbf{Ax}' - \mathbf{b} \geqslant \mathbf{0}$, $\mathbf{v} \geqslant \mathbf{0}$, and $\mathbf{x}' \geqslant \mathbf{0}$, the foregoing equation implies that $\mathbf{cx}' \geqslant \mathbf{cx}$. Since this is true for each feasible solution, then $\mathbf{x}$ is indeed an optimal solution of the problem. This shows that the Kuhn-Tucker conditions are sufficient for optimality.

Conversely suppose that $\mathbf{x}$ is an optimal (and hence feasible) solution of the problem. We shall now show that conditions (5.17) and (5.18) hold. Possibly after rearranging the columns and rows of $\mathbf{A}$, suppose that $x_1, x_2, \ldots, x_p$ are positive, $x_{p+1}, \ldots, x_n = 0$, $\mathbf{A}_1\mathbf{x} = \mathbf{b}_1$ and $\mathbf{A}_2\mathbf{x} > \mathbf{b}_2$ where $\mathbf{A} = \begin{bmatrix} \mathbf{A}_1 \\ \mathbf{A}_2 \end{bmatrix}$ and $\mathbf{b} = \begin{bmatrix} \mathbf{b}_1 \\ \mathbf{b}_2 \end{bmatrix}$, $\mathbf{A}_1$ is $k \times n$, $\mathbf{A}_2$ is $(m-k) \times n$, $\mathbf{b}_1$ is a k vector, and $\mathbf{b}_2$ is an $m-k$ vector. Note that $\mathbf{A}_1\mathbf{d} \geqslant \mathbf{0}$, $d_{p+1}, \ldots, d_n \geqslant 0$ and $\mathbf{cd} < 0$ has no solution. This follows, since otherwise it can be easily verified that $\mathbf{x} + \lambda\mathbf{d}$ is a feasible solution with a better objective value, for $\lambda > 0$ and sufficiently small, thus violating optimality of $\mathbf{x}$ (see Exercise 5.33). The system $\mathbf{A}_1\mathbf{d} \geqslant \mathbf{0}$, $d_{p+1}, \ldots, d_n \geqslant 0$ can be rewritten as $\mathbf{Td} \leqslant \mathbf{0}$, where

$$\mathbf{T} = \left[\begin{array}{c} -\mathbf{A}_1 \\ \hline \mathbf{0} \,|\, -\mathbf{I} \end{array} \right] \begin{array}{l} k \text{ rows} \\ n - p \text{ rows} \end{array}$$

and $\mathbf{I}$ is an $(n-p) \times (n-p)$ identity matrix. The system $\mathbf{Td} \leqslant \mathbf{0}$ and $\mathbf{cd} < 0$ therefore has no solution and applying Farkas's Theorem (see Section 2.7), there exists a nonnegative k vector $\mathbf{w}_1$ and a nonnegative $n-p$ vector $\mathbf{v}_2$ such that $\mathbf{c} - \mathbf{w}_1\mathbf{A}_1 - (\mathbf{0}, \mathbf{v}_2) = (\mathbf{0}, \mathbf{0})$. Letting $\mathbf{w}_2 = \mathbf{0}$ and $\mathbf{v}_1 = \mathbf{0}$, we have $\mathbf{c} - \mathbf{w}_1\mathbf{A}_1 - \mathbf{w}_2\mathbf{A}_2 - (\mathbf{v}_1, \mathbf{v}_2) = (\mathbf{0}, \mathbf{0})$, that is, $\mathbf{c} - \mathbf{wA} - \mathbf{v} = \mathbf{0}$, where $\mathbf{w}, \mathbf{v} \geqslant \mathbf{0}$. Furthermore, note that $x_j = 0$ for $j = p+1, \ldots, n$ and $v_j = 0$ for $j = 1, \ldots, p$, and hence $x_j v_j = 0$ for all j and

$$\mathbf{w}(\mathbf{Ax} - \mathbf{b}) = (\mathbf{w}_1, \mathbf{w}_2)\begin{bmatrix} \mathbf{A}_1\mathbf{x} - \mathbf{b}_1 \\ \mathbf{A}_2\mathbf{x} - \mathbf{b}_2 \end{bmatrix} = (\mathbf{w}_1, \mathbf{0})\begin{bmatrix} \mathbf{0} \\ \mathbf{A}_2\mathbf{x} - \mathbf{b}_2 \end{bmatrix} = 0$$

This shows that conditions (5.16), (5.17), and (5.18) hold and hence the Kuhn-Tucker conditions are also necessary for optimality.

The Kuhn-Tucker Conditions for Equality Constraints

Consider the following linear programming problem in equality form.

$$\begin{aligned} \text{Minimize} \quad & \mathbf{cx} \\ \text{Subject to} \quad & \mathbf{Ax} = \mathbf{b} \\ & \mathbf{x} \geqslant \mathbf{0} \end{aligned}$$

By changing the equality into two inequalities of the form $\mathbf{Ax} \geqslant \mathbf{b}$ and $-\mathbf{Ax} \geqslant -\mathbf{b}$, the Kuhn-Tucker conditions developed earlier would simplify to

$$\mathbf{Ax} = \mathbf{b}, \qquad \mathbf{x} \geqslant \mathbf{0} \tag{5.19}$$

$$\mathbf{c} - \mathbf{wA} - \mathbf{v} = \mathbf{0}, \qquad \mathbf{w} \text{ unrestricted}, \qquad \mathbf{v} \geqslant \mathbf{0} \tag{5.20}$$

$$\mathbf{vx} = 0 \tag{5.21}$$

The main difference between these conditions and the conditions for the inequality problem is that the Lagrangian multiplier vector (or dual vector) $\mathbf{w}$ corresponding to the constraint $\mathbf{Ax} = \mathbf{b}$ is unrestricted in sign.

Optimality at a Basic Feasible Solution

Consider the following problem.

$$\begin{aligned} \text{Minimize} \quad & \mathbf{cx} \\ \text{Subject to} \quad & \mathbf{Ax} = \mathbf{b} \\ & \mathbf{x} \geqslant \mathbf{0} \end{aligned}$$

Assume that rank $(\mathbf{A}) = m$, and let us reinvestigate how the simplex method recognizes an optimal basic feasible solution. Suppose that we have a basic feasible solution $\mathbf{x}$ with basis $\mathbf{B}$ and let us examine conditions (5.19), (5.20), and (5.21). Obviously (5.19) holds. The condition $\mathbf{c} - \mathbf{wA} - \mathbf{v} = \mathbf{0}$ can be rewritten as follows, where $\mathbf{v}$ is decomposed into $\mathbf{v}_B$ and $\mathbf{v}_N$:

$$(\mathbf{c}_B, \mathbf{c}_N) - \mathbf{w}(\mathbf{B}, \mathbf{N}) - (\mathbf{v}_B, \mathbf{v}_N) = (\mathbf{0}, \mathbf{0}) \tag{5.22}$$

If the complementary slackness condition $\mathbf{vx} = 0$ is to hold, and since $\mathbf{x}_N = \mathbf{0}$, it suffices to have $\mathbf{v}_B = \mathbf{0}$, in order to guarantee that $\mathbf{vx} = \mathbf{v}_B\mathbf{x}_B + \mathbf{v}_N\mathbf{x}_N = 0$. With $\mathbf{v}_B = \mathbf{0}$, Equation (5.22) reduces to the following two equations:

$$\mathbf{c}_B - \mathbf{wB} = \mathbf{0}$$

$$\mathbf{c}_N - \mathbf{wN} - \mathbf{v}_N = \mathbf{0}$$

From the first equation we get $\mathbf{w} = \mathbf{c}_B\mathbf{B}^{-1}$, and from the second equation we get $\mathbf{v}_N = \mathbf{c}_N - \mathbf{wN} = \mathbf{c}_N - \mathbf{c}_B\mathbf{B}^{-1}\mathbf{N}$.

To summarize, given a basic feasible solution, condition (5.19) automatically holds, (5.21) holds by letting $\mathbf{v}_B = \mathbf{0}$, and the condition $\mathbf{c} - \mathbf{wA} - \mathbf{v} = \mathbf{0}$ is satisfied by letting $\mathbf{w} = \mathbf{c}_B\mathbf{B}^{-1}$ and $\mathbf{v}_N = \mathbf{c}_N - \mathbf{c}_B\mathbf{B}^{-1}\mathbf{N}$. The only possible source of violation of the Kuhn-Tucker conditions is that $\mathbf{v}_N$ may violate the nonnegativity restrictions. Note, however, that $\mathbf{v}_N$ consists of the $c_j - z_j$ values for the nonbasic variables. Therefore the nonnegativity of $\mathbf{v}_N$ in Equation (5.20) is violated if $c_j - z_j < 0$ for some nonbasic variable. Of course, if $c_j - z_j \geqslant 0$ for each nonbasic variable, then $\mathbf{v}_N \geqslant \mathbf{0}$ and all the Kuhn-Tucker conditions are met. These are precisely the simplex termination criteria.

Reinterpretation of the Simplex Method

From the previous discussion, the simplex method can be interpreted as a systematic procedure for approaching an optimal extreme point satisfying the Kuhn-Tucker conditions. At each iteration, feasibility (called primal feasibility) is satisfied, and hence condition (5.19) always holds. Also complementary slackness is always satisfied since either a variable x_j is nonbasic and has value zero, or else $v_j = c_j - z_j = 0$, and hence $v_j x_j = 0$ for all j and $\mathbf{vx} = 0$. So condition (5.21) is always satisfied during the simplex method. Condition (5.20) (called dual feasibility; more on this in Chapter 6) is partially violated during the iterations of the simplex method. Condition (5.20) has two portions, namely $\mathbf{c} - \mathbf{wA} - \mathbf{v} = \mathbf{0}$, and $\mathbf{v} \geqslant \mathbf{0}$. The first portion always holds by letting $\mathbf{w} = \mathbf{c}_B\mathbf{B}^{-1}$ and $\mathbf{v} = (\mathbf{v}_B, \mathbf{v}_N) = (\mathbf{0}, \mathbf{c}_N - \mathbf{c}_B\mathbf{B}^{-1}\mathbf{N})$. However, the second portion, namely nonnegativity of $\mathbf{c}_N - \mathbf{c}_B\mathbf{B}^{-1}\mathbf{N}$ (called dual feasibility), is violated, until of course, the optimal solution is reached. To summarize, the simplex method always satisfies primal feasibility and complementary slackness. Dual feasibility is violated, and the violation is used to improve the objective function, by increasing the nonbasic variable with the most negative $c_j - z_j$.

Finding the Lagrangian Multipliers From the Simplex Tableau

We already know that $\mathbf{v}_B = \mathbf{0}$ and $\mathbf{v}_N = \mathbf{c}_N - \mathbf{c}_B\mathbf{B}^{-1}\mathbf{N}$. Therefore the Lagrangian multiplier v_j corresponding to the constraint $x_j \geqslant 0$ can be easily obtained from row 0 of the simplex tableau. More precisely, v_j is the negative of the $z_j - c_j$ entry in row 0 under the x_j column.

Now we turn to the Lagrangian vector $\mathbf{w} = \mathbf{c}_B\mathbf{B}^{-1}$ corresponding to the constraints $\mathbf{Ax} = \mathbf{b}$. The method for obtaining $\mathbf{w}$ from the simplex tableau was discussed in Chapter 3. We shall elaborate on this further. Recall that row 0 of the simplex method consists of $z_j - c_j$ for $j = 1, \ldots, n$, which are given by the vector $\mathbf{c}_B\mathbf{B}^{-1}\mathbf{A} - \mathbf{c}$. If the matrix $\mathbf{A}$ has an identity matrix as a portion of its columns, then in row 0 under these columns, we have $\mathbf{c}_B\mathbf{B}^{-1}\mathbf{I} - \hat{\mathbf{c}} = \mathbf{w} - \hat{\mathbf{c}}$, where $\hat{\mathbf{c}}$ is the part of the cost vector $\mathbf{c}$ corresponding to the identity columns in the original problem. By simply adding the vector $\hat{\mathbf{c}}$ to $\mathbf{w} - \hat{\mathbf{c}}$ in row 0, we get $\mathbf{w}$.

EXERCISES

5.1 Solve the following linear program by the revised simplex method.

$$
\begin{aligned}
\text{Minimize} \quad & -2x_2 + x_3 \\
\text{Subject to} \quad & x_1 - 2x_2 + x_3 \geqslant -4 \\
& x_1 + x_2 + x_3 \leqslant 9 \\
& 2x_1 - x_2 - x_3 \leqslant 5 \\
& x_1, \quad x_2, \quad x_3 \geqslant 0
\end{aligned}
$$

5.2 Solve the following problem using the revised simplex method.

$$
\begin{aligned}
\text{Minimize} \quad & x_1 + 6x_2 - 7x_3 + x_4 + 5x_5 \\
\text{Subject to} \quad & x_1 - \tfrac{3}{4}x_2 + 2x_3 - \tfrac{1}{4}x_4 = 5 \\
& -\tfrac{1}{4}x_2 + 3x_3 - \tfrac{3}{4}x_4 + x_5 = 5 \\
& x_1, \quad x_2, \quad x_3, \quad x_4, \quad x_5 \geqslant 0
\end{aligned}
$$

5.3 Verify the entries in Table 5.1.

5.4 Solve Exercise 5.2 using the product form of the inverse.

5.5 Solve the following problem by the revised simplex method using the product form of the inverse.

$$
\begin{aligned}
\text{Maximize} \quad & 3x_1 + 4x_2 + x_3 + 7x_4 \\
\text{Subject to} \quad & 8x_1 + 3x_2 + 4x_3 + x_4 \leqslant 7 \\
& 2x_1 + 6x_2 + x_3 + 5x_4 \leqslant 3 \\
& x_1 + 4x_2 + 5x_3 + 2x_4 \leqslant 8 \\
& x_1, \quad x_2, \quad x_3, \quad x_4 \geqslant 0
\end{aligned}
$$

5.6 Apply the revised simplex method, with and without the cycling prevention rule, to the following problem (this is, the example problem of cycling in Chapter 4).

$$\begin{array}{lllllll}
\text{Minimize} & & -\tfrac{3}{4}x_4 + 20x_5 - \tfrac{1}{2}x_6 + 6x_7 \\
\text{Subject to} & x_1 & +\tfrac{1}{4}x_4 - 8x_5 - x_6 + 9x_7 = 0 \\
& x_2 & +\tfrac{1}{2}x_4 - 12x_5 - \tfrac{1}{2}x_6 + 3x_7 = 0 \\
& x_3 & + x_6 = 1 \\
& x_1, x_2, x_3, & x_4, \quad x_5, \quad x_6, \quad x_7 \geq 0
\end{array}$$

5.7 In the revised simplex method with product form of the inverse, the number of elementary matrices increases by 1 at each iteration. If the number of elementary matrices becomes excessive, it would be necessary to reinvert **B**. Let **B** be a basis. Show how can **B** be reinverted such that $\mathbf{B}^{-1}$ is represented as the product of m elementary matrices. Illustrate by an example.

5.8 Determine the number of multiplications and additions needed per iteration of the revised simplex method using the product form of the inverse. How can we take advantage of sparsity of nonzero elements in the matrix **A**? Give a detailed comparison between the simplex method and the revised simplex method using the product form of the inverse.

5.9 Use the simplex method for bounded variables to solve the following problem.

$$\begin{array}{ll}
\text{Maximize} & x_1 + x_2 + 3x_3 \\
\text{Subject to} & x_1 + x_2 + x_3 \leq 12 \\
& -x_1 + x_2 \leq 5 \\
& x_2 + 2x_3 \leq 8 \\
& 0 \leq x_1 \leq 3 \\
& 0 \leq x_2 \leq 6 \\
& 0 \leq x_3 \leq 4
\end{array}$$

5.10 Use the simplex method for bounded variables to solve the following problem.

$$\begin{aligned}
\text{Minimize} \quad & x_1+2x_2+3x_3-x_4 \\
\text{Subject to} \quad & x_1-x_2+x_3-2x_4 \leqslant 6 \\
& -x_1+x_2-x_3+x_4 \leqslant 8 \\
& 2x_1+x_2-x_3 \geqslant 2 \\
& 0 \leqslant x_1 \leqslant 3 \\
& 1 \leqslant x_2 \leqslant 4 \\
& 0 \leqslant x_3 \leqslant 10 \\
& 2 \leqslant x_4 \leqslant 5
\end{aligned}$$

5.11 Consider the problem: Minimize $\mathbf{cx}$ subject to $\mathbf{Ax} = \mathbf{b}$, $\mathbf{l} \leqslant \mathbf{x} \leqslant \mathbf{u}$. Show in detail that the collection of extreme points and the collection of basic feasible solutions as defined in Section 5.2 are equal.

5.12 Consider the problem: Minimize $\mathbf{cx}$ subject to $\mathbf{Ax} = \mathbf{b}$, $\mathbf{0} \leqslant \mathbf{x} \leqslant \mathbf{u}$. Show that the basic feasible solutions defined in Section 5.2, and the basic feasible solutions that would be obtained if the constraint $\mathbf{x} \leqslant \mathbf{u}$ is transformed into $\mathbf{x} + \mathbf{x}_s = \mathbf{u}$ and $\mathbf{x}_s \geqslant \mathbf{0}$, are equivalent.

5.13 Solve the following problem by the simplex method for bounded variables.

$$\begin{aligned}
\text{Maximize} \quad & 2x_1+3x_2-2x_3 \\
\text{Subject to} \quad & x_1+x_2+x_3 \leqslant 8 \\
& 2x_1+x_2-x_3 \geqslant 3 \\
& x_1 \leqslant 4 \\
& -2 \leqslant x_2 \leqslant 6 \\
& x_3 \geqslant 2
\end{aligned}$$

5.14 Compare the simplex method of Chapter 3 with the lower-upper bound simplex method. Indicate the number of operations (additions, subtractions, and so on) when each of the two methods is applied to the same lower-upper bounded linear program.

5.15 Show in detail that the simplex method for bounded variables discussed in Section 5.2 converges in a finite number of steps. Discuss in detail the problem of degeneracy and devise a rule that prevents cycling. Give all proofs.

5.16 a. Solve the following (knapsack) problem.

$$\begin{aligned} \text{Maximize} \quad & 2x_1 + 3x_2 + 8x_3 + x_4 + x_5 \\ \text{Subject to} \quad & 3x_1 + 7x_2 + 12x_3 + 2x_4 + 3x_5 \leqslant 10 \\ & x_1, \quad x_2, \quad x_3, \quad x_4, \quad x_5 \geqslant 0 \end{aligned}$$

b. Give a generalized closed form solution for the following problem.

$$\begin{aligned} \text{Maximize} \quad & c_1x_1 + \cdots + c_nx_n \\ \text{Subject to} \quad & a_1x_1 + \cdots + a_nx_n \leqslant b \\ & x_1, \quad \cdots, \quad x_n \geqslant 0 \end{aligned}$$

where c_j and a_j are positive scalars for each j.

c. What is the form of the optimal solution if c_j and a_j are allowed to be any scalars in part (b)?

5.17 Let $Q = \{1, 2, \ldots, n\}$, $P_i \subset Q$ with $P_i \cap P_j = \Phi$ for $i, j = 1, 2, \ldots, r$ and $i \neq j$, and

$$\bigcup_{i=1}^{r} P_i = Q$$

Develop an efficient method to solve the following problem where $c_j \geqslant 0$ for each j.

$$\begin{aligned} \text{Maximize} \quad & \sum_{j \in Q} c_j x_j \\ \text{Subject to} \quad & b_0' \leqslant \sum_{j \in Q} x_j \leqslant b_0'' \\ & b_i' \leqslant \sum_{j \in P_i} x_j \leqslant b_i'' \qquad i = 1, \ldots, r \\ & 0 \leqslant x_j \leqslant u_j \qquad j \in Q \end{aligned}$$

Apply the method to the following problem.

$$\begin{aligned} \text{Maximize} \quad & 10x_1 + 6x_2 + 3x_3 + 5x_4 + 8x_5 \\ \text{Subject to} \quad & 30 \leqslant x_1 + x_2 + x_3 + x_4 + x_5 \leqslant 100 \\ & 2 \leqslant x_1 + x_2 \leqslant 50 \\ & 70 \leqslant x_3 + x_4 + x_5 \leqslant 80 \\ & 0 \leqslant x_1, x_4, x_5 \leqslant 30 \\ & 0 \leqslant x_2, x_3 \leqslant 25 \end{aligned}$$

5.18 Solve the following problem by the simplex method for bounded variables.

$$\begin{aligned}
\text{Maximize} \quad & 2x_1+6x_2-x_3-4x_4+x_5 \\
\text{Subject to} \quad & 2x_1+x_2+4x_3+x_4+x_5=10 \\
& 3x_1+8x_2-3x_3+x_4 = 7 \\
& 0\leqslant x_1\leqslant 3 \\
& 1\leqslant x_2\leqslant 4 \\
& 0\leqslant x_3\leqslant 8 \\
& 1\leqslant x_4\leqslant 2 \\
& 0\leqslant x_5\leqslant 20
\end{aligned}$$

5.19 Consider the following problem.

$$\begin{aligned}
\text{Minimize} \quad & x_1+3x_2+4x_3 \\
\text{Subject to} \quad & -x_1-2x_2-x_3\leqslant -12 \\
& -x_1-x_2+2x_3\leqslant -6 \\
& -2x_1-x_2-4x_3\leqslant -24 \\
& x_1, \quad x_2, \quad x_3\geqslant 0
\end{aligned}$$

Let x_a be an artificial variable with an activity vector $\hat{\mathbf{b}} \leqslant \mathbf{b}$. Introducing the restrictions $0 \leqslant x_a \leqslant 1$ and letting $x_a = 1$ would lead to a starting basic feasible solution of the new system. Use the bounded simplex method to find a basic feasible solution of the original system.

5.20 Solve the following problem by the simplex method for bounded variables.

$$\begin{aligned}
\text{Maximize} \quad & 6x_1+4x_2+2x_3 \\
\text{Subject to} \quad & 4x_1-3x_2+x_3\leqslant 8 \\
& x_1+2x_2+4x_3\leqslant 10 \\
& 0\leqslant x_1\leqslant 3 \\
& 0\leqslant x_2\leqslant 2 \\
& 0\leqslant x_3
\end{aligned}$$

5.21 Solve the following problem by the simplex method for bounded variables.

$$\begin{aligned} \text{Minimize} \quad & 6x_1 + 2x_2 \\ \text{Subject to} \quad & x_1 + 3x_2 \geqslant 3 \\ & 5x_1 + x_2 \geqslant 4 \\ & x_1 \geqslant 2 \\ & x_2 \geqslant 1 \end{aligned}$$

5.22 Solve the following problem by the simplex method for bounded variables.

$$\begin{aligned} \text{Maximize} \quad & 6x_1 + 4x_2 \\ \text{Subject to} \quad & 3x_1 + 2x_2 \leqslant 4 \\ & x_1 + 2x_2 \leqslant 9 \\ & 0 \leqslant x_1 \leqslant 3 \\ & 0 \leqslant x_2 \leqslant 4 \end{aligned}$$

5.23 Show that the following two problems are equivalent.

$$\begin{aligned} P_1: \text{Minimize} \quad & \mathbf{cx} \\ \text{Subject to} \quad & \mathbf{b}_1 \leqslant \mathbf{Ax} \leqslant \mathbf{b}_2 \\ & \mathbf{x} \geqslant \mathbf{0} \end{aligned}$$

$$\begin{aligned} P_2: \text{Minimize} \quad & \mathbf{cx} \\ \text{Subject to} \quad & \mathbf{Ax} + \mathbf{s} = \mathbf{b}_2 \\ & \mathbf{x} \geqslant \mathbf{0} \\ & \mathbf{0} \leqslant \mathbf{s} \leqslant \mathbf{b}_2 - \mathbf{b}_1 \end{aligned}$$

Use the lower-upper bound simplex method to solve the following.

$$\begin{aligned} \text{Minimize} \quad & 3x_1 - 4x_2 \\ \text{Subject to} \quad & 3 \leqslant x_1 + x_2 \leqslant 5 \\ & 2 \leqslant 2x_1 - 5x_2 \leqslant 8 \\ & x_1, \quad x_2 \geqslant 0. \end{aligned}$$

(Note that it will be necessary to use artificial variables to get started.)

5.24 A government has allocated \$1.5 billion of its budget for military purposes. Sixty percent of the military budget will be used to purchase tanks, planes, and missile batteries. These can be acquired at a unit cost of \$600,000, \$2 million, and \$800,000 respectively. It is decided that at least 200 tanks and 200 planes must be acquired. Because of the shortage of experienced pilots, it is also decided not to purchase more than 300 planes. For strategic purposes the ratio of the missile batteries to the planes purchased must fall in the range from $\frac{1}{4}$ to $\frac{1}{2}$. The objective is to maximize the overall utility of these weapons where the individual utilities are given as 1, 3, and 2 respectively. Find the optimal solution.

5.25 Solve Exercise 1.13.

5.26 A farmer who raises chickens would like to determine the amounts of the available ingredients that will meet certain nutritional requirements. The available ingredients and their cost, the nutrients in the ingredients, and the daily requirements are summarized below.

	INGREDIENT			MINIMUM DAILY
NUTRIENT	CORN	LIME	ALFALFA	REQUIREMENT
Protein	8	4	4	10
Carbohydrates	4	2	4	6
Vitamins	2	3	4	5
\$ cost	0.10	0.06	0.04	

Find the optimal mix using the revised simplex method with the product form of the inverse. Use only one artificial variable.

5.27 An automobile manufacturer has contracted to export 400 cars of model A and 500 cars of model B overseas. The model A car occupies a volume of 12 cubic meters, and the model B car occupies a volume of 15 cubic meters. Three ships for transporting the automobiles are available. These arrive at the port of destination at the beginning of January, the middle of February, and the end of March respectively. The first ship only transports model A cars at \$450 per automobile. The second and third ships transport both types at a cost of \$35 and \$40 per cubic meter respectively. The first ship can only accommodate 200 cars, and the second and third ships have available volumes of 4500 and 6000 cubic meters. If the manufacturer has contracted to deliver at least 250 and 200 cars of model A and B by the middle of February and the remainder by the end of March, what is the shipping pattern that minimizes the total cost? Use the revised simplex method.

5.28 A manufacturing firm would like to plan its production/inventory policy for the months of August, September, October, and November. The product under consideration is seasonable, and its demand over the particular months is estimated to be 500, 600, 800, and 1200 units respectively. Presently the monthly production capacity is 600 units with a unit cost of \$25. Management has decided to install a new production system with monthly capacity of 1100 units at a unit cost of \$30. However, the new system cannot be installed until the middle of November. Assume that the starting inventory is 250 units and that at most 400 units can be stored during any given month. If the holding inventory cost per month per item is \$3, find the production schedule that minimizes the total production and inventory cost using the bounded simplex method. Assume that demand must be satisfied and that 100 units are required in inventory at the end of November.

5.29 Consider the problem: Minimize $\mathbf{cx}$ subject to $\mathbf{Ax} = \mathbf{b}$, $\mathbf{x} \geqslant \mathbf{0}$. Let $\mathbf{B}$ be a basis. After adding the redundant constraints $\mathbf{x}_N - \mathbf{x}_N = \mathbf{0}$, the following equations represent all the variables in terms of the independent variables $\mathbf{x}_N$:

	$\mathbf{x}_N$	RHS
z	$\mathbf{c}_B\mathbf{B}^{-1}\mathbf{N} - \mathbf{c}_N$	$\mathbf{c}_B\bar{\mathbf{b}}$
$\mathbf{x}_B$	$\mathbf{B}^{-1}\mathbf{N}$	$\bar{\mathbf{b}}$
$\mathbf{x}_N$	$-\mathbf{I}$	$\mathbf{0}$

The simplex method proceeds by choosing the most positive $z_j - c_j$, say $z_k - c_k$. Then x_k enters the basis and the usual minimum ratio test indicates that x_{B_r} leaves the basis. The foregoing array can be updated by *column pivoting* at y_{rk} as follows.

1. Divide the kth column (pivot column) by $-y_{rk}$.
2. Multiply the kth column (pivot column) by y_{rj} and add to the jth column.
3. Multiply the kth column by $\bar{b}_r$ and add to the right-hand side.
4. Remove the variable x_k from the list of nonbasic variables and add x_{B_r} instead in its place. Note that no row designations are changed.

This method of displaying the tableau and updating it is usually called the *column simplex method*. Show that pivoting gives the representation of all the variables in terms of the new nonbasic variables. In particular, show that pivoting updates the tableau such that the new $\mathbf{c}_B\mathbf{B}^{-1}\mathbf{N} - \mathbf{c}_N$, $\mathbf{B}^{-1}\mathbf{N}$, $\mathbf{B}^{-1}\mathbf{b}$, and $\mathbf{c}_B\mathbf{B}^{-1}\mathbf{b}$ are immediately available.

5.30 Referring to Problem 5.29, solve the following problem using the column simplex method.

$$\begin{array}{ll} \text{Maximize} & x_1+2x_2+3x_3 \\ \text{Subject to} & 3x_1+2x_2+x_3 \leqslant 6 \\ & -x_1+2x_2+4x_3 \leqslant 8 \\ & 2x_1+x_2-x_3 \leqslant 2 \\ & x_1, \quad x_2, \quad x_3 \geqslant 0 \end{array}$$

5.31 Referring to Problem 5.29, is it possible to extract the inverse basis from a typical column simplex tableau? If so, how can this be done?

5.32 Write the optimality conditions for each of the following problems.

a. Maximize $\mathbf{cx}$
Subject to $\mathbf{Ax} \leqslant \mathbf{b}$
$\mathbf{x} \geqslant \mathbf{0}$

b. Maximize $\mathbf{cx}$
Subject to $\mathbf{Ax} \geqslant \mathbf{b}$
$\mathbf{x} \geqslant \mathbf{0}$

c. Minimize $\mathbf{cx}$
Subject to $\mathbf{Ax} \leqslant \mathbf{b}$
$\mathbf{x} \geqslant \mathbf{0}$

d. Minimize $\mathbf{cx}$
Subject to $\mathbf{A}_1\mathbf{x} = \mathbf{b}_1$
$\mathbf{A}_2\mathbf{x} \geqslant \mathbf{b}_2$
$\mathbf{x} \geqslant \mathbf{0}$

e. Minimize $\mathbf{cx}$
Subject to $\mathbf{Ax} = \mathbf{b}$
$\mathbf{l} \leqslant \mathbf{x} \leqslant \mathbf{u}$

5.33 Consider the problem: Minimize $\mathbf{cx}$ subject to $\mathbf{Ax} \geqslant \mathbf{b}$, $\mathbf{x} \geqslant \mathbf{0}$. Suppose that $\mathbf{x}$ is an optimal solution. Further suppose that $x_j > 0$ for $j = 1, 2, \ldots, p$, $x_j = 0$ for $j = p+1, \ldots, n$, $\mathbf{A}_1\mathbf{x} = \mathbf{b}_1$, and $\mathbf{A}_2\mathbf{x} > \mathbf{b}_2$ where $\mathbf{A} = \begin{bmatrix} \mathbf{A}_1 \\ \mathbf{A}_2 \end{bmatrix}$ and $\mathbf{b} = \begin{bmatrix} \mathbf{b}_1 \\ \mathbf{b}_2 \end{bmatrix}$. Show that the system $\mathbf{A}_1\mathbf{d} \geqslant \mathbf{0}$, $d_{p+1}, \ldots, d_n \geqslant 0$ and $\mathbf{cd} < 0$ has no solution $\mathbf{d}$ in E^n.

5.34 Consider the problem: Maximize $\mathbf{cx}$ subject to $\mathbf{Ax} \leqslant \mathbf{b}$, $\mathbf{x} \geqslant \mathbf{0}$. Introducing the slack vector $\mathbf{x}_s$, we get the equivalent problem: Maximize $\mathbf{cx}$ subject to $\mathbf{Ax} + \mathbf{x}_s = \mathbf{b}$, $\mathbf{x} \geqslant \mathbf{0}$, $\mathbf{x}_s \geqslant \mathbf{0}$. Write the Kuhn-Tucker optimality conditions for both problems. Show equivalence of the optimality conditions, and show that the Lagrangian multipliers corresponding to the nonnegativity $\mathbf{x}_s \geqslant \mathbf{0}$ are equal to the Lagrangian multipliers of the constraints $\mathbf{Ax} + \mathbf{x}_s = \mathbf{b}$ and $\mathbf{Ax} \leqslant \mathbf{b}$.

5.35 Consider the following problem.

$$\begin{aligned} \text{Maximize} \quad & 2x_1 + x_2 \\ \text{Subject to} \quad & x_1 + x_2 \leqslant 4 \\ & x_2 \leqslant 3 \\ & x_1, \ x_2 \geqslant 0 \end{aligned}$$

The optimal point is (4, 0). Verify this statement by the Kuhn-Tucker optimality conditions, and interpret your result geometrically.

5.36 Solve the following linear program by the simplex method, and show that the solution satisfies the Kuhn-Tucker conditions. At each iteration, point out the source of violating the optimality conditions.

$$\begin{aligned} \text{Maximize} \quad & 10x_1 + 15x_2 + 5x_3 \\ \text{Subject to} \quad & 2x_1 + x_2 \leqslant 6000 \\ & 3x_1 + 3x_2 + x_3 \leqslant 9000 \\ & x_1 + 2x_2 + 2x_3 \leqslant 4000 \\ & x_1, \quad x_2, \quad x_3 \geqslant 0 \end{aligned}$$

5.37 Consider the linear programming problem: Minimize $\mathbf{cx}$ subject to $\mathbf{Ax} = \mathbf{b}$, $\mathbf{x} \geqslant \mathbf{0}$. It is well known that a feasible point $\mathbf{x}$ is an optimal solution if and only if there exist vectors $\mathbf{w}$ and $\mathbf{v}$ such that

$$\mathbf{c} - \mathbf{wA} - \mathbf{v} = \mathbf{0}$$

$$\mathbf{v} \geqslant \mathbf{0}$$

$$\mathbf{vx} = 0$$

Is it possible that $\mathbf{x}$ be optimal if $v_i \geqslant 0$ for all $i \neq j$ whereas $v_j < 0$ and the corresponding $x_j = 0$? In other words, is it possible to have a degenerate optimal solution with one of the Lagrangian multipliers (shadow prices) of the nonnegativity constraints being negative? Explain why or why not. Illustrate by a numerical example. (*Hint.* Construct a linear program with an optimal degenerate basic feasible solution. Try different basis representations.)

5.38 Consider the problem: Minimize $\mathbf{cx}$ subject to $\mathbf{Ax} \geqslant \mathbf{b}$, $\mathbf{x} \geqslant \mathbf{0}$. Let $\mathbf{x}^*$ be an optimal solution. Suppose that $\mathbf{A}$ is decomposed into $\begin{bmatrix} \mathbf{A}_1 \\ \mathbf{A}_2 \end{bmatrix}$ and $\mathbf{b}$ is decomposed into $\begin{bmatrix} \mathbf{b}_1 \\ \mathbf{b}_2 \end{bmatrix}$ such that $\mathbf{A}_1\mathbf{x}^* = \mathbf{b}_1$ and $\mathbf{A}_2\mathbf{x}^* > \mathbf{b}_2$. Show that $\mathbf{x}^*$ is also an optimal solution of the problem: Minimize $\mathbf{cx}$ subject to $\mathbf{A}_1\mathbf{x} \geqslant \mathbf{b}_1$, $\mathbf{x} \geqslant \mathbf{0}$, and to the problem: Minimize $\mathbf{cx}$ subject to $\mathbf{A}_1\mathbf{x} = \mathbf{b}_1$, $\mathbf{x} \geqslant \mathbf{0}$.

5.39 A manufacturer produces two items with unit profits \$10.00 and \$15.00. Each unit of item 1 uses 4 man-hours and 3 machine-hours. Each unit of item 2 uses 7 man-hours and 6 machine-hours. If the total man-hours and machine-hours available are 300 and 500 respectively, find the optimal solution and verify optimality by the Kuhn-Tucker conditions. Interpret the optimality conditions geometrically. Give an economic interpretation of the Kuhn-Tucker conditions at the optimal point. (*Hint*. Recall the economic interpretation of w_1 and w_2.)

5.40 Can a linear program be solved by writing down the Kuhn-Tucker conditions and finding a solution to them? If yes, why is this not done?

5.41 Consider the problem: Minimize $\mathbf{cx}$ subject to $\mathbf{Ax} = \mathbf{b}$, $\mathbf{x} \geqslant \mathbf{0}$. Leaving the question of feasibility aside, show that starting from any point $\mathbf{x}$, the direction with norm 1, which best improves the objective function, is $-\mathbf{c}/\|\mathbf{c}\|$.

5.42 Consider the following problem.

$$\begin{aligned} \text{Maximize} \quad & x_1 - 2x_2 + x_3 \\ \text{Subject to} \quad & x_1 + x_2 + x_3 \leqslant 6 \\ & 2x_1 + x_2 \leqslant 4 \\ & -x_1 + 2x_2 - x_3 \leqslant 4 \\ & x_1, \quad x_2, \quad x_3 \geqslant 0 \end{aligned}$$

Solve the problem by the simplex method, illustrating at each iteration source of violation of the Kuhn-Tucker conditions.

5.43 Consider the following problem.

$$\begin{aligned} \text{Maximize} \quad & 2x_1 + 3x_2 \\ \text{Subject to} \quad & x_1 + x_2 \leqslant 8 \\ & -2x_1 + 3x_2 \leqslant 12 \\ & x_1, \quad x_2 \geqslant 0 \end{aligned}$$

a. Solve the problem geometrically. At each iteration identify the variable that enters and the variable that leaves the basis.
b. Write the Kuhn-Tucker conditions and show that they hold at optimality.

5.44 The following is an idea of a graphical example of the simplex method at work.

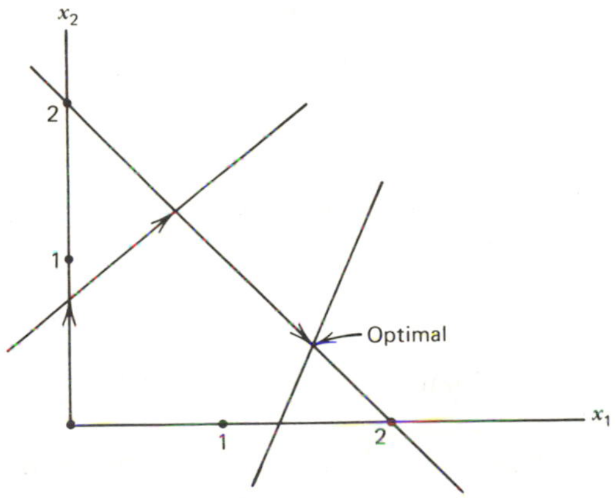

a. Give the starting basis and each succeeding basis until the optimal point is reached. Specify the entering and leaving vectors.
b. If the optimal point is unique, could the simplex method have gone in the direction it did assuming that the entering variable is that with the most positive $z_j - c_j$?

5.45 Consider the linear programming problem: Minimize $\mathbf{cx}$ subject to $\mathbf{Ax} = \mathbf{b}$, $\mathbf{x} \geqslant \mathbf{0}$. Let $\mathbf{x}$ be a basic feasible solution with basis $\mathbf{B}$. The simplex method proceeds by increasing the nonbasic variable with the most positive $z_j - c_j$.
a. Devise a procedure in which all nonbasic variables with positive $z_j - c_j$'s are increased. How are the basic variables modified? By how much would you increase the nonbasic variables? Interpret increasing several variables simultaneously.
b. At later iterations we may have more than m positive variables. How would you represent the corresponding point in a tableau format? (*Hint.* Let the basis consist of the largest m variables. Other nonbasic variables may be either positive or zero.)
c. At later iterations you may have a positive nonbasic variable with negative $z_j - c_j$. What happens if you decrease x_j?
d. Use the ideas of (a), (b), and (c) above to construct a complete algorithm for solving linear programs where several nonbasic variables

are simultaneously modified. How do you recognize reaching the optimal solution? What are the advantages and disadvantages of your procedure?

e. Illustrate your procedure by solving the following problem.

$$\begin{aligned} \text{Minimize} \quad & -x_1 - 2x_2 - x_3 \\ \text{Subject to} \quad & x_1 + x_2 + 3x_3 \leqslant 12 \\ & x_1 + 2x_2 \leqslant 6 \\ & x_1 + x_3 \leqslant 8 \\ & x_1, \ x_2, \ x_3 \geqslant 0 \end{aligned}$$

f. Consider the following alternative procedure for modifying the nonbasic variables. For each nonbasic variable x_j let

$$d_j = \begin{cases} z_j - c_j & \text{if } x_j > 0 \\ \text{Maximum } (0, z_j - c_j) & \text{if } x_j = 0 \end{cases}$$

Modify the nonbasic variables according to the d_j's above and the basic variables according to the relationship

$$\mathbf{x}_B = \hat{\mathbf{b}} - \lambda \sum_{j \in R} \mathbf{y}_j d_j$$

where the vector $\hat{\mathbf{b}}$ represents the current values of the basic variables and $\lambda \geqslant 0$ is to be determined. Interpret this method and compare it with the method in (d). Solve the problem in (e) by this procedure.

5.46 The accompanying diagram depicts the region given by $a_1x_1 + a_2x_2 \leqslant b$ and $x_1, x_2 \geqslant 0$. Let (x_1, x_2) be the shown point. The distance from the x_1 and x_2 axes give the values x_1 and x_2 respectively. Indicate on the diagram the value of the slack variable. How can you generalize the result to n variables?

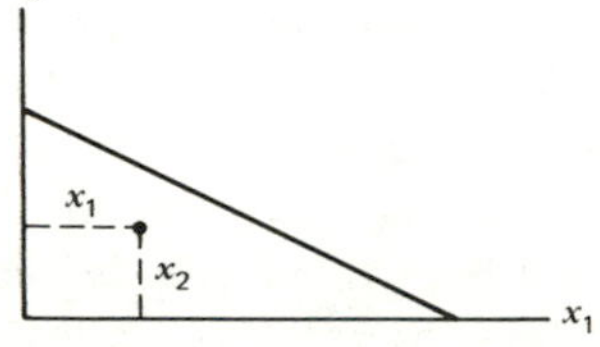

5.47 Suppose that we are given an extreme point $\mathbf{x}$ of a polyhedral set X. An extreme point $\mathbf{y} \neq \mathbf{x}$ is called *adjacent* to $\mathbf{x}$ if there exists a hyperplane that supports X and its intersection with X is the line segment joining $\mathbf{x}$ and $\mathbf{y}$. In the accompanying diagram, obviously $\mathbf{x}_2$ and $\mathbf{x}_5$ are extreme points of X that are adjacent to $\mathbf{x}_1$, whereas $\mathbf{x}_3$ and $\mathbf{x}_4$ are not adjacent to $\mathbf{x}_1$. Now let X consist of all points satisfying $\mathbf{Ax} = \mathbf{b}$ and $\mathbf{x} \geqslant \mathbf{0}$, where $\mathbf{A}$ is a $m \times n$ matrix with rank m. Further suppose that X is bounded. Let $\mathbf{x}$ be a nondegenerate basic feasible solution (extreme point). Characterize the collection of adjacent extreme points. What is their number? What happens if the nondegeneracy assumption is relaxed? In each case justify your answer.

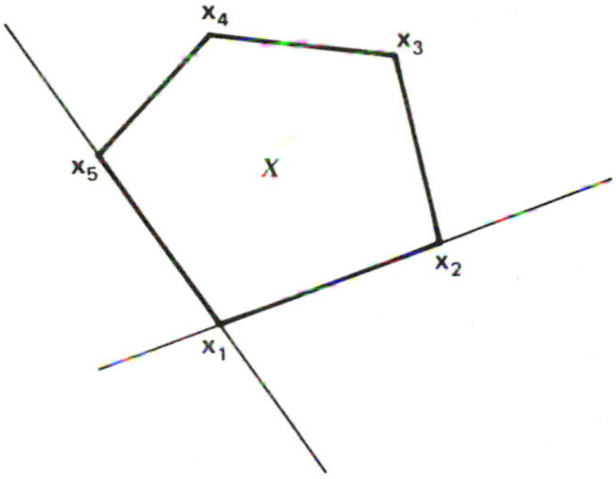

5.48 Referring to Exercise 5.47, show that the simplex method moves from an extreme point to an adjacent extreme point. (*Hint*. Suppose you have a basic feasible solution $\mathbf{x}$ with basis $\mathbf{B}$ consisting of the first m columns of $\mathbf{A}$. Further suppose that x_k entered the basis. Consider the hyperplane passing through $\mathbf{x}$ and whose normal vector is $(\mathbf{p}, \mathbf{pB}^{-1}\mathbf{N} + (1, 1, \ldots, 1, 0, 1, \ldots, 1))$, where $\mathbf{p}$ is an arbitrary m vector and the zero component of $(1, 1, \ldots, 1, 0, 1, 1, \ldots, 1)$ appears at position $k - m$.)

5.49 Consider the collection of points satisfying $\mathbf{x} \geqslant \mathbf{0}$ and $\mathbf{Ax} = \mathbf{b}$, where $\mathbf{A}$ is an $m \times n$ matrix with rank m. Further suppose that the region is bounded. Let $\mathbf{x}_0$ be an extreme point of the region, and let $\mathbf{x}_1, \mathbf{x}_2, \ldots, \mathbf{x}_k$ be the adjacent extreme points of the region (refer to Exercise 5.47). Let $\mathbf{x}$ be any point in the region. Show that $\mathbf{x}$ can be represented as

$$\mathbf{x} = \mathbf{x}_0 + \sum_{j=1}^{k} \mu_j(\mathbf{x}_j - \mathbf{x}_0) \quad \text{where} \quad \mu_j \geqslant 0 \quad \text{for} \quad j = 1, 2, \ldots, k$$

Interpret this result geometrically. [*Hint*. Let $\mathbf{x}_0 = \begin{bmatrix} \mathbf{0} \\ \mathbf{B}^{-1}\mathbf{b} \end{bmatrix} = \begin{bmatrix} \mathbf{0} \\ \bar{\mathbf{b}} \end{bmatrix}$ (the nonbasic variables are placed first for convenience). Show that

$$\mathbf{x}_j = \begin{bmatrix} 0 \\ \vdots \\ \lambda_j \\ \vdots \\ 0 \\ \hline \bar{\mathbf{b}} - \lambda_j \mathbf{y}_j \end{bmatrix}$$

for $j = 1, 2, \ldots, k = n - m$ where $\mathbf{y}_j = \mathbf{B}^{-1}\mathbf{a}_j$ and

$$\lambda_j = \underset{1 \leqslant i \leqslant m}{\text{Minimum}} \left\{ \frac{\bar{b}_i}{y_{ij}} : y_{ij} > 0 \right\} \Bigg]$$

5.50 Suppose that the boundedness restriction in Exercise 5.49 is dropped. Can you generalize the foregoing result? Interpret your result geometrically. (*Hint.* Introduce the notion of an adjacent direction. Then $\mathbf{x}$ in the region can be represented as

$$\mathbf{x} = \mathbf{x}_0 + \sum_{j \in I} \mu_j (\mathbf{x}_j - \mathbf{x}_0) + \sum_{j \in J} \mu_j \mathbf{d}_j$$

where $\mu_j \geqslant 0$ for $j \in I \cup J$, $\mathbf{x}_j$ for $j \in I$ are adjacent extreme points and $\mathbf{d}_j$ for $j \in J$ are adjacent extreme directions.)

5.51 Show that an extreme point of a bounded polyhedral set has a minimal objective if and only if it has an objective that is smaller than or equal to that of any adjacent extreme point. Can you generalize the result to the unbounded case? (*Hint.* Use Exercises 5.49 and 5.50)

5.52 Consider the problem: Minimize $\mathbf{cx}$ subject to $\mathbf{Ax} = \mathbf{b}$, $\mathbf{x} \geqslant \mathbf{0}$. Let $\mathbf{x}^*$ be an unique optimal extreme point. Show that the second best extreme point must be adjacent to $\mathbf{x}^*$. What happens if the uniqueness assumption is relaxed?

NOTES AND REFERENCES

1. The revised simplex method was devised by Dantzig and Orchard-Hays [114] in 1953 (also see Dantzig and Orchard-Hays [113] for the product form of the inverse). For further reading on this topic, refer to Dantzig [97].

2. In Chapter 7 we describe the use of the revised simplex method in solving large-scale problems in the context of decomposition by generating columns at each iteration.
3. The simplex method for bounded variables was published by Dantzig [93] in 1955. The method was first developed at the RAND corporation to provide a shortcut routine for solving a problem of assigning personnel. The method was independently developed by Charnes and Lemke [71] in 1954.
4. The Kuhn-Tucker optimality conditions for nonlinear programs were first developed by Kuhn and Tucker [295] in 1950. These conditions are necessary (and sufficient under suitable convexity assumptions) for optimality. Specialization of these conditions for linear programs is given in Section 5.3. For further reading on the Kuhn-Tucker conditions the reader may refer to Mangasarian [319] and Zangwill [486]. The Kuhn-Tucker conditions in linear programming and the subject of duality are very closely associated. This fact will become apparent after the reader studies Chapter 6.

SIX: DUALITY AND SENSITIVITY

For every linear program there is another associated linear program. This new linear program satisfies some very important properties. It may be used to obtain the solution to the original program. Its variables provide extremely useful information about the optimal solution to the original linear program. To distinguish points of view in this and subsequent chapters we shall call the original linear programming problem the *primal* (linear programming) problem.

We shall begin by formulating this new linear program, called the *dual* (linear programming) problem, and proceed to develop some of its important properties. These properties will lead to two new algorithms, the dual simplex method and the primal-dual algorithm, for solving linear programs. Finally, we shall discuss the effect of variation in the data, that is, the cost coefficients, the right-hand-side coefficients, and the constraint coefficients on the optimal solution to a linear program.

6.1 FORMULATION OF THE DUAL PROBLEM

Associated with each linear programming problem there is another linear programming problem called the dual. The dual linear program possesses many important properties relative to the original primal linear program. There are two important forms (definitions) of duality: the *canonical* form of duality and the *standard* form of duality. These two forms are completely equivalent.

Canonical Form of Duality

Suppose that the primal linear program is given in the form:

$$\begin{aligned} \text{P: Minimize} \quad & \mathbf{cx} \\ \text{Subject to} \quad & \mathbf{Ax} \geqslant \mathbf{b} \\ & \mathbf{x} \geqslant \mathbf{0} \end{aligned}$$

Then the *dual* linear program is defined by:

$$\begin{aligned} \text{D: Maximize} \quad & \mathbf{wb} \\ \text{Subject to} \quad & \mathbf{wA} \leqslant \mathbf{c} \\ & \mathbf{w} \geqslant \mathbf{0} \end{aligned}$$

Note that there is exactly one dual variable for each primal constraint and exactly one dual constraint for each primal variable. We shall say more about this later.

Example 6.1

Consider the following linear program and its dual.

$$\begin{aligned} \text{P: Minimize} \quad & 6x_1 + 8x_2 \\ \text{Subject to} \quad & 3x_1 + x_2 \geqslant 4 \\ & 5x_1 + 2x_2 \geqslant 7 \\ & x_1, \quad x_2 \geqslant 0 \end{aligned}$$

$$\begin{aligned} \text{D: Maximize} \quad & 4w_1 + 7w_2 \\ \text{Subject to} \quad & 3w_1 + 5w_2 \leqslant 6 \\ & w_1 + 2w_2 \leqslant 8 \\ & w_1, \quad w_2 \geqslant 0 \end{aligned}$$

Before proceeding further, try solving both problems and compare their optimal objective values. This will provide a hint of things to come.

In the canonical definition of duality it is important for problem P to have a "Minimization" objective with all "greater than or equal to" constraints and all "nonnegative" variables. In theory, to apply the canonical definition of duality we must first convert the primal linear program to the foregoing format. However, in practice it is possible to immediately write down the dual of any linear program. We shall discuss this shortly.

Standard Form of Duality

Another equivalent definition of duality applies when the constraints are equalities. Suppose that the primal linear program is given in the form:

$$\begin{aligned} P:\ \text{Minimize}\ & \mathbf{cx} \\ \text{Subject to}\ & \mathbf{Ax} = \mathbf{b} \\ & \mathbf{x} \geq \mathbf{0} \end{aligned}$$

Then the *dual* linear program is defined by:

$$\begin{aligned} D:\ \text{Maximize}\ & \mathbf{wb} \\ \text{Subject to}\ & \mathbf{wA} \leq \mathbf{c} \\ & \mathbf{w}\ \text{unrestricted} \end{aligned}$$

Example 6.2

Consider the following linear program and its dual.

$$\begin{aligned} P:\ \text{Minimize}\ & 6x_1 + 8x_2 \\ \text{Subject to}\ & 3x_1 + x_2 - x_3 = 4 \\ & 5x_1 + 2x_2 - x_4 = 7 \\ & x_1,\ x_2,\ x_3,\ x_4 \geq 0 \end{aligned}$$

$$\begin{aligned} D:\ \text{Maximize}\ & 4w_1 + 7w_2 \\ \text{Subject to}\ & 3w_1 + 5w_2 \leq 6 \\ & w_1 + 2w_2 \leq 8 \\ & -w_1 \leq 0 \\ & -w_2 \leq 0 \\ & w_1,\ w_2\ \text{unrestricted} \end{aligned}$$

Given one of the definitions, canonical or standard, it is easy to demonstrate that the other definition is valid. For example, suppose that we accept the standard form as a definition and wish to demonstrate that the canonical form is correct. By adding slack variables to the canonical form of a linear program, we may apply the standard form of duality to obtain the dual problem.

P: Minimize $\mathbf{cx}$	D: Maximize $\mathbf{wb}$
Subject to $\mathbf{Ax} - \mathbf{Ix}_s = \mathbf{b}$	Subject to $\mathbf{wA} \leqslant \mathbf{c}$
$\mathbf{x}, \quad \mathbf{x}_s \geqslant \mathbf{0}$	$-\mathbf{wI} \leqslant \mathbf{0}$
	$\mathbf{w}$ unrestricted

But since $-\mathbf{wI} \leqslant \mathbf{0}$ is the same as $\mathbf{w} \geqslant \mathbf{0}$, we obtain the canonical form of the dual problem.

Dual of the Dual

Since the dual linear program is itself a linear program, we may wonder what its dual might be. Consider the dual in canonical form:

$$\begin{aligned} &\text{Maximize } \mathbf{wb} \\ &\text{Subject to } \mathbf{wA} \leqslant \mathbf{c} \\ &\qquad\qquad \mathbf{w} \geqslant \mathbf{0} \end{aligned}$$

Applying the transformation techniques of Chapter 1, we may rewrite this problem in the form:

$$\begin{aligned} &\text{Minimize } (-\mathbf{b}^t)\mathbf{w}^t \\ &\text{Subject to } (-\mathbf{A}^t)\mathbf{w}^t \geqslant (-\mathbf{c}^t) \\ &\qquad\qquad \mathbf{w}^t \geqslant \mathbf{0} \end{aligned}$$

The dual linear program for this linear program is given by (letting $\mathbf{x}^t$ play the role of the row vector of dual variables):

$$\begin{aligned} &\text{Maximize } \mathbf{x}^t(-\mathbf{c}^t) \\ &\text{Subject to } \mathbf{x}^t(-\mathbf{A}^t) \leqslant (-\mathbf{b}^t) \\ &\qquad\qquad \mathbf{x}^t \geqslant \mathbf{0} \end{aligned}$$

But this is the same as

$$\begin{aligned} \text{Minimize} \quad & \mathbf{cx} \\ \text{Subject to } & \mathbf{Ax} \geqslant \mathbf{b} \\ & \mathbf{x} \geqslant \mathbf{0} \end{aligned}$$

which is precisely the original primal problem. Thus we have the following lemma.

Lemma 1

The dual of the dual is the primal.

This lemma indicates that the definitions may be applied in reverse. The terms "primal" and "dual" are relative to the frame of reference we choose.

Mixed Forms of Duality

In practice, many linear programs contain some constraints of the "less than or equal to" type, some of the "greater than or equal to" type and some of the "equal to" type. Also, variables may be "$\geqslant 0$," "$\leqslant 0$," or "unrestricted." In theory, this presents no problem since we may apply the transformation techniques of Chapter 1 to convert any "mixed" problem to one of the primal or dual forms discussed above, after which the dual can be readily obtained. In practice such conversions can be tedious. Fortunately, it is not necessary actually to make these conversions, and it is possible to give immediately the dual of any linear program.

Consider the following linear program.

$$\begin{aligned} \text{Minimize} \quad & \mathbf{cx} \\ \text{Subject to } & \mathbf{A}_1\mathbf{x} \geqslant \mathbf{b}_1 \\ & \mathbf{A}_2\mathbf{x} = \mathbf{b}_2 \\ & \mathbf{A}_3\mathbf{x} \leqslant \mathbf{b}_3 \\ & \mathbf{x} \geqslant \mathbf{0} \end{aligned}$$

Converting this problem to the standard format, we get

$$\begin{aligned} \text{Minimize} \quad & \mathbf{cx} \\ \text{Subject to} \quad & \mathbf{A}_1\mathbf{x} - \mathbf{I}\mathbf{x}_s = \mathbf{b}_1 \\ & \mathbf{A}_2\mathbf{x} = \mathbf{b}_2 \\ & \mathbf{A}_3\mathbf{x} + \mathbf{I}\mathbf{x}_t = \mathbf{b}_3 \\ & \mathbf{x}, \quad \mathbf{x}_s, \quad \mathbf{x}_t \geqslant \mathbf{0} \end{aligned}$$

The dual of this problem is

$$
\begin{aligned}
\text{Maximize} \quad & \mathbf{w}_1\mathbf{b}_1 + \mathbf{w}_2\mathbf{b}_2 + \mathbf{w}_3\mathbf{b}_3 \\
\text{Subject to} \quad & \mathbf{w}_1\mathbf{A}_1 + \mathbf{w}_2\mathbf{A}_2 + \mathbf{w}_3\mathbf{A}_3 \leqslant \mathbf{c} \\
& -\mathbf{w}_1\mathbf{I} \leqslant \mathbf{0} \\
& \mathbf{w}_3\mathbf{I} \leqslant \mathbf{0} \\
& \mathbf{w}_1, \quad \mathbf{w}_2, \quad \mathbf{w}_3 \quad \text{unrestricted}
\end{aligned}
$$

From this example we see that "greater than or equal to" constraints in the minimization problem give rise to "$\geqslant 0$" variables in the maximization problem. Also, "equal to" constraints in the minimization problem give rise to "unrestricted" variables in the maximization problem; and "less than or equal to" constraints in the minimization problem give rise to "$\leqslant 0$" variables in the maximization problem. In Exercise 6.5 we ask the reader to consider the various cases for variables and constraints in the minimization problem and their counterparts in the maximization problem. The complete results may be summarized in Table 6.1.

Table 6.1 Relationships Between Primal and Dual Problems

	MINIMIZATION PROBLEM		MAXIMIZATION PROBLEM	
Variables	$\geqslant 0$	$\longleftrightarrow$	$\leqslant$	Constraints
	$\leqslant 0$	$\longleftrightarrow$	$\geqslant$	
	Unrestricted	$\longleftrightarrow$	$=$	
Constraints	$\geqslant$	$\longleftrightarrow$	$\geqslant 0$	Variables
	$\leqslant$	$\longleftrightarrow$	$\leqslant 0$	
	$=$	$\longleftrightarrow$	Unrestricted	

We may utilize this table to develop the dual of any linear program without first transforming it to the standard or canonical forms.

Example 6.3

Consider the following linear program.

$$
\begin{aligned}
\text{Maximize} \quad & 8x_1 + 3x_2 \\
\text{Subject to} \quad & x_1 - 6x_2 \geqslant 2 \\
& 5x_1 + 7x_2 = -4 \\
& x_1 \leqslant 0 \\
& x_2 \geqslant 0
\end{aligned}
$$

Applying the results of the table, we can immediately write down the dual.

$$\begin{array}{lrl} \text{Minimize} & 2w_1 - 4w_2 & \\ \text{Subject to} & w_1 + 5w_2 & \leqslant 8 \\ & -6w_1 + 7w_2 & \geqslant 3 \\ & w_1 & \leqslant 0 \\ & w_2 & \text{unrestricted} \end{array}$$

6.2 PRIMAL-DUAL RELATIONSHIPS

The definition we have selected for the dual problem leads to many important relationships between the primal and dual linear programs.

The Relationship Between Objective Values

Consider the canonical form of duality and let $\mathbf{x}_0$ and $\mathbf{w}_0$ be feasible solutions to the primal and dual programs respectively. Then $\mathbf{Ax}_0 \geqslant \mathbf{b}$, $\mathbf{x}_0 \geqslant \mathbf{0}$, $\mathbf{w}_0\mathbf{A} \leqslant \mathbf{c}$, and $\mathbf{w}_0 \geqslant \mathbf{0}$. Multiplying $\mathbf{Ax}_0 \geqslant \mathbf{b}$ on the left by $\mathbf{w}_0 \geqslant \mathbf{0}$ and $\mathbf{w}_0\mathbf{A} \leqslant \mathbf{c}$ on the right by $\mathbf{x}_0 \geqslant \mathbf{0}$, we get

$$\mathbf{cx}_0 \geqslant \mathbf{w}_0\mathbf{Ax}_0 \geqslant \mathbf{w}_0\mathbf{b}$$

The result is the following.

Lemma 2

The objective function value for any feasible solution to the minimization problem is always greater than or equal to the objective function value for any feasible solution to the maximization problem. In particular, the objective value of any feasible solution of the minimization problem gives an upper bound on the optimal objective of the maximization problem. Similarly, the objective value of any feasible solution of the maximization problem is a lower bound on the optimal objective of the minimization problem.

As an illustration of the application of this lemma, suppose that in Example 6.1 we select the feasible primal and dual solutions $\mathbf{x}_0 = (\frac{7}{5}, 0)^t$ and $\mathbf{w}_0 = (2, 0)$. Then $\mathbf{cx}_0 = \frac{42}{5} = 8.4$ and $\mathbf{w}_0\mathbf{b} = 8$. Thus the optimal solution for either problem has objective value between 8 and 8.4. This allows us to stop a linear programming solution procedure with a near optimal solution.

The following corollaries are immediate consequences of Lemma 2.

Corollary 1

If $\mathbf{x}_0$ and $\mathbf{w}_0$ are feasible solutions to the primal and dual problems such that $\mathbf{c}\mathbf{x}_0 = \mathbf{w}_0\mathbf{b}$, then $\mathbf{x}_0$ and $\mathbf{w}_0$ are optimal solutions to their respective problems.

Corollary 2

If either problem has an unbounded objective value, then the other problem possesses no feasible solution.

This corollary indicates that unboundedness in one problem implies infeasibility in the other problem. Is this property symmetric? Does infeasibility in one problem imply unboundedness in the other? The answer is "not necessarily." This is best illustrated by the following example.

Example 6.4

Consider the following primal and dual problems.

$$\begin{aligned} \text{P: Minimize} \quad & -x_1 - x_2 \\ \text{Subject to} \quad & x_1 - x_2 \geqslant 1 \\ & -x_1 + x_2 \geqslant 1 \\ & x_1, \ x_2 \geqslant 0 \end{aligned}$$

$$\begin{aligned} \text{D: Maximize} \quad & w_1 + w_2 \\ \text{Subject to} \quad & w_1 - w_2 \leqslant -1 \\ & -w_1 + w_2 \leqslant -1 \\ & w_1, \ w_2 \geqslant 0 \end{aligned}$$

Upon graphing both problems (in Figure 6.1) we find that neither problem possesses a feasible solution.

Duality and the Kuhn-Tucker Optimality Conditions

Recall from Chapter 5 that the optimality conditions for a linear program state that a necessary and sufficient condition for $\mathbf{x}^*$ to be an optimal point to the linear program Minimize $\mathbf{cx}$ subject to $\mathbf{Ax} \geqslant \mathbf{b}$, $\mathbf{x} \geqslant \mathbf{0}$ is that there exists a vector

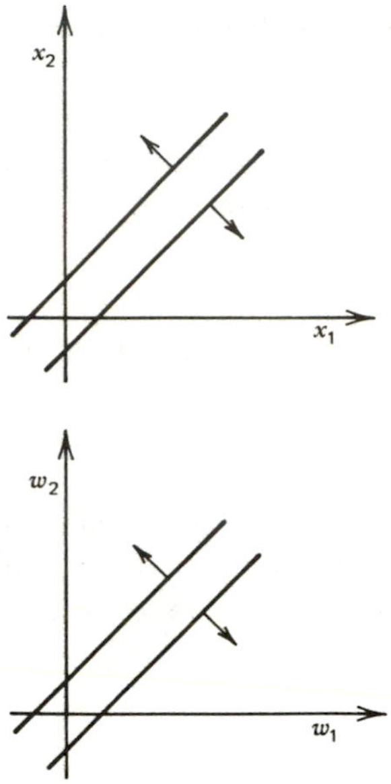

Figure 6.1. **An example of infeasible primal and dual problems.**

w* such that

1. $\mathbf{Ax}^* \geqslant \mathbf{b}$, $\mathbf{x}^* \geqslant \mathbf{0}$
2. $\mathbf{w}^*\mathbf{A} \leqslant \mathbf{c}$, $\mathbf{w}^* \geqslant \mathbf{0}$
3. $\mathbf{w}^*(\mathbf{Ax}^* - \mathbf{b}) = 0$
   $(\mathbf{c} - \mathbf{w}^*\mathbf{A})\mathbf{x}^* = 0$

Condition 1 above simply requires that the optimal point $\mathbf{x}^*$ must be feasible to the primal. In light of our discussion of duality we can now interpret condition 2. This condition indicates that the vector $\mathbf{w}^*$ must be a feasible point for the dual problem. From condition 3 above, we find that $\mathbf{cx}^* = \mathbf{w}^*\mathbf{b}$. Hence $\mathbf{w}^*$ must be an optimal solution to the dual problem. The Kuhn-Tucker optimality conditions for the dual problem imply the existence of a primal feasible solution whose objective is equal to that of the optimal dual (why?). This leads to the following lemma.

Lemma 3

If one problem possesses an optimal solution, then both problems possess optimal solutions and the two optimal objective values are equal.

It is also possible to see how the Kuhn-Tucker optimality conditions naturally give rise to the definition of the dual problem. Rather than solving for the optimal $\mathbf{x}^*$ directly, one might reasonably choose to search over values of $\mathbf{w}$ satisfying condition 2 above. Knowing that any feasible $\mathbf{w}_0$ (condition 2) satisfies $\mathbf{w}_0\mathbf{b} \leqslant \mathbf{c}\mathbf{x}^*$ and that the optimal $\mathbf{w}^*$ satisfies $\mathbf{w}^*\mathbf{b} = \mathbf{c}\mathbf{x}^*$, we would naturally be led to the maximization of the linear form $\mathbf{wb}$ over all feasible values of $\mathbf{w}$ satisfying condition 2.

By utilizing the foregoing results we obtain two important basic theorems of duality. These two theorems will permit us to use the dual problem to solve the primal problem and, also, to develop new algorithms to solve both problems.

The Fundamental Theorem of Duality

Combining the results of the lemmas, corollaries, and examples of the previous section we obtain the following.

Theorem 1 (Fundamental Theorem of Duality)

With regard to the primal and dual linear programming problems, exactly one of the following statements is true.

1. Both possess optimal solutions $\mathbf{x}^*$ and $\mathbf{w}^*$ with $\mathbf{c}\mathbf{x}^* = \mathbf{w}^*\mathbf{b}$.
2. One problem has unbounded objective value, in which case the other problem must be infeasible.
3. Both problems are infeasible.

From this theorem we see that duality is not completely symmetric. The best we can say is that (here optimal means finite optimal, and unbounded means having an unbounded optimal objective):

P	OPTIMAL	$\Leftrightarrow$	D	OPTIMAL
P	UNBOUNDED	$\Rightarrow$	D	INFEASIBLE
D	UNBOUNDED	$\Rightarrow$	P	INFEASIBLE
P	INFEASIBLE	$\Rightarrow$	D	UNBOUNDED OR INFEASIBLE
D	INFEASIBLE	$\Rightarrow$	P	UNBOUNDED OR INFEASIBLE

Complementary Slackness

Let $\mathbf{x}^*$ and $\mathbf{w}^*$ be any pair of optimal solutions to the primal and dual problems in canonical form respectively. Then

$$\mathbf{c}\mathbf{x}^* \geqslant \mathbf{w}^*\mathbf{A}\mathbf{x}^* \geqslant \mathbf{w}^*\mathbf{b}$$

But $\mathbf{c}\mathbf{x}^* = \mathbf{w}^*\mathbf{b}$ (why?). Thus

$$\mathbf{c}\mathbf{x}^* = \mathbf{w}^*\mathbf{A}\mathbf{x}^* = \mathbf{w}^*\mathbf{b}$$

This gives $\mathbf{w}^*(\mathbf{Ax}^* - \mathbf{b}) = 0$ and $(\mathbf{c} - \mathbf{w}^*\mathbf{A})\mathbf{x}^* = 0$. Since $\mathbf{w}^* \geqslant \mathbf{0}$ and $\mathbf{Ax}^* - \mathbf{b} \geqslant \mathbf{0}$, then $\mathbf{w}^*(\mathbf{Ax}^* - \mathbf{b}) = 0$ implies $w_i^*(\mathbf{a}^i\mathbf{x}^* - b_i) = 0$ for $i = 1, \ldots, m$. Similarly $(\mathbf{c} - \mathbf{w}^*\mathbf{A})\mathbf{x}^* = 0$ implies $(c_j - \mathbf{w}^*\mathbf{a}_j)x_j^* = 0$ for $j = 1, \ldots, n$.

Thus we have the following theorem.

***Theorem* 2** (Weak Theorem of Complementary Slackness)

If $\mathbf{x}^*$ and $\mathbf{w}^*$ are *any* optimal points to the primal and dual problems in the canonical form, then

$$(c_j - \mathbf{w}^*\mathbf{a}_j)x_j^* = 0 \qquad j = 1, \ldots, n$$

and

$$w_i^*\left(\mathbf{a}^i\mathbf{x}^* - b_i\right) = 0 \qquad i = 1, \ldots, m$$

This is a very important theorem relating the primal and dual problems. It obviously indicates that at least one of the two terms in each expression above must be zero. In particular,

$$x_j^* > 0 \;\Rightarrow\; \mathbf{w}^*\mathbf{a}_j = c_j$$

$$\mathbf{w}^*\mathbf{a}_j < c_j \;\Rightarrow\; x_j^* = 0$$

$$w_i^* > 0 \;\Rightarrow\; \mathbf{a}^i\mathbf{x}^* = b_i$$

$$\mathbf{a}^i\mathbf{x}^* > b_i \;\Rightarrow\; w_i^* = 0$$

The weak theorem of complementary slackness can also be stated as follows: at optimality "If a variable in one problem is positive, then the corresponding constraint in the other problem must be *tight*" and "If a constraint in one problem is not tight, then the corresponding variable in the other problem must be zero."

Suppose that we let $x_{n+i} = \mathbf{a}^i\mathbf{x} - b_i \geqslant 0$, $i = 1, \ldots, m$ be the m slack variables in the primal problem and let $w_{m+j} = c_j - \mathbf{w}\mathbf{a}_j \geqslant 0$, $j = 1, \ldots, n$ be the n slack variables in the dual problem (in Chapter 5 while stating the Kuhn-Tucker conditions, w_{m+j} was denoted by v_j). Then we may rewrite the complementary slackness conditions as follows:

$$x_j^* w_{m+j}^* = 0 \qquad j = 1, \ldots, n$$

$$w_i^* x_{n+i}^* = 0 \qquad i = 1, \ldots, m$$

This relates variables in one problem to slack variables in the other problem.

It should be noted that if $\mathbf{x}^*$ and $\mathbf{w}^*$ are feasible to their respective problems and satisfy the complementary slackness conditions, then they are optimal.

Using the Dual to Solve the Primal

We now have at hand powerful analysis tools, in the form of the two theorems of this section, to utilize the dual problem in solving the primal problem. Let us illustrate this potential usefulness by the following example.

Example 6.5

Consider the following primal and dual problems.

$$\begin{aligned}
\text{P: Minimize} \quad & 2x_1+3x_2+5x_3+2x_4+3x_5 \\
\text{Subject to} \quad & x_1+\ \ x_2+2x_3+x_4+3x_5 \geqslant 4 \\
& 2x_1-2x_2+3x_3+x_4+\ \ x_5 \geqslant 3 \\
& x_1,\quad x_2,\quad x_3\quad x_4,\quad x_5 \geqslant 0
\end{aligned}$$

$$\begin{aligned}
\text{D: Maximize} \quad & 4w_1+3w_2 \\
\text{Subject to} \quad & w_1+2w_2 \leqslant 2 \\
& w_1-2w_2 \leqslant 3 \\
& 2w_1+3w_2 \leqslant 5 \\
& w_1+\ \ w_2 \leqslant 2 \\
& 3w_1+\ \ w_2 \leqslant 3 \\
& w_1,\quad w_2 \geqslant 0
\end{aligned}$$

Since the dual has only two variables, we may solve it graphically as shown in Figure 6.2. The optimal solution to the dual is $w_1^* = \frac{4}{5}$, $w_2^* = \frac{3}{5}$ with objective 5. Right away we know that $z^* = 5$. Utilizing the weak theorem of complementary slackness, we further know that $x_2^* = x_3^* = x_4^* = 0$ since none of the corresponding complementary dual constraints are tight. Since $w_1^*, w_2^* > 0$, then $x_1^* + 3x_5^* = 4$ and $2x_1^* + x_5^* = 3$. From these two equations we get $x_1^* = 1$ and $x_5^* = 1$. Thus the primal optimal point is obtained from the duality theorems and the dual optimal point.

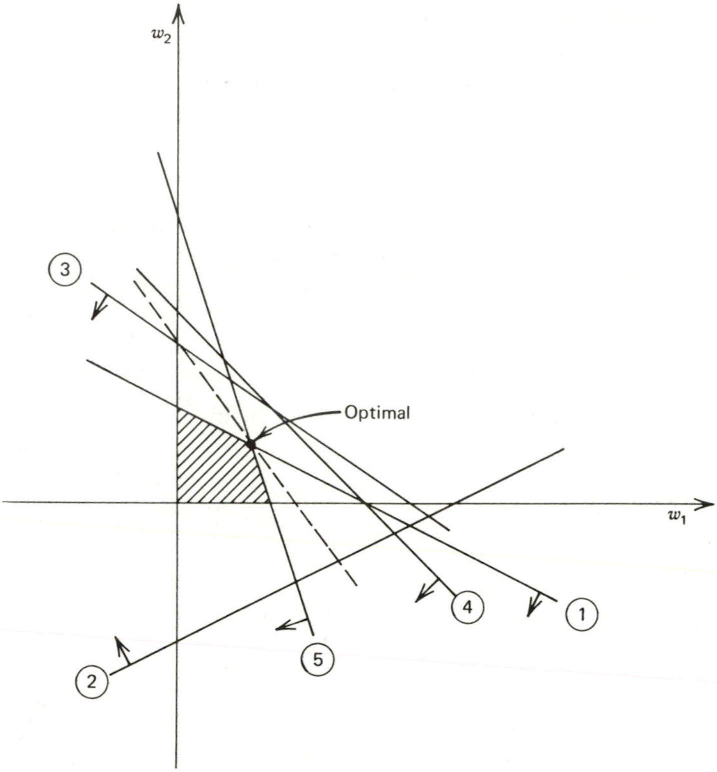

Figure 6.2. **Solving the dual problem graphically.**

6.3 ECONOMIC INTERPRETATION OF THE DUAL

Consider the following linear program and its dual.

P: Minimize $\mathbf{cx}$	D: Maximize $\mathbf{wb}$
Subject to $\mathbf{Ax} \geqslant \mathbf{b}$	Subject to $\mathbf{wA} \leqslant \mathbf{c}$
$\mathbf{x} \geqslant \mathbf{0}$	$\mathbf{w} \geqslant \mathbf{0}$

If $\mathbf{B}$ is the optimal basis for the primal problem and $\mathbf{c}_B$ is the basic cost vector, then we know that

$$z^* = \mathbf{c}_B \mathbf{B}^{-1} \mathbf{b} = \mathbf{w}^* \mathbf{b}$$

from which

$$\frac{\partial z^*}{\partial \mathbf{b}} = \mathbf{c}_B\mathbf{B}^{-1} = \mathbf{w}^*$$

Thus w_i^* is the rate of change of the optimal objective value with a unit increase in the ith right-hand-side value. Since $w_i^* \geqslant 0$, z^* will increase or stay constant as b_i increases.

Economically, we may think of $\mathbf{w}^*$ as a vector of *shadow prices* for the right-hand-side vector. To illustrate, if the ith constraint represents a demand for production of at least b_i units of the ith product and $\mathbf{cx}$ represents the total cost of production, then w_i^* is the *incremental cost* of producing one more unit of the ith product. Put another way, w_i^* is the *fair price* we would pay to have an extra unit of the ith product.

We may also interpret the entire dual problem economically. Suppose that you engage a firm to produce specified amounts $b_1, b_2, \ldots, b_m$ of m outputs or goods. The firm may engage in any of n activities at varying levels to produce the outputs. Each activity j has its own unit cost c_j, and you agree to pay the total cost of production. From your point of view you would like to have control over the firm's operations so that you can specify the mix and levels of activities that the firm will engage in so as to minimize the total production cost. If a_{ij} denotes the amount of product i generated by one unit of activity j, then $\sum_{j=1}^{n} a_{ij}x_j$ represents the units of output i that are produced. These must be greater than or equal to the required amount b_i. Therefore you wish to solve the following problem, which is precisely the primal problem.

$$\begin{aligned}
\text{Minimize} \quad & \sum_{j=1}^{n} c_j x_j \\
\text{Subject to} \quad & \sum_{j=1}^{n} a_{ij}x_j \geqslant b_i \qquad i = 1, 2, \ldots, m \\
& x_j \geqslant 0 \qquad j = 1, 2, \ldots, n
\end{aligned}$$

Instead of trying to control the operation of the firm to obtain the most desirable mix of activities, suppose that you agree to pay the firm unit prices $w_1, w_2, \ldots, w_m$ for each of the m outputs. However, you stipulate that these prices announced by the firm must be *fair*. Since a_{ij} is the number of units of output i produced by 1 unit of activity j, and since w_i is the unit price of output i, then $\sum_{i=1}^{m} a_{ij}w_i$ can be interpreted as the unit price of activity j consistent with the prices $w_1, w_2, \ldots, w_m$. Therefore you ask the firm that the implicit price of activity j, namely $\sum_{i=1}^{m} a_{ij}w_i$, does not exceed the actual price c_j. Therefore the firm must observe the constraints $\sum_{i=1}^{m} a_{ij} w_i \leq c_j$ for $j = 1, 2, \ldots, n$. Within these constraints the firm would like to choose a set of prices that maximize his return $\sum_{i=1}^{m} w_i b_i$. This leads to the following dual problem of the firm.

$$\text{Maximize} \quad \sum_{i=1}^{m} w_i b_i$$

$$\begin{aligned} \text{Subject to} \quad & \sum_{i=1}^{m} a_{ij} w_i \leq c_j \qquad j = 1, 2, \ldots, n \\ & w_i \geq 0 \qquad\qquad\quad i = 1, \ldots, m \end{aligned}$$

The main duality theorem states that there is an equilibrium set of activities and set of prices where the minimal production cost is equal to the maximal return. That the two objectives are equal at optimality becomes intuitively clear by noting that they represent the fair charge to the customer, where the primal objective is derived by cost considerations and the dual objective is arrived at by a pricing mechanism.

6.4 THE DUAL SIMPLEX METHOD

In this section we describe the *dual simplex* method, which solves the dual problem directly on the (primal) simplex tableau. At each iteration we move from a basic feasible solution of the dual problem to an improved basic feasible solution until optimality of the dual (and also the primal) is reached, or else until we conclude that the dual is unbounded and that the primal is infeasible.

Interpretation of Dual Feasibility on the Primal Simplex Tableau

Consider the following linear programming problem.

$$\begin{aligned} \text{Minimize} \quad & \mathbf{cx} \\ \text{Subject to} \quad & \mathbf{Ax} \geq \mathbf{b} \\ & \mathbf{x} \geq \mathbf{0} \end{aligned}$$

Let $\mathbf{B}$ be a basis that is not necessarily feasible and consider the following tableau.

	z	x_1	x_2	$\cdots$	x_n	x_{n+1} (SLACK VARIABLES)	$\cdots$	x_{n+m}	RHS
z	1	$z_1 - c_1$	$z_2 - c_2$	$\cdots$	$z_n - c_n$	$z_{n+1} - c_{n+1}$	$\cdots$	$z_{n+m} - c_{n+m}$	$\mathbf{c}_B\bar{\mathbf{b}}$
x_{B_1}	0	y_{11}	y_{12}	$\cdots$	y_{1n}	$y_{1,n+1}$	$\cdots$	$y_{1,n+m}$	$\bar{b}_1$
x_{B_2}	0	y_{21}	y_{22}	$\cdots$	y_{2n}	$y_{2,n+1}$	$\cdots$	$y_{2,n+m}$	$\bar{b}_2$
$\vdots$	$\vdots$	$\vdots$	$\vdots$		$\vdots$	$\vdots$		$\vdots$	$\vdots$
x_{B_m}	0	y_{m1}	y_{m2}	$\cdots$	y_{mn}	$y_{m,n+1}$	$\cdots$	$y_{m,n+m}$	$\bar{b}_m$

The tableau presents a primal feasible solution if $\bar{b}_i \geqslant 0$ for $i = 1, 2, \ldots, m$; that is, if $\bar{\mathbf{b}} = \mathbf{B}^{-1}\mathbf{b} \geqslant \mathbf{0}$. Furthermore, the tableau is optimal if $z_j - c_j \leqslant 0$ for $j = 1, 2, \ldots, n + m$. Define $\mathbf{w} = \mathbf{c}_B\mathbf{B}^{-1}$. For $j = 1, 2, \ldots, n$ we have

$$z_j - c_j = \mathbf{c}_B\mathbf{B}^{-1}\mathbf{a}_j - c_j = \mathbf{w}\mathbf{a}_j - c_j$$

Hence $z_j - c_j \leqslant 0$ for $j = 1, 2, \ldots, n$ implies that $\mathbf{w}\mathbf{a}_j - c_j \leqslant 0$ for $j = 1, 2, \ldots, n$, which in turn implies that $\mathbf{wA} \leqslant \mathbf{c}$. Furthermore, note that $\mathbf{a}_{n+i} = -\mathbf{e}_i$ and $c_{n+i} = 0$ for $i = 1, 2, \ldots, m$ and so we have

$$\begin{aligned} z_{n+i} - c_{n+i} &= \mathbf{w}\mathbf{a}_{n+i} - c_{n+i} \\ &= \mathbf{w}(-\mathbf{e}_i) - 0 \\ &= -w_i \qquad i = 1, 2, \ldots, m \end{aligned}$$

In addition, if $z_{n+i} - c_{n+i} \leqslant 0$ for $i = 1, 2, \ldots, m$, then $w_i \geqslant 0$ for $i = 1, 2, \ldots, m$ and so $\mathbf{w} \geqslant \mathbf{0}$. We have just shown that $z_j - c_j \leqslant 0$ for $j = 1, 2, \ldots, n + m$ implies that $\mathbf{wA} \leqslant \mathbf{c}$ and $\mathbf{w} \geqslant \mathbf{0}$, where $\mathbf{w} = \mathbf{c}_B\mathbf{B}^{-1}$. In other words, dual feasibility is precisely the simplex optimality criteria $z_j - c_j \leqslant 0$ for all j. At optimality $\mathbf{w}^* = \mathbf{c}_B\mathbf{B}^{-1}$ and the dual objective $\mathbf{w}^*\mathbf{b} = (\mathbf{c}_B\mathbf{B}^{-1})\mathbf{b} = \mathbf{c}_B(\mathbf{B}^{-1}\mathbf{b}) = \mathbf{c}_B\bar{\mathbf{b}} = z^*$; that is, the primal and dual objectives are equal. Thus we have the following result.

Lemma 4

At optimality of the primal minimization problem in the canonical form (that is, $z_j - c_j \leqslant 0$ for all j), $\mathbf{w}^* = \mathbf{c}_B\mathbf{B}^{-1}$ is an optimal solution to the dual problem. Furthermore $w_i^* = -(z_{n+i} - c_{n+i}) = -z_{n+i}$ for $i = 1, 2, \ldots, m$.

The Dual Simplex Method

Consider the following linear programming problem.

$$\begin{aligned} \text{Minimize} \quad & \mathbf{cx} \\ \text{Subject to} \quad & \mathbf{Ax}=\mathbf{b} \\ & \mathbf{x}\geqslant\mathbf{0} \end{aligned}$$

In certain instances it is difficult to find a starting basic solution that is feasible (that is, all $\bar{b}_i \geqslant 0$) to a linear program without adding artificial variables. In these same instances it is often possible to find a starting basic, but not necessarily feasible, solution that is dual feasible (that is, all $z_j - c_j \leqslant 0$ for a minimization problem). In such cases it is useful to develop a variant of the simplex method that would produce a series of simplex tableaux that maintain dual feasibility and complementary slackness and strive toward primal feasibility.

	z	x_1	$\cdots$	x_j	$\cdots$	x_k	$\cdots$	x_n	RHS
z	1	$z_1 - c_1$	$\cdots$	$z_j - c_j$	$\cdots$	$z_k - c_k$	$\cdots$	$z_n - c_n$	$\mathbf{c}_B\bar{\mathbf{b}}$
x_{B_1}	0	y_{11}	$\cdots$	y_{1j}	$\cdots$	y_{1k}	$\cdots$	y_{1n}	$\bar{b}_1$
x_{B_2}	0	y_{21}	$\cdots$	y_{2j}	$\cdots$	y_{2k}	$\cdots$	y_{2n}	$\bar{b}_2$
$\vdots$	$\vdots$	$\vdots$		$\vdots$	$\vdots$	$\vdots$		$\vdots$	$\vdots$
x_{B_r}	0	y_{r1}	$\cdots$	y_{rj}	$\cdots$	$\boxed{y_{rk}}$	$\cdots$	y_{rn}	$\bar{b}_r$
$\vdots$	$\vdots$	$\vdots$		$\vdots$		$\vdots$		$\vdots$	$\vdots$
x_{B_m}	0	y_{m1}	$\cdots$	y_{mj}	$\cdots$	y_{mk}	$\cdots$	y_{mn}	$\bar{b}_m$

Consider the above tableau representing a basic solution at some iteration. Suppose that the tableau is dual feasible (that is, $z_j - c_j \leqslant 0$ for a minimization problem). Then, if the tableau is also primal feasible (that is, all $\bar{b}_i \geqslant 0$) then we have the optimal solution. Otherwise, consider some $\bar{b}_r < 0$. By selecting row r as a pivot row and some column k such that $y_{rk} < 0$ as a pivot column we can make the new right-hand side $\bar{b}_r' > 0$. Through a series of such pivots we hope to make all $\bar{b}_i \geqslant 0$ while maintaining all $z_j - c_j \leqslant 0$ and thus achieve optimality. The question that remains is how do we select the pivot column so as to maintain dual feasibility after pivoting. The pivot column k is determined by the following minimum ratio test.

$$\frac{z_k - c_k}{y_{rk}} = \underset{j}{\text{minimum}} \left\{ \frac{z_j - c_j}{y_{rj}} : y_{rj} < 0 \right\} \tag{6.1}$$

Note that the new entries in row 0 after pivoting are given by:

$$(z_j - c_j)' = (z_j - c_j) - \frac{y_{rj}}{y_{rk}}(z_k - c_k)$$

If $y_{rj} \geq 0$, and since $z_k - c_k \leq 0$ and $y_{rk} < 0$, then $(y_{rj}/y_{rk})(z_k - c_k) \geq 0$ and hence $(z_j - c_j)' \leq z_j - c_j$. Since the previous solution was dual feasible, then $z_j - c_j \leq 0$ and hence $(z_j - c_j)' \leq 0$. Now consider the case where $y_{rj} < 0$. By 6.1 we have:

$$\frac{z_k - c_k}{y_{rk}} \leq \frac{z_j - c_j}{y_{rj}}$$

Multiplying both sides by $y_{rj} < 0$, we get $(z_j - c_j) - (y_{rj}/y_{rk})(z_k - c_k) \leq 0$, that is, $(z_j - c_j)' \leq 0$. To summarize, if the pivot column is chosen according to Equation (6.1), then the new basis obtained by pivoting at y_{rk} is still dual feasible. Moreover, the dual objective after pivoting is given by $\mathbf{c}_B\mathbf{B}^{-1}\mathbf{b} - (z_k - c_k)\bar{b}_r/y_{rk}$. Since $z_k - c_k \leq 0$, $\bar{b}_r < 0$, and $y_{rk} < 0$, then $-(z_k - c_k)\bar{b}_r/y_{rk} \geq 0$ and the dual objective improves over the current value of $\mathbf{c}_B\mathbf{B}^{-1}\mathbf{b} = \mathbf{wb}$.

We have just described a procedure that moves from a dual basic feasible solution to an improved (at least not worse) basic dual feasible solution. To complete the analysis we must consider the case when $y_{rj} \geq 0$ for all j and hence no column is eligible to be the pivot column. In this case the ith row reads: $\Sigma_j y_{rj} x_j = \bar{b}_r$. Since $y_{rj} \geq 0$ for all j and x_j is required to be nonnegative, then $\Sigma_j y_{rj} x_j \geq 0$ for any feasible solution. However, $\bar{b}_r < 0$. This contradiction shows that the primal is infeasible and the dual is unbounded (why?). In Exercise 6.31 we ask the reader to show directly that the dual is unbounded by constructing a direction satisfying the unboundedness criterion.

Summary of the Dual Simplex Method (Minimization Problem)

INITIALIZATION STEP

Find a basis $\mathbf{B}$ of the primal such that $z_j - c_j = \mathbf{c}_B\mathbf{B}^{-1}\mathbf{a}_j - c_j \leq 0$ for all j (in Section 6.6 we describe a procedure for finding such a basis if it is not immediately available).

MAIN STEP

1. If $\bar{\mathbf{b}} = \mathbf{B}^{-1}\mathbf{b} \geqslant \mathbf{0}$, stop; the current solution is optimal. Otherwise select the pivot row r with $\bar{b}_r < 0$, say $\bar{b}_r = \text{Minimum}\ \{\bar{b}_i\}$.
2. If $y_{rj} \geqslant 0$ for all j, stop; the dual is unbounded and the primal is infeasible. Otherwise select the pivot column k by the following minimum ratio test:

$$\frac{z_k - c_k}{y_{rk}} = \underset{j}{\text{Minimum}} \left\{ \frac{z_j - c_j}{y_{rj}} : y_{rj} < 0 \right\}$$

3. Pivot at y_{rk} and return to step 1.

Example 6.6

Consider the following problem.

$$\begin{aligned} \text{Minimize}\quad & 2x_1 + 3x_2 + 4x_3 \\ \text{Subject to}\quad & x_1 + 2x_2 + x_3 \geqslant 3 \\ & 2x_1 - x_2 + 3x_3 \geqslant 4 \\ & x_1,\ x_2,\ x_3 \geqslant 0 \end{aligned}$$

A starting basic solution that is dual feasible can be obtained by utilizing the slack variables x_4 and x_5. This results from the fact that the cost vector is nonnegative. Applying the dual simplex method, we obtain the following series of tableaux.

	z	x_1	x_2	x_3	x_4	x_5	RHS
z	1	-2	-3	-4	0	0	0
x_4	0	-1	-2	-1	1	0	-3
x_5	0	$\textcircled{-2}$	1	-3	0	1	-4

	z	x_1	x_2	x_3	x_4	x_5	RHS
z	1	0	-4	-1	0	-1	4
x_4	0	0	$\textcircled{-\frac{5}{2}}$	$\frac{1}{2}$	1	$-\frac{1}{2}$	-1
x_1	0	1	$-\frac{1}{2}$	$\frac{3}{2}$	0	$-\frac{1}{2}$	2

	z	x_1	x_2	x_3	x_4	x_5	RHS
z	1	0	0	$-\frac{9}{5}$	$-\frac{8}{5}$	$-\frac{1}{5}$	$\frac{28}{5}$
x_2	0	0	1	$-\frac{1}{5}$	$-\frac{2}{5}$	$\frac{1}{5}$	$\frac{2}{5}$
x_1	0	1	0	$\frac{7}{5}$	$-\frac{1}{5}$	$-\frac{2}{5}$	$\frac{11}{5}$

Since $\bar{\mathbf{b}} \geqslant \mathbf{0}$ and $z_j - c_j \leqslant 0$ for all j, the optimal primal and dual solutions are at hand. In particular,

$$(x_1^*, x_2^*, x_3^*, x_4^*, x_5^*) = (\tfrac{11}{5}, \tfrac{2}{5}, 0, 0, 0)$$

$$(w_1^*, w_2^*) = (\tfrac{8}{5}, \tfrac{1}{5})$$

Note that w_1^* and w_2^* are respectively the negatives of the $z_j - c_j$ entries under the slack variables x_4 and x_5. Also note that in each subsequent tableau the value of the objective function is increasing, as it should, for the dual (maximization) problem.

The Complementary Basic Dual Solution

Consider the following pair of primal and dual problems in standard form.

P: Minimize $\mathbf{cx}$	D: Maximize $\mathbf{wb}$
Subject to $\mathbf{Ax} = \mathbf{b}$	Subject to $\mathbf{wA} \leqslant \mathbf{c}$
$\mathbf{x} \geqslant \mathbf{0}$	$\mathbf{w}$ unrestricted

Given any primal basis $\mathbf{B}$, there is an associated complementary dual basis. To illustrate, introduce the dual slack vector $\mathbf{w}_s$ so that $\mathbf{wA} + \mathbf{w}_s = \mathbf{c}$. The dual constraints can be rewritten in the following more convenient form:

$$\mathbf{A}^t\mathbf{w}^t + \mathbf{I}\mathbf{w}_s^t = \mathbf{c}^t \tag{6.2}$$

$$\mathbf{w}^t \text{ unrestricted}$$

$$\mathbf{w}_s^t \geqslant \mathbf{0}$$

Given the primal basis $\mathbf{B}$, recall that $\mathbf{w} = \mathbf{c}_B\mathbf{B}^{-1}$. Substituting in Equation (6.2), we get

$$\begin{aligned}
\mathbf{w}_s^t &= \mathbf{c}^t - \mathbf{A}^t\mathbf{w}^t \\
&= \mathbf{c}^t - \mathbf{A}^t(\mathbf{B}^{-1})^t\mathbf{c}_B^t \\
&= \begin{pmatrix} \mathbf{c}_B^t \\ \mathbf{c}_N^t \end{pmatrix} - \begin{pmatrix} \mathbf{B}^t \\ \mathbf{N}^t \end{pmatrix}(\mathbf{B}^{-1})^t\mathbf{c}_B^t \\
&= \begin{pmatrix} \mathbf{0} \\ \mathbf{c}_N^t - \mathbf{N}^t(\mathbf{B}^{-1})^t\mathbf{c}_B^t \end{pmatrix}
\end{aligned} \tag{6.3}$$

Note that $\mathbf{w} = \mathbf{c}_B\mathbf{B}^{-1}$ and Equation (6.3) lead naturally to a dual basis. Since both $\mathbf{c}_B\mathbf{B}^{-1}$ and $\mathbf{c}_N^t - \mathbf{N}^t(\mathbf{B}^{-1})^t\mathbf{c}_B^t$ are not necessarily zero, then the vector $\mathbf{w}$ and the last $n - m$ components of $\mathbf{w}_s$ form the dual basis. In particular the dual basis corresponding to the primal basis $\mathbf{B}$ is given by

$$\left[\begin{array}{c|c} \mathbf{B}^t & \mathbf{0} \\ \hline \mathbf{N}^t & \mathbf{I}_{n-m} \end{array}\right]$$

The rank of the preceding matrix is n. The primal basis is feasible if $\mathbf{B}^{-1}\mathbf{b} \geqslant \mathbf{0}$ and the dual basis is feasible if $\mathbf{w}_s \geqslant \mathbf{0}$; that is, if $\mathbf{c}_N - \mathbf{c}_B\mathbf{B}^{-1}\mathbf{N} \geqslant \mathbf{0}$ (see Equation 6.3). Even if these conditions do not hold, the primal and dual bases are *complementary* in the sense that the complementary slackness condition $(\mathbf{wA} - \mathbf{c})\mathbf{x} = 0$ holds as shown below:

$$(\mathbf{wA} - \mathbf{c})\mathbf{x} = \mathbf{w}_s\mathbf{x} = (\mathbf{0}, \mathbf{c}_N - \mathbf{c}_B\mathbf{B}^{-1}\mathbf{N})\begin{pmatrix} \mathbf{B}^{-1}\mathbf{b} \\ \mathbf{0} \end{pmatrix} = 0$$

To summarize, during any dual simplex iteration we have a primal basis that is not necessarily feasible, and a complementary dual feasible basis. At termination primal feasibility is attained, and so all the Kuhn-Tucker optimality conditions hold.

Finite Convergence of the Dual Simplex Method in the Absence of Dual Degeneracy

Note that the dual simplex method moves among dual feasible bases. Also recall that the difference in the dual objective between two successive iterations is $-(z_k - c_k)\bar{b}_r/y_{rk}$. Note that $\bar{b}_r < 0$, $y_{rk} < 0$, and $z_k - c_k \leqslant 0$ and hence $-(z_k - c_k)\bar{b}_r/y_{rk} \geqslant 0$. In particular, if $z_k - c_k < 0$, then the dual objective strictly increases and hence no basis can be repeated and the algorithm must converge in a finite number of steps. By the foregoing characterization of the complementary dual basis, and since x_k is a nonbasic primal variable, then the dual slack of the constraint $\mathbf{wa}_k \leqslant c_k$ is basic. Assuming dual nondegeneracy, this dual slack variable must be positive so that $\mathbf{wa}_k < c_k$, that is $z_k - c_k < 0$. As discussed above, this would prevent cycling since the dual objective strictly increases at each iteration. In Exercise 6.33 we ask the reader to prove finite convergence in the presence of degeneracy provided that a special rule for choosing the pivot column is adopted.

6.5 THE PRIMAL-DUAL METHOD

Recall that in the dual simplex method we begin with a basic (not necessarily feasible) solution to the primal problem and a complementary basic feasible solution to the dual problem. The dual simplex method proceeds, by pivoting, through a series of dual basic feasible solutions until the associated complementary primal basic solution is feasible, thus satisfying all of the Kuhn-Tucker conditions for optimality.

In this section we describe a method, called the *primal-dual* algorithm, similar to the dual simplex method, which begins with dual feasibility, and proceeds to obtain primal feasibility while maintaining complementary slackness. An important difference between the dual simplex method and the primal-dual method is that the primal-dual algorithm does not require a dual feasible solution to be basic. Given a dual feasible solution, the primal variables that correspond to tight dual constraints (so that complementary slackness is satisfied) are determined. Using phase I of the simplex method, we attempt to attain primal feasibility using only these variables. If we are unable to obtain primal feasibility, we change the dual feasible solution in such a way as to admit at least one new variable to the phase I problem. This is continued until either the primal becomes feasible or the dual becomes unbounded.

Development of the Primal-Dual Method

Consider the following primal and dual problems in standard form where $\mathbf{b} \geqslant \mathbf{0}$.

P: Minimize $\mathbf{cx}$	D: Maximize $\mathbf{wb}$
Subject to $\mathbf{Ax}=\mathbf{b}$	Subject to $\mathbf{wA} \leqslant \mathbf{c}$
$\mathbf{x} \geqslant \mathbf{0}$	$\mathbf{w}$ unrestricted

Let $\mathbf{w}$ be an initial dual feasible solution, that is, $\mathbf{wa}_j \leqslant c_j$ for all j. By complementary slackness, if $\mathbf{wa}_j = c_j$, then x_j is allowed to be positive and we attempt to attain primal feasibility from among these variables. Let $Q = \{j : \mathbf{wa}_j - c_j = 0\}$, that is, the set of indices of primal variables allowed to be positive. Then the phase I problem that attempts to find a feasible solution to the primal problem among variables in the set Q becomes:

$$\text{Minimize} \quad \sum_{j \in Q} 0x_j + \mathbf{1x}_a$$

$$\text{Subject to} \quad \sum_{j \in Q} \mathbf{a}_j x_j + \mathbf{Ix}_a = \mathbf{b}$$

$$x_j \geq 0 \text{ for } j \in Q$$

$$\mathbf{x}_a \geq \mathbf{0}$$

We utilize the artificial vector $\mathbf{x}_a$ to obtain a starting basic feasible solution to the phase I problem. The phase I problem is sometimes called the *restricted primal* problem.

Denote the optimal objective value of the foregoing problem by x_0. At optimality of the phase I problem either $x_0 = 0$ or $x_0 > 0$. When $x_0 = 0$, we have a feasible solution to the primal problem since all artificials are zeros. Furthermore, we have a dual feasible solution, and the complementary slackness condition $(\mathbf{wa}_j - c_j)x_j = 0$ holds because either $j \in Q$ in which case $\mathbf{wa}_j - c_j = 0$, or else $j \notin Q$ in which case $x_j = 0$. Therefore we have an optimal solution of the overall problem whenever $x_0 = 0$. If $x_0 > 0$, primal feasibility is not achieved and we must construct a new dual solution that would admit a new variable to the restricted primal problem in such a way that x_0 might be decreased. We shall modify the dual vector $\mathbf{w}$ such that all the basic primal variables in the restricted problem remain in the new restricted primal problem, and in addition, at least one primal variable that did not belong to the set Q would get passed to the restricted primal problem. Furthermore, this variable would reduce x_0 if introduced in the basis. In order to construct such a dual vector, consider the following dual of the phase I problem.

$$\text{Maximize} \quad \mathbf{vb}$$

$$\text{Subject to} \quad \mathbf{va}_j \leq 0 \qquad j \in Q$$

$$\mathbf{v} \leq \mathbf{1}$$

$$\mathbf{v} \quad \text{unrestricted}$$

Let $\mathbf{v}^*$ be an optimal solution to the foregoing problem. Then, if a real variable x_j is a member of the optimal basis for the restricted primal, the associated dual constraint must be tight, that is, $\mathbf{v}^*\mathbf{a}_j = 0$. Also the criterion for basis entry in the restricted primal problem is that the associated dual constraint be violated, that is, $\mathbf{v}^*\mathbf{a}_j > 0$. However, no variable presently in the restricted primal has this property since the restricted primal is optimal. For $j \notin Q$, compute $\mathbf{v}^*\mathbf{a}_j$. If $\mathbf{v}^*\mathbf{a}_j > 0$, then if x_j could be passed to the restricted primal problem it would be a candidate to enter the basis with the potential of a further decrease in x_0. Therefore we must find a way to force some variable x_j with $\mathbf{v}^*\mathbf{a}_j > 0$ into the set Q.

Construct the following dual vector $\mathbf{w}'$, where $\theta > 0$:

$$\mathbf{w}' = \mathbf{w} + \theta\mathbf{v}^*$$

Then

$$\begin{aligned}\mathbf{w}'\mathbf{a}_j - c_j &= (\mathbf{w} + \theta\mathbf{v}^*)\mathbf{a}_j - c_j \\ &= (\mathbf{w}\mathbf{a}_j - c_j) + \theta(\mathbf{v}^*\mathbf{a}_j) \qquad (6.4)\end{aligned}$$

Note that $\mathbf{w}\mathbf{a}_j - c_j = 0$ and $\mathbf{v}^*\mathbf{a}_j \leqslant 0$ for $j \in Q$. Thus Equation (6.4) implies that $\mathbf{w}'\mathbf{a}_j - c_j \leqslant 0$ for $j \in Q$. In particular, if x_j with $j \in Q$ is a basic variable in the restricted primal, then $\mathbf{v}^*\mathbf{a}_j = 0$ and $\mathbf{w}'\mathbf{a}_j - c_j = 0$, permitting j in the new restricted primal problem. If $j \notin Q$ and $\mathbf{v}^*\mathbf{a}_j \leqslant 0$, then from Equation (6.4) and noting that $\mathbf{w}\mathbf{a}_j - c_j < 0$, we have $\mathbf{w}'\mathbf{a}_j - c_j \leqslant 0$. Finally consider $j \notin Q$ with $\mathbf{v}^*\mathbf{a}_j > 0$. Examining Equation (6.4), and noting that $\mathbf{w}\mathbf{a}_j - c_j < 0$ for $j \notin Q$, it is evident that we can choose a $\theta > 0$ such that $\mathbf{w}'\mathbf{a}_j - c_j \leqslant 0$ for $j \notin Q$ with at least one component equal to zero. In particular, define θ as follows:

$$\theta = \frac{-(\mathbf{w}\mathbf{a}_k - c_k)}{\mathbf{v}^*\mathbf{a}_k} = \underset{j}{\text{Minimum}} \left\{ \frac{-(\mathbf{w}\mathbf{a}_j - c_j)}{\mathbf{v}^*\mathbf{a}_j} : \mathbf{v}^*\mathbf{a}_j > 0 \right\} > 0 \qquad (6.5)$$

By definition of θ above and from Equation (6.4), we see that $\mathbf{w}'\mathbf{a}_k - c_k = 0$. Furthermore, for each j with $\mathbf{v}^*\mathbf{a}_j > 0$, and noting Equations (6.4) and (6.5), we have $\mathbf{w}'\mathbf{a}_j - c_j \leqslant 0$.

To summarize, modifying the dual vector as detailed above leads to a new feasible dual solution where $\mathbf{w}'\mathbf{a}_j - c_j \leqslant 0$ for all j. Furthermore, all the variables that belonged to the restricted primal basis are passed to the new restricted primal. In addition, a new variable x_k that is a candidate to enter the basis, is passed to the restricted primal problem. Hence we continue from the present restricted primal basis by entering x_k, which leads to a potential reduction in x_0.

Case of Unbounded Dual

The foregoing process is continued until either $x_0 = 0$ in which case we have an optimal solution, or else $x_0 > 0$ and $\mathbf{v}^*\mathbf{a}_j \leqslant 0$ for all $j \notin Q$. In this case consider

$\mathbf{w}' = \mathbf{w} + \theta\mathbf{v}^*$. Since $\mathbf{wa}_j - c_j \leqslant 0$ for all j, and by assumption $\mathbf{v}^*\mathbf{a}_j \leqslant 0$ for all j, then from Equation (6.4) $\mathbf{w}'$ is a dual feasible solution for all $\theta > 0$. Furthermore, the dual objective is

$$\mathbf{w}'\mathbf{b} = (\mathbf{w} + \theta\mathbf{v}^*)\mathbf{b} = \mathbf{wb} + \theta\mathbf{v}^*\mathbf{b}$$

Since $\mathbf{v}^*\mathbf{b} = x_0$ (why?), and the latter is positive, then $\mathbf{w}'\mathbf{b}$ can be increased indefinitely by choosing θ arbitrarily large. Therefore the dual is unbounded and hence the primal is infeasible.

Summary of the Primal-Dual Algorithm (Minimization Problem)

INITIALIZATION STEP

Choose a vector $\mathbf{w}$ such that $\mathbf{wa}_j - c_j \leqslant 0$ for all j.

MAIN STEP

1. Let $Q = \{j : \mathbf{wa}_j - c_j = 0\}$ and solve the following restricted primal problem.

$$\begin{aligned} \text{Minimize} \quad & \sum_{j \in Q} 0x_j + \mathbf{1x}_a \\ \text{Subject to} \quad & \sum_{j \in Q} \mathbf{a}_j x_j + \mathbf{x}_a = \mathbf{b} \\ & x_j \geqslant 0 \quad \text{for } j \in Q \\ & \mathbf{x}_a \geqslant \mathbf{0} \end{aligned}$$

   Denote the optimal objective by x_0. If $x_0 = 0$, stop; an optimal solution is obtained. Otherwise let $\mathbf{v}^*$ be the optimal dual solution to the foregoing restricted primal problem.
2. If $\mathbf{v}^*\mathbf{a}_j \leqslant 0$ for all j, then stop; the dual is unbounded and the primal is infeasible. Otherwise let

$$\theta = \underset{j}{\text{Minimum}} \left\{ \frac{-(\mathbf{wa}_j - c_j)}{\mathbf{v}^*\mathbf{a}_j} : \mathbf{v}^*\mathbf{a}_j > 0 \right\} > 0$$

   and replace $\mathbf{w}$ by $\mathbf{w} + \theta\mathbf{v}^*$. Repeat step 1.

Example 6.7

Consider the following problem.

$$\begin{aligned}
\text{Minimize} \quad & 3x_1+4x_2+6x_3+7x_4+x_5 \\
\text{Subject to} \quad & 2x_1-x_2+x_3+6x_4-5x_5-x_6 = 6 \\
& x_1+x_2+2x_3+x_4+2x_5 - x_7=3 \\
& x_1, \quad x_2, \quad x_3, \quad x_4, \quad x_5, \; x_6, \; x_7 \geqslant 0
\end{aligned}$$

The dual problem is given by the following.

$$\begin{aligned}
\text{Maximize} \quad & 6w_1+3w_2 \\
\text{Subject to} \quad & 2w_1+w_2 \leqslant 3 \\
& -w_1+w_2 \leqslant 4 \\
& w_1+2w_2 \leqslant 6 \\
& 6w_1+w_2 \leqslant 7 \\
& -5w_1+2w_2 \leqslant 1 \\
& -w_1 \leqslant 0 \\
& -w_2 \leqslant 0 \\
& w_1, \quad w_2 \quad \text{unrestricted}
\end{aligned}$$

An initial dual feasible solution is given by $\mathbf{w} = (w_1, w_2) = (0, 0)$. Substituting $\mathbf{w}$ in each dual constraint, we find that the last two dual constraints are tight so that $Q = \{6, 7\}$. Denoting the artificial variables by x_8 and x_9, the restricted primal problem becomes as follows.

$$\begin{aligned}
\text{Minimize} \quad & x_8+x_9 \\
\text{Subject to} \quad & -x_6 + x_8 = 6 \\
& -x_7 + x_9 = 3 \\
& x_6, \; x_7, \; x_8, \; x_9 \geqslant 0
\end{aligned}$$

The optimal solution to this restricted primal is clearly $(x_6, x_7, x_8, x_9) = (0, 0, 6, 3)$ and the optimal objective $x_0 = 9$. The dual of the foregoing restricted primal is the following.

$$\begin{aligned}
\text{Maximize} \quad & 6v_1+3v_2 \\
\text{Subject to} \quad & -v_1 \leqslant 0 \\
& -v_2 \leqslant 0 \\
& v_1 \leqslant 1 \\
& v_2 \leqslant 1 \\
& v_1, \quad v_2 \quad \text{unrestricted}
\end{aligned}$$

Utilizing complementary slackness, we see that since x_8 and x_9 are basic, the last two dual constraints must be tight and $\mathbf{v}^* = (v_1^*, v_2^*) = (1, 1)$. Computing $\mathbf{v}^*\mathbf{a}_j$ for each column j, we have $\mathbf{v}^*\mathbf{a}_1 = 3$, $\mathbf{v}^*\mathbf{a}_2 = 0$, $\mathbf{v}^*\mathbf{a}_3 = 3$, $\mathbf{v}^*\mathbf{a}_4 = 7$, and $\mathbf{v}^*\mathbf{a}_5 = -3$. Then θ is determined as follows:

$$\theta = \text{Minimum}\left\{-(-\tfrac{3}{3}),\ -(-\tfrac{6}{3}),\ -(-\tfrac{7}{7})\right\} = 1$$

and $\mathbf{w}' = (0, 0) + 1(1, 1) = (1, 1)$.

With the new dual solution $\mathbf{w}'$ we recompute Q and obtain $Q = \{1, 4\}$, giving the following restricted primal:

$$\begin{aligned}
\text{Minimize} \quad & x_8+x_9 \\
\text{Subject to} \quad & 2x_1+6x_4+x_8 \qquad = 6 \\
& x_1+\ x_4 \qquad +x_9 = 3 \\
& x_1, \quad x_4, \quad x_8, \quad x_9 \geqslant 0
\end{aligned}$$

This time an optimal solution to the restricted problem is given by

$$(x_1, x_4, x_8, x_9) = (3, 0, 0, 0)$$

with $x_0 = 0$. Thus we have an optimal solution to the original problem with the optimal primal and dual solutions being

$$(x_1^*, x_2^*, x_3^*, x_4^*, x_5^*, x_6^*, x_7^*) = (3, 0, 0, 0, 0, 0, 0)$$

and

$$(w_1^*, w_2^*) = (1, 1).$$

Tableau Form of the Primal-Dual Method

Let $z_j - c_j$ be the row zero coefficients for the original primal problem and let $\hat{z}_j - \hat{c}_j$ be the row zero coefficients for the restricted primal problem. Then for each real variable x_j we have

$$z_j - c_j = \mathbf{wa}_j - c_j \qquad \text{and} \quad \hat{z}_j - \hat{c}_j = \mathbf{va}_j - 0 = \mathbf{va}_j$$

We also have

$$\frac{\mathbf{wa}_j - c_j}{\mathbf{va}_j} = \frac{z_j - c_j}{\hat{z}_j - \hat{c}_j}$$

and

$$(\mathbf{wa}_j - c_j) + \theta \mathbf{va}_j = (z_j - c_j) + \theta(\hat{z}_j - \hat{c}_j)$$

We can carry out all of the necessary operations directly in one tableau. In this tableau we have two objective rows; the first gives the $z_j - c_j$'s and the second gives the $\hat{z}_j - \hat{c}_j$'s. We shall apply this tableau method to the foregoing problem. The initial tableau is displayed below. In this example $\mathbf{w}$ is initially (0, 0), so that $z_j - c_j = \mathbf{wa}_j - c_j = -c_j$ and the right-hand-side entry in the z-row is zero. When $\mathbf{w} \neq \mathbf{0}$, we still compute $z_j - c_j = \mathbf{wa}_j - c_j$, but also initialize the RHS entry of the z-row to $\mathbf{wb}$ instead of zero. [Try starting the tableau with $\mathbf{w} = (1, 0)$.]

x_1	x_2	x_3	x_4	x_5	x_6	x_7	x_8	x_9	RHS
−3	−4	−6	−7	−1	0	0	0	0	0
0	0	0	0	0	0	0	−1	−1	0
2	−1	1	6	−5	−1	0	1	0	6
1	1	2	1	2	0	−1	0	1	3

Since we begin with x_8 and x_9 in the basis for the restricted primal, we must perform some preliminary pivoting to zero their cost coefficients in the phase I objective. We do this by adding the first and second constraint rows to the restricted primal objective row. Then $\hat{z}_j - \hat{c}_j = 0$ for the two basic variables x_8 and x_9. Let □ indicate the variables in the restricted primal problem, that is, those for which $z_j - c_j = 0$. As the restricted primal problem is solved, only the variables signaled with □ are allowed to enter the basis.

	x_1	x_2	x_3	x_4	x_5	□ x_6	□ x_7	□ x_8	□ x_9	RHS
z	−3	−4	−6	−7	−1	0	0	0	0	0
x_0	3	0	3	7	−3	−1	−1	0	0	9
x_8	2	−1	1	6	−5	−1	0	1	0	6
x_9	1	1	2	1	2	0	−1	0	1	3

Since $\hat{z}_j - \hat{c}_j \leqslant 0$ for all variables in the restricted problem, we have an optimal solution for phase I. Then θ is given by

$$\theta = \quad \text{Minimum}\left\{ \frac{-(z_j - c_j)}{\hat{z}_j - \hat{c}_j} : \hat{z}_j - \hat{c}_j > 0 \right\}$$

$$\text{Minimum}\{ -(-\tfrac{3}{3}), -(-\tfrac{6}{3}), -(-\tfrac{7}{7})\} = 1$$

Thus we add 1 times the phase I objective row to the original objective row. This leads to the following tableau. The phase I problem is solved by only utilizing the variables in the set Q, that is, those with $z_j - c_j = 0$.

	□ x_1	x_2	x_3	□ x_4	x_5	x_6	x_7	□ x_8	□ x_9	RHS
z	0	−4	−3	0	−4	−1	−1	0	0	9
x_0	3	0	3	7	−3	−1	−1	0	0	9
x_8	2	−1	1	(6)	−5	−1	0	1	0	6
x_9	1	1	2	1	2	0	−1	0	1	3

	□ x_1	x_2	x_3	□ x_4	x_5	x_6	x_7	□ x_8	□ x_9	RHS
z	0	−4	−3	0	−4	−1	−1	0	0	9
x_0	$\frac{4}{6}$	$\frac{7}{6}$	$\frac{11}{6}$	0	$\frac{17}{6}$	$\frac{1}{6}$	−1	$-\frac{7}{6}$	0	2
x_4	$\frac{2}{6}$	$-\frac{1}{6}$	$\frac{1}{6}$	1	$-\frac{5}{6}$	$-\frac{1}{6}$	0	$\frac{1}{6}$	0	1
x_9	($\frac{4}{6}$)	$\frac{7}{6}$	$\frac{11}{6}$	0	$\frac{17}{6}$	$\frac{1}{6}$	−1	$-\frac{1}{6}$	1	2

	□ x_1	x_2	x_3	□ x_4	x_5	x_6	x_7	□ x_8	□ x_9	RHS
z	0	−4	−3	0	−4	−1	−1	0	0	9
x_0	0	0	0	0	0	0	0	−1	−1	0
x_4	0	$-\frac{3}{4}$	$-\frac{3}{4}$	1	$-\frac{9}{4}$	$-\frac{1}{4}$	$\frac{2}{4}$	$\frac{1}{4}$	$-\frac{2}{4}$	0
x_1	1	$\frac{7}{4}$	$\frac{11}{4}$	0	$\frac{17}{4}$	$\frac{1}{4}$	$-\frac{6}{4}$	$-\frac{1}{4}$	$\frac{6}{4}$	3

Since $x_0 = 0$, then the optimal solution is found, namely,

$$(x_1^*, x_2^*, x_3^*, x_4^*, x_5^*, x_6^*, x_7^*) = (3, 0, 0, 0, 0, 0, 0)$$

whose objective is 9.

Finite Convergence of the Primal-Dual Method in the Absence of Degeneracy

Recall that at each iteration an improving variable is added to the restricted primal problem. Therefore, in the absence of degeneracy in the restricted primal problem, the optimal objective x_0 strictly decreases at each iteration. This means that the set Q generated at any iteration is distinct from all those generated at previous iterations. Since there is only a finite number of sets of the form Q (recall $Q \subset \{1, 2, \ldots, n\}$) and none of them can be repeated, then the algorithm terminates in a finite number of steps. In Exercise 6.48 we ask the reader to consider the case of degeneracy of the restricted primal problem.

6.6 FINDING AN INITIAL DUAL FEASIBLE SOLUTION: THE ARTIFICIAL CONSTRAINT TECHNIQUE

Both the dual simplex method and the primal-dual method require an initial dual feasible solution. In the primal tableau this requirement of dual feasibility translates to $z_j - c_j \leqslant 0$ for all j for a minimization problem. We shall now see that this can be accommodated by adding a single new primal constraint.

Suppose that the first m columns constitute the initial basis and consider adding the constraint $\sum_{j=m+1}^{n} x_j \leqslant M$, where $M > 0$ is large. The initial tableau is displayed below where x_{n+1} is the slack variable of the additional constraint.

	z	x_1	x_2	$\cdots$	x_m	x_{m+1}	$\cdots$	x_n	x_{n+1}	RHS
z	1	0	0		0	$z_{m+1} - c_{m+1}$	$\cdots$	$z_n - c_n$	0	$\mathbf{c}_B\bar{\mathbf{b}}$
x_{n+1}	0	0	0	$\cdots$	0	1	$\cdots$	1	1	M
x_1	0	1	0	$\cdots$	0	$y_{1,m+1}$	$\cdots$	y_{1n}	0	$\bar{b}_1$
x_2	0	0	1	$\cdots$	0	$y_{2,m+1}$	$\cdots$	y_{2n}	0	$\bar{b}_2$
$\vdots$	$\vdots$	$\vdots$	$\vdots$		$\vdots$	$\vdots$		$\vdots$	$\vdots$	$\vdots$
x_m	0	0	0	$\cdots$	1	$y_{m,m+1}$	$\cdots$	y_{mn}	0	$\bar{b}_m$

This additional constraint bounds the nonbasic variables and thus indirectly bounds the basic variables and thereby the overall primal problem. To obtain a dual feasible solution in the new tableau we let

$$z_k - c_k = \underset{j}{\text{Maximum}} \{z_j - c_j\}$$

Once column k has been selected, perform a single pivot with column k as an entry column and column $n + 1$ as an exit column. In particular, to zero $z_k - c_k$ we shall subtract $z_k - c_k$ times the new row from the objective function row. Note that the choice of k and the single pivot described above ensure that all new entries in row 0 are nonpositive, and thus we have a (basic) feasible dual solution. With this solution available either the dual simplex method or the primal-dual simplex method can be applied, eventually leading to one of the following three cases.

1. The dual solution is unbounded.
2. The optimal primal and dual solutions are obtained with $x^*_{n+1} > 0$.
3. The optimal primal and dual solutions are obtained with $x^*_{n+1} = 0$.

In case 1 the primal problem is infeasible. In case 2 we have the optimal solution to the primal problem. However, case 3 indicates that the new bounding constraint is limiting the primal solution (recall that M is arbitrarily large) and therefore the primal problem is itself unbounded. In Exercise 6.41 we ask the reader to give a formal proof of this conclusion.

In Exercise 6.40 we ask the reader to show that applying the artificial constraint technique to the primal problem is equivalent to applying the single artificial variable technique (described in Chapter 4) with the big-M method to the dual problem and vice versa.

Example 6.8

Suppose that we wish to apply the dual simplex method to the following tableau.

	z	x_1	x_2	x_3	x_4	x_5	RHS
z	1	0	1	5	−1	0	0
x_1	0	1	2	−1	1	0	4
x_5	0	0	3	4	−1	1	3

Adding the artificial constraint $x_2 + x_3 + x_4 \leqslant M$ whose slack is x_6, we get the following tableau.

	z	x_1	x_2	x_3	x_4	x_5	x_6	RHS
z	1	0	1	5	−1	0	0	0
x_6	0	0	1	(1)	1	0	1	M
x_1	0	1	2	−1	1	0	0	4
x_5	0	0	3	4	−1	1	0	3

From the tableau we find that Maximum $z_j - c_j = z_3 - c_3 = 5$. Pivoting in the x_3 column and the x_6 row, we get a new tableau that is dual feasible. The dual simplex method can now be carried out in standard form.

	z	x_1	x_2	x_3	x_4	x_5	x_6	RHS
z	1	0	-4	0	-6	0	-5	$-5M$
x_3	0	0	1	1	1	0	1	M
x_1	0	1	3	0	2	0	1	$M+4$
x_5	0	0	-1	0	-5	1	-4	$-4M+3$

6.7 SENSITIVITY ANALYSIS

In most practical applications, some of the problem data are not known exactly and hence are estimated as well as possible. It is important to be able to find the new optimal solution of the problem as other estimates of some of the data become available, without the expensive task of resolving the problem from scratch. Also at early stages of problem formulation some factors may be overlooked. It is important to update the current solution in a way that takes care of these factors. Furthermore, in many situations the constraints are not very rigid. For example, a constraint may reflect the availability of some resource. This availability can be increased by extra purchase, overtime, buying new equipment, and the like. It is desirable to examine the effect of relaxing some of the constraints on the value of the optimal objective without having to resolve the problem. These and other related topics constitute *sensitivity analysis*. We shall discuss some methods for updating the optimal solution under different problem variations.

Consider the following problem.

$$\begin{aligned} \text{Minimize} \quad & \mathbf{cx} \\ \text{Subject to} \quad & \mathbf{Ax} = \mathbf{b} \\ & \mathbf{x} \geqslant \mathbf{0} \end{aligned}$$

Suppose that the simplex method produced an optimal basis **B**. We shall describe how to make use of the optimality conditions (primal-dual relationships) in order to find the new optimal solution, if some of the problem data change, without resolving the problem from scratch. In particular, the following variations in the problem will be considered.

Change in the cost vector **c**.
Change in the right-hand-side vector **b**.
Change in the constraint matrix **A**.
Addition of a new activity.
Addition of a new constraint.

Change in the Cost Vector

Given an optimal basic feasible solution, suppose that the cost coefficient of one (or more) of the variables is changed from c_k to c'_k. The effect of this change on the final tableau will occur in the cost row; that is, dual feasibility may be lost. Consider the following two cases.

Case I: x_k is nonbasic

In this case $\mathbf{c}_B$ is not affected, and hence $z_j = \mathbf{c}_B\mathbf{B}^{-1}\mathbf{a}_j$ is not changed for any j. Thus $z_k - c_k$ is replaced by $z_k - c'_k$. Note that $z_k - c_k \leqslant 0$ since the current point was an optimal solution of the original problem. If $z_k - c'_k = (z_k - c_k) + (c_k - c'_k)$ is positive, then x_k must be introduced into the basis and the (primal) simplex method is continued as usual. Otherwise the old solution is still optimal with respect to the new problem.

Case II: x_k is basic, say $x_k \equiv x_{B_t}$

Here c_{B_t} is replaced by c'_{B_t}. Let the new value of z_j be z'_j. Then $z'_j - c_j$ is calculated as follows:

$$z'_j - c_j = \mathbf{c}'_B\mathbf{B}^{-1}\mathbf{a}_j - c_j = \left(\mathbf{c}_B\mathbf{B}^{-1}\mathbf{a}_j - c_j\right) + (0, 0, \ldots, c'_{B_t} - c_{B_t}, 0, \ldots, 0)\mathbf{y}_j$$

$$= (z_j - c_j) + \left(c'_{B_t} - c_{B_t}\right)y_{tj} \qquad \text{for all } j$$

In particular for $j = k$, $z_k - c_k = 0$, and $y_{tk} = 1$, and hence $z'_k - c_k = c'_k - c_k$. As we should expect, $z'_k - c'_k$ is still equal to zero. Therefore the cost row can be updated by adding the net change in the cost of $x_{B_t} \equiv x_k$ times the current t row of the final tableau, to the original cost row. Then $z'_k - c_k$ is updated to $z'_k - c'_k = 0$. Of course the new objective value $\mathbf{c}'_B\mathbf{B}^{-1}\mathbf{b} = \mathbf{c}_B\mathbf{B}^{-1}\mathbf{b} + (c'_{B_t} - c_{B_t})\bar{b}_t$ will be obtained in the process.

Example 6.9

Consider the following problem.

$$\begin{array}{lrl}
\text{Minimize} & -2x_1 + x_2 - x_3 & \\
\text{Subject to} & x_1 + x_2 + x_3 & \leqslant 6 \\
& -x_1 + 2x_2 \quad\;\; & \leqslant 4 \\
& x_1, \quad x_2, \quad x_3 & \geqslant 0
\end{array}$$

The optimal tableau is given by the following.

	z	x_1	x_2	x_3	x_4	x_5	RHS
z	1	0	−3	−1	−2	0	−12
x_1	0	1	1	1	1	0	6
x_5	0	0	3	1	1	1	10

Suppose that $c_2 = 1$ is replaced by -3. Since x_2 is nonbasic, then $z_2 - c_2' = (z_2 - c_2) + (c_2 - c_2') = -3 + 4 = 1$, and all other $z_j - c_j$ are unaffected. Hence x_2 enters the basis.

	z	x_1	x_2	x_3	x_4	x_5	RHS
z	1	0	1	−1	−2	0	−12
x_1	0	1	1	1	1	0	6
x_5	0	0	(3)	1	1	1	10

The subsequent tableaux are not shown. Next suppose that $c_1 = -2$ is replaced by zero. Since x_1 is basic, then the new cost row, except $z_1 - c_1$, is obtained by multiplying the row of x_1 by the net change in c_1 [that is, $0 - (-2) = 2$] and adding to the old cost row. The new $z_1 - c_1$ remains zero. Note that the new $z_3 - c_3$ is now positive and so x_3 enters the basis.

	z	x_1	x_2	x_3	x_4	x_5	RHS
z	1	0	−1	1	0	0	0
x_1	0	1	1	(1)	1	0	6
x_5	0	0	3	1	1	1	10

The subsequent tableaux are not shown.

Change in the Right-Hand-Side

If the right-hand-side vector $\mathbf{b}$ is replaced by $\mathbf{b}'$, then $\mathbf{B}^{-1}\mathbf{b}$ will be replaced by $\mathbf{B}^{-1}\mathbf{b}'$. The new right-hand side can be calculated without explicitly evaluating $\mathbf{B}^{-1}\mathbf{b}'$. This is evident by noting that $\mathbf{B}^{-1}\mathbf{b}' = \mathbf{B}^{-1}\mathbf{b} + \mathbf{B}^{-1}(\mathbf{b}' - \mathbf{b})$. If the first m columns originally form the identity, then $\mathbf{B}^{-1}(\mathbf{b}' - \mathbf{b}) = \Sigma_{j=1}^{m}\mathbf{y}_j(b_j' - b_j)$ and hence $\mathbf{B}^{-1}\mathbf{b}' = \bar{\mathbf{b}} + \Sigma_{j=1}^{m}\mathbf{y}_j(b_j' - b_j)$. Since $z_j - c_j \leqslant 0$ for all nonbasic variables (for a minimum problem), the only possible violation of optimality is that the new vector $\mathbf{B}^{-1}\mathbf{b}'$ may have some negative entries. If $\mathbf{B}^{-1}\mathbf{b}' \geqslant \mathbf{0}$, then the same basis remains optimal, and the values of the basic variables are $\mathbf{B}^{-1}\mathbf{b}'$ and the objective has value $\mathbf{c}_B\mathbf{B}^{-1}\mathbf{b}'$. Otherwise the dual simplex method is used to find the new optimal solution by restoring feasibility.

Example 6.10

Suppose that the right-hand side of Example 6.9 is replaced by $\begin{pmatrix} 3 \\ 4 \end{pmatrix}$. Note that $\mathbf{B}^{-1} = \begin{bmatrix} 1 & 0 \\ 1 & 1 \end{bmatrix}$ and hence $\mathbf{B}^{-1}\mathbf{b}' = \begin{bmatrix} 1 & 0 \\ 1 & 1 \end{bmatrix}\begin{bmatrix} 3 \\ 4 \end{bmatrix} = \begin{pmatrix} 3 \\ 7 \end{pmatrix}$. Then $\mathbf{B}^{-1}\mathbf{b}' \geqslant \mathbf{0}$ and hence the new optimal solution is $x_1 = 3, x_5 = 7, x_2 = x_3 = x_4 = 0$.

Change in the Constraint Matrix

We now discuss the effect of changing some of the entries of the constraint matrix $\mathbf{A}$. Two cases are possible, namely, changes involving nonbasic columns, and changes involving basic columns.

Case I: Changes in Activity Vectors for Nonbasic Columns

Suppose that the nonbasic column $\mathbf{a}_j$ is modified to $\mathbf{a}_j'$. Then the new updated column is $\mathbf{B}^{-1}\mathbf{a}_j'$ and $z_j' - c_j = \mathbf{c}_B\mathbf{B}^{-1}\mathbf{a}_j' - c_j$. If $z_j' - c_j \leqslant 0$, then the old solution is optimal; otherwise the simplex method is continued, after column j of the tableau is updated, by introducing the nonbasic variable x_j.

Case II: Changes in Activity Vectors for Basic Columns

Suppose that a basic column $\mathbf{a}_j$ is modified to $\mathbf{a}_j'$. This case can cause considerable trouble. It is possible that the current set of basic vectors no longer form a basis after the change. Even if this does not occur, a change in the activity vector for a single basic column will change $\mathbf{B}^{-1}$ and thus the entries in every column.

Assume that the basic columns are ordered from 1 to m. Let the activity vector for basic column j change from $\mathbf{a}_j$ to $\mathbf{a}_j'$. Compute $\mathbf{y}_j' = \mathbf{B}^{-1}\mathbf{a}_j'$ where $\mathbf{B}^{-1}$ is the current basis inverse. There are two possibilities. If $y_{jj}' = 0$, the current set of basic vectors no longer forms a basis (why?). In this case it is probably best to add an artificial variable to take the place of x_j in the basis and resort to the two-phase method or the big-M method. However, if $y_{jj}' \neq 0$, we may replace column j, which is currently a unit vector, by $\mathbf{y}_j'$ and pivot on y_{jj}'. The current basis continues to be a basis (why?). However, upon pivoting we may have destroyed both primal and dual feasibility and, if so, must resort to one of the artificial variable (primal or dual) techniques.

Example 6.11

Suppose that in Example 6.9, $\mathbf{a}_2$ is changed from $\begin{pmatrix} 1 \\ 2 \end{pmatrix}$ to $\begin{pmatrix} 2 \\ 5 \end{pmatrix}$. Then

$$\mathbf{y}_2' = \mathbf{B}^{-1}\mathbf{a}_2' = \begin{pmatrix} 1 & 0 \\ 1 & 1 \end{pmatrix}\begin{pmatrix} 2 \\ 5 \end{pmatrix} = \begin{pmatrix} 2 \\ 7 \end{pmatrix}$$

$$\mathbf{c}_B\mathbf{B}^{-1}\mathbf{a}_2' - c_2 = (-2, 0)\begin{pmatrix} 2 \\ 7 \end{pmatrix} - 1 = -5$$

Thus the current optimal tableau remains optimal with column x_2 replaced by $(-5, 2, 7)^t$.

Next suppose that column $\mathbf{a}_1$ is changed from $\begin{pmatrix} 1 \\ -1 \end{pmatrix}$ to $\begin{pmatrix} 0 \\ -1 \end{pmatrix}$. Then

$$\mathbf{y}_1' = \mathbf{B}^{-1}\mathbf{a}_1' = \begin{pmatrix} 1 & 0 \\ 1 & 1 \end{pmatrix}\begin{pmatrix} 0 \\ -1 \end{pmatrix} = \begin{pmatrix} 0 \\ -1 \end{pmatrix}$$

$$\mathbf{c}_B\mathbf{B}^{-1}\mathbf{a}_1' - c_1 = (-2, 0)\begin{pmatrix} 0 \\ -1 \end{pmatrix} - (-2) = 2$$

Here the entry in the x_1 row of $\mathbf{y}_1'$ is zero, and so the current basic columns no longer span the space. Replacing column x_1 by $(2, 0, -1)^t$ and adding the artificial variable x_6 to replace x_1 in the basis, we get the following tableau.

	z	x_1	x_2	x_3	x_4	x_5	x_6	RHS
z	1	2	−3	−1	−2	0	$-M$	−12
x_6	0	0	1	1	1	0	(1)	6
x_5	0	−1	3	1	1	1	0	10

After preliminary pivoting at row x_6 and column x_6 to get $z_6 - c_6 = 0$, that is, to get the tableau in basic form, we may proceed with the big-M method.

Finally, suppose that column $\mathbf{a}_1$ is changed from $\begin{pmatrix} 1 \\ -1 \end{pmatrix}$ to $\begin{pmatrix} 3 \\ 6 \end{pmatrix}$. Then

$$\mathbf{y}_1' = \mathbf{B}^{-1}\mathbf{a}_1' = \begin{pmatrix} 1 & 0 \\ 1 & 1 \end{pmatrix}\begin{pmatrix} 3 \\ 6 \end{pmatrix} = \begin{pmatrix} 3 \\ 9 \end{pmatrix}$$

$$\mathbf{c}_B\mathbf{B}^{-1}\mathbf{a}_1' - c_1 = (-2, 0)\begin{pmatrix} 3 \\ 9 \end{pmatrix} - (-2) = -4$$

In this case the entry in the x_1 row of $\mathbf{y}_1'$ is nonzero and so we replace column x_1 by $(-4, 3, 9)^t$, pivot in the x_1 column and x_1 row, and proceed.

	z	x_1	x_2	x_3	x_4	x_5	RHS
z	1	−4	−3	−1	−2	0	−12
x_1	0	(3)	1	1	1	0	6
x_5	0	9	3	1	1	1	10

The subsequent tableaux are not shown.

Adding a New Activity

Suppose that a new activity x_{n+1} with unit cost c_{n+1} and *consumption* column $\mathbf{a}_{n+1}$ is considered for possible production. Without resolving the problem, we can easily determine whether producing x_{n+1} is worthwhile. First calculate $z_{n+1} - c_{n+1}$. If $z_{n+1} - c_{n+1} \leq 0$ (for a minimization problem), then $x^*_{n+1} = 0$ and the current solution is optimal. On the othe hand, if $z_{n+1} - c_{n+1} > 0$, then x_{n+1} is introduced into the basis and the simplex method continues to find the new optimal solution.

Example 6.12

Consider Example 6.9. We wish to find the new optimal solution if a new activity $x_6 \geq 0$ with $c_6 = 1$, and $\mathbf{a}_6 = \begin{pmatrix} -1 \\ 2 \end{pmatrix}$ is introduced. First we calculate $z_6 - c_6$:

$$z_6 - c_6 = \mathbf{w}\mathbf{a}_6 - c_6$$

$$= (-2, 0)\begin{pmatrix} -1 \\ 2 \end{pmatrix} - 1 = 1$$

$$\mathbf{y}_6 = \mathbf{B}^{-1}\mathbf{a}_6 = \begin{bmatrix} 1 & 0 \\ 1 & 1 \end{bmatrix}\begin{bmatrix} -1 \\ 2 \end{bmatrix} = \begin{bmatrix} -1 \\ 1 \end{bmatrix}$$

Therefore x_6 is introduced in the basis by pivoting at the x_5 row and the x_6 column.

	z	x_1	x_2	x_3	x_4	x_5	x_6	RHS
z	1	0	-3	-1	-2	0	1	-12
x_1	0	1	1	1	1	0	-1	6
x_5	0	0	3	1	1	1	(1)	10

The subsequent tableaux are not shown.

Adding a New Constraint

Suppose that a new constraint is added to the problem. If the optimal solution to the original problem satisfies the added constraint, it is then obvious that the point is also an optimal solution of the new problem (why?). If, on the other hand, the point does not satisfy the new constraint, that is, if the constraint "cuts away" the optimal point, we can use the dual simplex method to find the new optimal solution. These two cases are illustrated in Figure 6.3.

Suppose that $\mathbf{B}$ is the optimal basis before the constraint $\mathbf{a}^{m+1}\mathbf{x} \leq b_{m+1}$ is added. The corresponding tableau is shown below.

$$z + (\mathbf{c}_B\mathbf{B}^{-1}\mathbf{N} - \mathbf{c}_N)\mathbf{x}_N = \mathbf{c}_B\mathbf{B}^{-1}\mathbf{b}$$

$$\mathbf{x}_B + \mathbf{B}^{-1}\mathbf{N}\mathbf{x}_N = \mathbf{B}^{-1}\mathbf{b} \tag{6.6}$$

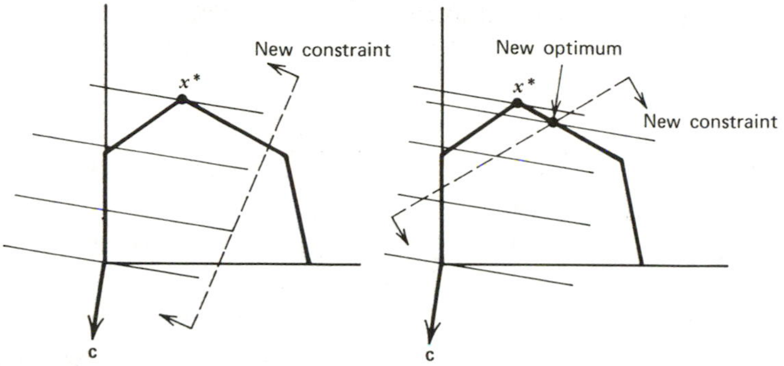

Figure 6.3. **Addition of a new constraint.**

The constraint $\mathbf{a}^{m+1}\mathbf{x} \leqslant b_{m+1}$ is rewritten as $\mathbf{a}_B^{m+1}\mathbf{x}_B + \mathbf{a}_N^{m+1}\mathbf{x}_N + x_{n+1} = b_{m+1}$, where $\mathbf{a}^{m+1}$ is decomposed into $(\mathbf{a}_B^{m+1}, \mathbf{a}_N^{m+1})$ and x_{n+1} is a nonnegative slack variable. Multiplying Equation (6.6) by $\mathbf{a}_B^{m+1}$ and subtracting from the new constraint gives the following system:

$$z + (\mathbf{c}_B\mathbf{B}^{-1}\mathbf{N} - \mathbf{c}_N)\mathbf{x}_N = \mathbf{c}_B\mathbf{B}^{-1}\mathbf{b}$$

$$\mathbf{x}_B + \mathbf{B}^{-1}\mathbf{N}\mathbf{x}_N = \mathbf{B}^{-1}\mathbf{b}$$

$$(\mathbf{a}_N^{m+1} - \mathbf{a}_B^{m+1}\mathbf{B}^{-1}\mathbf{N})\mathbf{x}_N + x_{n+1} = b_{m+1} - \mathbf{a}_B^{m+1}\mathbf{B}^{-1}\mathbf{b}$$

These equations give us a basic solution of the new system (why?). The only possible violation of optimality of the new problem is the sign of $b_{m+1} - \mathbf{a}_B^{m+1}\mathbf{B}^{-1}\mathbf{b}$. So if $b_{m+1} - \mathbf{a}_B^{m+1}\mathbf{B}^{-1}\mathbf{b} \geqslant 0$, then the current solution is optimal. Otherwise, if $b_{m+1} - \mathbf{a}_B^{m+1}\mathbf{B}^{-1}\mathbf{b} < 0$, then the dual simplex method is used to restore feasibility.

Example 6.13

Consider Example 6.9 with the added restriction that $-x_1 + 2x_3 \geqslant 2$. Clearly the optimal point $(x_1, x_2, x_3) = (6, 0, 0)$ does not satisfy this constraint. The constraint $-x_1 + 2x_3 \geqslant 2$ is rewritten as $x_1 - 2x_3 + x_6 = -2$, where x_6 is a nonnegative slack variable. This row is added to the optimal simplex tableau of Example 6.9 to obtain the following tableau.

	z	x_1	x_2	x_3	x_4	x_5	x_6	RHS
z	1	0	−3	−1	−2	0	0	−12
x_1	0	1	1	1	1	0	0	6
x_5	0	0	3	1	1	1	0	10
x_6	0	1	0	−2	0	0	1	−2

Multiply row 1 by -1 and add to row 3 in order to restore column x_1 to a unit vector. The dual simplex method can then be applied to the resulting tableau below.

	z	x_1	x_2	x_3	x_4	x_5	x_6	RHS
z	1	0	-3	-1	-2	0	0	-12
x_1	0	1	1	1	1	0	0	6
x_5	0	0	3	1	1	1	0	10
x_6	0	0	-1	(-3)	-1	0	1	-8

Subsequent tableaux are not shown. Note that adding a new constraint in the primal problem is equivalent to adding a new variable in the dual problem and vice versa.

An Application of Adding Constraints in Integer Programming

The linear integer programming problem may be stated as follows.

$$\begin{aligned} \text{Minimize} \quad & \mathbf{cx} \\ \text{Subject to} \quad & \mathbf{Ax}=\mathbf{b} \\ & \mathbf{x} \geqslant \mathbf{0} \\ & \mathbf{x} \quad \text{integer} \end{aligned}$$

A natural method to solve this problem is to ignore the last condition, **x** integer, and solve the problem as a linear program. At optimality, if all of the variables have integer values, then we have the optimal solution to the original integer program (why?). Otherwise consider adding a new constraint to the linear program. This additional constraint should "cut off" the current optimal noninteger linear programming solution without cutting-off any feasible integer solution. Adding the new constraint to the optimal tableau, we apply the dual simplex to reoptimize the new linear program. The new solution is either integer or not. The procedure of adding constraints is repeated until either an all integer solution is found or infeasibility results (indicating no integer solution). How, then, can such a cutting constraint be generated?

Consider the optimal simplex tableau when a noninteger solution results. Let $\bar{b}_r$ be noninteger. Assume that the basic variables are indexed from 1 to m. The equation associated with $\bar{b}_r$ is

$$x_r + \sum_{j=m+1}^{n} y_{rj} x_j = \bar{b}_r$$

Let I_{rj} be the greatest integer that is less than or equal to y_{rj} (I_{rj} is called the integer part of y_{rj}). Similarly, let I_r be the integer part of $\bar{b}_r$. Let F_{rj} and F_r be the respective fractional parts, that is,

$$F_{rj} = y_{rj} - I_{rj} \qquad \text{and} \qquad F_r = \bar{b}_r - I_r$$

Then $0 \leqslant F_{rj} < 1$ and $0 < F_r < 1$ (why?). Using this, we may rewrite the basic equation for x_r as

$$x_r + \sum_{j=m+1}^{n} (I_{rj} + F_{rj})x_j = I_r + F_r$$

Rearranging terms, we get

$$x_r + \sum_{j=m+1}^{n} I_{rj}x_j - I_r = F_r - \sum_{j=m+1}^{n} F_{rj}x_j$$

Now the left-hand side of this equation will be integer for any feasible integer solution (why?). The right-hand side is strictly less than 1, since $F_r < 1$, $F_{rj} \geqslant 0$, and $x_j \geqslant 0$. But since the right-hand side must also be integer, because it equals the left-hand side, we may conclude that it must be less than or equal to zero (there are no integers greater than zero and less than one). Thus we may write

$$F_r - \sum_{j=m+1}^{n} F_{rj}x_j \leqslant 0$$

However, since x_j is currently nonbasic (and hence $x_j = 0$) for $j = m + 1, \ldots, n$ and $F_r > 0$, the current optimal (noninteger) linear programming solution does not satisfy this additional constraint. In other words, this new constraint will cut-off the current optimal solution if added to the current optimal tableau. The dual simplex method is then applied to obtain a new optimal linear programming solution, which is again tested for integrality. Such a procedure is called a *cutting plane* algorithm.

Example 6.14

Consider the following integer program.

Minimize $\quad 3x_1 + 4x_2$

Subject to $\quad 3x_1 + x_2 \geqslant 4$

$\qquad x_1 + 2x_2 \geqslant 4$

$\qquad x_1, \quad x_2 \geqslant 0$

$\qquad x_1, \quad x_2$ integer

In Figure 6.4 we show the optimal linear programming and integer programming solutions respectively.

Ignoring the integer conditions, the following tableau gives the optimal linear programming solution.

	z	x_1	x_2	x_3	x_4	RHS
z	1	0	0	$-\frac{2}{5}$	$-\frac{9}{5}$	$\frac{44}{5}$
x_1	0	1	0	$-\frac{2}{5}$	$\frac{1}{5}$	$\frac{4}{5}$
x_2	0	0	1	$\frac{1}{5}$	$-\frac{3}{5}$	$\frac{8}{5}$

Since this solution is noninteger, we may select a noninteger variable for generating a cut (including z). Select x_2. (*Note.* Selecting different variables may generate different cuts.) The equation for the basic variable x_2 is

$$x_2 + \tfrac{1}{5}x_3 - \tfrac{3}{5}x_4 = \tfrac{8}{5}$$

From this we get

$$I_{23} = 0, \quad I_{24} = -1, \quad I_2 = 1$$
$$F_{23} = \tfrac{1}{5}, \quad F_{24} = \tfrac{2}{5}, \quad F_2 = \tfrac{3}{5}$$

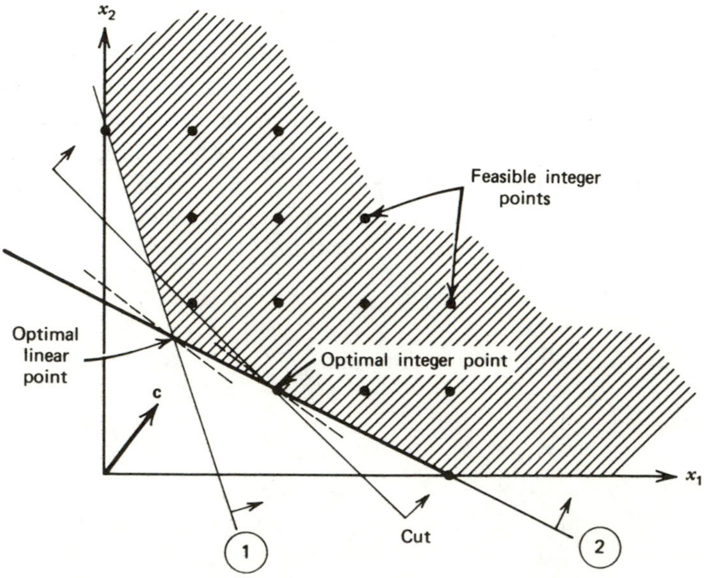

Figure 6.4. The graphical solution.

and the additional constraint becomes

$$\tfrac{1}{5}x_3 + \tfrac{2}{5}x_4 \geqslant \tfrac{3}{5} \qquad \text{(cut)}$$

Adding this constraint with slack variable x_5 to the tableau and applying the dual simplex method, we get the following tableaux.

	z	x_1	x_2	x_3	x_4	x_5	RHS
z	1	0	0	$-\frac{2}{5}$	$-\frac{9}{5}$	0	$\frac{44}{5}$
x_1	0	1	0	$-\frac{2}{5}$	$\frac{1}{5}$	0	$\frac{4}{5}$
x_2	0	0	1	$\frac{1}{5}$	$-\frac{3}{5}$	0	$\frac{8}{5}$
x_5	0	0	0	$\left(-\frac{1}{5}\right)$	$-\frac{2}{5}$	1	$-\frac{3}{5}$

	z	x_1	x_2	x_3	x_4	x_5	RHS
z	1	0	0	0	-1	-2	10
x_1	0	1	0	0	1	-2	2
x_2	0	0	1	0	-1	1	1
x_3	0	0	0	1	2	-5	3

Hence we have obtained the optimal integer solution $\mathbf{x}^* = (2, 1)^t$ with only one cut. In other integer programs we might have to repeat the cutting plane process many times. If in the foregoing tableau some variable had turned out noninteger, we would have generated a new cut and continued.

It is interesting to examine the cut in terms of the original variables. Substituting $x_3 = 3x_1 + x_2 - 4$ and $x_4 = x_1 + 2x_2 - 4$ into $\frac{1}{5}x_3 + \frac{2}{5}x_4 \geqslant \frac{3}{5}$ and simplifying, we get

$$x_1 + x_2 \geqslant 3 \qquad \text{(cut in terms of } x_1 \text{ and } x_2\text{)}$$

It can easily be seen that the addition of this constraint to Figure 6.4 will yield the required integer optimum.

6.8 PARAMETRIC ANALYSIS

Parametric analysis is used quite often in large-scale optimization and nonlinear optimization, where one often finds a direction along which the objective function or the constraints are perturbed, and then seeks to move along this direction. So we seek the optimal solutions to a class of problems by perturbing either the objective vector or the RHS vector along a fixed direction.

Perturbation of the Cost Vector

Consider the following problem.

$$\begin{aligned} \text{Minimize} \quad & \mathbf{cx} \\ \text{Subject to} \quad & \mathbf{Ax} = \mathbf{b} \\ & \mathbf{x} \geqslant \mathbf{0} \end{aligned}$$

Assume that $\mathbf{B}$ is an optimal basis. Suppose that the cost vector $\mathbf{c}$ is perturbed along the cost direction $\mathbf{c}'$, that is, $\mathbf{c}$ is replaced by $\mathbf{c} + \lambda\mathbf{c}'$ where $\lambda \geqslant 0$. We are interested in finding the optimal points and corresponding objective values as a function of $\lambda \geqslant 0$. Decomposing $\mathbf{A}$ into $[\mathbf{B}, \mathbf{N}]$, $\mathbf{c}$ into $(\mathbf{c}_B, \mathbf{c}_N)$, and $\mathbf{c}'$ into $(\mathbf{c}'_B, \mathbf{c}'_N)$, we get

$$\begin{aligned} z - (\mathbf{c}_B + \lambda\mathbf{c}'_B)\mathbf{x}_B - (\mathbf{c}_N + \lambda\mathbf{c}'_N)\mathbf{x}_N &= 0 \\ \mathbf{B}\mathbf{x}_B + \mathbf{N}\mathbf{x}_N &= \mathbf{b} \end{aligned}$$

Updating the tableau and denoting $\mathbf{c}'_B\mathbf{y}_j$ by z'_j, we get

$$z + \sum_{j \in R} \left[(z_j - c_j) + \lambda(z'_j - c'_j) \right] x_j = \mathbf{c}_B\bar{\mathbf{b}} + \lambda\mathbf{c}'_B\bar{\mathbf{b}}$$

$$\mathbf{x}_B + \sum_{j \in R} \mathbf{y}_j x_j = \bar{\mathbf{b}}$$

where R is the set of current indices associated with the nonbasic variables. The current tableau has $\lambda = 0$ and gives an optimal basic feasible solution of the original problem without perturbation. We would like to find out how far we can move in the direction $\mathbf{c}'$ while still maintaining optimality of the current point. Let $S = \{j : (z'_j - c'_j) > 0\}$. If $S = \phi$, then the current solution is optimal for all values of $\lambda \geqslant 0$ (why?). Otherwise, calculate $\hat{\lambda}$ as follows:

$$\hat{\lambda} = \underset{j \in S}{\text{Minimum}} \left\{ \frac{-(z_j - c_j)}{z'_j - c'_j} \right\} = \frac{-(z_k - c_k)}{z'_k - c'_k} \tag{6.7}$$

Let $\lambda_1 = \hat{\lambda}$. For $\lambda \in [0, \lambda_1]$ the current solution is optimal and the optimal objective value is given by $\mathbf{c}_B\bar{\mathbf{b}} + \lambda\mathbf{c}'_B\bar{\mathbf{b}} = \mathbf{c}_B\mathbf{B}^{-1}\mathbf{b} + \lambda\mathbf{c}'_B\mathbf{B}^{-1}\mathbf{b}$. For $\lambda \in [0, \lambda_1]$, the shadow prices in the simplex tableau are replaced by $(z_j - c_j) + \lambda(z'_j - c'_j)$. At $\lambda = \lambda_1$, x_k is introduced into the basis (if a blocking variable exists). After the tableau is updated, the process is repeated by recalculating S and $\hat{\lambda}$ and letting $\lambda_2 = \hat{\lambda}$. For $\lambda \in [\lambda_1, \lambda_2]$ the new current solution is optimal and its objective value is given by $\mathbf{c}_B\bar{\mathbf{b}} + \lambda\mathbf{c}'_B\bar{\mathbf{b}} = \mathbf{c}_B\mathbf{B}^{-1}\mathbf{b} + \lambda\mathbf{c}'_B\mathbf{B}^{-1}\mathbf{b}$ where $\mathbf{B}$ is the current basis. The process is repeated until S becomes empty. If there is no blocking variable when x_k enters the basis, then the problem is unbounded for all values of λ greater than the current value.

Example 6.15

Consider the following problem.

$$\begin{aligned} \text{Minimize} \quad & -x_1 - 3x_2 \\ \text{Subject to} \quad & x_1 + x_2 \leqslant 6 \\ & -x_1 + 2x_2 \leqslant 6 \\ & x_1, \quad x_2 \geqslant 0 \end{aligned}$$

It is desired to find the optimal solutions and optimal objective values of the class of problems whose objective function is $(-1 + 2\lambda, -3 + \lambda)$ for $\lambda \geqslant 0$; that is, we perturb the cost vector along the vector (2, 1). First we solve the problem with $\lambda = 0$ where x_3 and x_4 are the slack variables. The optimal tableau for $\lambda = 0$ is given by the following.

	z	x_1	x_2	x_3	x_4	RHS
z	1	0	0	$-\frac{5}{3}$	$-\frac{2}{3}$	-14
x_1	0	1	0	$\frac{2}{3}$	$-\frac{1}{3}$	2
x_2	0	0	1	$\frac{1}{3}$	$\frac{1}{3}$	4

In order to find the range over which this tableau is optimal, first find $\mathbf{c}'_B\mathbf{B}^{-1}\mathbf{N} - \mathbf{c}'_N$:

$$\begin{aligned} \mathbf{c}'_B\mathbf{B}^{-1}\mathbf{N} - \mathbf{c}'_N &= \mathbf{c}'_B(\mathbf{y}_3, \mathbf{y}_4) - (c'_3, c'_4) \\ &= (2, 1)\begin{bmatrix} \frac{2}{3} & -\frac{1}{3} \\ \frac{1}{3} & \frac{1}{3} \end{bmatrix} - (0, 0) \\ &= \left(\tfrac{5}{3}, -\tfrac{1}{3}\right) \end{aligned} \tag{6.8}$$

Therefore $S = \{3\}$ and from Equation (6.7) $\hat{\lambda}$ is given by

$$\hat{\lambda} = \frac{-(z_3 - c_3)}{z'_3 - c'_3} = \frac{-(-\frac{5}{3})}{\frac{5}{3}} = 1$$

Therefore $\lambda_1 = 1$ and for $\lambda \in [0, 1]$ the basis $[\mathbf{a}_1, \mathbf{a}_2]$ remains optimal. The optimal objective value $z(\lambda)$ in this interval is given by

$$z(\lambda) = \mathbf{c}_B\bar{\mathbf{b}} + \lambda \mathbf{c}'_B\bar{\mathbf{b}}$$

$$= -14 + \lambda(2, 1)\begin{pmatrix} 2 \\ 4 \end{pmatrix} = -14 + 8\lambda$$

Noting Equation (6.8), the shadow prices of the nonbasic variables x_3 and x_4 are given by

$$(z_3 - c_3) + \lambda(z'_3 - c'_3) = -\tfrac{5}{3} + \tfrac{5}{3}\lambda$$

$$(z_4 - c_4) + \lambda(z'_4 - c'_4) = -\tfrac{2}{3} - \tfrac{1}{3}\lambda$$

Hence the optimal solution for any λ in the interval [0, 1] is given by the following tableau.

	z	x_1	x_2	x_3	x_4	RHS
z	1	0	0	$-\frac{5}{3} + \frac{5}{3}\lambda$	$-\frac{2}{3} - \frac{1}{3}\lambda$	$-14 + 8\lambda$
x_1	0	1	0	$\frac{2}{3}$	$-\frac{1}{3}$	2
x_2	0	0	1	$\frac{1}{3}$	$\frac{1}{3}$	4

At $\lambda = 1$ the coefficient of x_3 in row 0 is equal to 0 and x_3 is introduced in the basis leading to the following new tableau.

	z	x_1	x_2	x_3	x_4	RHS
z	1	0	0	0	-1	-6
x_3	0	$\frac{3}{2}$	0	1	$-\frac{1}{2}$	3
x_2	0	$-\frac{1}{2}$	1	0	$\frac{1}{2}$	3

We would now like to find the interval $[1, \lambda_2]$ over which the foregoing tableau is optimal. Note that

$$z_1 - c_1 = \mathbf{c}_B\mathbf{y}_1 - c_1 = (0, -3)\begin{pmatrix} \frac{3}{2} \\ -\frac{1}{2} \end{pmatrix} + 1 = \tfrac{5}{2}$$

$$z_4 - c_4 = \mathbf{c}_B\mathbf{y}_4 - c_4 = (0, -3)\begin{pmatrix} -\frac{1}{2} \\ \frac{1}{2} \end{pmatrix} - 0 = -\tfrac{3}{2}$$

$$z'_1 - c'_1 = \mathbf{c}'_B\mathbf{y}_1 - c'_1 = (0, 1)\begin{pmatrix} \frac{3}{2} \\ -\frac{1}{2} \end{pmatrix} - 2 = -\tfrac{5}{2}$$

$$z'_4 - c'_4 = \mathbf{c}'_B\mathbf{y}_4 - c'_4 = (0, 1)\begin{pmatrix} -\frac{1}{2} \\ \frac{1}{2} \end{pmatrix} - 0 = \tfrac{1}{2}$$

Therefore the shadow prices for the nonbasic variables x_1 and x_4 are given by

$$(z_1 - c_1) + \lambda(z'_1 - c'_1) = \tfrac{5}{2} - \tfrac{5}{2}\lambda$$

$$(z_4 - c_4) + \lambda(z'_4 - c'_4) = -\tfrac{3}{2} + \tfrac{1}{2}\lambda$$

Therefore for λ in the interval [1, 3] the shadow prices are nonpositive and the basis consisting of $\mathbf{a}_3$ and $\mathbf{a}_2$ is optimal. Note that $\lambda = 3$ can also be determined as follows:

$$S = \{4\} \text{ and } \lambda = \frac{-(z_4 - c_4)}{z'_4 - c'_4} = \frac{\frac{3}{2}}{\frac{1}{2}} = 3$$

Over the interval [1, 3] the objective function $z(\lambda)$ is given by

$$z(\lambda) = \mathbf{c}_B\bar{\mathbf{b}} + \lambda\mathbf{c}'_B\bar{\mathbf{b}}$$

$$= (0, -3)\binom{3}{3} + \lambda(0, 1)\binom{3}{3} = -9 + 3\lambda$$

Hence the optimal tableau for $\lambda \in [1, 3]$ is given below.

	z	x_1	x_2	x_3	x_4	RHS
z	1	$\frac{5}{2} - \frac{5}{2}\lambda$	0	0	$-\frac{3}{2} + \frac{1}{2}\lambda$	$-9 + 3\lambda$
x_3	0	$\frac{3}{2}$	0	1	$-\frac{1}{2}$	3
x_2	0	$-\frac{1}{2}$	1	0	$\frac{1}{2}$	3

At $\lambda = 3$ the coefficient of x_4 in row 0 is equal to zero, and x_4 is introduced in the basis leading to the following tableau.

	z	x_1	x_2	x_3	x_4	RHS
z	1	-5	0	0	0	0
x_3	0	1	1	1	0	6
x_4	0	-1	2	0	1	6

We would like to calculate the interval over which the foregoing tableau is optimal. First calculate

$$z'_1 - c'_1 = \mathbf{c}'_B\mathbf{y}_1 - c'_1 = -2$$

$$z'_2 - c'_2 = \mathbf{c}'_B\mathbf{y}_2 - c'_2 = -1$$

Therefore $S = \phi$ and hence the basis $[\mathbf{a}_3, \mathbf{a}_4]$ is optimal for all $\lambda \in [3, \infty)$. Figure 6.5 shows the optimal points and the optimal objective as a function of λ. Note

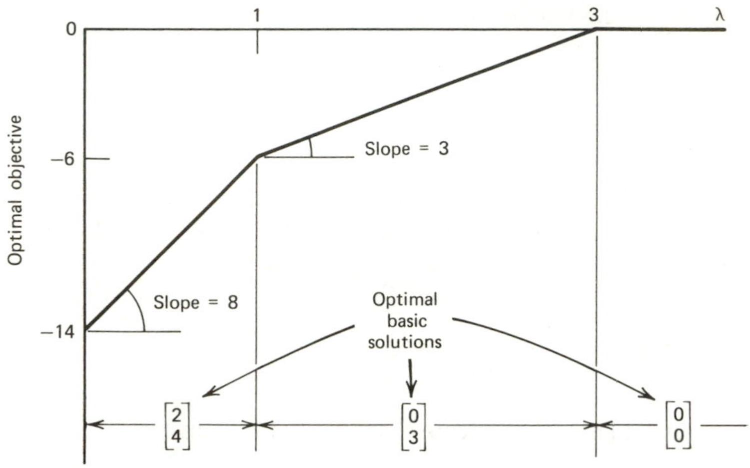

Figure 6.5. **Optimal objectives and optimal points as a function of λ.**

that this function is piecewise-linear and concave. In Exercise 6.60 the reader is asked to show that this is always true. The break points correspond to the value of λ at which alternative optimal solutions exist.

Perturbation of the Right-Hand-Side

Suppose that the right-hand-side vector $\mathbf{b}$ is replaced by $\mathbf{b} + \lambda\mathbf{b}'$ where $\lambda \geqslant 0$. This means that the right-hand side is perturbed along the vector $\mathbf{b}'$. Since the right-hand side of the primal problem is the objective of the dual problem, perturbing the right-hand side can be analyzed as perturbing the objective function of the dual problem. We shall now handle the perturbation directly by considering the primal problem. Suppose that we have an optimal basis $\mathbf{B}$ of the original problem, that is, with $\lambda = 0$. The corresponding tableau is given by

$$z + (\mathbf{c}_B\mathbf{B}^{-1}\mathbf{N} - \mathbf{c}_N)\mathbf{x}_N = \mathbf{c}_B\mathbf{B}^{-1}\mathbf{b}$$

$$\mathbf{x}_B + \mathbf{B}^{-1}\mathbf{N}\mathbf{x}_N = \mathbf{B}^{-1}\mathbf{b}$$

where $\mathbf{c}_B\mathbf{B}^{-1}\mathbf{N} - \mathbf{c}_N \leqslant \mathbf{0}$. If $\mathbf{b}$ is replaced by $\mathbf{b} + \lambda\mathbf{b}'$, the vector $\mathbf{c}_B\mathbf{B}^{-1}\mathbf{N} - \mathbf{c}_N$ will not be affected; that is, dual feasibility will not be affected. The only change is that $\mathbf{B}^{-1}\mathbf{b}$ will be replaced by $\mathbf{B}^{-1}(\mathbf{b} + \lambda\mathbf{b}')$, and accordingly the objective becomes $\mathbf{c}_B\mathbf{B}^{-1}(\mathbf{b} + \lambda\mathbf{b}')$. As long as $\mathbf{B}^{-1}(\mathbf{b} + \lambda\mathbf{b}')$ is nonnegative, the current basis remains optimal. The value of λ at which another basis becomes optimal, can therefore, be determined as follows. Let $S = \{i : \bar{b}'_i < 0\}$ where $\bar{\mathbf{b}}' = \mathbf{B}^{-1}\mathbf{b}'$.

If $S = \phi$, then the current basis is optimal for all values of $\lambda \geqslant 0$. Otherwise let

$$\hat{\lambda} = \underset{i \in S}{\text{Minimum}} \left\{ \frac{\bar{b}_i}{-\bar{b}'_i} \right\} = \frac{\bar{b}_r}{-\bar{b}'_r} \tag{6.9}$$

Let $\lambda_1 = \hat{\lambda}$. For $\lambda \in [0, \lambda_1]$ the current basis is optimal, where $\mathbf{x}_B = \mathbf{B}^{-1}(\mathbf{b} + \lambda\mathbf{b}')$ and the optimal objective is $\mathbf{c}_B\mathbf{B}^{-1}(\mathbf{b} + \lambda\mathbf{b}')$. At λ_1 the right-hand side is replaced by $\mathbf{B}^{-1}(\mathbf{b} + \lambda_1\mathbf{b}')$, x_{B_r} is removed from the basis, and an appropriate variable (according to dual simplex method criterion) enters the basis. After the tableau is updated, the process is repeated in order to find the range $[\lambda_1, \lambda_2]$ over which the new basis is optimal, where $\lambda_2 = \hat{\lambda}$ from Equation (6.9). The process is terminated when either S is empty, in which case the current basis is optimal for all values of λ greater than or equal to the last value of λ, or else when all the entries in the row whose right-hand side dropped to zero, are nonnegative. In this latter case no feasible solutions exist for all values of λ greater than the current value (why?).

Example 6.16

Consider the following problem.

$$\begin{array}{lrl}
\text{Minimize} & -x_1 - 3x_2 & \\
\text{Subject to} & x_1 + x_2 & \leqslant 6 \\
& -x_1 + 2x_2 & \leqslant 6 \\
& x_1, \quad x_2 & \geqslant 0
\end{array}$$

It is desired to find the optimal solution and optimal basis as the right-hand-side is perturbed along direction $\begin{pmatrix} -1 \\ 1 \end{pmatrix}$, that is, if $\mathbf{b} = \begin{pmatrix} 6 \\ 6 \end{pmatrix}$ is replaced by $\mathbf{b} + \lambda\mathbf{b}' = \begin{pmatrix} 6 \\ 6 \end{pmatrix} + \lambda\begin{pmatrix} -1 \\ 1 \end{pmatrix}$ for $\lambda \geqslant 0$. The optimal solution with $\lambda = 0$ is shown below where x_3 and x_4 are the slack variables.

	z	x_1	x_2	x_3	x_4	RHS
z	1	0	0	$-\frac{5}{3}$	$-\frac{2}{3}$	-14
x_1	0	1	0	$\frac{2}{3}$	$-\frac{1}{3}$	2
x_2	0	0	1	$\frac{1}{3}$	$\frac{1}{3}$	4

In order to find the range over which the above basis is optimal, first calculate $\bar{\mathbf{b}}'$:

$$\bar{\mathbf{b}}' = \mathbf{B}^{-1}\mathbf{b}'$$

$$= \begin{bmatrix} \frac{2}{3} & -\frac{1}{3} \\ \frac{1}{3} & \frac{1}{3} \end{bmatrix} \begin{bmatrix} -1 \\ 1 \end{bmatrix} = \begin{bmatrix} -1 \\ 0 \end{bmatrix}$$

Therefore $S = \{1\}$ and from Equation (6.9) λ_1 is given by

$$\lambda_1 = \frac{\bar{b}_1}{-\bar{b}'_1} = \frac{2}{-(-1)} = 2$$

Therefore the basis $[\mathbf{a}_1, \mathbf{a}_2]$ remains optimal over the interval $[0, 2]$. In particular, for any $\lambda \in [0, 2]$ the objective value and the right-hand-side are given by

$$z(\lambda) = \mathbf{c}_B\bar{\mathbf{b}} + \lambda\mathbf{c}_B\bar{\mathbf{b}}'$$

$$= (-1, -3)\binom{2}{4} + \lambda(-1, -3)\binom{-1}{0}$$

$$= -14 + \lambda$$

$$\bar{\mathbf{b}} + \lambda\bar{\mathbf{b}}' = \binom{2}{4} + \lambda\binom{-1}{0}$$

$$= \binom{2-\lambda}{4}$$

	z	x_1	x_2	x_3	x_4	RHS
z	1	0	0	$-\frac{5}{3}$	$-\frac{2}{3}$	$-14+\lambda$
x_1	0	1	0	$\frac{2}{3}$	$-\frac{1}{3}$	$2-\lambda$
x_2	0	0	1	$\frac{1}{3}$	$\frac{1}{3}$	4

At $\lambda = 2$, $x_{B_r} = x_1$ drops to zero. A dual simplex pivot is performed so that x_1 leaves the basis and x_4 enters the basis leading to the following tableau.

	z	x_1	x_2	x_3	x_4	RHS
z	1	-2	0	-3	0	-12
x_4	0	-3	0	-2	1	0
x_2	0	1	1	1	0	4

In order to find the range $[2, \lambda_2]$ over which this tableau is optimal, first find $\bar{\mathbf{b}}$ and $\bar{\mathbf{b}}'$:

$$\bar{\mathbf{b}} = \mathbf{B}^{-1}\mathbf{b} = \begin{bmatrix} -2 & 1 \\ 1 & 0 \end{bmatrix}\begin{bmatrix} 6 \\ 6 \end{bmatrix} = \begin{bmatrix} -6 \\ 6 \end{bmatrix}$$

$$\bar{\mathbf{b}}' = \mathbf{B}^{-1}\mathbf{b}' = \begin{bmatrix} -2 & 1 \\ 1 & 0 \end{bmatrix}\begin{bmatrix} -1 \\ 1 \end{bmatrix} = \begin{bmatrix} 3 \\ -1 \end{bmatrix}$$

Therefore $S = \{2\}$ and λ_2 is given by

$$\lambda_2 = \frac{\bar{b}_2}{-\bar{b}'_2} = \frac{6}{-(-1)} = 6$$

For λ in the interval [2, 6] the optimal objective value and the right-hand-side are given by

$$z(\lambda) = \mathbf{c}_B\bar{\mathbf{b}} + \lambda\mathbf{c}_B\bar{\mathbf{b}}'$$

$$= (0, -3)\begin{pmatrix} -6 \\ 6 \end{pmatrix} + \lambda(0, -3)\begin{pmatrix} 3 \\ -1 \end{pmatrix}$$

$$= -18 + 3\lambda$$

$$\bar{\mathbf{b}} + \lambda\bar{\mathbf{b}}' = \begin{bmatrix} -6 \\ 6 \end{bmatrix} + \lambda\begin{bmatrix} 3 \\ -1 \end{bmatrix}$$

$$= \begin{bmatrix} -6 + 3\lambda \\ 6 - \lambda \end{bmatrix}$$

The optimal tableau over the interval [2, 6] is depicted below.

	z	x_1	x_2	x_3	x_4	RHS
z	1	-2	0	-3	0	$-18 + 3\lambda$
x_4	0	-3	0	-2	1	$-6 + 3\lambda$
x_2	0	1	1	1	0	$6 - \lambda$

At $\lambda = 6$, x_2 drops to zero. Since all entries in the x_2 row are nonnegative, we stop with the conclusion that for $\lambda > 6$ there exist no feasible solutions. Figure 6.6 summarizes the optimal bases and corresponding objectives for $\lambda \geqslant 0$. Note that the optimal objective as a function of λ is piecewise-linear and convex. In Exercise 6.61 we ask the reader to show that this is always true. The break points correspond to the values of λ for which alternative optimal dual solutions exist.

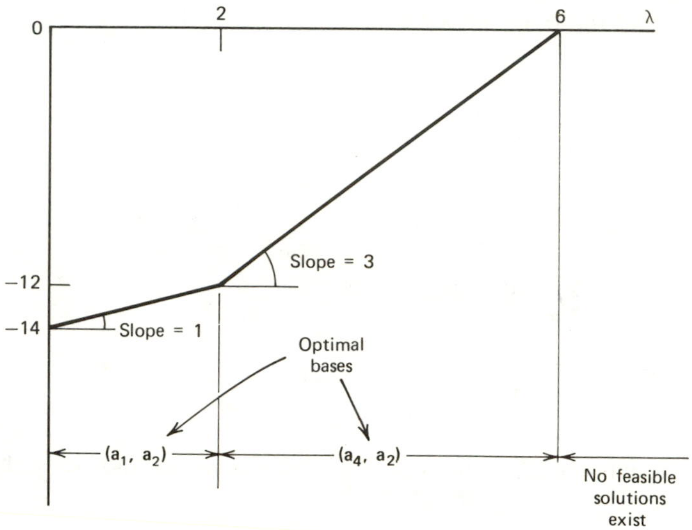

Figure 6.6. **Optimal objectives and bases as a function of λ.**

EXERCISES

6.1 Consider the following problem.

$$\begin{aligned} \text{Maximize} \quad & -x_1+2x_2 \\ \text{Subject to} \quad & 3x_1+4x_2 \leqslant 12 \\ & 2x_1- \ \ x_2 \geqslant 2 \\ & x_1, \quad x_2 \geqslant 0 \end{aligned}$$

a. Solve the problem graphically.
b. State the dual and solve it graphically. Utilize the theorems of duality to obtain the values of all the primal variables from the optimal dual solution.

6.2 Consider the following problem.

$$\begin{aligned} \text{Minimize} \quad & 2x_1+3x_2+5x_3+6x_4 \\ \text{Subject to} \quad & x_1+2x_2+3x_3+ \ \ x_4 \geqslant \ \ 2 \\ & -2x_1+ \ \ x_2- \ \ x_3+3x_4 \leqslant -3 \\ & x_1, \quad x_2, \quad x_3, \quad x_4 \geqslant \ \ 0 \end{aligned}$$

a. Give the dual linear program.
b. Solve the dual geometrically.
c. Utilize information about the dual linear program and the theorems of duality to solve the primal problem.

6.3 Solve the following linear program by a graphical method.

$$\begin{aligned} \text{Maximize} \quad & 3x_1 + x_2 + 4x_3 \\ \text{Subject to} \quad & 6x_1 + 3x_2 + 5x_3 \leqslant 25 \\ & 3x_1 + 4x_2 + 5x_3 \leqslant 20 \\ & x_1, \quad x_2, \quad x_3 \geqslant 0 \end{aligned}$$

(*Hint*. Utilize the dual problem.)

6.4 Give the dual of the following problem.

$$\begin{aligned} \text{Minimize} \quad & 2x_1 + 3x_2 - 5x_3 \\ \text{Subject to} \quad & x_1 + x_2 - x_3 + x_4 \geqslant 5 \\ & 2x_1 + x_3 \leqslant 4 \\ & x_2 + x_3 + x_4 = 6 \\ & x_1 \leqslant 0 \\ & x_2, \ x_3 \geqslant 0 \\ & x_4 \ \text{unrestricted} \end{aligned}$$

6.5 Use the standard form or the canonical form of duality to obtain the dual of the following problem. Also verify the relationships in Table 6.1.

$$\begin{aligned} \text{Minimize} \quad & \mathbf{c}_1\mathbf{x}_1 + \mathbf{c}_2\mathbf{x}_2 + \mathbf{c}_3\mathbf{x}_3 \\ \text{Subject to} \quad & \mathbf{A}_{11}\mathbf{x}_1 + \mathbf{A}_{12}\mathbf{x}_2 + \mathbf{A}_{13}\mathbf{x}_3 \geqslant \mathbf{b}_1 \\ & \mathbf{A}_{21}\mathbf{x}_1 + \mathbf{A}_{22}\mathbf{x}_2 + \mathbf{A}_{23}\mathbf{x}_3 \leqslant \mathbf{b}_2 \\ & \mathbf{A}_{31}\mathbf{x}_1 + \mathbf{A}_{32}\mathbf{x}_2 + \mathbf{A}_{33}\mathbf{x}_3 = \mathbf{b}_3 \\ & \mathbf{x}_1 \geqslant \mathbf{0} \\ & \mathbf{x}_2 \leqslant \mathbf{0} \\ & \mathbf{x}_3 \ \text{unrestricted} \end{aligned}$$

6.6 Consider the following problem.

$$\begin{aligned} \text{Maximize} \quad & 10x_1+24x_2+20x_3+20x_4+25x_5 \\ \text{Subject to} \quad & x_1+\ \ x_2+2x_3+3x_4+5x_5 \leqslant 19 \\ & 2x_1+4x_2+3x_3+2x_4+\ \ x_5 \leqslant 57 \\ & x_1,\ \ x_2,\ \ x_3,\ \ x_4,\ \ x_5 \geqslant 0 \end{aligned}$$

a. Write the dual problem and verify that $(w_1, w_2) = (4, 5)$ is a feasible solution.
b. Use the information in part (a) to derive an optimal solution to both the primal and the dual problems.

6.7 Consider the following linear program.

$$\begin{aligned} \text{P: Minimize} \quad & 6x_1+2x_2 \\ \text{Subject to} \quad & x_1+2x_2 \geqslant 3 \\ & x_2 \geqslant 0 \\ & x_1 \quad \text{unrestricted} \end{aligned}$$

a. State the dual of P.
b. Draw the set of feasible solutions for the dual of part (a).
c. Convert P to canonical form by replacing x_1 by $x_1' - x_1''$ with x_1', $x_1'' \geqslant 0$. Give the dual of this converted problem.
d. Draw the set of feasible solutions for the dual of part (c).
e. Compare parts (b) and (d). What did the transformation of part (c) do to the dual of part (a)?

6.8 The following simplex tableau shows the optimal solution of a linear programming problem. It is known that x_4 and x_5 are the slack variables in the first and second constraints of the original problem. The constraints are of the $\leqslant$ type.

	z	x_1	x_2	x_3	x_4	x_5	RHS
z	1	0	-4	0	-4	-2	-40
x_3	0	0	$\frac{1}{2}$	1	$\frac{1}{2}$	0	$\frac{5}{2}$
x_1	0	1	$-\frac{1}{2}$	0	$-\frac{1}{6}$	$\frac{1}{3}$	$\frac{5}{2}$

a. Write the original problem.
b. What is the dual of the original problem?
c. Obtain the optimal solution of the dual problem from the above tableau.

6.9 Consider the following linear programming problem.

$$\text{Maximize} \quad 2x_1 + 3x_2 + 6x_3$$

$$\begin{aligned} \text{Subject to} \quad x_1 + 2x_2 + \ x_3 &\leqslant 10 \\ x_1 - \ x_2 + 3x_3 &\leqslant 6 \\ x_1, \quad x_2, \quad x_3 &\geqslant 0 \end{aligned}$$

a. Write the dual problem.
b. Solve the foregoing problem by the simplex method. At each iteration, identify the dual variables, and show which dual constraints are violated.
c. At each iteration, identify the dual basis that goes with the simplex iteration. Identify the dual basic and nonbasic variables.
d. Show that at each iteration of the simplex method, the dual objective is "worsened."
e. Verify that at termination, feasible solutions of both problems are at hand, with equal objectives, and with complementary slackness.

6.10 Consider the problem: Minimize $\mathbf{cx}$ subject to $\mathbf{Ax} = \mathbf{b}$, $\mathbf{x} \geqslant \mathbf{0}$ where $m = n$, $\mathbf{c} = \mathbf{b}^t$ and $\mathbf{A} = \mathbf{A}^t$. Show that if there exists an $\mathbf{x}_0$ such that $\mathbf{Ax}_0 = \mathbf{b}, \mathbf{x}_0 \geqslant \mathbf{0}$, then $\mathbf{x}_0$ is an optimal point. (*Hint.* Use duality.)

6.11 Consider the problem: Minimize z subject to $z - \mathbf{cx} = 0$, $\mathbf{Ax} = \mathbf{b}$, $\mathbf{x} \geqslant \mathbf{0}$.
a. State the dual.
b. At optimality what will be the value of the first dual variable? Explain!

6.12 Consider the problem: Minimize $\mathbf{cx}$ subject to $\mathbf{Ax} = \mathbf{b}$, $\mathbf{x} \geqslant \mathbf{0}$. Let $\mathbf{B}$ be a basis that is neither primal nor dual feasible. Indicate how one can solve this problem starting with the basis $\mathbf{B}$.

6.13 Prove that if a given basic feasible solution to some linear programming problem is optimal, the same basic vectors will yield an optimal solution for any requirements vector that lies in the cone spanned by these basic vectors.

6.14 The Sewel Manufacturing Company produces two types of reclining chairs for sale in the Northeast. Two basic types of skilled labor are involved —assembly and finishing. One unit of the top-of-the-line recliner requires $\frac{3}{2}$ hours of assembly, 1 hour of finishing, and sells for a profit of \$20. A unit of the second-line recliner requires $\frac{1}{2}$ hour of assembly, $\frac{1}{2}$ hour of finishing, and sells for a profit of \$12. Presently there are 100 assembly hours and 80 finishing hours available to the company. The company is involved in labor negotiations concerning salary modifications for the coming year. You are asked to provide the company with indications of the worth of an hour of assembly worker's time and an hour of finishing worker's time.

6.15 Show by duality that if the problem Minimize $\mathbf{cx}$ subject to $\mathbf{Ax} = \mathbf{b}$ and $\mathbf{x} \geqslant \mathbf{0}$ has a finite optimal solution, then the new problem Minimize $\mathbf{cx}$ subject to $\mathbf{Ax} = \mathbf{b}'$ and $\mathbf{x} \geqslant \mathbf{0}$ cannot be unbounded, no matter what value the vector $\mathbf{b}'$ might take.

6.16 Show that if the primal problem Minimize $\mathbf{cx}$ subject to $\mathbf{Ax} \geqslant \mathbf{b}$ and $\mathbf{x} \geqslant \mathbf{0}$ has no feasible solutions, and if the dual problem has a feasible solution, then the dual problem is unbounded. (*Hint.* Use Farkas's lemma. If the system $\mathbf{Ax} \geqslant \mathbf{b}$ and $\mathbf{x} \geqslant \mathbf{0}$ has no solutions, then the system $\mathbf{wA} \leqslant \mathbf{0}$, $\mathbf{w} \geqslant \mathbf{0}$, $\mathbf{wb} > 0$ has a solution.)

6.17 Consider the problem: Minimize $\mathbf{cx}$ subject to $\mathbf{Ax} = \mathbf{b}$, $\mathbf{l} \leqslant \mathbf{x} \leqslant \mathbf{u}$.
a. Give the dual.
b. Show that the dual always possesses a feasible solution.
c. If the primal problem possesses a feasible solution, what conclusions would you reach?

6.18 Consider the problem: Maximize $\mathbf{cx}$ subject to $\mathbf{Ax} = \mathbf{b}$, $\mathbf{x} \geqslant \mathbf{0}$. Let $z_j - c_j$, y_{ij}, and $\bar{b}_i$ be the updated entries at some iteration of the simplex algorithm. Indicate whether each of the following statements is true or false. Discuss.
a. $y_{ij} = -\dfrac{\partial x_{B_i}}{\partial x_j}$
b. $z_j - c_j = -\dfrac{\partial z}{\partial x_j}$
c. Dual feasibility is the same as primal optimality.
d. Performing row operations on inequality systems yields equivalent systems.
e. Adding artificial variables to the primal serves to restrict variables that are really unrestricted in the dual.
f. Linear programming by the simplex method is essentially a gradient search.
g. A linear problem can be solved by the two-phase method if it can be solved by the big-M method.
h. There is a *duality gap* (difference in optimal objective values) when both the primal and the dual programs have no feasible solutions.
i. Converting a maximization problem to a minimization problem changes the sign of the dual variables.
j. If w_i is a dual variable, then

$$w_i = -\frac{\partial z}{\partial b_i}$$

k. A linear program with some variables required to be greater than or equal to zero can always be converted into one where all variables are unrestricted, without adding any new constraints.

6.19 Use the main duality theorem to prove Farkas's theorem. (*Hint*. Consider the following pair of primal and dual problems.)

Minimize	$\mathbf{0x}$	Maximize	$\mathbf{wb}$
Subject to	$\mathbf{Ax} = \mathbf{b}$	Subject to	$\mathbf{wA} \leqslant \mathbf{0}$
	$\mathbf{x} \geqslant \mathbf{0}$		$\mathbf{w}$ unrestricted

6.20 Consider a pair of primal and dual linear programs in standard form.

a. What happens to the dual solution if the kth primal constraint is multiplied by a nonzero scalar λ?

b. What happens to the dual solution if a scalar multiple of one primal constraint is added to another primal constraint?

c. What happens to the primal and dual solutions if a scalar multiple of one primal column is added to another primal column?

6.21 Show that if a set of constraints is redundant, then the corresponding dual variables can only be specified within a constant of addition (that is, if one dual variable in the set is changed by an amount θ, then all dual variables in the set would change by appropriate multiples of θ).

6.22 Two players are involved in a competitive game. One player, called the row player, has two strategies available; the other player, called the column player, has three strategies available. If the row player selects strategy i and the column player selects strategy j, the payoff to the row player is c_{ij} and the payoff to the column player is $-c_{ij}$. Thus the column player loses what the row player wins and vice versa—a *two-person zero-sum game*. The following matrix gives the payoffs to the row player.

	1	2	3
1	2	−1	0
2	−3	1	1

Let x_1, x_2, and x_3 be probabilities with which the column player will select the various strategies over many plays of the game. Thus $x_1 + x_2 + x_3 = 1$, $x_1, x_2, x_3 \geqslant 0$. If the column player applies these probabilities to the selection of his strategy for any play of the game, consider the row player's options. If the row player selects row 1, then his expected payoff is $2x_1 - x_2$. If the row player selects row 2, his payoff is $-3x_1 + x_2 + x_3$. Wishing to minimize the maximum expected payoff to the row player, the column player should solve the following linear program.

$$\begin{aligned}
\text{Minimize} \quad & z \\
\text{Subject to} \quad & x_1+x_2+x_3=1 \\
& 2x_1-x_2 \leqslant z \\
& -3x_1+x_2+x_3 \leqslant z \\
& x_1,\ x_2,\ x_3 \geqslant 0 \\
& z \quad \text{unrestricted}
\end{aligned}$$

Transposing the variable z to the left-hand side, we get the column player's problem:

$$\begin{aligned}
\text{Minimize} \quad & z \\
\text{Subject to} \quad & x_1+x_2+x_3=1 \\
& z-2x_1+x_2 \geqslant 0 \\
& z+3x_1-x_2-x_3 \geqslant 0 \\
& x_1,\ x_2,\ x_3 \geqslant 0 \\
& z \quad \text{unrestricted}
\end{aligned}$$

a. Give the dual of this linear program.
b. Interpret the dual problem in part (a). (*Hint*. Consider the row player's problem.)
c. Solve the dual problem of part (a). (*Hint*. This problem may be solved graphically.)
d. Use the optimal dual solution of part (c) to compute the column player's probabilities.
e. Interpret the complementary slackness conditions for this two-person zero-sum game.

6.23 Show that discarding a redundant constraint is equivalent to setting the corresponding dual variable to zero.

6.24 Let $\mathbf{x}^*$ be an optimal solution to the problem: Minimize $\mathbf{cx}$ subject to $\mathbf{a}^i\mathbf{x} = b_i$, $i = 1, \ldots, m$, $\mathbf{x} \geqslant \mathbf{0}$. Let $\mathbf{w}^*$ be an optimal dual solution. Show that $\mathbf{x}^*$ is also an optimal to the problem: Minimize $(\mathbf{c} - w_k^*\mathbf{a}^k)\mathbf{x}$ subject to $\mathbf{a}^i\mathbf{x} = b_i$, $i = 1, \ldots, m$, $i \neq k$, $\mathbf{x} \geqslant \mathbf{0}$, where w_k^* is the kth component of $\mathbf{w}^*$. Discuss!

6.25 The following are the initial and current tableaux of a linear programming problem.

	z	x_1	x_2	x_3	x_4	x_5	x_6	x_7	RHS
z	1	1	6	-7	a	5	0	0	0
x_6	0	5	-4	13	b	1	1	0	20
x_7	0	1	-1	5	c	1	0	1	8

	z	x_1	x_2	x_3	x_4	x_5	x_6	x_7	RHS
z	1	$\frac{72}{7}$	0	0	$\frac{11}{7}$	$\frac{8}{7}$	$\frac{23}{7}$	$-\frac{50}{7}$	$\frac{60}{7}$
x_3	0	$-\frac{1}{7}$	0	1	$-\frac{2}{7}$	$\frac{3}{7}$	$-\frac{1}{7}$	$\frac{4}{7}$	$\frac{12}{7}$
x_2	0	$-\frac{12}{7}$	1	0	$-\frac{3}{7}$	$\frac{8}{7}$	$-\frac{5}{7}$	$\frac{13}{7}$	$\frac{4}{7}$

a. Find a, b, and c.
b. Find $\mathbf{B}^{-1}$.
c. Find $\partial x_2/\partial x_5$.
d. Find $\partial x_3/\partial b_2$.
e. Find $\partial z/\partial x_6$.
f. Find the complementary dual solution.

6.26 The following is an optimal simplex tableau (maximization and all $\leqslant$ constraints).

	z	x_1	x_2	x_3	SLACKS x_4	SLACKS x_5	SLACKS x_6	RHS
z	1	0	0	0	4	0	9	5
x_1	0	1	1	0	2	0	1	2
x_3	0	0	0	1	1	0	4	$\frac{3}{2}$
x_5	0	0	-2	0	1	1	6	1

a. Give the optimal solution.
b. Give the optimal dual solution.
c. Find $\partial z/\partial b_1$. Interpret this number.
d. Find $\partial x_1/\partial x_6$. Interpret this number.
e. If you could buy an additional unit of the first resource for a cost of $\frac{5}{2}$ would you do this? Why?
f. Another firm wishes to purchase one unit of the third resource from you. How much is such a unit worth to you? Why?
g. Are there any alternate optimal solutions? If not, why not? If so, give one.

6.27 Solve the following problem by the dual simplex method.

$$\begin{array}{ll} \text{Maximize} & -4x_1 - 6x_2 - 18x_3 \\ \text{Subject to} & x_1 \qquad + 3x_3 \geqslant 3 \\ & \qquad x_2 + 2x_3 \geqslant 5 \\ & x_1, \quad x_2, \quad x_3 \geqslant 0 \end{array}$$

Give the optimal values of all the primal and dual variables. Demonstrate that complementary slackness holds.

6.28 Consider the following linear programming problem.

$$\begin{array}{ll} \text{Maximize} & 2x_1 - 3x_2 \\ \text{Subject to} & x_1 + x_2 \geqslant 3 \\ & 3x_1 + x_2 \leqslant 6 \\ & x_1, \quad x_2 \geqslant 0 \end{array}$$

You are told that the optimal solution is $x_1 = \frac{3}{2}$ and $x_2 = \frac{3}{2}$. Verify this statement by duality. Describe two procedures for modifying the problem in such a way that the dual simplex method can be used. Use one of these procedures for solving the problem by the dual simplex method.

6.29 Solve the following linear program by the dual simplex method.

$$\begin{array}{ll} \text{Minimize} & 2x_1 + 3x_2 + 5x_3 + 6x_4 \\ \text{Subject to} & x_1 + 2x_2 + 3x_3 + x_4 \geqslant 2 \\ & -2x_1 + x_2 - x_3 + 3x_4 \leqslant -3 \\ & x_1, \quad x_2, \quad x_3, \quad x_4 \geqslant 0 \end{array}$$

6.30 Consider the following problem.

$$\begin{array}{ll} \text{Minimize} & 3x_1 + 5x_2 - x_3 + 2x_4 - 4x_5 \\ \text{Subject to} & x_1 + x_2 + x_3 + 3x_4 + x_5 \leqslant 6 \\ & -x_1 - x_2 + 2x_3 + x_4 - x_5 \geqslant 3 \\ & x_1, \quad x_2, \quad x_3, \quad x_4, \quad x_5 \geqslant 0 \end{array}$$

a. Give the dual problem.
b. Solve the dual problem using the artificial constraint technique.
c. Find the primal solution from the dual solution.

6.31 Show that, in the dual simplex, if $\bar{b}_r < 0$ and $y_{rj} \geqslant 0$ for $j = 1, \ldots, n$, then the dual is unbounded (and the primal is infeasible) by constructing a suitable direction. (*Hint*. Consider $\mathbf{w} = \mathbf{c}_B\mathbf{B}^{-1} + \lambda\ \mathbf{B}^r$, where $\mathbf{B}^r$ is the rth row of $\mathbf{B}^{-1}$.)

6.32 Show that the dual simplex algorithm is precisely the primal simplex algorithm applied to the dual problem. Be explicit.

6.33 Indicate how the lexicographic method can be implemented in the dual simplex method to guarantee finiteness. (*Hint*. Here we are interested in columns starting and continuing to be lexicographically negative.)

6.34 In Section 6.5 we showed that the complementary dual basis matrix is given by

$$\left[\begin{array}{c|c} \mathbf{B}^t & \mathbf{0} \\ \hline \mathbf{N}^t & I_{n-m} \end{array}\right]$$

a. Give the complete starting dual tableau.
b. Give the inverse of this basis matrix.
c. Use the result of part (b) to develop the dual tableau associated with this basis matrix.

6.35 Suppose that an optimal solution to the problem Minimize $\mathbf{cx}$ subject to $\mathbf{Ax} \geqslant \mathbf{b}$, $\mathbf{x} \geqslant \mathbf{0}$, exists. Prove the following complementarity theorem.
a. A variable is zero for all primal optimal solutions if and only if its complementary dual variable is positive for some dual optimal solution.
b. A variable is unbounded in the primal feasible set if and only if its complementary dual variable is bounded in the dual feasible set.
c. A variable is unbounded in the primal *optimal* set if and only if its complementary dual variable is zero for all dual feasible solutions.
d. A variable is positive for some primal feasible solution if and only if its complementary dual variable is bounded in the dual *optimal* set.

6.36 Consider the following problem.

$$\begin{array}{lrl} \text{P: Minimize} & 2x_1 - 3x_2 & \\ \text{Subject to} & x_1 + x_2 \geqslant & 2 \\ & - x_2 \geqslant & -4 \\ & x_1, \quad x_2 \geqslant & 0 \end{array}$$

a. Solve P graphically.
b. Give the dual of P. Solve the dual graphically.
c. Illustrate the theorem of the previous exercise for each primal and dual variable, including slacks.

6.37 Consider the following problem.

$$\begin{aligned} \text{P: Minimize} \quad & -x_2 \\ \text{Subject to} \quad & x_1 + x_2 \geqslant 2 \\ & -x_2 \geqslant -4 \\ & x_1, \ x_2 \geqslant 0 \end{aligned}$$

a. Solve P graphically.
b. Give the dual of P. Solve the dual graphically.
c. Illustrate the complementarity theorem of Exercise 6.35 for each primal and dual variable, including slacks.

6.38 Apply the dual simplex method to the following problem.

$$\begin{aligned} \text{Minimize} \quad & w_3 \\ \text{Subject to} \quad & -\tfrac{1}{4}w_1 - \tfrac{1}{2}w_2 \leqslant -\tfrac{3}{4} \\ & 8w_1 + 12w_2 \leqslant 20 \\ & w_1 + \tfrac{1}{2}w_2 - w_3 \leqslant -\tfrac{1}{2} \\ & -9w_1 - 3w_2 \leqslant 6 \\ & w_1, \ w_2, \ w_3 \geqslant 0 \end{aligned}$$

6.39 Apply the perturbation technique to the dual simplex method to ensure finiteness. Specifically consider the perturbed problem

$$\begin{aligned} \text{Minimize} \quad & (\mathbf{c} + \boldsymbol{\epsilon})\mathbf{x} \\ \text{Subject to} \quad & \mathbf{A}\mathbf{x} = \mathbf{b} \\ & \mathbf{x} \geqslant \mathbf{0} \end{aligned}$$

where $\boldsymbol{\epsilon} = (\epsilon^1, \epsilon^2, \ldots, \epsilon^n)$ is a vector formed from the powers of a positive scalar ϵ. Show that the derived procedure is not exactly the same as the one produced by the lexicographic method developed in Exercise 6.33.

6.40 Show that the artificial constraint technique applied to the primal problem is precisely the single artificial variable technique (of Chapter 4) with the big-M method applied to the dual problem and vice versa. (*Hint*. Consider the dual of minimize $\mathbf{0x}_B + (\mathbf{c}_N - \mathbf{c}_B\mathbf{B}^{-1}\mathbf{N})\mathbf{x}_N$ subject to $\mathbf{x_B} + \mathbf{B}^{-1}\mathbf{N}\mathbf{x}_N = \mathbf{B}^{-1}\mathbf{b}$, $\mathbf{x}_B, \mathbf{x}_N \geqslant \mathbf{0}$.)

6.41 Suppose that the artificial constraint technique is utilized to find a starting dual solution. Show that if this constraint is tight at optimality, then the primal problem is unbounded. How large should M be in order to reach this conclusion?

6.42 Solve the following problem by the primal-dual algorithm.

$$\begin{array}{ll} \text{Maximize} & x_1 + 6x_2 \\ \text{Subject to} & x_1 + x_2 \geqslant 2 \\ & x_1 + 3x_2 \leqslant 3 \\ & x_1, \quad x_2 \geqslant 0 \end{array}$$

6.43 Apply the primal-dual method to the following problem.

$$\begin{array}{lrrrrrrl} \text{Minimize} & 9x_1 + & 7x_2 + & 4x_3 + & 2x_4 + & 6x_5 + & 10x_6 & \\ \text{Subject to} & x_1 + & x_2 + & x_3 & & & & = 8 \\ & & & & x_4 + & x_5 + & x_6 & = 5 \\ & x_1 & & + & x_4 & & & = 6 \\ & & x_2 & & + & x_5 & & = 4 \\ & & & x_3 & & + & x_6 & = 3 \\ & x_1, & x_2, & x_3, & x_4, & x_5, & x_6 & \geqslant 0 \end{array}$$

6.44 Solve the following problem by the primal-dual algorithm.

$$\begin{array}{ll} \text{Minimize} & x_1 \qquad + 2x_3 - x_4 \\ \text{Subject to} & x_1 + x_2 + x_3 + x_4 \leqslant 6 \\ & 2x_1 - x_2 + 3x_3 - 3x_4 \geqslant 5 \\ & x_1, \quad x_2, \quad x_3, \quad x_4 \geqslant 0 \end{array}$$

6.45 Suppose that at the end of the restricted primal problem we have $x_0 > 0$

and $\mathbf{v}^*\mathbf{a}_j \leqslant 0$ for all j. Show directly that the primal problem has no feasible solutions.

6.46 When is the primal-dual algorithm preferred to the dual simplex alogrithm, and vice versa?

6.47 Apply the primal-dual algorithm to the following problem.

$$\begin{aligned} \text{Maximize} \quad & 7x_1+2x_2+x_3+4x_4+6x_5 \\ \text{Subject to} \quad & 3x_1+5x_2-6x_3+2x_4+4x_5=27 \\ & x_1+2x_2+3x_3-7x_4+6x_5 \geqslant 2 \\ & 9x_1-4x_2+2x_3+5x_4-2x_5=16 \\ & x_1, \quad x_2, \quad x_3, \quad x_4, \quad x_5 \geqslant 0 \end{aligned}$$

6.48 We have shown that the primal-dual algorithm converges in a finite number of steps in the absence of degeneracy. What happens in the degenerate case? How can we guarantee finite convergence? (*Hint*. Consider applying the lexicographic simplex or the perturbation method to the restricted primal problem.)

6.49 Consider the following linear programming problem and its optimal final tableau shown below.

$$\begin{aligned} \text{Maximize} \quad & 2x_1+x_2-x_3 \\ \text{Subject to} \quad & x_1+2x_2+x_3 \leqslant 8 \\ & -x_1+x_2-2x_3 \leqslant 4 \\ & x_1, \quad x_2, \quad x_3 \geqslant 0 \end{aligned}$$

Final Tableau

	z	x_1	x_2	x_3	x_4	x_5	RHS
z	1	0	3	3	2	0	16
x_1	0	1	2	1	1	0	8
x_5	0	0	3	−1	1	1	12

a. Write the dual problem and find the optimal dual variables from the foregoing tableau.
b. Using sensitivity analysis, find the new optimal solution if the coefficient of x_2 in the objective function is changed from 1 to 5.
c. Suppose that the coefficient of x_3 in the second constraint is changed from −2 to 1. Using sensitivity, find the new optimal solution.

d. Suppose that the following constraint is added to the problem: $x_2 + x_3 \geq 2$. Using sensitivity, find the new optimal solution.
e. If you were to choose between increasing the right-hand side of the first and second constraints, which one would you choose? Why? What is the effect of this increase on the optimal value of the objective function?
f. Suppose that a new activity x_6 is proposed with unit return 4 and *consumption* vector $\mathbf{a}_6 = (1, 2)'$. Find the new optimal solution.

6.50 Consider the following optimal tableau of a maximization problem where the constraints are of the $\leq$ type.

	z	x_1	x_2	x_3	x_4	x_5	x_6 (SLACKS)	x_7 (SLACKS)	x_8 (SLACKS)	RHS
z	1	0	0	0	2	0	2	$\frac{1}{10}$	2	17
x_1	0	1	0	0	-1	0	$\frac{1}{2}$	$\frac{1}{5}$	-1	3
x_2	0	0	1	0	2	1	-1	0	$\frac{1}{2}$	1
x_3	0	0	0	1	-1	-2	5	$-\frac{3}{10}$	2	7

a. Would the solution be altered if a new activity x_9 with coefficients $(2, 0, 3)'$ in the constraints, and price of 5, were added to the problem?
b. How large can b_1 (the first constraint resource) be made without violating feasibility?

6.51 Consider the tableau of exercise 6.50. Suppose that we add the constraint $x_1 - x_2 + 2x_3 \leq 10$ to the problem. Is the solution still optimal? If not, find the new optimal solution.

6.52 A farmer has 500 acres of land and wishes to determine the acreage allocated to the following three crops: wheat, corn, and soybeans. The man-days, preparation cost, and profit per acre of the three crops are summarized below.

CROP	MAN-DAYS	PREPARATION COST \$	PROFIT \$
Wheat	6	100	60
Corn	8	150	100
Soybeans	10	120	80

Suppose that the maximum number of man-days available are 5000 and that the farmer has \$60,000 for preparation.
a. Find the optimal solution.
b. Assuming an 8 hour work day, would it be profitable to the farmer to acquire additional help at \$3 per hour? Why or why not?

c. Suppose that the farmer has contracted to deliver at least the equivalent of 100 acres of wheat. Use sensitivity analysis to find the new optimal solution.

6.53 A product is assembled from three parts that can be manufactured on two machines A and B. Neither machine can process different parts at the same time. The number of parts processed by each machine per hour are summarized below.

	MACHINE A	MACHINE B
Part 1	12	6
Part 2	15	12
Part 3	—	25

Management seeks a daily schedule of the machines so that the number of assemblies is maximized. Currently the company has three machines of type A and five machines of type B.

a. Solve the problem.
b. If only one machine can be acquired, which type would you recommend and why?
c. Management is contemplating the purchase of a type A machine at a cost of $100,000. Suppose that the life of the machine is 10 years and that each year is equivalent to 2000 working hours. Would you recommend the purchase if the unit profit from each assembly is $1? Why or why not?

6.54 The tourism department of a certain country would like to decide which projects to fund during the comming year. The projects were divided into three main categories: religious, historical, and construction (hotels, roads, nightclubs, and so on). Three proposals A, B, and C for restoring religious sites are submitted, with estimated costs of $5, $7, and $3 million respectively. Four proposals D, E, F, and G for the restoration of historical sites are submitted with estimated costs of $15, $12, $5, and $7 million. Finally, five proposals H, I, J, K, and L for constructing new facilities are submitted. These cost $2, $15, $22, $8, and $10 million respectively. In order to determine relative priority of these projects, experts from the tourism department developed a scoring model with the following scores for proposals A, B, C, D, E, F, G, H, I, J, K, L: 5, 6, 2, 8, 11, 1, 7, 2, 10, 9, 5, and 4 respectively. The department decides that at least one project of each category must be funded. Projects E and F represent a continuation of a plan that started during the previous year, and at least one of them must be funded. Furthermore, at most two historical and three construction projects can be chosen. Which projects should the tourism department fund in order to maximize the total score and not to exceed \$80 million? (*Hint*. Project j is chosen if $x_j = 1$ and is not chosen if $x_j = 0$. First solve the continuous linear program by the bounded simplex method and then add appropriate cuts.)

6.55 An airline company wishes to assign two types of its aircraft to three routes. Each aircraft can make at most two daily trips. Furthermore, 3 and 4 aircraft of types A and B are available respectively. The capacity of type A aircraft is 140 passengers and that of type B aircraft is 100 passengers. The expected number of daily passengers on the three routes is 300, 700, and 220 respectively.

The operating costs per trip on the different routes are summarized below.

	OPERATING COST FOR A GIVEN ROUTE		
AIRCRAFT TYPE	1	2	3
A	3000	2500	2000
B	2400	2000	1800

a. Find the optimal solution of the continuous linear programming problem. Does this solution make any sense?
b. Using the cutting plane algorithm of Section 6.9, find the optimal integer solution.

6.56 Consider the following optimal tableau of a maximization problem where the constraints are of the $\leqslant$ type.

	z	x_1	x_2	x_3	x_4	x_5	x_6 (SLACKS)	x_7 (SLACKS)	x_8 (SLACKS)	RHS
z	1	0	0	0	2	0	2	$\frac{1}{10}$	2	17
x_1	0	1	0	0	-1	0	$\frac{1}{2}$	$\frac{1}{5}$	-1	3
x_2	0	0	1	0	2	1	-1	0	$\frac{1}{2}$	1
x_3	0	0	0	1	-1	-2	5	$-\frac{3}{10}$	2	7

Construct the sequence of optimal solutions for $b_1' = b_1 - \theta$ where θ varies between 0 and ∞.

6.57 Consider Exercise 6.1. Suppose that the cost vector is modified in the direction $(-1, -1)$. Using parametric analysis on the cost vector, find the sequence of optimal solutions.

6.58 Consider the following problem.

$$\begin{aligned} \text{Minimize} \quad & -x_1 + x_2 - 2x_3 \\ \text{Subject to} \quad & x_1 + x_2 + x_3 \leqslant 6 \\ & -x_1 + 2x_2 + 3x_3 \leqslant 9 \\ & x_1, \quad x_2, \quad x_3 \geqslant 0 \end{aligned}$$

a. Solve the problem by the simplex method.
b. Suppose that the vector $\mathbf{c} = (-1, 1, -2)$ is replaced by $(-1, 1, -2) + \lambda(2, 1, 1)$ where λ is a real number. Find optimal solutions for all values of λ.

6.59 Consider the following problem.

$$\begin{aligned} \text{Maximize} \quad & 2x_1 + 3x_2 + 5x_3 \\ \text{Subject to} \quad & x_1 + x_2 + 2x_3 \leqslant 8 \\ & x_1 - x_2 + x_3 \leqslant 4 \\ & x_1, \quad x_2, \quad x_3 \geqslant 0 \end{aligned}$$

a. Find the optimal solution.
b. Find the new optimal solution if the cost coefficient c_2 changes from 3 to -4.
c. Find the optimal solution if the cost coefficient c_2 varies over the entire real line $(-\infty, \infty)$.

6.60 Prove that parametric analysis on the cost vector in a minimization problem always produces a piecewise-linear and concave function $z(\lambda)$.

6.61 Prove that parametric analysis on the RHS vector in a minimization problem always produces a piecewise-linear and convex function $z(\lambda)$.

6.62 Prove that if $\mathbf{K}$ is a skew symmetric matrix (that is, $\mathbf{K} = -\mathbf{K}'$) then the system

$$\mathbf{Kx} \geqslant \mathbf{0}, \quad \mathbf{x} \geqslant \mathbf{0}$$

possesses at least one solution $\bar{\mathbf{x}}$ such that $\mathbf{K}\bar{\mathbf{x}} + \bar{\mathbf{x}} > \mathbf{0}$. (*Hint.* Apply Farkas's theorem to the system $\mathbf{Kx} \geqslant \mathbf{0}$, $\mathbf{x} \geqslant \mathbf{0}$, $\mathbf{e}_j\mathbf{x} > 0$. Repeat for each j and combine solutions by summing.)

6.63 Apply the results of the previous problem to the system

$$\begin{aligned} \mathbf{Ax} - r\mathbf{b} &\geqslant \mathbf{0}, & \mathbf{x} &\geqslant \mathbf{0} \\ -\mathbf{wA} + r\mathbf{c} &\geqslant \mathbf{0}, & \mathbf{w} &\geqslant \mathbf{0} \\ \mathbf{wb} - \mathbf{cx} &\geqslant 0, & r &\geqslant 0 \end{aligned}$$

a. Use the results to derive the fundamental theorem of duality.
b. Use the results to prove that at optimality there exists at least one pair of primal and dual optimal points with the property that if a variable in one problem is zero, then the complementary variable in the other problem is positive. This is called the *strong theorem of complementary slackness*.
c. Illustrate the strong theorem of complimentary slackness geometrically. (*Hint*. Consider a linear program where the objective function is parallel to one of the constraints and alternate optimal solutions result.)
d. When must the solution to the strong duality theorem not occur at an extreme point? (*Hint*. Consider part (c) above.)

6.64 Consider the following primal problem.

$$\begin{aligned} \text{Minimize} \quad & \mathbf{cx} \\ \text{Subject to} \quad & \mathbf{Ax} \geqslant \mathbf{b} \\ & \mathbf{x} \in X \end{aligned}$$

where X is a polyhedral set. (Often the set X consists of constraints which are easy to handle.) Associated with the foregoing primal problem is the following *Lagrangian dual* problem.

$$\begin{aligned} \text{Maximize} \quad & f(\mathbf{w}) \\ \text{Subject to} \quad & \mathbf{w} \geqslant \mathbf{0} \end{aligned}$$

where $f(\mathbf{w}) = \mathbf{wb} + \underset{\mathbf{x} \in X}{\text{Minimum}} \; (\mathbf{c} - \mathbf{wA})\mathbf{x}$.

a. Show that if $\mathbf{x}_0$ is feasible to the primal problem, that is, $\mathbf{Ax}_0 \geqslant \mathbf{b}$ and $\mathbf{x}_0 \in X$, and $\mathbf{w}_0$ is feasible to the Lagrangian dual problem, that is, $\mathbf{w}_0 \geqslant \mathbf{0}$, then $\mathbf{cx}_0 \geqslant f(\mathbf{w}_0)$.
b. Suppose that X is nonempty and bounded and that the primal problem

possesses a finite optimal solution. Show that

$$\begin{array}{ll} \text{Minimum } \mathbf{cx} = & \text{Maximum} f(\mathbf{w}) \\ \mathbf{Ax} \geqslant \mathbf{b} & \mathbf{w} \geqslant \mathbf{0} \\ \mathbf{x} \in X & \end{array}$$

6.65 Consider the problem: Minimize $x_1 + 2x_2$ subject to $3x_1 + x_2 \geqslant 6$, $-x_1 + x_2 \leqslant 2$, $x_1 + x_2 \leqslant 8$, and $x_1, x_2 \geqslant 0$. Let $X = \{\mathbf{x} : -x_1 + x_2 \leqslant 2, \ \ x_1 + x_2 \leqslant 8, \ \ x_1, x_2 \geqslant 0\}$.

a. Formulate the Lagrangian dual problem.
b. Show that $f(w) = 6w + \text{Minimum } \{0, 4 - 2w, 13 - 14w, 8 - 24w\}$. (*Hint*. Examine the second term in $f(w)$ in Exercise 6.64 and enumerate the extreme points of X graphically.)
c. Plot $f(w)$ for each value of w.
d. From part (c) locate the optimal solution to the Lagrangian dual problem.
e. From part (d) find the optimal solution to the primal problem.

NOTES AND REFERENCES

1. John Von Neumann is credited with having first postulated the existence of a dual linear program. His insights came through his work in game theory and economics together with a strong mathematical capability. Many individuals have continued to develop and extend the basic duality theorems, notably Tucker [438] and A. C. Williams [470].
2. The dual simplex method was first developed by Lemke [308].
3. The primal-dual algorithm was developed by Dantzig, Ford, and Fulkerson [102]. This development was fostered out of the work of Kuhn [294] on the assignment problem.

SEVEN: THE DECOMPOSITION PRINCIPLE

In practice, many linear programming problems are simply too large to fit into today's computers. It is not unusual in a corporate management model or in a logistics model to produce a linear program with many thousands of rows and a seemingly unlimited number of columns. In such problems some method must be applied to convert the large problems into one or more smaller problems of manageable size. Fortunately, there is a technique, called the *decomposition principle*, that does exactly this.

Even if a linear program is of manageable size, certain of its constraints may possess special structure that would permit efficient handling. In such cases we would like to separate the linear program into one with general structure and one with special structure where a more efficient method may be applied. Again, the decomposition principle can be applied to such a linear program to achieve the desired effect.

The decomposition principle is a systematic procedure for solving large-scale linear programs or linear programs that contain constraints of special structure. The constraints are divided into two sets: general constraints (or *complicating* constraints) and constraints with special structure. It will become apparent that it is not necessary for either set to have special structure; however, special structure, when available, enhances the efficiency of the decomposition principle.

The strategy of the decomposition procedure is to operate on two separate linear programs: one over the set of general constraints and one over the set of special constraints. Information is passed back and forth between the two linear programs until a point is reached where the solution to the original problem is achieved. The linear program over the general constraints is called the *master problem*, and the linear program over the special constraints is called the *subproblem*. The master problem passes down a new set of cost coefficients to the subproblem and receives a new column based on these cost coefficients.

We shall begin by assuming that the special constraint set is bounded. Once the decomposition principle is developed for this case and we have discussed how to get started, we shall relax the boundedness assumption and also extend the procedure to multiple subproblems.

7.1 THE DECOMPOSITION ALGORITHM

Consider the following linear program, where X is a polyhedral set representing constraints of special structure, $\mathbf{A}$ is an $m \times n$ matrix, and $\mathbf{b}$ is an m vector.

$$\begin{aligned} \text{Minimize} \quad & \mathbf{cx} \\ \text{Subject to} \quad & \mathbf{Ax} = \mathbf{b} \\ & \mathbf{x} \in X \end{aligned}$$

To simplify the presentation, assume that X is bounded (this assumption will be relaxed in Section 7.4). Since X is a bounded polyhedral set, then any point $\mathbf{x} \in X$ can be represented as a convex combination of the finite number of extreme points of X. Denoting these points by $\mathbf{x}_1, \mathbf{x}_2, \ldots, \mathbf{x}_t$, any $\mathbf{x} \in X$ can be represented as

$$\mathbf{x} = \sum_{j=1}^{t} \lambda_j \mathbf{x}_j$$

$$\sum_{j=1}^{t} \lambda_j = 1$$

$$\lambda_j \geq 0 \qquad j = 1, 2, \ldots, t$$

Substituting for $\mathbf{x}$, the foregoing optimization problem can be transformed into the following problem in the variables $\lambda_1, \lambda_2, \ldots, \lambda_t$.

$$\begin{aligned} \text{Minimize} \quad & \sum_{j=1}^{t} (\mathbf{cx}_j)\lambda_j \\ \text{Subject to} \quad & \sum_{j=1}^{t} (\mathbf{Ax}_j)\lambda_j = \mathbf{b} && (7.1) \\ & \sum_{j=1}^{t} \lambda_j = 1 && (7.2) \\ & \lambda_j \geqslant 0 \quad j = 1, 2, \ldots, t && (7.3) \end{aligned}$$

Since t, the number of extreme points of the set X, is usually very large, attempting to explicitly enumerate all the extreme points $\mathbf{x}_1, \mathbf{x}_2, \ldots, \mathbf{x}_t$, and explicitly solving this problem is a very difficult task. Rather, we shall attempt to find an optimal solution of the problem (and hence the original problem) without explicitly enumerating all the extreme points.

Application of the Revised Simplex Method

Consider solving the foregoing problem by the revised simplex method. Suppose that we have a basic feasible solution $\boldsymbol{\lambda} = (\boldsymbol{\lambda}_B, \boldsymbol{\lambda}_N)$. Further suppose that the $(m + 1) \times (m + 1)$ basis inverse $\mathbf{B}^{-1}$ is known (the process of initialization is discussed in detail in Section 7.3). Denoting the dual variables corresponding to Equations (7.1) and (7.2) by $\mathbf{w}$ and α, we get $(\mathbf{w}, \alpha) = \hat{\mathbf{c}}_B\mathbf{B}^{-1}$, where $\hat{\mathbf{c}}_B$ is the cost of the basic variables with $\hat{c}_j = \mathbf{cx}_j$ for each basic variable λ_j. The basis inverse, the dual variables, the values of the basic variables, and the objective function are displayed below, where $\bar{\mathbf{b}} = \mathbf{B}^{-1}\binom{\mathbf{b}}{1}$.

BASIS INVERSE	RHS
$(\mathbf{w}, \alpha)$	$\hat{\mathbf{c}}_B\bar{\mathbf{b}}$
$\mathbf{B}^{-1}$	$\bar{\mathbf{b}}$

The revised simplex method proceeds by concluding that the current solution is optimal or else by deciding to increase a nonbasic variable. This is done by first calculating

$$\begin{aligned} z_k - \hat{c}_k &= \underset{1 \leqslant j \leqslant t}{\text{Maximum}}\ z_j - \hat{c}_j = \underset{1 \leqslant j \leqslant t}{\text{maximum}}\ (\mathbf{w}, \alpha)\begin{bmatrix} \mathbf{Ax}_j \\ 1 \end{bmatrix} - \mathbf{cx}_j \\ &= \underset{1 \leqslant j \leqslant t}{\text{maximum}}\ \mathbf{wAx}_j + \alpha - \mathbf{cx}_j \qquad (7.4) \end{aligned}$$

Since $z_j - \hat{c}_j = 0$ for basic variables, then the foregoing maximum is $\geqslant 0$. Thus if $z_k - \hat{c}_k = 0$, then $z_j - \hat{c}_j \leqslant 0$ for all nonbasic variables and the optimal solution is at hand. On the other hand, if $z_k - \hat{c}_k > 0$, then the nonbasic variable x_k is increased.

Determining the index k using Equation (7.4) is computationally infeasible because t is very large and the extreme points $\mathbf{x}_j$'s corresponding to the nonbasic λ_j's are not explicitly known. Therefore an alternative scheme must be devised. Since X is a bounded polyhedral set, the maximum of any linear objective can be achieved at one of the extreme points. Therefore

$$\underset{1 \leqslant j \leqslant t}{\text{Maximum}} (\mathbf{wA} - \mathbf{c})\mathbf{x}_j + \alpha = \underset{\mathbf{x} \in X}{\text{Maximum}} (\mathbf{wA} - \mathbf{c})\mathbf{x} + \alpha$$

To summarize, given a basic feasible solution $(\boldsymbol{\lambda}_B, \boldsymbol{\lambda}_N)$ with dual variables $(\mathbf{w}, \alpha)$, solve the following linear *subproblem*, which is "easy" because of the special structure of X.

$$\begin{aligned} \text{Maximize} \quad & (\mathbf{wA} - \mathbf{c})\mathbf{x} + \alpha \\ \text{Subject to} \quad & \mathbf{x} \in X \end{aligned}$$

Note that the objective function contains a constant. This is easily handled by initializing the RHS value for z to α instead of the normal value of 0. Let $\mathbf{x}_k$ be an optimal solution to the foregoing subproblem with objective value $z_k - \hat{c}_k$. If $z_k - \hat{c}_k = 0$, then the basic feasible solution $(\boldsymbol{\lambda}_B, \boldsymbol{\lambda}_N)$ is optimal. Otherwise if $z_k - \hat{c}_k > 0$, then the variable λ_k enters the basis. As in the revised simplex method the corresponding column $\begin{pmatrix} \mathbf{Ax}_k \\ 1 \end{pmatrix}$ is updated by premultiplying it by $\mathbf{B}^{-1}$ giving $\mathbf{y}_k = \mathbf{B}^{-1}\begin{pmatrix} \mathbf{Ax}_k \\ 1 \end{pmatrix}$. Note that $\mathbf{y}_k \leqslant \mathbf{0}$ cannot occur since X was assumed bounded, producing a bounded master problem. The updated column $\begin{pmatrix} z_k - \hat{c}_k \\ \mathbf{y}_k \end{pmatrix}$ is adjoined to the foregoing above array. The variable λ_{B_r} leaving the basis is determined by the usual minimum ratio test. The basis inverse, dual variables, and right RHS are updated by pivoting at y_{rk}. After updating, the process is repeated.

Now we have all the ingredients of the decomposition algorithm, a summary of which is given below. Note that the master step gives an improved feasible solution of the overall problem, and the subproblem checks whether $z_j - \hat{c}_j \leqslant 0$ for all λ_j, or else determines the most positive $z_k - \hat{c}_k$.

Summary of the Decomposition Algorithm

INITIALIZATION STEP

Find an initial basic feasible solution of the system defined by Equations (7.1), (7.2), and (7.3) (getting an initial basic feasible solution is discussed in detail in Section 7.3). Let the basis be $\mathbf{B}$ and form the following *master array* where $(\mathbf{w}, \alpha) = \hat{\mathbf{c}}_B\mathbf{B}^{-1}$ (recall that $\hat{c}_j = \mathbf{c}\mathbf{x}_j$), and $\bar{\mathbf{b}} = \mathbf{B}^{-1}\begin{bmatrix}\mathbf{b}\\1\end{bmatrix}$.

BASIS INVERSE	RHS
$(\mathbf{w}, \alpha)$	$\hat{\mathbf{c}}_B\bar{\mathbf{b}}$
$\mathbf{B}^{-1}$	$\bar{\mathbf{b}}$

MAIN STEP

1. Solve the following *subproblem*.

$$\begin{aligned}&\text{Maximize} && (\mathbf{wA} - \mathbf{c})\mathbf{x} + \alpha\\ &\text{Subject to} && \mathbf{x} \in X\end{aligned}$$

Let $\mathbf{x}_k$ be an optimal basic feasible solution with objective value of $z_k - \hat{c}_k$. If $z_k - \hat{c}_k = 0$ stop; the basic feasible solution of the last master step is an optimal solution of the overall problem. Otherwise go to step 2 below.

2. Let $\mathbf{y}_k = \mathbf{B}^{-1}\begin{bmatrix}\mathbf{Ax}_k\\1\end{bmatrix}$ and adjoin the updated column $\begin{pmatrix}z_k - \hat{c}_k\\ \mathbf{y}_k\end{pmatrix}$ to the master array. Pivot at y_{rk} where the index r is determined as follows:

$$\frac{\bar{b}_r}{y_{rk}} = \underset{1 \leq i \leq m+1}{\text{Minimum}}\left\{\frac{\bar{b}_i}{y_{ik}} : y_{ik} > 0\right\}$$

This updates the dual variables, the basis inverse, and the right-hand side. After pivoting, the column of λ_k is deleted and step 1 is repeated.

Some Remarks on the Decomposition Algorithm

1. Note that the foregoing algorithm is a direct implementation of the revised simplex method except that the calculation $z_k - \hat{c}_k$ is performed by solving a subproblem. Therefore the algorithm converges in a finite number of iterations provided that a cycling prevention rule is used in both the master step and the subproblem in the presence of degeneracy.

2. At each iteration the master step provides a new improved basic feasible solution of the system given by Equations (7.1), (7.2), and (7.3) by introducing the nonbasic variable λ_k, which is generated by the subproblem. At each iteration the subproblem provides an extreme point $\mathbf{x}_k$, which corresponds to an updated column $\begin{bmatrix} z_k - \hat{c}_k \\ \mathbf{y}_k \end{bmatrix}$, and hence this procedure is sometimes referred to as a *column generation scheme*.
3. At each iteration a different dual vector is passed from the master step to the subproblem. Rather than solving the subproblem anew at each iteration, the optimal basis of the last iteration could be utilized by modifying the cost row.
4. If the master constraints are of the inequality type, then we must check the $z_j - \hat{c}_j$ for nonbasic slack variables in addition to solving the subproblem. For a master constraint i of the $\leqslant$ type with associated slack variables s_i we get

$$z_{s_i} - c_{s_i} = (\mathbf{w}, \alpha)\begin{pmatrix} \mathbf{e}_i \\ 0 \end{pmatrix} - 0 = w_i$$

   Thus, for a minimization problem a slack variable associated with a $\leqslant$ constraint is eligible to enter the basis if $w_i > 0$. (Note that the entry criterion is $w_i < 0$ for constraints of the $\geqslant$ type.)
5. At each iteration, the subproblem need not be completely optimized. It is only necessary that the current extreme point $\mathbf{x}_k$ satisfies $z_k - \hat{c}_k = (\mathbf{wA} - \mathbf{c})\mathbf{x}_k + \alpha > 0$. In this case λ_k is a candidate to enter the basis of the master problem.

Calculation and Use of Lower Bounds

Recall that the decomposition algorithm stops when Maximum $z_j - \hat{c}_j = 0$. Because of the large number of variables $\lambda_1, \lambda_2, \ldots, \lambda_t$, continuing the computations until this condition is satisfied may be time-consuming for large problems.

We shall develop a lower bound on the objective of any feasible solution of the overall problem, and hence a lower bound on the optimal objective. Since the decomposition algorithm generates feasible points with improving objective values, we may stop when the difference between the objective of the current feasible point and the lower bound is within an acceptable tolerance. This may not give the true optimal point, but will guarantee good feasible solutions, within any desirable accuracy from the optimal. Consider the following subproblem.

$$\begin{aligned} &\text{Maximize} \quad (\mathbf{wA} - \mathbf{c})\mathbf{x} + \alpha \\ &\text{Subject to} \quad \mathbf{x} \in X \end{aligned}$$

where **w** is the dual vector passed from the master step. Let the optimal objective of the foregoing subproblem be $z_k - \hat{c}_k$. Now let **x** be any feasible solution of the overall problem, that is, $\mathbf{Ax} = \mathbf{b}$ and $\mathbf{x} \in X$. By definition of $z_k - \hat{c}_k$ and since $\mathbf{x} \in X$, we have

$$(\mathbf{wA} - \mathbf{c})\mathbf{x} + \alpha \leqslant (z_k - \hat{c}_k)$$

Since $\mathbf{Ax} = \mathbf{b}$, then the above inequality implies that

$$\mathbf{cx} \geqslant \mathbf{wAx} - (z_k - \hat{c}_k) + \alpha = \mathbf{wb} + \alpha - (z_k - \hat{c}_k) = \hat{\mathbf{c}}_B\bar{\mathbf{b}} - (z_k - \hat{c}_k)$$

Since this is true for each $\mathbf{x} \in X$ with $\mathbf{Ax} = \mathbf{b}$, then

$$\underset{\substack{\mathbf{Ax}=\mathbf{b} \\ \mathbf{x}\in X}}{\text{Minimum}} \quad \mathbf{cx} \geqslant \hat{\mathbf{c}}_B\bar{\mathbf{b}} - (z_k - \hat{c}_k)$$

In other words, $\hat{\mathbf{c}}_B\bar{\mathbf{b}} - (z_k - \hat{c}_k)$ is a lower bound on the optimal objective value of the overall problem.

7.2 NUMERICAL EXAMPLE

Consider the following problem.

$$\begin{array}{llcl}
\text{Minimize} & -2x_1 - x_2 - x_3 + x_4 & & \\
\text{Subject to} & x_1 + x_3 & \leqslant & 2 \\
& x_1 + x_2 + 2x_4 & \leqslant & 3 \\
& x_1 & \leqslant & 2 \\
& x_1 + 2x_2 & \leqslant & 5 \\
& - x_3 + x_4 & \leqslant & 2 \\
& 2x_3 + x_4 & \leqslant & 6 \\
& x_1, \quad x_2, \quad x_3, \quad x_4 & \geqslant & 0
\end{array}$$

Note that the third and fourth constraints involve only x_1 and x_2, whereas the fifth and sixth constraints involve only x_3 and x_4 (we shall have more to say about this special structure later). If we let X consist of the last four constraints, in addition to the nonnegativity restrictions, then minimizing a linear function over X becomes a simple process, since the subproblem can be decomposed into two subproblems. Therefore we shall handle the first two constraints as $\mathbf{Ax} \leqslant \mathbf{b}$, where $\mathbf{A} = \begin{bmatrix} 1 & 0 & 1 & 0 \\ 1 & 1 & 0 & 2 \end{bmatrix}$, $\mathbf{b} = \begin{bmatrix} 2 \\ 3 \end{bmatrix}$, and the remaining constraints as X. Note that any point (x_1, x_2, x_3, x_4) in X must have its first two components and its last two components in the sets X_1 and X_2 shown in Figure 7.1.

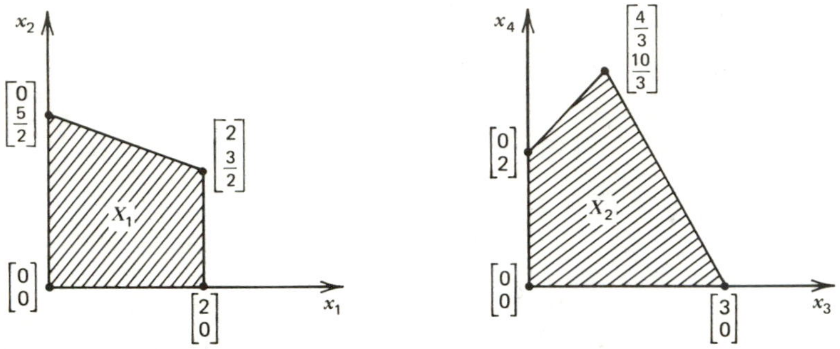

Figure 7.1. **Representation of X by two sets.**

Initialization Step

The problem is reformulated as follows, where $\mathbf{x}_1, \mathbf{x}_2, \ldots, \mathbf{x}_t$ are the extreme points of X, $\hat{c}_j = \mathbf{c}\mathbf{x}_j$ for $j = 1, 2, \ldots, t$, and $\mathbf{s} \geqslant \mathbf{0}$ is the slack vector.

$$\text{Minimize} \quad \sum_{j=1}^{t} \hat{c}_j \lambda_j$$

$$\text{Subject to} \quad \sum_{j=1}^{t} (\mathbf{A}\mathbf{x}_j)\,\lambda_j + \mathbf{s} = \mathbf{b}$$

$$\sum_{j=1}^{t} \lambda_j = 1$$

$$\lambda_j \geqslant 0 \qquad j = 1, 2, \ldots, t$$

$$\mathbf{s} \geqslant \mathbf{0}$$

We need a starting basis with known $\mathbf{B}^{-1}$. Let the starting basis consist of $\mathbf{s}$ and λ_1 where $\mathbf{x}_1 = (0, 0, 0, 0)$ is an extreme point of X with $\mathbf{c}\mathbf{x}_1 = 0$. Therefore

$$\mathbf{B} = \begin{bmatrix} 1 & 0 & 0 \\ 0 & 1 & 0 \\ 0 & 0 & 1 \end{bmatrix}$$

The vector $(\mathbf{w}, \alpha) = \hat{\mathbf{c}}_B\mathbf{B}^{-1} = \mathbf{0}\mathbf{B}^{-1} = \mathbf{0}$, and $\bar{\mathbf{b}} = \mathbf{B}^{-1}\begin{bmatrix} \mathbf{b} \\ 1 \end{bmatrix} = \begin{bmatrix} \mathbf{b} \\ 1 \end{bmatrix}$. This gives the following tableau. Note that the first three columns give (w_1, w_2, α) in row 0 and $\mathbf{B}^{-1}$ in the remaining rows.

	BASIS INVERSE			RHS
z	0	0	0	0
s_1	1	0	0	2
s_2	0	1	0	3
λ_1	0	0	1	1

Iteration 1

SUBPROBLEM

Solve the following subproblem.

$$\begin{aligned} &\text{Maximize} \quad (\mathbf{wA} - \mathbf{c})\mathbf{x} + \alpha \\ &\text{Subject to} \quad \mathbf{x} \in X \end{aligned}$$

Here $(w_1, w_2) = (0, 0)$ from the foregoing array. Therefore the subproblem is as follows:

$$\begin{aligned} &\text{Maximize} \quad 2x_1 + x_2 + x_3 - x_4 + 0 \\ &\text{Subject to} \quad \mathbf{x} \in X \end{aligned}$$

This problem is separable in the vectors (x_1, x_2) and (x_3, x_4) and can be solved geometrically. Using Figure 7.1, it is easily verified that the optimal solution is $\mathbf{x}_2 = (2, \frac{3}{2}, 3, 0)$ with objective $z_2 - \hat{c}_2 = \frac{17}{2}$. Since $z_2 - \hat{c}_2 = \frac{17}{2} > 0$, then λ_2 corresponding to $\mathbf{x}_2$ is introduced. The lower bound $= \hat{\mathbf{c}}_B\bar{\mathbf{b}} - (z_2 - \hat{c}_2) = 0 - \frac{17}{2}$. Recall that the best objective so far is 0.

MASTER STEP

$$z_2 - \hat{c}_2 = \frac{17}{2}$$

$$\mathbf{A}\mathbf{x}_2 = \begin{bmatrix} 1 & 0 & 1 & 0 \\ 1 & 1 & 0 & 2 \end{bmatrix} \begin{bmatrix} 2 \\ \frac{3}{2} \\ 3 \\ 0 \end{bmatrix} = \begin{bmatrix} 5 \\ \frac{7}{2} \end{bmatrix}$$

Then

$$\begin{bmatrix} \mathbf{Ax}_2 \\ 1 \end{bmatrix} = \begin{bmatrix} 5 \\ \frac{7}{2} \\ 1 \end{bmatrix}$$

is updated by premultiplying by $\mathbf{B}^{-1}$. So

$$\mathbf{y}_2 = \mathbf{B}^{-1}\begin{bmatrix} 5 \\ \frac{7}{2} \\ 1 \end{bmatrix} = \mathbf{I}\begin{bmatrix} 5 \\ \frac{7}{2} \\ 1 \end{bmatrix} = \begin{bmatrix} 5 \\ \frac{7}{2} \\ 1 \end{bmatrix}$$

Insert the column

$$\begin{bmatrix} z_2 - \hat{c}_2 \\ \mathbf{y}_2 \end{bmatrix} = \begin{bmatrix} \frac{17}{2} \\ 5 \\ \frac{7}{2} \\ 1 \end{bmatrix}$$

into the foregoing array and pivot. This leads to the following two tableaux (the λ_2 column is deleted after pivoting).

	BASIS INVERSE			RHS	λ_2
z	0	0	0	0	$\frac{17}{2}$
s_1	1	0	0	2	(5)
s_2	0	1	0	3	$\frac{7}{2}$
λ_1	0	0	1	1	1

	BASIS INVERSE			RHS
z	$-\frac{17}{10}$	0	0	$-\frac{17}{5}$
λ_2	$\frac{1}{5}$	0	0	$\frac{2}{5}$
s_2	$-\frac{7}{10}$	1	0	$\frac{8}{5}$
λ_1	$-\frac{1}{5}$	0	1	$\frac{3}{5}$

The best-known feasible solution of the overall problem is given by

$$\mathbf{x} = \lambda_1\mathbf{x}_1 + \lambda_2\mathbf{x}_2$$
$$= \tfrac{3}{5}(0, 0, 0, 0) + \tfrac{2}{5}(2, \tfrac{3}{2}, 3, 0) = (\tfrac{4}{5}, \tfrac{3}{5}, \tfrac{6}{5}, 0)$$

The objective is $-\frac{17}{5}$. Also $(w_1, w_2, \alpha) = (-\frac{17}{10}, 0, 0)$.

Iteration 2

Since $w_1 < 0$, s_1 is not eligible to enter the basis at this time.

SUBPROBLEM

Solve the following problem.

$$\begin{array}{ll}\text{Maximize} & (\mathbf{wA} - \mathbf{c})\mathbf{x} + \alpha \\ \text{Subject to} & \mathbf{x} \in X\end{array}$$

$$\mathbf{wA} - \mathbf{c} = (-\tfrac{17}{10}, 0)\begin{bmatrix} 1 & 0 & 1 & 0 \\ 1 & 1 & 0 & 2 \end{bmatrix} - (-2, -1, -1, 1) = (\tfrac{3}{10}, 1, -\tfrac{7}{10}, -1)$$

Therefore the subproblem is

$$\begin{array}{ll}\text{Maximize} & \tfrac{3}{10}x_1 + x_2 - \tfrac{7}{10}x_3 - x_4 + 0 \\ \text{Subject to} & \mathbf{x} \in X\end{array}$$

The problem decomposes into two problems involving (x_1, x_2) and (x_3, x_4). Using Figure 7.1, the optimal solution is $\mathbf{x}_3 = (0, \frac{5}{2}, 0, 0)$ with objective $z_3 - \hat{c}_3 = \frac{5}{2}$. Since $z_3 - \hat{c}_3 > 0$, then λ_3 is introduced.

The lower bound is $\hat{\mathbf{c}}_B\bar{\mathbf{b}} - (z_3 - \hat{c}_3) = -\frac{17}{5} - \frac{5}{2} = -5.9$. (Recall that the best-known objective so far is -3.4.)

MASTER STEP

$$z_3 - \hat{c}_3 = \tfrac{5}{2}$$

$$\mathbf{A}\mathbf{x}_3 = \begin{bmatrix} 1 & 0 & 1 & 0 \\ 1 & 1 & 0 & 2 \end{bmatrix}\begin{bmatrix} 0 \\ \frac{5}{2} \\ 0 \\ 0 \end{bmatrix} = \begin{bmatrix} 0 \\ \frac{5}{2} \end{bmatrix}$$

$$\mathbf{y}_3 = \mathbf{B}^{-1}\begin{bmatrix} \mathbf{A}\mathbf{x}_3 \\ 1 \end{bmatrix} = \begin{bmatrix} \frac{1}{5} & 0 & 0 \\ -\frac{7}{10} & 1 & 0 \\ -\frac{1}{5} & 0 & 1 \end{bmatrix}\begin{bmatrix} 0 \\ \frac{5}{2} \\ 1 \end{bmatrix} = \begin{bmatrix} 0 \\ \frac{5}{2} \\ 1 \end{bmatrix}$$

Insert the column $\begin{bmatrix} z_3 - \hat{c}_3 \\ \mathbf{y}_3 \end{bmatrix}$ into the foregoing array and pivot. This leads to the following two tableaux (the λ_3 column is deleted after pivoting).

	BASIS INVERSE			RHS		λ_3
z	$-\frac{17}{10}$	0	0	$-\frac{17}{5}$		$\frac{5}{2}$
λ_2	$\frac{1}{5}$	0	0	$\frac{2}{5}$		0
s_2	$-\frac{7}{10}$	1	0	$\frac{8}{5}$		$\frac{5}{2}$
λ_1	$-\frac{1}{5}$	0	1	$\frac{3}{5}$		(1)

	BASIS INVERSE			RHS
z	$-\frac{6}{5}$	0	$-\frac{5}{2}$	$-\frac{49}{10}$
λ_2	$\frac{1}{5}$	0	0	$\frac{2}{5}$
s_2	$-\frac{1}{5}$	1	$-\frac{5}{2}$	$\frac{1}{10}$
λ_3	$-\frac{1}{5}$	0	1	$\frac{3}{5}$

The best-known feasible solution of the overall problem is given by

$$\mathbf{x} = \lambda_2\mathbf{x}_2 + \lambda_3\mathbf{x}_3$$
$$= \tfrac{2}{5}(2, \tfrac{3}{2}, 3, 0) + \tfrac{3}{5}(0, \tfrac{5}{2}, 0, 0) = (\tfrac{4}{5}, \tfrac{21}{10}, \tfrac{6}{5}, 0)$$

The objective is -4.9. Also $(w_1, w_2, \alpha) = (-\frac{6}{5}, 0, -\frac{5}{2})$.

Iteration 3

Since $w_1 < 0$, s_1 is not eligible to enter the basis at this time.

SUBPROBLEM

Solve the following subproblem.

Maximize $(\mathbf{wA} - \mathbf{c})\mathbf{x} + \alpha$
Subject to $\mathbf{x} \in X$

$$\mathbf{wA} - \mathbf{c} = (-\tfrac{6}{5}, 0)\begin{bmatrix} 1 & 0 & 1 & 0 \\ 1 & 1 & 0 & 2 \end{bmatrix} - (-2, -1, -1, 1) = (\tfrac{4}{5}, 1, -\tfrac{1}{5}, -1)$$

Therefore the subproblem is as follows.

Maximize $\frac{4}{5}x_1 + x_2 - \frac{1}{5}x_3 - x_4 - \frac{5}{2}$
Subject to $\mathbf{x} \in X$

Using Figure 7.1, the optimal solution is $\mathbf{x}_4 = (2, \frac{3}{2}, 0, 0)$ with objective $z_4 - \hat{c}_4 = \frac{3}{5}$, and so λ_4 is introduced.

The lower bound is given by $\hat{\mathbf{c}}_B\bar{\mathbf{b}} - (z_4 - \hat{c}_4) = -\frac{49}{10} - \frac{3}{5} = -5.5$. Recall that the best-known objective so far is -4.9. If we are interested only in an approximate solution, we could have stopped here with the feasible solution $\mathbf{x} = (\frac{4}{5}, \frac{21}{10}, \frac{6}{5}, 0)$ whose objective is -4.9.

MASTER STEP

$z_4 - \hat{c}_4 = \frac{3}{5}$

$$\mathbf{A}\mathbf{x}_4 = \begin{bmatrix} 1 & 0 & 1 & 0 \\ 1 & 1 & 0 & 2 \end{bmatrix} \begin{bmatrix} 2 \\ \frac{3}{2} \\ 0 \\ 0 \end{bmatrix} = \begin{bmatrix} 2 \\ \frac{7}{2} \end{bmatrix}$$

The updated column $\mathbf{y}_4$ is given by

$$\mathbf{y}_4 = \mathbf{B}^{-1}\begin{bmatrix} \mathbf{A}\mathbf{x}_4 \\ 1 \end{bmatrix} = \begin{bmatrix} \frac{1}{5} & 0 & 0 \\ -\frac{1}{5} & 1 & -\frac{5}{2} \\ -\frac{1}{5} & 0 & 1 \end{bmatrix} \begin{bmatrix} 2 \\ \frac{7}{2} \\ 1 \end{bmatrix} = \begin{bmatrix} \frac{2}{5} \\ \frac{3}{5} \\ \frac{3}{5} \end{bmatrix}$$

Insert the column $\begin{pmatrix} z_4 - \hat{c}_4 \\ \mathbf{y}_4 \end{pmatrix}$ in the foregoing array and pivot. This leads to the following two tableaux (the λ_4 column is deleted after pivoting).

	BASIS INVERSE			RHS	λ_4
z	$-\frac{6}{5}$	0	$-\frac{5}{2}$	$\frac{49}{10}$	$\frac{3}{5}$
λ_2	$\frac{1}{5}$	0	0	$\frac{2}{5}$	$\frac{2}{5}$
s_2	$-\frac{1}{5}$	1	$-\frac{5}{2}$	$\frac{1}{10}$	$\textcircled{\frac{3}{5}}$
λ_3	$-\frac{1}{5}$	0	1	$\frac{3}{5}$	$\frac{3}{5}$

	BASIS INVERSE			RHS
z	-1	-1	0	-5
λ_2	$\frac{1}{3}$	$-\frac{2}{3}$	$\frac{5}{3}$	$\frac{1}{3}$
λ_4	$-\frac{1}{3}$	$\frac{5}{3}$	$-\frac{25}{6}$	$\frac{1}{6}$
λ_3	0	-1	$\frac{7}{2}$	$\frac{1}{2}$

The best-known feasible solution of the overall problem is given by

$$\mathbf{x} = \lambda_2\mathbf{x}_2 + \lambda_3\mathbf{x}_3 + \lambda_4\mathbf{x}_4$$
$$= \tfrac{1}{3}(2, \tfrac{3}{2}, 3, 0) + \tfrac{1}{2}(0, \tfrac{5}{2}, 0, 0) + \tfrac{1}{6}(2, \tfrac{3}{2}, 0, 0) = (1, 2, 1, 0)$$

The objective is -5. Also $(w_1, w_2, \alpha) = (-1, -1, 0)$.

Iteration 4

Since $w_1 < 0$ and $w_2 < 0$, s_1 and s_2 are not eligible to enter the basis at this time.

SUBPROBLEM

Maximize $(\mathbf{wA} - \mathbf{c})\mathbf{x} + \alpha$
Subject to $\mathbf{x} \in X$

$$\mathbf{wA} - \mathbf{c} = (-1, -1)\begin{bmatrix} 1 & 0 & 1 & 0 \\ 1 & 1 & 0 & 2 \end{bmatrix} - (-2, -1, -1, 1) = (0, 0, 0, -3)$$

Therefore the subproblem is as follows.

Maximize $0x_1 + 0x_2 + 0x_3 - 3x_4 + 0$
Subject to $\mathbf{x} \in X$

Using Figure 7.1, an optimal solution is $\mathbf{x}_5 = (0, 0, 0, 0)$ with objective $z_5 - \hat{c}_5 = 0$, which is the termination criterion. Also note that the lower bound is $\hat{\mathbf{c}}_B\bar{\mathbf{b}} - (z_5 - \hat{c}_5) = -5 - 0 = -5$, which is equal to the best (and therefore optimal) solution known so far.

To summarize, the optimal solution $(x_1, x_2, x_3, x_4) = (1, 2, 1, 0)$ with objective -5 is at hand. The progress of the lower bounds and the objective values of the primal feasible solutions generated by the decomposition algorithm is shown in Figure 7.2. Optimality is reached at iteration 4. If we were interested in an approximate solution, we could have stopped at iteration 3, since we have a feasible solution with an objective value equal to -4.9, and meanwhile are assured (by the lower bound) that there exist no feasible solutions with an objective less than -5.5.

The optimal point $(x_1, x_2, x_3, x_4) = (1, 2, 1, 0)$ is shown in Figure 7.3 in the two sets X_1 and X_2. Note that $(1, 2)$ is not an extreme point of X_1 and $(1, 0)$ is not an extreme point of X_2. Note, however, that we can map the master

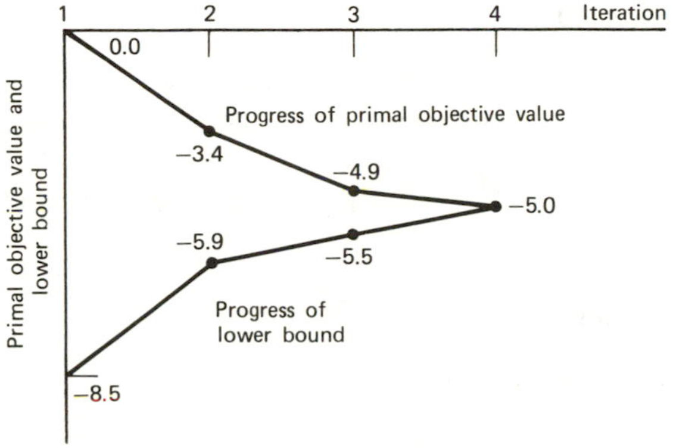

Figure 7.2. **Progress of the primal objective value and the lower bound.**

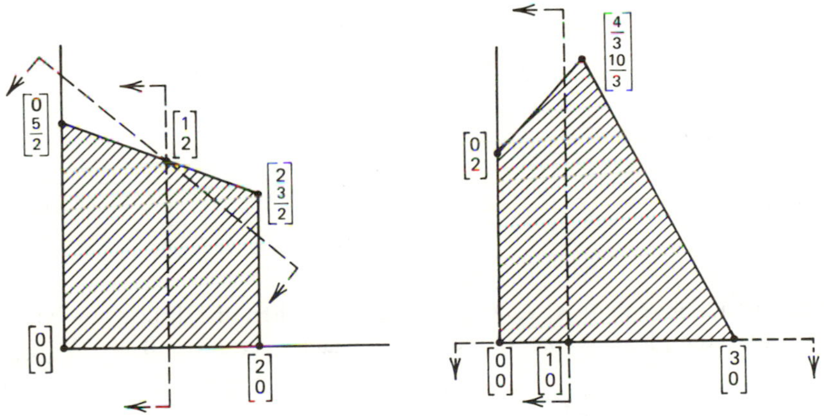

Figure 7.3. **Illustration of the optimal point.**

constraints

$$x_1 \qquad + x_3 \qquad \leqslant 2$$

$$x_1 + x_2 \qquad + 2x_4 \leqslant 3$$

into the (x_1, x_2) space by substituting $x_3 = 1$ and $x_4 = 0$. This leads to the two restrictions $x_1 \leqslant 1$ and $x_1 + x_2 \leqslant 3$, which are shown in Figure 7.3. We see that $(1, 2)$ is an extreme point of X_1 intersected with these two additional constraints. Similarly, in the (x_3, x_4) space, by substituting the values $x_1 = 1$ and $x_2 = 2$, the master constraints reduce to $x_3 \leqslant 1$ and $2x_4 \leqslant 0$. Again $(1, 0)$ is an extreme point of X_2 intersected with these additional constraints. It is worthwhile noting that the decomposition algorithm may not provide an optimal extreme point of the overall problem if alternative optima exist. The reader may refer to Exercise 7.19.

7.3 GETTING STARTED

In this section we describe a method to obtain a starting basic feasible solution for the master problem using artificial variables if necessary. These artificial variables are eliminated by the use of phase I or by the big-M method. If at termination there is a positive artificial variable, then the overall problem has no feasible solution.

Inequality Constraints

Consider the following problem.

$$\begin{aligned} \text{Minimize} \quad & \sum_{j=1}^{t} (\mathbf{c}\mathbf{x}_j)\lambda_j \\ \text{Subject to} \quad & \sum_{j=1}^{t} (\mathbf{A}\mathbf{x}_j)\lambda_j \leqslant \mathbf{b} \\ & \sum_{j=1}^{t} \lambda_j = 1 \\ & \lambda_j \geqslant 0 \qquad j = 1, 2, \ldots, t \end{aligned}$$

If there is a convenient $\mathbf{x}_1 \in X$ with $\mathbf{A}\mathbf{x}_1 \leqslant \mathbf{b}$, then the following basis is at hand, where the identity corresponds to the slack vector $\mathbf{s} \geqslant \mathbf{0}$.

$$\mathbf{B} = \left[\begin{array}{c|c} \mathbf{I} & \mathbf{A}\mathbf{x}_1 \\ \hline \mathbf{0} & 1 \end{array}\right], \qquad \mathbf{B}^{-1} = \left[\begin{array}{c|c} \mathbf{I} & -\mathbf{A}\mathbf{x}_1 \\ \hline \mathbf{0} & 1 \end{array}\right]$$

The initial array is given by the following tableau.

	BASIS INVERSE		RHS
z	$\mathbf{0}$	$\mathbf{c}\mathbf{x}_1$	$\mathbf{c}\mathbf{x}_1$
$\mathbf{s}$	$\mathbf{I}$	$-\mathbf{A}\mathbf{x}_1$	$\mathbf{b} - \mathbf{A}\mathbf{x}_1$
λ_1	$\mathbf{0}$	1	1

Now suppose that there is no obvious $\mathbf{x} \in X$ with $\mathbf{A}\mathbf{x} \leqslant \mathbf{b}$. In this case after converting the master problem to equality form by adding appropriate slack variables, the constraints are manipulated so that the RHS values are nonnegative. Then artificial variables are added, as needed, to create an identity matrix. This identity matrix constitutes the starting basis. The two-phase or big-M methods can be used to drive the artificial variables out of the basis.

Equality Constraints

In this case $m + 1$ artificial variables can be introduced to form the initial basis. The artificial variables are eliminated by the two-phase or by the big-M method.

7.4 THE CASE OF UNBOUNDED REGION X

For an unbounded set X, the decomposition algorithm must be slightly modified. In this case points in X can no longer be represented as a convex combination of the extreme points, but rather as a convex combination of the extreme points plus a nonnegative combination of the extreme directions. In other words, $\mathbf{x} \in X$ if and only if

$$\mathbf{x} = \sum_{j=1}^{t} \lambda_j \mathbf{x}_j + \sum_{j=1}^{l} \mu_j \mathbf{d}_j$$

$$\sum_{j=1}^{t} \lambda_j = 1$$

$$\lambda_j \geqslant 0 \qquad j = 1, 2, \ldots, t$$

$$\mu_j \geqslant 0 \qquad j = 1, 2, \ldots, l$$

where $\mathbf{x}_1, \mathbf{x}_2, \ldots, \mathbf{x}_t$ are the extreme points of X and $\mathbf{d}_1, \mathbf{d}_2, \ldots, \mathbf{d}_l$ are the extreme directions of X. The primal problem can be transformed into a problem in the variables $\lambda_1, \lambda_2, \ldots, \lambda_t$ and $\mu_1, \mu_2, \ldots, \mu_l$ as follows:

$$\text{Minimize} \quad \sum_{j=1}^{t} (\mathbf{c}\mathbf{x}_j)\lambda_j + \sum_{j=1}^{l} (\mathbf{c}\mathbf{d}_j)\mu_j$$

$$\text{Subject to} \quad \sum_{j=1}^{t} (\mathbf{A}\mathbf{x}_j)\lambda_j + \sum_{j=1}^{l} (\mathbf{A}\mathbf{d}_j)\mu_j = \mathbf{b} \tag{7.5}$$

$$\sum_{j=1}^{t} \lambda_j = 1 \tag{7.6}$$

$$\lambda_j \geqslant 0 \qquad j = 1, 2, \ldots, t$$

$$\mu_j \geqslant 0 \qquad j = 1, 2, \ldots, l$$

Since t and l are usually very large, we shall attempt to solve the foregoing problem by the revised simplex method. Suppose that we have a basic feasible solution of the foregoing system with basis $\mathbf{B}$, and let $\mathbf{w}$ and α be the dual variables corresponding to constraints (7.5) and (7.6) above. Further suppose that $\mathbf{B}^{-1}$, $(\mathbf{w}, \alpha) = \hat{\mathbf{c}}_B\mathbf{B}^{-1}$ ($\hat{\mathbf{c}}_B$ is the cost of the basic variables), and $\bar{\mathbf{b}} = \mathbf{B}^{-1}\binom{\mathbf{b}}{1}$

are known and displayed below.

BASIS INVERSE	RHS
$(\mathbf{w}, \alpha)$	$\hat{\mathbf{c}}_B\bar{\mathbf{b}}$
$\mathbf{B}^{-1}$	$\bar{\mathbf{b}}$

Recall that the current solution is optimal to the overall problem if $z_j - \hat{c}_j \leqslant 0$ for each variable. In particular, the following conditions must hold at optimality:

$$\lambda_j \text{ nonbasic} \Rightarrow 0 \geqslant z_j - \hat{c}_j = (\mathbf{w}, \alpha)\begin{pmatrix} \mathbf{Ax}_j \\ 1 \end{pmatrix} - \mathbf{cx}_j = \mathbf{wAx}_j + \alpha - \mathbf{cx}_j \quad (7.7)$$

$$\mu_j \text{ nonbasic} \Rightarrow 0 \geqslant z_j - \hat{c}_j = (\mathbf{w}, \alpha)\begin{pmatrix} \mathbf{Ad}_j \\ 0 \end{pmatrix} - \mathbf{cd}_j = \mathbf{wAd}_j - \mathbf{cd}_j \quad (7.8)$$

Since the number of nonbasic variables is very large, checking conditions (7.7) and (7.8) by generating the corresponding extreme points and directions is computationally infeasible. However, we may determine whether or not these conditions hold by solving the following subproblem. More importantly, as the subproblem is solved, if conditions (7.7) or (7.8) do not hold, a nonbasic variable with a positive $z_k - \hat{c}_k$, and hence eligible to enter the basis, is found.

$$\begin{array}{ll} \text{Maximize} & (\mathbf{wA} - \mathbf{c})\mathbf{x} + \alpha \\ \text{Subject to} & \mathbf{x} \in X \end{array}$$

First, suppose that the optimal solution of the subproblem is unbounded. Recall that this is only possible if an extreme direction $\mathbf{d}_k$ is found such that $(\mathbf{wA} - \mathbf{c})\mathbf{d}_k > 0$. This means that condition (7.8) is violated. Moreover, $z_k - \hat{c}_k = (\mathbf{wA} - \mathbf{c})\mathbf{d}_k > 0$ and μ_k is eligible to enter the basis. In this case $\begin{pmatrix} \mathbf{Ad}_k \\ 0 \end{pmatrix}$ is updated by premultiplying by $\mathbf{B}^{-1}$ and the resulting column $\begin{pmatrix} z_k - \hat{c}_k \\ \mathbf{y}_k \end{pmatrix}$ is inserted in the foregoing array and the revised simplex method is continued. Now consider the case where the optimal solution is bounded. A necessary and sufficient condition for boundedness is that $(\mathbf{wA} - \mathbf{c})\mathbf{d}_j \leqslant 0$ for all extreme directions, and so Equation (7.8) holds. Now we check whether (7.7) holds. Let $\mathbf{x}_k$ be an optimal extreme point and consider the optimal objective, $z_k - \hat{c}_k$, to the subproblem. If $z_k - \hat{c}_k \leqslant 0$, then by optimality of $\mathbf{x}_k$, for each extreme point $\mathbf{x}_j$, we have

$$(\mathbf{wA} - \mathbf{c})\mathbf{x}_j + \alpha \leqslant (\mathbf{wA} - \mathbf{c})\mathbf{x}_k + \alpha = z_k - \hat{c}_k \leqslant 0$$

and hence condition (7.7) holds and we stop with an optimal solution of the overall problem. If, on the other hand, $z_k - \hat{c}_k > 0$, then λ_k is introduced in the basis. This is done by inserting the column $\begin{pmatrix} z_k - \hat{c}_k \\ \mathbf{y}_k \end{pmatrix}$ into the foregoing array and pivoting, where $\mathbf{y}_k = \mathbf{B}^{-1}\begin{pmatrix} \mathbf{Ax}_k \\ 1 \end{pmatrix}$. Note that, as in the bounded case, if the master constraints are of the inequality type, then the $z_j - \hat{c}_j$ values for the slack variables must be checked before deducing optimality.

To summarize, solving the foregoing subproblem leads us either to terminate the algorithm with an optimal solution, or else to identify an improving variable to enter the basis. We now have all the ingredients for a decomposition algorithm in the case of an unbounded set X. Example 7.1 below illustrates such an algorithm. We ask the reader in Exercise 7.15 to write a step-by-step procedure for this case.

Example 7.1

$$\begin{aligned} \text{Minimize} \quad & -x_1 - 2x_2 - x_3 \\ \text{Subject to} \quad & x_1 + x_2 + x_3 \leqslant 12 \\ & -x_1 + x_2 \leqslant 2 \\ & -x_1 + 2x_2 \leqslant 8 \\ & x_3 \leqslant 3 \\ & x_1, \quad x_2, \quad x_3 \geqslant 0 \end{aligned}$$

The first constraint is handled as $\mathbf{Ax} \leqslant \mathbf{b}$ and the rest of the constraints are treated by X. Note that X decomposes into the two sets of Figure 7.4. The problem is transformed into $\lambda_1, \ldots, \lambda_t$ and $\mu_1, \ldots, \mu_l$ as follows.

$$\begin{aligned} \text{Minimize} \quad & \sum_{j=1}^{t} (\mathbf{cx}_j)\lambda_j + \sum_{j=1}^{l} (\mathbf{cd}_j)\,\mu_j \\ & \sum_{j=1}^{t} (\mathbf{Ax}_j)\lambda_j + \sum_{j=1}^{l} (\mathbf{Ad}_j)\,\mu_j \leqslant \mathbf{b} \\ & \sum_{j=1}^{t} \lambda_j = 1 \\ & \lambda_j \geqslant 0 \qquad j = 1, 2, \ldots, t \\ & \mu_j \geqslant 0 \qquad j = 1, 2, \ldots, l \end{aligned}$$

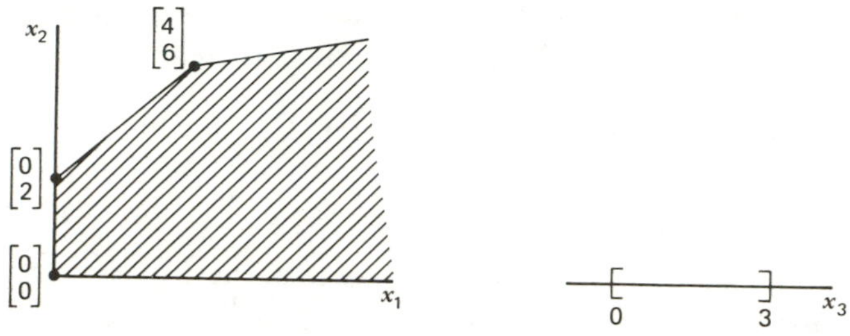

Figure 7.4. **Illustration of the unbounded** X.

Note that $\mathbf{x}_1 = (0, 0, 0)$ belongs to X and $\mathbf{A}\mathbf{x}_1 = 0 + 0 + 0 \leqslant 12$. Therefore the initial basis consists of λ_1 (corresponding to $\mathbf{x}_1$) plus the slack variable s. This leads to the following array, where $w = \alpha = 0$.

	BASIS INVERSE		RHS
z	0	0	0
s	1	0	12
λ_1	0	1	1

Iteration 1

SUBPROBLEM

$$\begin{array}{ll} \text{Maximize} & (\mathbf{wA} - \mathbf{c})\mathbf{x} + \alpha \\ \text{Subject to} & \mathbf{x} \in X \end{array}$$

Since $w = \alpha = 0$ and $\mathbf{A} = (1, 1, 1)$, this problem reduces to the following.

$$\text{Maximize} \quad x_1 + 2x_2 + x_3 + 0$$

$$\begin{array}{lrl} \text{Subject to} & -x_1 + x_2 & \leqslant 2 \\ & -x_1 + 2x_2 & \leqslant 8 \\ & x_3 & \leqslant 3 \\ & x_1, \quad x_2, \quad x_3 & \geqslant 0 \end{array}$$

Note that the problem decomposes into two problems in (x_1, x_2) and x_3. The optimal value of x_3 is 3. The other part of the problem can be solved geometrically or by the simplex method below, where x_4 and x_5 are the slack variables.

	z	x_1	x_2	x_4	x_5	RHS
z	1	−1	−2	0	0	0
x_4	0	−1	(1)	1	0	2
x_5	0	−1	2	0	1	8

The optimal is unbounded by noting the x_1 column. Suppose that we continue by introducing x_2 as the simplex method would normally proceed.

	z	x_1	x_2	x_4	x_5	RHS
z	1	−3	0	2	0	4
x_2	0	−1	1	1	0	2
x_5	0	(1)	0	−2	1	4

	z	x_1	x_2	x_4	x_5	RHS
z	1	0	0	−4	3	16
x_2	0	0	1	−1	1	6
x_1	0	1	0	−2	1	4

Since the most negative entry in row 0 corresponds to negative entries in rows 1 and 2, then the optimal is unbounded. As x_4 increases by 1 unit, x_1 increases by 2 units and x_2 increases by 1 unit; that is, in the (x_1, x_2) space we have found a direction $\mathbf{d}_1 = \binom{2}{1}$ leading to an unbounded solution. In the (x_1, x_2, x_3) space, $\mathbf{d}_1$ is given by $(2, 1, 0)^t$ (why?). Also $(\mathbf{wA} - \mathbf{c})\mathbf{d}_1 = 4$ (the negative of -4 in row 0 under x_4) and so μ_1 is introduced in the basis.

MASTER STEP

$$z_1 - \hat{c}_1 = 4$$

$$\mathbf{Ad}_1 = (1, 1, 1)\begin{bmatrix} 2 \\ 1 \\ 0 \end{bmatrix} = 3$$

$$\mathbf{y}_1 = \mathbf{B}^{-1}\begin{pmatrix} \mathbf{Ad}_1 \\ 0 \end{pmatrix}$$

where $\mathbf{B}^{-1}$ is obtained from the initial array of the master problem. In particular,

$$\mathbf{y}_1 = \begin{bmatrix} 1 & 0 \\ 0 & 1 \end{bmatrix}\begin{bmatrix} 3 \\ 0 \end{bmatrix} = \begin{bmatrix} 3 \\ 0 \end{bmatrix}$$

Introduce the column $\begin{pmatrix} z_1 - \hat{c}_1 \\ \mathbf{y}_1 \end{pmatrix}$ in the master array and pivot. The μ_1 column is eliminated after pivoting.

	BASIS INVERSE		RHS		μ_1
z	0	0	0		4
s	1	0	12		(3)
λ_1	0	1	1		0

	BASIS INVERSE		RHS
z	$-\frac{4}{3}$	0	-16
μ_1	$\frac{1}{3}$	0	4
λ_1	0	1	1

Iteration 2

$w = -\frac{4}{3}$ and $\alpha = 0$

Since $w < 0$, s is not a candidate to enter the basis.

SUBPROBLEM

Maximize $(\mathbf{wA} - \mathbf{c})\mathbf{x} + \alpha$
Subject to $\mathbf{x} \in X$

This reduces to the following.

Maximize $-\frac{1}{3}x_1 + \frac{2}{3}x_2 + 0$

Subject to
$$-x_1 + x_2 \leqslant 2$$
$$-x_1 + 2x_2 \leqslant 8$$
$$x_1, \quad x_2 \geqslant 0$$

Maximize $-\frac{1}{3}x_3$

Subject to $0 \leqslant x_3 \leqslant 3$

Here the value $\alpha = 0$ is added to only one of the subproblems. Obviously $x_3 = 0$. The new problem in (x_1, x_2) is solved by utilizing the corresponding tableau of the last iteration, deleting row 0, and introducing the new costs.

	z	x_1	x_2	x_4	x_5	RHS
z	1	$\frac{1}{3}$	$-\frac{2}{3}$	0	0	0
x_2	0	0	1	-1	1	6
x_1	0	1	0	-2	1	4

Multiply row 1 by $\frac{2}{3}$ and row 2 by $-\frac{1}{3}$ and add to row 0.

	z	x_1	x_2	x_4	x_5	RHS
z	1	0	0	0	$\frac{1}{3}$	$\frac{8}{3}$
x_2	0	0	1	-1	1	6
x_1	0	1	0	-2	1	4

The foregoing tableau is optimal (not unique). The optimal objective of the subproblem is $z_2 - \hat{c}_2 = \frac{8}{3} > 0$ and so λ_2 corresponding to $\mathbf{x}_2 = (x_1, x_2, x_3) = (4, 6, 0)$ is introduced.

MASTER STEP

$z_2 - \hat{c}_2 = \frac{8}{3}$

$\mathbf{Ax}_2 = 10$

$$\mathbf{y}_2 = \mathbf{B}^{-1}\begin{pmatrix} \mathbf{Ax}_2 \\ 1 \end{pmatrix} = \begin{bmatrix} \frac{1}{3} & 0 \\ 0 & 1 \end{bmatrix}\begin{bmatrix} 10 \\ 1 \end{bmatrix} = \begin{bmatrix} \frac{10}{3} \\ 1 \end{bmatrix}$$

Introduce $\begin{pmatrix} z_2 - \hat{c}_2 \\ \mathbf{y}_2 \end{pmatrix}$ into the master array and pivot. The λ_2 column is eliminated after pivoting.

	BASIS INVERSE		RHS	λ_2
z	$-\frac{4}{3}$	0	-16	$\frac{8}{3}$
μ_1	$\frac{1}{3}$	0	4	$\frac{10}{3}$
λ_1	0	1	1	(1)

	BASIS INVERSE		RHS
z	$-\frac{4}{3}$	$-\frac{8}{3}$	$-\frac{56}{3}$
μ_1	$\frac{1}{3}$	$-\frac{10}{3}$	$\frac{2}{3}$
λ_2	0	1	1

Iteration 3

Note that $w = -\frac{4}{3}$ did not alter from the last iteration. So s is still not a candidate to enter the basis. Also the optimal solution of the last subproblem remains the same (see iteration 2). The objective value of $\frac{8}{3}$ was for the previous dual solution with $\alpha = 0$. For $\alpha = -\frac{8}{3}$ we have $z_3 - \hat{c}_3 = \frac{8}{3} - \frac{8}{3} = 0$, which is the termination criterion, and the optimal solution is at hand. More specifically, the optimal $\mathbf{x}^*$ is given by

$$\mathbf{x}^* = \lambda_2\mathbf{x}_2 + \mu_1\mathbf{d}_1$$

$$= 1\begin{bmatrix} 4 \\ 6 \\ 0 \end{bmatrix} + \frac{2}{3}\begin{bmatrix} 2 \\ 1 \\ 0 \end{bmatrix} = \begin{bmatrix} \frac{16}{3} \\ \frac{20}{3} \\ 0 \end{bmatrix}$$

The objective is $-\frac{56}{3}$.

7.5 BLOCK DIAGONAL STRUCTURE

In this section we discuss the important special case when the set X has a block diagonal structure. In this case, X can itself be decomposed into several sets $X_1, X_2, \ldots, X_T$, each involving a subset of the variables, which do not appear in any other set. If we decompose the vector $\mathbf{x}$ accordingly into the vectors $\mathbf{x}_1, \mathbf{x}_2, \ldots, \mathbf{x}_T$, the vector $\mathbf{c}$ into $\mathbf{c}_1, \mathbf{c}_2, \ldots, \mathbf{c}_T$, and the matrix $\mathbf{A}$ of the master constraints $\mathbf{Ax} = \mathbf{b}$ into the matrices $\mathbf{A}_1, \mathbf{A}_2, \ldots, \mathbf{A}_T$, we get the following problem.

$$\begin{array}{llllll}
\text{Minimize} & \mathbf{c}_1\mathbf{x}_1 + & \mathbf{c}_2\mathbf{x}_2 + & \cdots & + \mathbf{c}_T\mathbf{x}_T & \\
\text{Subject to} & \mathbf{A}_1\mathbf{x}_1 + & \mathbf{A}_2\mathbf{x}_2 + & \cdots & + \mathbf{A}_T\mathbf{x}_T & = \mathbf{b} \\
 & \mathbf{B}_1\mathbf{x}_1 & & & & \leqslant \mathbf{b}_1 \\
 & & \mathbf{B}_2\mathbf{x}_2 & & & \leqslant \mathbf{b}_2 \\
 & & & \ddots & & \vdots \\
 & & & & \mathbf{B}_T\mathbf{x}_T & \leqslant \mathbf{b}_T \\
 & \mathbf{x}_1, & \mathbf{x}_2, & \ldots, & \mathbf{x}_T & \geqslant \mathbf{0}
\end{array}$$

where $X_i = \{\mathbf{x}_i : \mathbf{B}_i\mathbf{x}_i \leqslant \mathbf{b}_i, \mathbf{x}_i \geqslant \mathbf{0}\}$ for $i = 1, 2, \ldots, T$.

Problems with the foregoing structure arise frequently in network flows with several commodities and in the allocation of scarce resources among competing activities. Problems of this structure can be solved by the decomposition algorithm of this chapter (see Section 7.2 and Example 7.1). However, this block diagonal structure of X can be utilized further, as will be discussed in this section.

For subproblem i, $\mathbf{x}_i \in X_i$ if and only if

$$\mathbf{x}_i = \sum_{j=1}^{t_i} \lambda_{ij}\mathbf{x}_{ij} + \sum_{j=1}^{l_i} \mu_{ij}\mathbf{d}_{ij}$$

$$\sum_{j=1}^{t_i} \lambda_{ij} = 1$$

$$\lambda_{ij} \geqslant 0 \qquad j = 1, 2, \ldots, t_i$$

$$\mu_{ij} \geqslant 0 \qquad j = 1, 2, \ldots, l_i$$

where the $\mathbf{x}_{ij}$'s and the $\mathbf{d}_{ij}$'s are the extreme points and the extreme directions (if any) of X_i. Replacing each $\mathbf{x}_i$ by the foregoing representation, the original problem can be reformulated as follows.

$$\text{Minimize} \quad \sum_{i=1}^{T} \sum_{j=1}^{t_i} (\mathbf{c}_i\mathbf{x}_{ij})\lambda_{ij} + \sum_{i=1}^{T} \sum_{j=1}^{l_i} (\mathbf{c}_i\mathbf{d}_{ij})\,\mu_{ij}$$

$$\text{Subject to} \quad \sum_{i=1}^{T} \sum_{j=1}^{t_i} (\mathbf{A}_i\mathbf{x}_{ij})\lambda_{ij} + \sum_{i=1}^{T} \sum_{j=1}^{l_i} (\mathbf{A}_i\mathbf{d}_{ij})\,\mu_{ij} = \mathbf{b} \tag{7.9}$$

$$\sum_{j=1}^{t_i} \lambda_{ij} = 1 \qquad i = 1, 2, \ldots, T \tag{7.10}$$

$$\lambda_{ij} \geqslant 0 \qquad j = 1, 2, \ldots, t_i \quad i = 1, 2, \ldots, T$$

$$\mu_{ij} \geqslant 0 \qquad j = 1, 2, \ldots, l_i \quad i = 1, 2, \ldots, T$$

Note the difference in the foregoing formulation and that of Section 7.2. Here we allow different convex combinations and linear combinations for each subproblem i since we have T *convexity constraints* [the constraints of Equation (7.10)]. This adds more flexibility but at the same time increases the number of constraints from $m + 1$ to $m + T$.

Suppose that we have a basic feasible solution of the foregoing system with an $(m + T) \times (m + T)$ basis $\mathbf{B}$. Note that each basis must contain at least one variable λ_{ij} from each block i (see Exercise 7.17). Further suppose that $\mathbf{B}^{-1}$, $\bar{\mathbf{b}} = \mathbf{B}^{-1}\binom{\mathbf{b}}{\mathbf{1}}$, $(\mathbf{w}, \boldsymbol{\alpha}) = (w_1, \ldots, w_m, \alpha_1, \ldots, \alpha_T) = \hat{\mathbf{c}}_B\mathbf{B}^{-1}$ are known, where $\hat{\mathbf{c}}_B$ is the cost of the basic variables ($\hat{c}_{ij} = \mathbf{c}_i\mathbf{x}_{ij}$ for λ_{ij} and $\hat{c}_{ij} = \mathbf{c}_i\mathbf{d}_{ij}$ for μ_{ij}). These are displayed below.

BASIS INVERSE	RHS
$(\mathbf{w}, \boldsymbol{\alpha})$	$\hat{\mathbf{c}}_B\bar{\mathbf{b}}$
$\mathbf{B}^{-1}$	$\bar{\mathbf{b}}$

This solution is optimal if $z_{ij} - \hat{c}_{ij} \leqslant 0$ for each variable (naturally $z_{ij} - \hat{c}_{ij} = 0$ for each basic variable). In particular the following conditions must hold at optimality:

$$\lambda_{ij} \text{ nonbasic} \Rightarrow 0 \geqslant z_{ij} - \hat{c}_{ij} = \mathbf{w}\mathbf{A}_i\mathbf{x}_{ij} + \alpha_i - \mathbf{c}_i\mathbf{x}_{ij} \tag{7.11}$$

$$\mu_{ij} \text{ nonbasic} \Rightarrow 0 \geqslant z_{ij} - \hat{c}_{ij} = \mathbf{w}\mathbf{A}_i\mathbf{d}_{ij} - \mathbf{c}_i\mathbf{d}_{ij} \tag{7.12}$$

Whether conditions (7.11) and (7.12) hold or not, can be easily verified by solving the following subproblems.

$$\begin{array}{ll} \text{Maximize} & (\mathbf{w}\mathbf{A}_i - \mathbf{c}_i)\mathbf{x}_i + \alpha_i \\ \text{Subject to} & \mathbf{x}_i \in X_i \end{array}$$

If the optimal solution is unbounded, then an extreme direction $\mathbf{d}_{ik}$ is found such that $(\mathbf{w}\mathbf{A}_i - \mathbf{c}_i)\mathbf{d}_{ik} > 0$; that is, condition (7.12) is violated and introducing μ_{ik} will improve the objective function since $z_{ik} - \hat{c}_{ik} = (\mathbf{w}\mathbf{A}_i - \mathbf{c}_i)\mathbf{d}_{ik} > 0$. If the optimal is bounded, then automatically condition (7.12) holds for subproblem i. Let $\mathbf{x}_{ik}$ be an optimal extreme point. If the optimal objective value $z_{ik} - \hat{c}_{ik} = \mathbf{w}\mathbf{A}_i\mathbf{x}_{ik} + \alpha_i - \mathbf{c}_i\mathbf{x}_{ik} \leqslant 0$, then condition (7.11) holds for subproblem i. Otherwise λ_{ik} can be introduced in the basis. When all subproblems have $z_{ik} - \hat{c}_{ik} \leqslant 0$, then the optimal solution to the original problem is obtained. If the master constraints are of the inequality type, then we must also check the $z_j - \hat{c}_j$ values for the nonbasic slack variables (as we did in Section 7.1) before terminating.

To summarize, each subproblem i is solved in turn. If subproblem i yields an unbounded solution, then an extreme direction $\mathbf{d}_{ik}$ is found that is a candidate to enter the master basis. If subproblem i yields a bounded optimal point and $\mathbf{w}\mathbf{A}_i\mathbf{x}_{ik} + \alpha_i - \mathbf{c}_i\mathbf{x}_{ik} > 0$, then the extreme point is a candidate to enter the master basis. If neither of these conditions occurs, then there is currently no candidate column to enter the master basis from subproblem i. If no subproblem yields a candidate to enter the master basis, then we are optimal. Otherwise, we must select one from among the various candidates to enter the master basis. We may use the rule of the most positive $z_{ik} - \hat{c}_{ik}$, the first positive $z_{ik} - \hat{c}_{ik}$, and so on. If we use the rule of the first positive $z_{ik} - \hat{c}_{ik}$, then we may stop solving the subproblems after the first candidate comes available. On selecting the candidate, we update the entering column, pivot on the master array, and repeat the process.

Calculation of lower bounds for the case of bounded subproblems

Let $\mathbf{x}_1, \mathbf{x}_2, \ldots, \mathbf{x}_T$ represent a feasible solution of the overall problem so that $\mathbf{x}_i \in X_i$ for each i and $\Sigma_i \mathbf{A}_i \mathbf{x}_i = \mathbf{b}$. By definition of $z_{ik} - \hat{c}_{ik}$ we have

$$(\mathbf{w}\mathbf{A}_i - \mathbf{c}_i)\mathbf{x}_i + \alpha_i \leqslant (z_{ik} - \hat{c}_{ik})$$

or

$$\mathbf{c}_i\mathbf{x}_i \geqslant \mathbf{w}\mathbf{A}_i\mathbf{x}_i + \alpha_i - (z_{ik} - \hat{c}_{ik}).$$

Summing on i we get

$$\sum_i \mathbf{c}_i\mathbf{x}_i \geqslant \mathbf{w}\sum_i \mathbf{A}_i\mathbf{x}_i + \sum_i \alpha_i - \sum_i (z_{ik} - \hat{c}_{ik}).$$

But $\Sigma_i \mathbf{c}_i\mathbf{x}_i = \mathbf{c}\mathbf{x}$ and $\Sigma_i \mathbf{A}_i\mathbf{x}_i = \mathbf{b}$. Thus we get

$$\mathbf{c}\mathbf{x} \geqslant \mathbf{w}\mathbf{b} + \boldsymbol{\alpha}\mathbf{1} - \sum_i (z_{ik} - \hat{c}_{ik})$$

or

$$\mathbf{c}\mathbf{x} \geqslant \hat{\mathbf{c}}_B\bar{\mathbf{b}} - \sum_i (z_{ik} - \hat{c}_{ik}).$$

This is a natural extension of the case for one bounded subproblem presented earlier.

Example 7.2

Minimize $-2x_1 - x_2 - 3x_3 - x_4$

Subject to

$$\begin{aligned}
x_1 + x_2 + x_3 + x_4 &\leqslant 6\\
x_2 + 2x_3 + x_4 &\leqslant 4\\
x_1 + x_2 &\leqslant 6\\
x_2 &\leqslant 2\\
-\ x_3 + x_4 &\leqslant 3\\
x_3 + x_4 &\leqslant 5\\
x_1,\ x_2,\ x_3,\ x_4 &\geqslant 0
\end{aligned}$$

The first two constraints are handled by $\mathbf{Ax} \leqslant \mathbf{b}$, and the rest of the constraints are treated by X. Note that X decomposes into two sets as shown in Figure 7.5. The problem is transformed into the following.

$$\text{Minimize} \quad \sum_{j=1}^{t_1} (\mathbf{c}_1\mathbf{x}_{1j})\lambda_{1j} + \sum_{j=1}^{t_2} (\mathbf{c}_2\mathbf{x}_{2j})\,\lambda_{2j}$$

$$\begin{aligned} \text{Subject to} \quad \sum_{j=1}^{t_1} (\mathbf{A}_1\mathbf{x}_{1j})\lambda_{1j} + \sum_{j=1}^{t_2} (\mathbf{A}_2\mathbf{x}_{2j})\lambda_{2j} &\leqslant \mathbf{b} \\ \sum_{j=1}^{t_1} \lambda_{1j} &= 1 \\ \sum_{j=1}^{t_2} \lambda_{2j} &= 1 \\ \lambda_{1j} &\geqslant 0 \quad j = 1, 2, \ldots, t_1 \\ \lambda_{2j} &\geqslant 0 \quad j = 1, 2, \ldots, t_2 \end{aligned}$$

where $\mathbf{c}_1 = (-2, -1)$, $\mathbf{c}_2 = (-3, -1)$, $\mathbf{A}_1 = \begin{bmatrix} 1 & 1 \\ 0 & 1 \end{bmatrix}$, and $\mathbf{A}_2 = \begin{bmatrix} 1 & 1 \\ 2 & 1 \end{bmatrix}$. Note that $\mathbf{x}_{11} = (x_1, x_2) = (0, 0)$ and $\mathbf{x}_{21} = (x_3, x_4) = (0, 0)$ belong to X_1 and X_2 and satisfy the master constraints. Therefore we have a basic feasible solution of the overall system where the basis consists of s_1, s_2, λ_{11}, and λ_{21} (s_1 and s_2 are the slacks). This leads to the following master array.

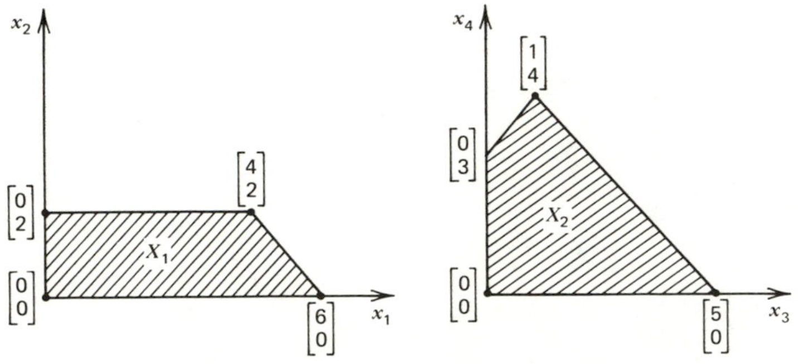

Figure 7.5. **The region** X.

	BASIS INVERSE				RHS
z	0	0	0	0	0
s_1	1	0	0	0	6
s_2	0	1	0	0	4
λ_{11}	0	0	1	0	1
λ_{21}	0	0	0	1	1

The first four entries of row 0 give w_1, w_2, α_1, and α_2 respectively. Under these entries $\mathbf{B}^{-1}$ is stored.

Iteration 1

Solve the following two subproblems.

SUBPROBLEM 1

Maximize $(\mathbf{w}\mathbf{A}_1 - \mathbf{c}_1)\mathbf{x}_1 + \alpha_1$

Subject to $\mathbf{x}_1 \in X_1$

SUBPROBLEM 2

Maximize $(\mathbf{w}\mathbf{A}_2 - \mathbf{c}_2)\mathbf{x}_2 + \alpha_2$

Subject to $\mathbf{x}_2 \in X_2$

Since $\mathbf{w} = (0, 0)$ and $\boldsymbol{\alpha} = (0, 0)$, these reduce to maximizing $2x_1 + x_2 + 0$ and maximizing $3x_3 + x_4 + 0$ over the two regions of Figure 7.5. The optimal solutions are respectively $\mathbf{x}_{12} = (x_1, x_2) = (6, 0)$ with objective 12 and $\mathbf{x}_{22} = (x_3, x_4) = (5, 0)$ with objective 15. Then $(\mathbf{w}\mathbf{A}_1 - \mathbf{c}_1)\mathbf{x}_{12} + \alpha_1 = 12$ and $(\mathbf{w}\mathbf{A}_2 - \mathbf{c}_2)\mathbf{x}_{22} + \alpha_2 = 15$. Therefore λ_{12} and λ_{22} are both candidates to enter. Select λ_{22} since $z_{22} - \hat{c}_{22} = 15$ is the most positive.

MASTER PROBLEM

$z_{22} - \hat{c}_{22} = 15$

Form the column

$$\begin{bmatrix} \mathbf{A}_2\mathbf{x}_{22} \\ 0 \\ 1 \end{bmatrix}$$

Note that:

$$\mathbf{A}_2\mathbf{x}_{22} = \begin{bmatrix} 1 & 1 \\ 2 & 1 \end{bmatrix}\begin{bmatrix} 5 \\ 0 \end{bmatrix} = \begin{bmatrix} 5 \\ 10 \end{bmatrix}, \qquad \begin{bmatrix} \mathbf{A}_2\mathbf{x}_{22} \\ 0 \\ 1 \end{bmatrix} = \begin{bmatrix} 5 \\ 10 \\ 0 \\ 1 \end{bmatrix}$$

This column is updated by premultiplying by $\mathbf{B}^{-1} = \mathbf{I}$. Insert this column along with $z_{22} - \hat{c}_{22}$ in the master array. Update the current basic feasible solution by pivoting. After completion, throw column λ_{22} away.

	BASIS INVERSE				RHS		λ_{22}
z	0	0	0	0	0		15
s_1	1	0	0	0	6		5
s_2	0	1	0	0	4		(10)
λ_{11}	0	0	1	0	1		0
λ_{21}	0	0	0	1	1		1

	BASIS INVERSE				RHS
z	0	$-\frac{3}{2}$	0	0	-6
s_1	1	$-\frac{1}{2}$	0	0	4
λ_{22}	0	$\frac{1}{10}$	0	0	$\frac{2}{5}$
λ_{11}	0	0	1	0	1
λ_{21}	0	$-\frac{1}{10}$	0	1	$\frac{3}{5}$

Note that $w_1 = 0$, $w_2 = -\frac{3}{2}$, $\alpha_1 = \alpha_2 = 0$.

Iteration 2

Since s_2 just left the basis, it will not be a candidate to immediately reenter.

Solve the following two subproblems.

SUBPROBLEM 1

Maximize $(\mathbf{w}\mathbf{A}_1 - \mathbf{c}_1)\mathbf{x}_1 + \alpha_1$

Subject to $\mathbf{x}_1 \in X_1$

SUBPROBLEM 2

Maximize $(\mathbf{w}\mathbf{A}_2 - \mathbf{c}_2)\mathbf{x}_2 + \alpha_2$

Subject to $\mathbf{x}_2 \in X_2$

These problems reduce to the following.

Maximize $2x_1 - \frac{1}{2}x_2 + 0$

Subject to $(x_1, x_2) \in X_1$

Maximize $0x_3 - \frac{1}{2}x_4 + 0$

Subject to $(x_3, x_4) \in X_2$

The optimal solutions are respectively $\mathbf{x}_{13} = (x_1, x_2) = (6, 0)$ with objective $z_{13} - \hat{c}_{13} = (\mathbf{w}\mathbf{A}_1 - \mathbf{c}_1)\mathbf{x}_{13} + \alpha_1 = 12$ and $\mathbf{x}_{23} = (x_3, x_4) = (5, 0)$ with objective $z_{23} - \hat{c}_{23} = (\mathbf{w}\mathbf{A}_2 - \mathbf{c}_2)\mathbf{x}_{23} + \alpha_2 = 0$. Thus there is no candidate from subproblem 2 at this time and λ_{13} is a candidate to enter the master basis.

MASTER PROBLEM

$z_{13} - \hat{c}_{13} = 12$

Form the column

$$\begin{bmatrix} \mathbf{A}_1\mathbf{x}_{13} \\ 1 \\ 0 \end{bmatrix}.$$

Note that

$$\mathbf{A}_1\mathbf{x}_{13} = \begin{bmatrix} 1 & 1 \\ 0 & 1 \end{bmatrix}\begin{bmatrix} 6 \\ 0 \end{bmatrix} = \begin{bmatrix} 6 \\ 0 \end{bmatrix}, \qquad \begin{bmatrix} \mathbf{A}_1\mathbf{x}_{13} \\ 1 \\ 0 \end{bmatrix} = \begin{bmatrix} 6 \\ 0 \\ 1 \\ 0 \end{bmatrix}$$

Updating this column, we get

$$\mathbf{y}_{13} = \mathbf{B}^{-1}\begin{bmatrix} \mathbf{A}_1\mathbf{x}_{13} \\ 1 \\ 0 \end{bmatrix} = \begin{bmatrix} 6 \\ 0 \\ 1 \\ 0 \end{bmatrix}$$

Insert this column along with $z_{13} - \hat{c}_{13} = 12$ into the master array. Update the current basic feasible solution by pivoting and then discard column λ_{13}.

	BASIS INVERSE				RHS	λ_{13}
z	0	$-\frac{3}{2}$	0	0	-6	12
s_1	1	$-\frac{1}{2}$	0	0	4	(6)
λ_{22}	0	$\frac{1}{10}$	0	0	$\frac{2}{5}$	0
λ_{11}	0	0	1	0	1	1
λ_{21}	0	$-\frac{1}{10}$	0	1	$\frac{3}{5}$	0

	BASIS INVERSE				RHS
z	-2	$-\frac{1}{2}$	0	0	-14
λ_{13}	$\frac{1}{6}$	$-\frac{1}{12}$	0	0	$\frac{2}{3}$
λ_{22}	0	$\frac{1}{10}$	0	0	$\frac{2}{5}$
λ_{11}	$-\frac{1}{6}$	$\frac{1}{12}$	1	0	$\frac{1}{3}$
λ_{21}	0	$-\frac{1}{10}$	0	1	$\frac{3}{5}$

Note that $w_1 = -2$, $w_2 = -\frac{1}{2}$, $\alpha_1 = \alpha_2 = 0$.

Iteration 3

Since $w_1 < 0$ and $w_2 < 0$, neither s_1 nor s_2 are candiates to enter the basis. Solve the following two subproblems.

SUBPROBLEM 1		SUBPROBLEM 2	
Maximize	$(\mathbf{wA}_1 - \mathbf{c}_1)\mathbf{x}_1 + \alpha_1$	Maximize	$(\mathbf{wA}_2 - \mathbf{c}_2)\mathbf{x}_2 + \alpha_2$
Subject to	$\mathbf{x}_1 \in X_1$	Subject to	$\mathbf{x}_2 \in X_2$

These problems reduce to the following.

Maximize	$0x_1 - \frac{3}{2}x_2 + 0$	Maximize	$0x_3 - \frac{3}{2}x_4 + 0$
Subject to	$(x_1, x_2) \in X_1$	Subject to	$(x_3, x_4) \in X_2$

From Figure 7.5, $\mathbf{x}_{14} = (x_1, x_2) = (0, 0)$ with objective 0 and $\mathbf{x}_{24} = (x_3, x_4) = (0, 0)$ with objective 0 are optimal solutions. Then $(\mathbf{wA}_1 - \mathbf{c}_1)\mathbf{x}_{14} + \alpha_1 = (\mathbf{wA}_2 - \mathbf{c}_2)\mathbf{x}_{24} + \alpha_2 = 0$ and the optimal solution is reached. From the master problem the optimal point $\mathbf{x}^*$ is given by

$$\begin{pmatrix} x_1 \\ x_2 \end{pmatrix} = \lambda_{11}\mathbf{x}_{11} + \lambda_{13}\mathbf{x}_{13}$$

$$= \tfrac{1}{3}\begin{pmatrix} 0 \\ 0 \end{pmatrix} + \tfrac{2}{3}\begin{pmatrix} 6 \\ 0 \end{pmatrix} = \begin{pmatrix} 4 \\ 0 \end{pmatrix}$$

$$\begin{pmatrix} x_3 \\ x_4 \end{pmatrix} = \lambda_{21}\mathbf{x}_{21} + \lambda_{22}\mathbf{x}_{22}$$

$$= \tfrac{3}{5}\begin{pmatrix} 0 \\ 0 \end{pmatrix} + \tfrac{2}{5}\begin{pmatrix} 5 \\ 0 \end{pmatrix} = \begin{pmatrix} 2 \\ 0 \end{pmatrix}$$

Therefore $\mathbf{x}^* = (x_1, x_2, x_3, x_4) = (4, 0, 2, 0)$ with objective -14.

Economic Interpretation

The decomposition algorithm has an interesting economic interpretation. Consider the case of a large system that is composed of smaller subsystems $1, 2, \ldots, T$. Each subsystem i has its own objective, and the objective function of the overall system is the sum of the objective functions of the subsystems. Each subsystem has its constraints designated by the set X_i, which is assumed to be bounded for the purpose of simplification. In addition, all the subsystems share a few common resources, and hence the consumption of these resources by all the subsystems must not exceed the availability given by the vector **b**.

Recall the following economic interpretation of the dual variables (Lagrangian multipliers). Here w_i is the rate of change of the objective as a function of b_i; that is, if b_i is replaced by $b_i + \Delta$, then the objective is modified by adding $w_i\Delta$. Hence $-w_i$ can be thought of as the price of consuming one unit of the ith common resource. Similarly, $-\alpha_i$ can be thought of as the price of consuming a portion of the ith convexity constraint.

With this in mind, the decomposition algorithm can be interpreted as follows. With the current proposals of the subsystems, the superordinate (total system) obtains the optimal weights of these proposals and announces a set of prices for using the common resources. These prices are passed down to the subsystems, which modify their proposals according to these new prices. A typical subsystem i solves the following subproblem.

$$\begin{array}{ll} \text{Maximize} & (\mathbf{wA}_i - \mathbf{c}_i)\mathbf{x}_i + \alpha_i \\ \text{Subject to} & \mathbf{x}_i \in X_i \end{array}$$

or equivalently

$$\begin{array}{ll} \text{Minimize} & (\mathbf{c}_i - \mathbf{wA}_i)\mathbf{x}_i - \alpha_i \\ \text{Subject to} & \mathbf{x}_i \in X_i \end{array}$$

The original objective of subsystem i is $\mathbf{c}_i\mathbf{x}_i$. The term $-\mathbf{wA}_i\mathbf{x}_i$ reflects the indirect price of using the common resources. Note that $\mathbf{A}_i\mathbf{x}_i$ is the amount of the common resources consumed by the $\mathbf{x}_i$ proposal. Since the price of using these resources is $-\mathbf{w}$, then the indirect cost of using them is $-\mathbf{wA}_i\mathbf{x}_i$, and the total cost is $(\mathbf{c}_i - \mathbf{wA}_i)\mathbf{x}_i$. Note that the term $-\mathbf{wAx}_i$ makes proposals that use much of the common resources unattractive from a cost point of view. Subsystem i announces an optimal proposal $\mathbf{x}_{ik}$. If this proposal is to be considered, then the weight of the older proposals $\mathbf{x}_{ij}$'s must decrease in order to "make room" for this proposal; that is, $\Sigma_j\lambda_{ij}$ must decrease from its present level of 1. The resulting saving is precisely α_i. If the cost of introducing the proposal $\mathbf{x}_{ik}$ is less than the saving realized; that is, if $(\mathbf{c}_i - \mathbf{wA}_i)\mathbf{x}_{ik} - \alpha_i < 0$, or $(\mathbf{wA}_i - \mathbf{c}_i)\mathbf{x}_{ik} + \alpha_i > 0$, then the superordinate would consider this new proposal. After all the subsystems introduce their new proposals, the superordinate calculates the optimum mix of these proposals and passes down new prices. The process is repeated until none of the subsystems has a new attractive proposal; that is, when $(\mathbf{c}_i - \mathbf{wA}_i)\mathbf{x}_{ik} - \alpha_i \geqslant 0$ for each i.

EXERCISES

7.1 Solve the following problem using the decomposition technique with one convexity constraint. Show the progress of the lower bound and primal objective.

$$
\begin{array}{lrl}
\text{Minimize} & -x_1 - x_2 - 2x_3 - x_4 & \\
\text{Subject to} & x_1 + 2x_2 + 2x_3 + x_4 & \leq 40 \\
& -x_1 + x_2 + x_3 + x_4 & \leq 10 \\
& x_1 + 3x_2 & \leq 30 \\
& 2x_1 + x_2 & \leq 20 \\
& x_3 & \leq 10 \\
& x_4 & \leq 10 \\
& x_3 + x_4 & \leq 15 \\
& x_1, \quad x_2, \quad x_3, \quad x_4 & \geq 0
\end{array}
$$

7.2 Solve the following linear programming problem by the decomposition method using one convexity constraint. Show the progress of the lower bound and primal objective.

$$
\begin{array}{lrl}
\text{Minimize} & -x_1 - 3x_2 + x_3 - x_4 & \\
\text{Subject to} & x_1 + x_2 + x_3 + x_4 & \leq 8 \\
& x_1 + x_2 & \leq 6 \\
& x_3 + 2x_4 & \leq 10 \\
& -x_3 + x_4 & \leq 4 \\
& x_1, \quad x_2, \quad x_3, \quad x_4 & \geq 0
\end{array}
$$

7.3 Consider the problem: Minimize $\mathbf{cx}$ subject to $\mathbf{Ax} = \mathbf{b}$, $\mathbf{x} \in X$. Let $\mathbf{x}^*$ be an optimal solution of the overall problem. Is it possible that $\mathbf{x}^*$ belongs to the interior of X? Interpret your answer geometrically. (*Hint*. Consider the case $\mathbf{wA} - \mathbf{c} = \mathbf{0}$ where $\mathbf{w}$ is the Lagrangian multiplier vector corresponding to $\mathbf{Ax} = \mathbf{b}$.)

7.4 Consider the problem: Minimize $\mathbf{cx}$ subject to $\mathbf{Ax} = \mathbf{b}$, $\mathbf{Dx} = \mathbf{d}$, $\mathbf{x} \geq \mathbf{0}$. Suppose that $\mathbf{w}_1^*$ and $\mathbf{w}_2^*$ are the optimal dual vectors of the constraints $\mathbf{Ax} = \mathbf{b}$ and $\mathbf{Dx} = \mathbf{d}$ respectively. Consider the problem: Maximize $(\mathbf{w}_1^*\mathbf{A} - \mathbf{c})\mathbf{x}$ subject to $\mathbf{Dx} = \mathbf{d}$, $\mathbf{x} \geq \mathbf{0}$. Let $\mathbf{x}_1^*, \mathbf{x}_2^*, \ldots, \mathbf{x}_k^*$ be optimal solutions of the foregoing problem that are extreme points of the set $\{\mathbf{x} : \mathbf{Dx} = \mathbf{d}, \mathbf{x} \geqslant \mathbf{0}\}$. Show that an optimal solution of the original problem can be represented as a convex combination of $\mathbf{x}_1^*, \mathbf{x}_2^*, \ldots, \mathbf{x}_k^*$, that is,

$$\mathbf{x}^* = \sum_{j=1}^{k} \lambda_j \mathbf{x}_j^*$$
$$\sum_{j=1}^{k} \lambda_j = 1$$
$$\lambda_j \geqslant 0 \qquad j = 1, 2, \ldots, k$$

7.5 Consider the following problem.

$$\begin{array}{lrl} \text{Minimize} & 2x_1 - 3x_2 - 4x_3 & \\ \text{Subject to} & x_1 + x_2 + x_3 & \leqslant 6 \\ & x_1 + x_2 - 2x_3 & \geqslant 2 \\ & 0 \leqslant x_1, \quad x_2, \quad x_3 & \leqslant 3 \end{array}$$

a. Set up the problem so that it can be solved by the decomposition algorithm.
b. Find a starting basis in the λ-space.
c. Find the optimal solution by the decomposition algorithm and compare with the bounded simplex method.

7.6 Consider the following (transportation) problem.

$$\text{Minimize} \quad 2x_{11} + 3x_{12} + 4x_{13} + 5x_{14} + 2x_{21} + 6x_{22} + 3x_{23} + 5x_{24}$$

$$\begin{array}{lllllllllll} \text{Subject to} & x_{11} + & x_{12} + & x_{13} + & x_{14} & & & & & = 500 \\ & & & & & x_{21} + & x_{22} + & x_{23} + & x_{24} & = 700 \\ & x_{11} & & & & + x_{21} & & & & = 200 \\ & & x_{12} & & & & + x_{22} & & & = 300 \\ & & & x_{13} & & & & + x_{23} & & = 400 \\ & & & & x_{14} & & & & + x_{24} & = 300 \\ & x_{11}, & x_{12}, & x_{13}, & x_{14}, & x_{21}, & x_{22}, & x_{23}, & x_{24} & \geqslant 0 \end{array}$$

a. Set up the problem so that it can be solved by the decomposition algorithm using several convexity constraints.
b. Find an optimal solution using the decomposition algorithm.

7.7 Use the decomposition principle to solve the following problem.

$$\begin{aligned}
\text{Maximize} \quad & x_1+8x_2+5x_3+6x_4 \\
\text{Subject to} \quad & x_1+4x_2+5x_3+2x_4 \leqslant 7 \\
& 2x_1+3x_2 \leqslant 6 \\
& 5x_1+x_2 \leqslant 5 \\
& 3x_3+4x_4 \geqslant 12 \\
& x_3 \leqslant 4 \\
& x_4 \leqslant 3 \\
& x_1, \quad x_2, \quad x_3, \quad x_4 \geqslant 0
\end{aligned}$$

7.8 In the text we developed a lower bound on the optimal objective value when the subproblem is bounded. Examine the case when this assumption is relaxed.

7.9 Solve the following problem by the decomposition technique using two convexity constraints.

$$\begin{aligned}
\text{Maximize} \quad & x_1+x_2+3x_3-x_4 \\
\text{Subject to} \quad & x_1+x_2+x_3+x_4 \leqslant 12 \\
& -x_1+x_2 \leqslant 2 \\
& 3x_1-4x_3 \leqslant 5 \\
& x_3+x_4 \leqslant 4 \\
& -x_3+x_4 \leqslant 5 \\
& x_1, \quad x_2, \quad x_3, \quad x_4 \geqslant 0
\end{aligned}$$

7.10 Solve the following problem by the decomposition technique.

$$\begin{aligned}
\text{Minimize} \quad & -x_1-2x_2+x_3+x_4 \\
\text{Subject to} \quad & x_1+2x_2+2x_3+x_4 \geqslant 40 \\
& x_1+x_2 \leqslant 2 \\
& -x_1+2x_2 \leqslant 2 \\
& x_3+x_4 \leqslant 6 \\
& x_1, \quad x_2, \quad x_3, \quad x_4 \geqslant 0
\end{aligned}$$

7.11 Consider the problem: Minimize $\mathbf{cx}$ subject to $\mathbf{Ax} = \mathbf{b}$, $\mathbf{x} \in X$. Suppose that X has a block diagonal structure. The decomposition algorithm can be applied by using either one convexity constraint or several convexity constraints, one for each subproblem. Discuss the advantages and disadvantages of both strategies. Which one would you prefer and why?

7.12 Apply the decomposition algorithm to the following problem.

$$\begin{aligned} \text{Minimize} \quad & -2x_1 + x_2 - 5x_3 \\ \text{Subject to} \quad & 4x_1 - 2x_2 + 3x_3 \leqslant 4 \\ & 0 \leqslant x_1, \quad x_2, \quad x_3 \leqslant 1 \end{aligned}$$

7.13 Suppose that the columns added during each master step of the decomposition algorithm are not deleted. In particular suppose that the master problem at iteration p is as follows.

$$\begin{aligned} \text{Minimize} \quad & \sum_{j=1}^{p} (\mathbf{cx}_j)\lambda_j \\ \text{Subject to} \quad & \sum_{j=1}^{p} (\mathbf{Ax}_j)\lambda_j = \mathbf{b} \\ & \sum_{j=1}^{p} \lambda_j = 1 \\ & \lambda_j \geqslant 0 \qquad j = 1, 2, \ldots, p \end{aligned}$$

where $\mathbf{x}_1, \ldots, \mathbf{x}_p$ are the extreme points generated so far. Discuss the details of such a decomposition procedure and compare with that of Section 7.1. Illustrate by solving the problem of Section 7.2.

7.14 In reference to Exercise 7.13 above, consider the following master problem.

$$\begin{aligned} \text{Minimize} \quad & \sum_{j=1}^{p} (\mathbf{cx}_j)\lambda_j \\ \text{Subject to} \quad & \sum_{j=1}^{p} (\mathbf{Ax}_j)\lambda_j = \mathbf{b} \\ & \sum_{j=1}^{p} \lambda_j = 1 \\ & \lambda_j \geqslant 0 \qquad j = 1, 2, \ldots, p \end{aligned}$$

Write the dual of this master problem. Suppose that the dual problem has been solved instead of this problem; how does the decomposition algorithm proceed? Interpret the dual master problem and show how to update it. Show convergence of the procedure and interpret it geometrically. (*Note.* The decomposition algorithm where the dual master problem replaces the original master problem is usually called a *cutting plane algorithm*, since a constraint that cuts away the previous optimal point of the dual master problem is added at each iteration.)

7.15 Construct both a flow chart and detailed steps of the decomposition algorithm for solving the problem: Minimize $\mathbf{cx}$ subject to $\mathbf{Ax} = \mathbf{b}$, $\mathbf{x} \in X$, where X is not necessarily bounded. Then code the decomposition algorithm in FORTRAN or another computer language. Use your code to solve some of the numerical problems given here.

7.16 Solve the following problem by the decomposition algorithm.

$$\begin{array}{lrl}
\text{Minimize} & -x_1 - 2x_2 - x_3 - 2x_4 & \\
\text{Subject to} & x_1 + x_2 + x_3 + x_4 & \leqslant 12 \\
 & -x_1 + x_2 & \leqslant 4 \\
 & -x_1 + 2x_2 & \leqslant 12 \\
 & x_3 + 2x_4 & \leqslant 8 \\
 & x_1, \quad x_2, \quad x_3, \quad x_4 & \geqslant 0
\end{array}$$

7.17 Consider the following problem.

$$\text{Minimize} \quad \sum_{i=1}^{T} \mathbf{c}_i \mathbf{x}_i$$

$$\text{Subject to} \quad \sum_{i=1}^{T} \mathbf{A}_i \mathbf{x}_i = \mathbf{b}$$

$$\mathbf{x}_i \in X_i \qquad i = 1, 2, \ldots, T$$

Show that any basis in the λ_{ij} space must contain at least one λ_{ij} for each $i = 1, 2, \ldots, T$.

7.18 Solve the following problem by the decomposition algorithm. Use phase I to get started in the λ-space.

$$
\begin{array}{lrl}
\text{Maximize} & x_1+3x_2+x_3 & \\
\text{Subject to} & x_1+\ x_2+x_3 & \leqslant 6 \\
& x_1+\ x_2 \qquad & \geqslant 4 \\
& -x_1+\ x_2 \qquad & \geqslant 2 \\
& x_3 & \geqslant 3 \\
& x_1,\quad x_2,\ x_3 & \geqslant 0
\end{array}
$$

7.19 Is it possible that the decomposition algorithm would generate an optimal nonextreme point of the overall problem in case of alternative optimal solutions? Discuss.
(*Hint*. Consider the following problem.

$$
\begin{array}{ll}
\text{Maximize} & x_1+x_2 \\
\text{Subject to} & x_1+x_2 \leqslant \tfrac{3}{2} \\
& 0 \leqslant x_1,\ x_2 \leqslant 1)
\end{array}
$$

7.20 Many options are available while solving the subproblem(s) and the master problem. These include the following.

a. The subproblem is terminated if an extreme point $\mathbf{x}_k$ is found with $z_k - \hat{c}_k > 0$. Then λ_k is introduced in the master problem.
b. Several columns can be generated from the subproblem(s) at each iteration.
c. At least one additional column is added to the master problem without discarding any of the previously generated columns. In this case the master problem reduces to finding the optimal mix of all columns generated so far.

Discuss in detail the foregoing options and compare and contrast them. Elaborate on the advantages and disadvantages of each.

7.21 Give a detailed analysis of the cases that may be encountered as artificial variables are used to find a starting basis of the master problem. Discuss both the two-phase method and the big-M method.

7.22 Consider the following problem.

$$\begin{array}{llllll}
\text{Minimize} & \mathbf{c}_0\mathbf{x}_0 + \mathbf{c}_1\mathbf{x}_1 + \mathbf{c}_2\mathbf{x}_2 \cdots + \mathbf{c}_T\mathbf{x}_T & & & & \\
\text{Subject to} & \mathbf{D}_0\mathbf{x}_0 + \mathbf{A}_1\mathbf{x}_1 + \mathbf{A}_2\mathbf{x}_2 \cdots + \mathbf{A}_T\mathbf{x}_T = \mathbf{b}_0 \\
& \mathbf{D}_1\mathbf{x}_0 + \mathbf{B}_1\mathbf{x}_1 \qquad\qquad\qquad\qquad = \mathbf{b}_1 \\
& \mathbf{D}_2\mathbf{x}_0 \qquad + \mathbf{B}_2\mathbf{x}_2 \qquad\qquad\quad = \mathbf{b}_2 \\
& \vdots \qquad\qquad\qquad \ddots \qquad\qquad\qquad \vdots \\
& \mathbf{D}_T\mathbf{x}_0 \qquad\qquad\qquad\qquad + \mathbf{B}_T\mathbf{x}_T = \mathbf{b}_T \\
& \quad \mathbf{x}_0, \quad \mathbf{x}_1, \quad \mathbf{x}_2, \ldots, \quad \mathbf{x}_T \geqslant \mathbf{0}
\end{array}$$

Describe in detail how the decomposition technique can be used to solve problems of the foregoing structure. (*Hint*. Let the first set of constraints be the constraints of the master problem. The subproblem consists of the remaining constraints. Take the dual of the subproblem and solve it by decomposition. This becomes a "three-level" algorithm.)

7.23 Consider the following problem.

$$\begin{array}{ll}
\text{Minimize} & \mathbf{c}_1\mathbf{x}_1 + \mathbf{c}_2\mathbf{x}_2 \cdots + \mathbf{c}_T\mathbf{x}_T \\
\text{Subject to} & \mathbf{A}_1\mathbf{x}_1 + \mathbf{A}_2\mathbf{x}_2 \cdots + \mathbf{A}_T\mathbf{x}_T \leqslant \mathbf{b} \\
& \mathbf{B}_1\mathbf{x}_1 \qquad\qquad\qquad\qquad \leqslant \mathbf{b}_1 \\
& \qquad\quad \mathbf{B}_2\mathbf{x}_2 \qquad\qquad\quad \leqslant \mathbf{b}_2 \\
& \qquad\qquad\qquad \ddots \qquad\qquad \vdots \\
& \qquad\qquad\qquad\qquad \mathbf{B}_T\mathbf{x}_T \leqslant \mathbf{b}_T \\
& \quad \mathbf{x}_1, \quad \mathbf{x}_2 \cdots, \quad \mathbf{x}_T \geqslant \mathbf{0}
\end{array}$$

The following implementation of the decomposition principle is a possibility. The subproblem constraints are

$$\begin{array}{c}
\mathbf{A}_1\mathbf{x}_1 + \mathbf{A}_2\mathbf{x}_2 \ldots + \mathbf{A}_T\mathbf{x}_T \leqslant \mathbf{b} \\
\mathbf{x}_1, \quad \mathbf{x}_2 \ldots, \quad \mathbf{x}_T \geqslant \mathbf{0}
\end{array}$$

and the master constraints are $\mathbf{B}_1\mathbf{x}_1 \leqslant \mathbf{b}_1$, $\mathbf{B}_2\mathbf{x}_2 \leqslant \mathbf{b}_2, \ldots, \mathbf{B}_T\mathbf{x}_T \leqslant \mathbf{b}_T$. Describe the details of such an algorithm. What are the advantages and the disadvantages of this procedure? Does this scheme have an economic interpretation? Use the scheme to solve Exercise 7.10.

7.24 Consider the following problem.

$$\text{Minimize} \quad \sum_{j=1}^{T} \mathbf{c}_j\mathbf{x}_j + \sum_{j=1}^{T} \mathbf{d}_j\mathbf{y}_j$$

$$\begin{aligned}
\text{Subject to} \quad \mathbf{x}_{j-1} - \mathbf{x}_j + \mathbf{A}_j\mathbf{y}_j &= \mathbf{b}_j \qquad j = 1, 2, \ldots, T \\
\mathbf{x}_T &= \mathbf{b} \\
\mathbf{0} \leqslant \mathbf{x}_j &\leqslant \mathbf{u}_j \qquad j = 1, 2, \ldots, T \\
\mathbf{0} \leqslant \mathbf{y}_j &\leqslant \mathbf{u}'_j \qquad j = 1, 2, \ldots, T
\end{aligned}$$

where $\mathbf{x}_0$ is a known vector.

a. What class of problems lend themselves to this general structure? What is the interpretation of the vectors $\mathbf{x}_j$ and $\mathbf{y}_j$? (*Hint*. Examine a discrete control system.)
b. How can the decomposition algorithm be applied to this system? (*Hint*. Choose every other constraint to form the master constraints.)
c. Apply the procedure in (b) to solve the following problem.

$$\begin{aligned}
\text{Minimize} \quad & x_1 + x_2 + x_3 + x_4 + 3y_1 + 5y_2 + 4y_3 + 6y_4 \\
& x_0 - x_1 + y_1 = 40 \\
& x_1 - x_2 + y_2 = 50 \\
& x_2 - x_3 + y_3 = 60 \\
& x_3 - x_4 + y_4 = 40 \\
& x_4 = 30 \\
& 0 \leqslant x_1,\ x_2,\ x_3,\ x_4 \leqslant 40 \\
& 0 \leqslant y_1,\ y_2 \leqslant 45 \\
& 0 \leqslant y_3,\ y_4 \leqslant 50
\end{aligned}$$

7.25 Consider the following *cutting stock problem*. We have standard rolls of length l and an order is placed requiring b_i units of length l_i where $i = 1, 2, \ldots, m$. It is desired to find the minimum number of rolls which satisfy the order.

a. Formulate the problem.
b. Apply the decomposition algorithm to solve the problem. Discuss in detail. (*Hint*. Consider the column $\mathbf{a}_j$ representing the jth cutting pattern. Here $\mathbf{a}_j$ is a vector of nonnegative integers; a_{ij}, the ith component, is the number of rolls of length l_i in the jth cutting pattern.

Develop a scheme for generating these $\mathbf{a}_j$ columns. What is the master problem and the subproblem?)

7.26 An agricultural mill produces cattle feed and chicken feed. These products are composed of three main ingredients, namely corn, lime, and fish meal. The ingredients contain two main types of nutrients, namely protein and calcium. The following table gives the nutrients' contents per pound of each ingredient.

	INGREDIENT		
NUTRIENT	CORN	LIME	FISH MEAL
Protein	25	15	25
Calcium	15	30	20

The protein content must lie in the interval [18, 22] per pound of cattle feed. Also the calcium content must be greater than or equal to 20 per pound of the cattle feed. Similarly, the protein content and the calcium content must be in the intervals [20, 23] and [20, 25], respectively, per pound of the chicken feed. Suppose that 3000, 2500, and 1000 pounds of corn, lime, and fish meal are available. Also suppose that it is required to produce 4000 and 2000 pounds of the cattle and chicken feed respectively. Let the price per pound of the corn, lime, fish meal be respectively \$0.10, \$0.10, and \$0.08. Formulate the blending problem with an objective of minimizing the cost. Solve the problem by the decomposition algorithm using two convexity constraints. Extra corn and fish meal can be obtained but, because of shortages, at the higher prices of \$0.12 and \$0.10 per pound. Would you advise the mill to consider extra corn and fish meal and modify their blending at these prices? Why or why not?

7.27 A company owns two refineries in Dallas and New York. The company can purchase two types of crude oil, light crude oil and heavy crude oil at the prices of \$11 and \$9 per barrel respectively. Because of shortages, the maximum amounts of these crudes that can be purchased are 2 million and 1 million barrels respectively. The following quantities of gasoline, kerosene, and jet fuel are produced per barrel of each type of oil.

	GASOLINE	KEROSENE	JET FUEL
Light crude oil	0.40	0.20	0.35
Heavy crude oil	0.32	0.40	0.20

Note that 5% and 8% of the crude are lost during the refining process, respectively. The company has contracted to deliver these products to three consumers in Kansas City, Los Angeles, and Detroit. The demands of these products are given below.

	GASOLINE	KEROSENE	JET FUEL
Kansas City	200,000	400,000	—
Los Angeles	300,000	200,000	600,000
Detroit	500,000	100,000	300,000

It is desired to find the amounts that must be purchased by the company of each crude at each of its refining facilities, and the shipping pattern of the products to Kansas City, Los Angeles, and Detroit, that satisfy the demands and minimize the total cost (purchase + shipping). The shipping and handling of a barrel of any finished product from the refineries to the consumers is given below.

	KANSAS CITY	LOS ANGELES	DETROIT
Dallas Refinery	\$0.60	\$0.40	\$0.50
New York Refinery	\$0.35	\$0.80	\$0.30

a. Formulate the problem.
b. Suggest a decomposition scheme for solving the problem.
c. Solve the problem using your scheme in (b).

7.28 A company has two manufacturing facilities in Atlanta and Los Angeles. The two facilities produce refrigerators and washer/dryers. The production capacities of these items in Atlanta are 5000 and 7000 respectively. Similarly the capacity of the Los Angeles facility is 8000 refrigerators and 4000 washer/dryers. The company delivers these products to three major customers in New York City, Seattle, and Miami. The customers' demand is given below.

DEMAND/CUSTOMER	NEW YORK	SEATTLE	MIAMI
Refrigerators	4000	5000	4000
Washer/dryers	3000	3000	4000

The items are transported from the manufacturing facilities to the customers via a railroad network. The unit transportation costs (no distinction is made between the two items) are summarized below. Also, because of limited space, the maximum number of refrigerators and/or washer/dryers that can be transported from a facility to a customer is given below.

FACILITY		CUSTOMER: NEW YORK	SEATTLE	MIAMI
Atlanta	Unit shipping cost $	6	14	7
	Maximum number of units	6000	3000	8000
Los Angeles	Unit shipping cost $	10	8	15
	Maximum number of units	3000	9000	3000

It is desired to find the shipping pattern that minimizes the total transportation cost.

a. Formulate the problem.

b. Use the decomposition technique with two convexity constraints to solve the problem.

(*Note*. This problem is called a *multicommodity transportation problem*. The subproblem decomposes into two transportation problems. If you are familiar with the transportation algorithm, you can use it to solve the subproblems. Otherwise use the simplex method to solve the subproblems.)

7.29 In this chapter the constraints were decomposed into special and general constraints. Now consider the following problem where the variables are decomposed.

$$\begin{aligned} \text{Minimize} \quad & \mathbf{cx} + f(\mathbf{y}) \\ \text{Subject to} \quad & \mathbf{Ax} + \mathbf{By} = \mathbf{b} \\ & \mathbf{x} \geqslant \mathbf{0} \\ & \mathbf{y} \in Y \end{aligned}$$

where $\mathbf{c}$, $\mathbf{b}$, $\mathbf{x}$, $\mathbf{y}$ are vectors, $\mathbf{A}$, $\mathbf{B}$ are matrices, f is an arbitrary function, and Y is an arbitrary set.

a. Show that the problem can be reformulated as follows.

$$\begin{aligned} \text{Minimize} \quad & z \\ \text{Subject to} \quad & z \geqslant f(\mathbf{y}) + \mathbf{w}(\mathbf{b} - \mathbf{By}) \\ & \mathbf{wA} \leqslant \mathbf{c} \\ & \mathbf{w} \quad \text{unrestricted} \\ & \mathbf{y} \in Y \end{aligned}$$

(*Hint*. The original problem can be reformulated as follows.

$$\underset{\mathbf{y}\in Y}{\text{Minimize}}\left(f(\mathbf{y}) + \underset{\substack{\mathbf{Ax}=\mathbf{b}-\mathbf{By}\\ \mathbf{x}\geqslant\mathbf{0}}}{\text{Minimum}}\,\mathbf{cx}\right)$$

Take the dual of

$$\begin{array}{ll} \text{Minimize} & \mathbf{cx} \\ \text{Subject to} & \mathbf{Ax} = \mathbf{b} - \mathbf{By} \\ & \mathbf{x} \geqslant \mathbf{0}) \end{array}$$

b. Show that $z \geqslant f(\mathbf{y}) + \mathbf{w}(\mathbf{b} - \mathbf{By})$ for each $\mathbf{wA} \leqslant \mathbf{c}$ if and only if $z \geqslant f(\mathbf{y}) + \mathbf{w}_j(\mathbf{b} - \mathbf{By})$ and $\mathbf{d}_j(\mathbf{b} - \mathbf{By}) \leqslant 0$ for each extreme point $\mathbf{w}_j$ and every extreme direction $\mathbf{d}_j$ of the region $\{\mathbf{w} : \mathbf{wA} \leqslant \mathbf{c}\}$.

c. Make use of (b) to reformulate the problem in (a) as follows:

$$\begin{array}{lll} \text{Minimize} & z & \\ \text{Subject to} & z \geqslant f(\mathbf{y}) + \mathbf{w}_j(\mathbf{b} - \mathbf{By}) & j = 1, 2, \ldots, t \\ & \mathbf{d}_j(\mathbf{b} - \mathbf{By}) \leqslant 0 & j = 1, 2, \ldots, l \\ & \mathbf{y} \in Y & \end{array}$$

where $\mathbf{w}_1, \ldots, \mathbf{w}_t$ and $\mathbf{d}_1, \ldots, \mathbf{d}_l$ are the extreme points and the extreme directions of $\{\mathbf{w} : \mathbf{wA} \leqslant \mathbf{c}\}$.

d. Without explicitly enumerating $\mathbf{w}_1, \ldots, \mathbf{w}_t$ and $\mathbf{d}_1, \ldots, \mathbf{d}_l$ beforehand, devise a decomposition algorithm for solving the problem in (c). (*Hint*. Master Problem:

$$\begin{array}{lll} \text{Minimize} & z & \\ \text{Subject to} & z \geqslant f(\mathbf{y}) + \mathbf{w}_j(\mathbf{b} - \mathbf{By}) & j = 1, 2, \ldots, t' \\ & \mathbf{d}_j(\mathbf{b} - \mathbf{By}) \leqslant 0 & j = 1, 2, \ldots, l' \\ & \mathbf{y} \in Y & \end{array}$$

where $\mathbf{w}_1, \ldots, \mathbf{w}_{t'}$ and $\mathbf{d}_1, \ldots, \mathbf{d}_{l'}$ are the extreme points and extreme directions generated so far.
Subproblem:

$$\begin{array}{ll} \text{Maximize} & \mathbf{w}(\mathbf{b} - \mathbf{By}) \\ \text{Subject to} & \mathbf{wA} \leqslant \mathbf{c} \end{array}$$

where $\mathbf{y}$ is obtained from the optimal of the master problem.)

e. How would you obtain the optimal $(\mathbf{x}, \mathbf{y})$ at termination of the decomposition algorithm in (d)?

(Note. This algorithm is referred to as *Benders's partitioning procedure*. Note that the set Y can be discrete and so the procedure can be used for solving mixed integer problems. In this case the master problem in (d) is a pure integer programming problem and the subproblem is a linear program.)

7.30 Apply Benders's partitioning procedure of Exercise 7.29 to solve the following problem [let (x_1, x_2) be $\mathbf{x}$ and (x_3, x_4) be $\mathbf{y}$].

$$\begin{aligned}
\text{Minimize} \quad & -x_1 - 2x_2 - 3x_3 - x_4 \\
\text{Subject to} \quad & x_1 + x_2 + 2x_3 + x_4 \leq 12 \\
& -x_1 + x_2 \leq 4 \\
& 2x_1 + x_2 \leq 6 \\
& x_3 + x_4 \leq 8 \\
& x_1, \quad x_2, \quad x_3, \quad x_4 \geq 0
\end{aligned}$$

7.31 A company is planning to build several warehouses for storing a certain product. These warehouses would serve two major customers with monthly demands of 3000 and 5000 units. Three candidate warehouses with capacities 4000, 5000, and 6000 can be constructed. Using the estimated construction cost of the warehouses, their useful life, and time value of money, the construction cost per month for the three warehouses is estimated as \$8000, \$12,000, and \$7000. The unit transportation cost from the three candidate warehouses to the customers is given below.

	CUSTOMER	
WAREHOUSE	1	2
1	1.50	2.00
2	2.00	1.50
3	2.50	2.25

Use Benders's partitioning procedure of Exercise 7.29 to determine which warehouses to construct and the corresponding shipping pattern.

7.32 Solve the following linear program entirely graphically using decomposition.

$$\begin{aligned}
\text{Minimize} \quad & 2x_1 + 5x_2 + x_3 - 2x_4 + 3x_5 \\
\text{Subject to} \quad & x_1 + x_2 + x_3 + x_4 \geq 2 \\
& 3x_1 + x_2 + 5x_3 + x_4 - 2x_5 \geq 5 \\
& -x_1 + 2x_3 + x_4 \geq 2 \\
& x_1, \quad x_2, \quad x_3, \quad x_4, \quad x_5 \geq 0
\end{aligned}$$

(*Hint*. Let the first constraint denote $\mathbf{Ax} \geqslant \mathbf{b}$ and the next two constraints represent X. Then take the dual of each set.)

7.33 Indicate how the results of the previous problem may be generalized to any number m of constraints. (*Hint*. Let the first constraint denote $\mathbf{Ax} = \mathbf{b}$ and the remaining $m - 1$ constraints be part of the subproblem. Then reapply decomposition to the subproblem in a similar way.)

7.34 Assume that a linear program requires $3m/2$ iterations for a solution. Also, assume that standard techniques of pivoting are used to update the basis inverse and RHS vector [in total an $(m + 1) \times (m + 1)$ matrix if we ignore the z column]. If there is no special structure to the constraint matrix, then is there an optimal split for decomposition? That is, find $m_1 + m_2 = m$ such that the first m_1 constraints form the master problem and the next m_2 constraints are subproblem constraints, and the total "effort" is minimized. Let the "effort" be defined by the number of elementary operations (additions, subtractions, multiplications, and divisions).

7.35 In the previous problem suppose that m_1 and m_2 are given and that the second m_2 constraints are of a special structure. Specifically, suppose that the subproblem requires only 5% of the normal effort to yield a solution when treated by itself.

a. Should the problem be decomposed for efficiency?
b. Is there a critical value of the percentage effort required?

7.36 Solve the following problem by decomposition.

$$
\begin{array}{lrl}
\text{Minimize} & -2x_1 + 5x_2 - 4x_3 & \\
\text{Subject to} & x_1 + 2x_2 + a_1x_3 & \leqslant 6 \\
 & 3x_1 - 6x_2 + a_2x_3 & \leqslant 5 \\
 & 2a_1 + 3a_2 & = 4 \\
 & x_1, x_2, x_3, a_1, a_2 & \geqslant 0
\end{array}
$$

[*Hint*. Let $X = \{(a_1, a_2) : 2a_1 + 3a_2 = 4, a_1, a_2 \geqslant 0\}$. This is called a *generalized linear programming problem*.]

NOTES AND REFERENCES

1. The decomposition algorithm of this chapter is an adaptation of the Dantzig-Wolfe decomposition principle [116]. The latter was inspired by the suggestions of Ford and Fulkerson [157] for solving the special case of multicommodity network flow problems.
2. The decomposition method presented in this chapter is closely associated

with the concepts of generalized Lagrangian multipliers, tangential approximation of the Lagrangian dual function, and the dual cutting plane algorithm. For further reading on these topics the reader may refer to Everett [139], Geoffrion [189], Lasdon [305], Kelley [278], and Zangwill [486].

3. In addition to the Dantzig-Wolfe and similar decomposition algorithms, the literature has a great deal of other decomposition methods. These can be classified as *price-directive* and *resource-directive* algorithms. In the former, a direction for modifying the Lagrangian multipliers of the coupling constraints is found and then a suitable step size is taken along this direction. See, for example, Geoffrion [188], Lasdon [305], Grinold [220], Balas [13], Held, Wolfe, and Crowder [235], and Bazaraa and Goode [24]. The resource-directive algorithms proceed by finding a direction for modifying the shares of the common resources among the subproblems and then determining the step size. The reader may refer to Geoffrion [188], Lasdon [305], and Abadie [1].
4. In Exercise 7.29 we describe the partitioning scheme of Benders [33]. This scheme is particularly suited for solving mixed integer programming problems. The relationship between Benders's scheme and the decomposition algorithm of this chapter becomes apparent upon studying Exercise 7.29.

EIGHT: THE TRANSPORTATION AND ASSIGNMENT PROBLEMS

An important special class of linear programming problems is that of transportation problems. This class and its extension, the class of network flow problems, possess a special structure that (1) permits the development of simple and efficient algorithms and (2) facilitates a greater intuition for and understanding of the techniques of linear programming and the simplex method. As a result these special problems deserve and receive our attention in this and later chapters.

8.1 DEFINITION OF THE TRANSPORTATION PROBLEM

Consider m origin points located on a map, where origin i has a supply of a_i units of a particular item (commodity). In addition, there are located n destination points, where destination j requires b_j units of the commodity. We assume that $a_i, b_j > 0$. Associated with each link (i, j), from origin i to destination j, there is a unit cost c_{ij} for transportation. The problem is to determine the feasible shipping pattern from origins to destinations that minimizes the total transportation cost.

Let x_{ij} be the number of units shipped along link (i, j) from origin i to destination j. Further assume that the total supply equals the total demand, that is,

$$\sum_{i=1}^{m} a_i = \sum_{j=1}^{n} b_j$$

If the total supply exceeds the total demand, then a dummy destination can be created with demand $b_{n+1} = \Sigma_i a_i - \Sigma_j b_j$, and $c_{i,n+1} = 0$ for $i = 1, \ldots, m$. Assuming that the total supply equals the total demand, the linear programming model for the transportation problem becomes as follows.

Minimize

$$c_{11}x_{11} + \cdots + c_{1n}x_{1n} + c_{21}x_{21} + \cdots + c_{2n}x_{2n} + \cdots + c_{m1}x_{m1} + \cdots + c_{mn}x_{mn}$$

Subject to

$$\begin{array}{llllllll}
x_{11} + \cdots + & x_{1n} & & & & & & = a_1 \\
& & x_{21} + \cdots & + x_{2n} & & & & = a_2 \\
& & & & \ddots & & & \vdots \\
& & & & & x_{m1} + \cdots & + x_{mn} & = a_m \\
x_{11} & & + x_{21} & & \cdots & + x_{m1} & & = b_1 \\
\quad \ddots & & \quad \ddots & & & \quad \ddots & & \vdots \\
& x_{1n} & & + x_{2n} & & \vdots & + x_{mn} & = b_n \\
x_{11}, \ldots, & x_{1n}, & x_{21}, \ldots, & x_{2n}, \cdots, & & x_{m1}, \ldots, & & x_{mn} \geqslant 0
\end{array}$$

The transportation problem is graphically illustrated in Figure 8.1.

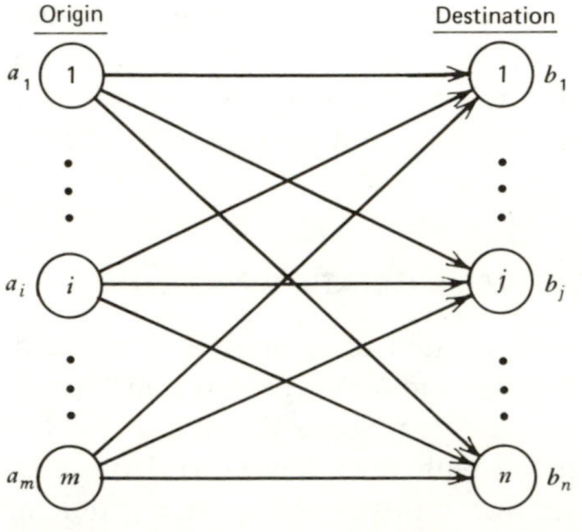

Figure 8.1. **Illustration of the graph of a transportation problem.**

We can cast the transportation problem in matrix form if we let

$$\begin{aligned}
\mathbf{x} &= (x_{11}, x_{12}, \ldots x_{1n}, x_{21}, \ldots x_{2n}, \ldots x_{mn})^t \\
\mathbf{c} &= (c_{11}, c_{12}, \ldots c_{1n}, c_{21}, \ldots c_{2n}, \ldots c_{mn}) \\
\mathbf{b} &= (a_1, a_2, \ldots, a_m, b_1, b_2, \ldots, b_n)^t \\
\mathbf{A} &= (\mathbf{a}_{11}, \mathbf{a}_{12}, \ldots, \mathbf{a}_{1n}, \mathbf{a}_{21}, \ldots, \mathbf{a}_{2n}, \ldots, \mathbf{a}_{mn})
\end{aligned}$$

where

$$\mathbf{a}_{ij} = \mathbf{e}_i + \mathbf{e}_{m+j}$$

and $\mathbf{e}_i$ and $\mathbf{e}_{m+j}$ are unit vectors in E^{m+n}, with ones in the ith and $(m + j)$th positions respectively. The reader should note that in this chapter a_i is a scalar, $\mathbf{a}_{ij}$ is a vector, and that these two terms should not be confused with one another. With these definitions the problem takes the following form.

$$\begin{aligned}
\text{Minimize} \quad & \mathbf{cx} \\
\text{Subject to} \quad & \mathbf{Ax} = \mathbf{b} \\
& \mathbf{x} \geqslant \mathbf{0}
\end{aligned}$$

The **A** matrix, with dimension $(m + n) \times mn$ has the following special form.

$$\overset{mn \text{ columns}}{\mathbf{A} = \begin{bmatrix} \mathbf{1} & \mathbf{0} & \cdots & \mathbf{0} \\ \mathbf{0} & \mathbf{1} & \cdots & \mathbf{0} \\ \vdots & \vdots & & \vdots \\ \mathbf{0} & \mathbf{0} & \cdots & \mathbf{1} \\ \mathbf{I} & \mathbf{I} & \cdots & \mathbf{I} \end{bmatrix}} \; m+n \text{ rows}$$

where **1** is an n row vector of all 1's and **I** is an $n \times n$ identity matrix. It is the **A** matrix that gives the transportation problem its special structure.

As an example of a transportation problem consider a 2-origin, 3-destination transportation problem with data as indicated below.

		Destination 1	Destination 2	Destination 3	a_i
Origin	1	$c_{11} = 4$	$c_{12} = 7$	$c_{13} = 5$	30
	2	$c_{21} = 2$	$c_{22} = 4$	$c_{23} = 3$	20
	b_j	15	10	25	

For this problem

$$\mathbf{A} = \begin{bmatrix} 1 & 1 & 1 & 0 & 0 & 0 \\ 0 & 0 & 0 & 1 & 1 & 1 \\ 1 & 0 & 0 & 1 & 0 & 0 \\ 0 & 1 & 0 & 0 & 1 & 0 \\ 0 & 0 & 1 & 0 & 0 & 1 \end{bmatrix}$$

Feasibility of the Transportation Problem

Under the assumption that supply equals demand, the transportation problem always possesses a feasible solution. For example, it is easy to show that

$$x_{ij} = \frac{a_i b_j}{d} \qquad i = 1, \ldots, m, j = 1, \ldots, n$$

where $d = \Sigma_i a_i = \Sigma_j b_j$, is a feasible solution. Note that for each feasible vector $\mathbf{x}$, every component x_{ij} is bounded as follows:

$$0 \leqslant x_{ij} \leqslant \text{Minimum}\{a_i, b_j\}$$

We know that a bounded linear program with a feasible solution possesses an optimal solution. Thus it is now only a matter of finding the optimal solution.

8.2 PROPERTIES OF THE A MATRIX

We shall examine some of the properties of the **A** matrix that give the transportation problem its special structure. As we shall see, these properties permit a simple and efficient application of the simplex method for transportation problems.

Rank of the A Matrix

Assuming $m, n \geqslant 2$, we have $m + n \leqslant mn$ so that rank $(\mathbf{A})$ is less than or equal to $m + n$. Clearly, rank $(\mathbf{A}) \neq m + n$ since the sum of the first m rows equals the sum of the last n rows and hence the $m + n$ rows of **A** are linearly dependent. Thus rank $(\mathbf{A}) \leqslant m + n - 1$.

To demonstrate that rank $(\mathbf{A}) = m + n - 1$ we need only to find an $(m + n - 1) \times (m + n - 1)$ submatrix from **A** which is nonsingular. Ignoring the last row of **A**, consider the submatrix

$$\mathbf{A}' = (\mathbf{a}_{1n}, \mathbf{a}_{2n}, \ldots, \mathbf{a}_{mn}, \mathbf{a}_{11}, \mathbf{a}_{12}, \ldots, \mathbf{a}_{1, n-1})$$

which is an upper triangular matrix of the form

$$\mathbf{A}' = \begin{bmatrix} \mathbf{I}_m & \mathbf{Q} \\ \mathbf{0} & \mathbf{I}_{n-1} \end{bmatrix}$$

Obviously $\mathbf{A}'$ has rank $m + n - 1$ and thus the matrix $\mathbf{A}$ has rank $m + n - 1$. In the previous example problem we have (deleting the last row)

$$\mathbf{A}' = \begin{array}{c} \begin{matrix} x_{13} & x_{23} & x_{11} & x_{12} \end{matrix} \\ \begin{bmatrix} 1 & 0 & 1 & 1 \\ 0 & 1 & 0 & 0 \\ 0 & 0 & 1 & 0 \\ 0 & 0 & 0 & 1 \end{bmatrix} \end{array}$$

Knowing that the rank of $\mathbf{A}$ is $m + n - 1$, we are left with two choices for a basis—we can either delete the last row or any row leaving $m + n - 1$ linearly independent constraints for which a basis exists, or we can add an artificial vector for one constraint. When applying the simplex method, we shall select the latter approach and augment $\mathbf{A}$ with a new artificial variable x_a with column vector $\mathbf{e}_{m+n}$.

Total Unimodularity of the A Matrix

The single most important property that the transportation matrix possesses is the total unimodularity property. The $\mathbf{A}$ matrix is *totally unimodular* if the determinant of every square submatrix formed from it has value -1, 0 or $+1$. In the case of the transportation matrix, since all entries are 0 or 1, every 1×1 submatrix has determinant of value 0 or 1. In addition, any $(m + n) \times (m + n)$ submatrix has determinant of value 0 since rank $(\mathbf{A}) = m + n - 1$. It remains to show that any $k \times k$ submatrix $(1 < k < m + n)$ has the required property.

Let $\mathbf{A}_k$ be any $k \times k$ submatrix from $\mathbf{A}$. We must show that det $\mathbf{A}_k = \pm 1$ or 0. By induction on k, suppose that the property is true for $\mathbf{A}_{k-1}$ (we know it is true for $\mathbf{A}_1$). Recall that each column of $\mathbf{A}_k$ has either no 1's, a single 1, or two 1's. If any column of $\mathbf{A}_k$ has no 1's, then det $\mathbf{A}_k = 0$. If, on the other hand, every column of $\mathbf{A}_k$ has two 1's, then one of the 1's will occur in an origin row and the other 1 will occur in a destination row. In this case the sum of the origin rows of $\mathbf{A}_k$ equals the sum of the destination rows of $\mathbf{A}_k$. Thus the rows of $\mathbf{A}_k$ are linearly dependent and det $\mathbf{A}_k = 0$. Finally, if some column of $\mathbf{A}_k$ contains a single 1, expanding det $\mathbf{A}_k$ by the minors of that column we get:

$$\det \mathbf{A}_k = \pm \det \mathbf{A}_{k-1}$$

where $\mathbf{A}_{k-1}$ is a $(k - 1) \times (k - 1)$ submatrix. But, by the induction hypothesis, det $\mathbf{A}_{k-1} = \pm 1$ or 0. Thus the property is true for $\mathbf{A}_k$ and the result is shown.

Triangularity of the Basis Matrix

We have previously demonstrated that rank $(\mathbf{A}) = m + n - 1$ by selecting a particular $(m + n - 1) \times (m + n - 1)$ submatrix that was nonsingular and, therefore, was a basis for $\mathbf{A}$ (ignoring the last row). That matrix was (upper) triangular. It can be demonstrated that every basis matrix of $\mathbf{A}$ enjoys this triangularity property.

Suppose that $\mathbf{B}$ is a basis matrix from $\mathbf{A}$. From previous considerations (recall the preceding discussion on total unimodularity) we know that there must be at least one column of $\mathbf{B}$ containing a single 1; otherwise, det $\mathbf{B} = 0$. Permuting the rows and columns of $\mathbf{B}$, we obtain

$$\mathbf{B} = \begin{bmatrix} 1 & \mathbf{q} \\ \mathbf{0} & \mathbf{B}_{m+n-2} \end{bmatrix}$$

Now considering $\mathbf{B}_{m+n-2}$ we again argue that it must contain at least one column with a single 1. Permuting its rows and columns, we get

$$\mathbf{B}_{m+n-2} = \begin{bmatrix} 1 & \mathbf{p} \\ \mathbf{0} & \mathbf{B}_{m+n-3} \end{bmatrix}$$

Letting $\mathbf{q} = (q_1, \mathbf{q}_2)$, $\mathbf{B}$ can be represented as follows:

$$\mathbf{B} = \begin{bmatrix} 1 & q_1 & \mathbf{q}_2 \\ 0 & 1 & \mathbf{p} \\ \mathbf{0} & \mathbf{0} & \mathbf{B}_{m+n-3} \end{bmatrix}$$

Continuing this procedure, we obtain the result that $\mathbf{B}$ is a triangular matrix.

Since $\mathbf{B}$ is a triangular matrix, there is a simple method of solving the basic system $\mathbf{B}\mathbf{x}_B = \mathbf{b}$. Suppose that we permute the rows and columns of $\mathbf{B}$ so that $\mathbf{B}$ is upper triangular. Then since the last row of $\mathbf{B}$ contains a single entry in the last column, we may use that row to solve for the last basic variable. Again, as a result of the (upper) triangularity of $\mathbf{B}$ we may substitute the value of the last basic variable into the next-to-last equation and solve for the next-to-last basic variable. This process of back substitution continues until the values of all basic variables have been determined.

In the example problem with the basis consisting of x_{13}, x_{23}, x_{11}, and x_{12} we append the last row and artificial column to get the following:

$$\begin{array}{cc} & \begin{array}{ccccc} x_{13} & x_{23} & x_{11} & x_{12} & x_a \end{array} \\ \mathbf{A}' = & \left[\begin{array}{cccc|c} 1 & 0 & 1 & 1 & 0 \\ 0 & 1 & 0 & 0 & 0 \\ 0 & 0 & 1 & 0 & 0 \\ 0 & 0 & 0 & 1 & 0 \\ 1 & 1 & 0 & 0 & 1 \end{array}\right] \end{array} \qquad \begin{array}{c} \mathbf{b} \\ \begin{bmatrix} 30 \\ 20 \\ 15 \\ 10 \\ 25 \end{bmatrix} \end{array}$$

Upon permuting the rows and columns of $\mathbf{A}'$, we get

$$\mathbf{A}' = \begin{array}{c} \begin{array}{c|cccc} x_a & x_{13} & x_{23} & x_{11} & x_{12} \end{array} \\ \left[\begin{array}{c|cccc} 1 & 1 & 1 & 0 & 0 \\ 0 & 1 & 0 & 1 & 1 \\ 0 & 0 & 1 & 0 & 0 \\ 0 & 0 & 0 & 1 & 0 \\ 0 & 0 & 0 & 0 & 1 \end{array}\right] \end{array} \qquad \begin{array}{c} \mathbf{b} \\ \left[\begin{array}{c} 25 \\ 30 \\ 20 \\ 15 \\ 10 \end{array}\right] \end{array}$$

Solving this triangular system, we get

$$x_{12} = 10$$

$$x_{11} = 15$$

$$x_{23} = 20$$

$$x_{13} = 30 - x_{11} - x_{12} = 30 - 15 - 10 = 5$$

$$x_a = 25 - x_{13} - x_{23} = 25 - 5 - 20 = 0$$

Integrality of Basic Solutions

Since each basis consists entirely of integer entries and is triangular with all diagonal elements equal to 1, we conclude that the values of all basic variables will be integral if the supplies and demands are integers. In particular, we may conclude that the optimal basic feasible solution will be all integer. This is a property not enjoyed by general linear programs.

Properties of the $\mathbf{y}_{ij}$ Vectors in the Simplex Tableau

Since a basis $\mathbf{B}$ consists entirely of zeros and ones and is triangular with 1's on the diagonal, then the elements of $\mathbf{B}^{-1}$ are all ± 1 or 0 (why?). Each vector, $\mathbf{y}_{ij}$, in the simplex tableau for a transportation problem is given by

$$\mathbf{B}\mathbf{y}_{ij} = \mathbf{a}_{ij}$$

which is a system of equations in the unknown elements of $\mathbf{y}_{ij}$. One method for solving such a system is by Cramer's rule (see Section 2.2). Utilizing Cramer's rule, the kth unknown element of $\mathbf{y}_{ij}$ is given by

$$y_{ijk} = \frac{\det \mathbf{B}_k}{\det \mathbf{B}}$$

where $\mathbf{B}_k$ is obtained from $\mathbf{B}$ by replacing the kth column by $\mathbf{a}_{ij}$. Then $\mathbf{B}_k$ is a square submatrix from **A**, and since the latter is totally unimodular, then det $\mathbf{B}_k = \pm 1$ or 0. Since det $\mathbf{B} = \pm 1$, it follows that $y_{ijk} = \pm 1$ or 0.

This shows that a typical updated simplex column $\mathbf{y}_{ij}$ consists of ± 1's and 0's. It also shows that any vector $\mathbf{a}_{ij}$ can be obtained by the simple addition and subtraction of basic vectors. This simplicity suggests that there may be a convenient method of obtaining the (unique) representation $\mathbf{B}\mathbf{y}_{ij} = \mathbf{a}_{ij}$ and, thereby, constructing the entire simplex tableau associated with a basic solution. In particular, in the representation of the nonbasic vector $\mathbf{a}_{ij} = \mathbf{e}_i + \mathbf{e}_{m+j}$ in terms of basic vectors there must be a basic vector of the form $\mathbf{a}_{ik} = \mathbf{e}_i + \mathbf{e}_{m+k}$ with a coefficient of $+1$. Then there must exist a basic vector of the form $\mathbf{a}_{lk} = \mathbf{e}_l + \mathbf{e}_{m+k}$ with a coefficient of -1 in the representation. This process continues until finally there must exist a vector of the form $\mathbf{a}_{uj} = \mathbf{e}_u + \mathbf{e}_{m+j}$ with a coefficient of $+1$ in the representation. A typical representation of $\mathbf{a}_{ij}$ is

$$\begin{aligned}\mathbf{a}_{ij} &= \mathbf{a}_{ik} - \mathbf{a}_{lk} + \mathbf{a}_{ls} - \mathbf{a}_{us} + \mathbf{a}_{uj} \\ &= (\mathbf{e}_i + \mathbf{e}_{m+k}) - (\mathbf{e}_l + \mathbf{e}_{m+k}) + (\mathbf{e}_l + \mathbf{e}_{m+s}) \\ &\quad - (\mathbf{e}_u + \mathbf{e}_{m+s}) + (\mathbf{e}_u + \mathbf{e}_{m+j}) \\ &= \mathbf{e}_i + \mathbf{e}_{m+j}\end{aligned}$$

Representing the nonbasic vector $\mathbf{a}_{ij}$ in terms of the basic vectors is illustrated on the transportation matrix (tableau) in Figure 8.2. Note that the cell (i, j) together with the cells (i, k), (l, k), (l, s), (u, s), and (u, j) form a *cycle* in the matrix. The cells (i, k), (l, k), (l, s), (u, s), and (u, j) form a *chain* in the matrix between cell (i, k) and cell (u, j). Other basic cells which do not appear in the representation of $\mathbf{a}_{ij}$ are deleted from Figure 8.2. Also, note that the signs of the coefficients alternate throughout the chain.

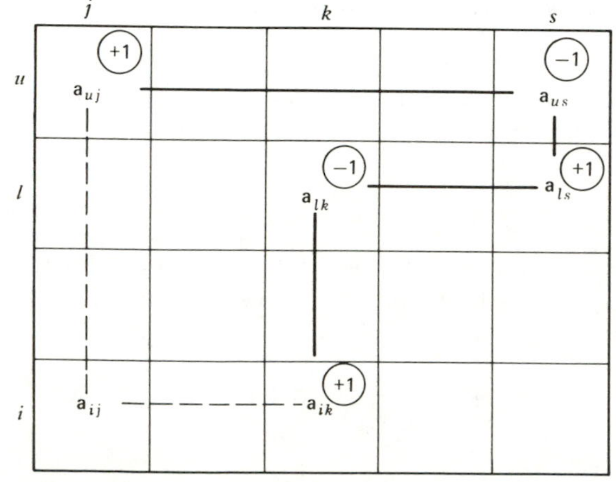

Figure 8.2. Illustration of the representation of $\mathbf{a}_{ij}$ in terms of the basic vectors.

Characterization of a Basis on a Transportation Tableau

We shall first show that the basic vectors cannot form a cycle on the transportation tableau. In reference to Figure 8.3, suppose by contradiction that cells (p, q), (r, q), (r, s), (u, s), (u, v), and (p, v) which form a cycle were basic. Consider the following linear combination:

$$\mathbf{a}_{pq} - \mathbf{a}_{rq} + \mathbf{a}_{rs} - \mathbf{a}_{us} + \mathbf{a}_{uv} - \mathbf{a}_{pv} =$$

$$(\mathbf{e}_p + \mathbf{e}_{m+q}) - (\mathbf{e}_r + \mathbf{e}_{m+q}) + (\mathbf{e}_r + \mathbf{e}_{m+s})$$

$$-(\mathbf{e}_u + \mathbf{e}_{m+s}) + (\mathbf{e}_u + \mathbf{e}_{m+v}) - (\mathbf{e}_p + \mathbf{e}_{m+v}) = \mathbf{0}$$

This means that the vectors $\mathbf{a}_{pq}$, $\mathbf{a}_{rq}$, $\mathbf{a}_{rs}$, $\mathbf{a}_{us}$, $\mathbf{a}_{uv}$, and $\mathbf{a}_{pv}$ are linearly dependent and could not have been in the basis. The conclusion is that no basis, as represented by a set of cells in the tableau, can contain a cycle.

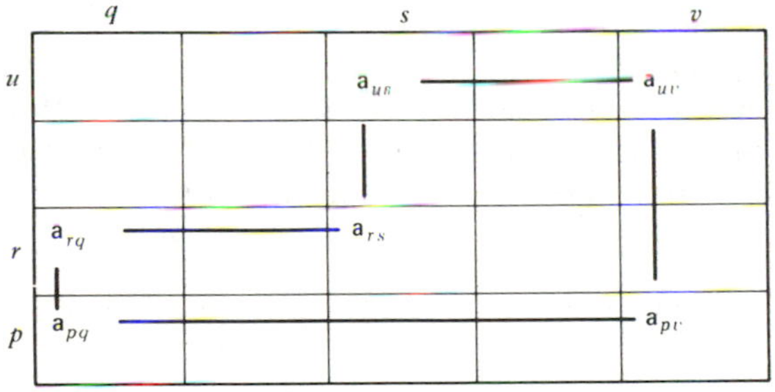

Figure 8.3. **A basis cannot contain a cycle.**

Figure 8.4 illustrates the basis used to show that the rank of $\mathbf{A}$ is $m + n - 1$ (the artificial variable used to complete the rank to $m + n$ is not shown). The B's indicate the basic cells while the lines connecting the B's indicate those basic cells in the same row or column as other basic cells. This structure, the basic cells and lines connecting them, has some very interesting properties. Ignoring the matrix for a moment, such a structure of cells and lines is called a *graph*. The basis graph of Figure 8.4 is *connected*; that is, every two basic cells in the tableau are connected via basic cells and lines (these cells and lines form a *chain*). The basis graph of Figure 8.4 is a *tree*, that is, a connected graph with no cycles. Finally, the basis tree in Figure 8.4 is *spanning*; that is, there is a cell of the basis tree in every row and column of the matrix.

Assuming that the artificial vector is present, we shall show that any basis on the transportation tableau can be characterized as a (single) connected spanning tree with $m + n - 1$ cells. To show that any basis is a spanning tree, first recall

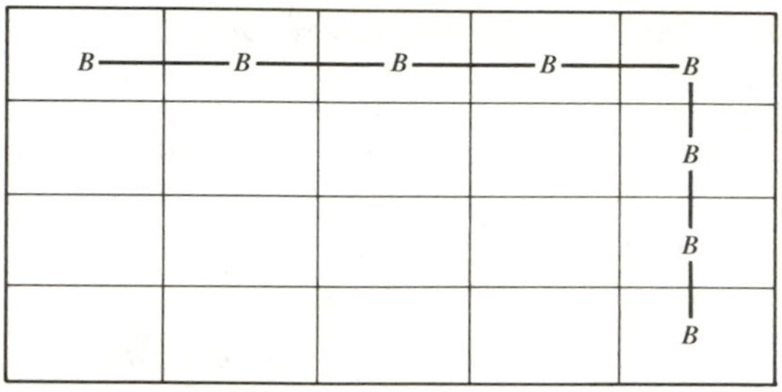

Figure 8.4. **Illustration of a basis.**

that a set of basic cells cannot contain a cycle. Thus the basis must be a single tree or several trees. It is also apparent that the basis tree or trees (*forest*) must be spanning. For suppose that the basis graph did not contain a cell in some row, i. Then the ith row of the associated matrix consists entirely of zeros, disqualifying it as a basis. Thus a basis contains at least one cell in each row. Similarly, it must contain at least one cell in each column. Finally, we must show that the basis graph is connected, and therefore, is a single tree. Consider the two basic cells (i, j) and (k, l) in Figure 8.5. If cell (k, j) is basic, then cells (i, j) and (k, l) are connected via the basic chain $\{(i, j), (k, j), (k, l)\}$. Suppose, on the other hand that cell (k, j) is not basic. We have already demonstrated that $\mathbf{a}_{kj}$ could be represented as the following linear combination of basic vectors.

$$\mathbf{a}_{kj} = \mathbf{a}_{rj} - \mathbf{a}_{rs} + \mathbf{a}_{ts} + \dots + \mathbf{a}_{vu} - \mathbf{a}_{vw} + \mathbf{a}_{kw}$$

In particular, the two basic cells (i, j) and (k, l) are connected by the basic chain $\{(i, j), (r, j), (r, s), (t, s), \dots, (v, u), (v, w), (k, w), (k, l)\}$. We have thus demonstrated that any two basic cells are connected by a chain in the basis graph and hence the basis graph is connected.

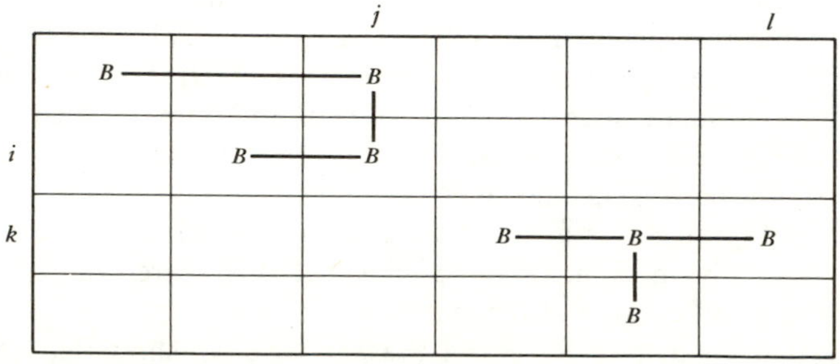

Figure 8.5. **Example of a structure not a single tree.**

To show the converse, we must demonstrate that a spanning tree with $m + n - 1$ points together with an artificial vector is a basis. We shall do this by showing that the matrix of vectors associated with a spanning tree has an $(m + n - 1) \times (m + n - 1)$ (upper) triangular submatrix with 1's on the diagonal.

We first note that a tree always possesses at least one *end*, that is, a point that has at most one line touching it. [The tree in Figure 8.4 has exactly two ends (1, 1) and (m, n).] Considering an end of the tree, it must either be the only point in the particular row or column. Suppose that the end is the only point in its particular row. In Figure 8.6, cell (i, j) is an example of such an end. In this case vector $\mathbf{a}_{ij} = \mathbf{e}_i + \mathbf{e}_{m+j}$ is the only tree vector with a nonzero entry in row i.

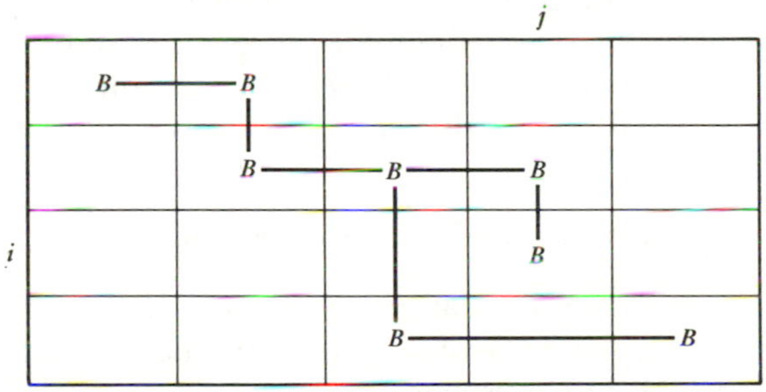

Figure 8.6. **Example of a tree with many ends.**

Perform row and column permutation on the $(m + n) \times (m + n - 1)$ matrix $\mathbf{T}$ of tree vectors so that this nonzero entry will be in the last column and last row of the matrix. Thus $\mathbf{T}$ is of the form:

$$\mathbf{T} = \left[\begin{array}{c|c} \mathbf{T}_1 & \mathbf{q} \\ \hline \mathbf{0} & 1 \end{array}\right]$$

where $\mathbf{T}_1$ is the matrix associated with the tree vectors when row i and vector $\mathbf{a}_{ij}$ are deleted. Graphically, the endpoint and line joining it are deleted to obtain a new connected tree. This is illustrated in Figure 8.7. Repeating the process, we locate an end of the reduced tree in Figure 8.7. Consider the end of cell (k, l). In this case the end is the only point in column l, so $\mathbf{a}_{kl} = \mathbf{e}_k + \mathbf{e}_{m+l}$ is the only vector with a nonzero entry in row $m + l$. Performing row and column permutations so that this nonzero entry is in the last row and column of $\mathbf{T}_1$, we obtain

$$\mathbf{T} = \left[\begin{array}{c|c|c} \mathbf{T}_2 & \mathbf{p} & \mathbf{q}_2 \\ \hline \mathbf{0} & 1 & q_1 \\ \hline \mathbf{0} & 0 & 1 \end{array}\right]$$

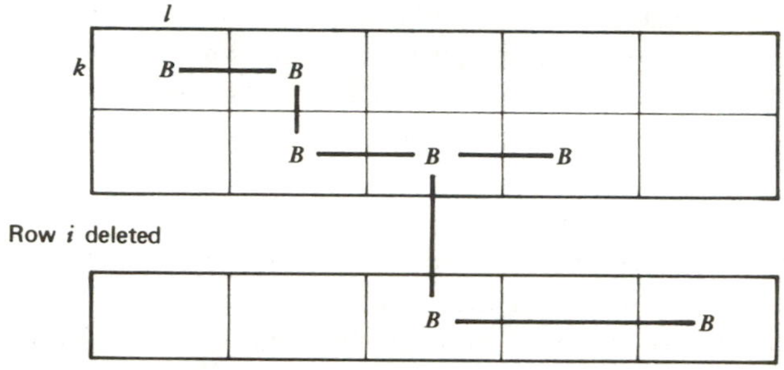

Figure 8.7. **The reduced tree of Figure 8.6.**

This process is continued. Eventually we obtain an $(m + n - 1) \times (m + n - 1)$ (upper) triangular matrix with 1's on the diagonal, in addition to an extra row with $(m + n - 1)$ components. This shows that the rank of $\mathbf{T}$ is $(m + n - 1)$, and hence $\mathbf{T}$ together with the artificial vector forms a basis of $\mathbf{A}$.

Representation of the Basis on the Transportation Graph

We have seen that a basis of the transportation problem consists of a tree on the transportation tableau plus an artificial variable, and conversely. Each tree on the transportation tableau corresponds uniquely to a tree on the transportation graph. In particular each basic cell (i, j) corresponds to the basic link (i, j) on the graph, and each line connecting two basic cells in the transportation tableau corresponds to the node (source or sink) connecting the associated links on the graph. The artificial vector $\mathbf{e}_{m+n}$ is represented on the graph by a link leaving destination n and ending nowhere. In Figure 8.8 we illustrate a basis of the transportation problem on both the transportation tableau and the transportation graph. The structure on the graph is called a *rooted spanning tree* where the root denotes the artificial variable.

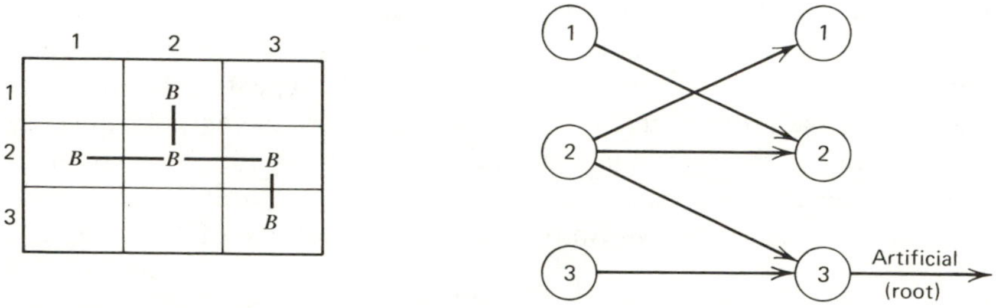

Figure 8.8. **Illustration of a basis on the transportation tableau and the transportation graph.**

From now on we shall carry all the computations on the transportation tableau. The reader should have no difficulty in drawing a parallel development directly on the graph. The graphical view will be taken in Chapter 9 on network flow problems.

8.3 REPRESENTATION OF A NONBASIC VECTOR IN TERMS OF THE BASIC VECTORS

We have determined that each nonbasic cell together with a subset of the basic cells forms a cycle and that the basic cells in this cycle provide the required representation for the nonbasic cell. Then we found that the set of basic cells formed a spanning tree on the transportation matrix. We further know that there is a unique chain between every pair of cells in the tree; otherwise cycles would be created. All of this suggests that to find the representation of a given nonbasic cell (i, j) we use the chain in the basic tree between some basic cell in row i and some basic cell in column j. This is essentially accurate except that not all basic cells in this chain (in the transportation matrix) are in the representation.

To produce the proper representation for a given nonbasic cell (variable) we simply locate the unique cycle, in the basis graph, containing the arc associated with the particular nonbasic cell. Then all of the basic cells of the transportation matrix associated with the arcs of the cycle in the graph are required for the representation of the nonbasic cell. The process of locating the representation directly on the transportation matrix is essentially the same except that not all basic cells in the unique cycle are used. In this case we use only those cells of the chain for which there is another cell of the chain in the same row and another cell of the chain in the same column.

To illustrate, consider Figure 8.9. Suppose that we want to represent $\mathbf{a}_{14}$ in terms of the basic vectors. In Figure 8.10 the unique cycle of the graph is given by (1, 4), (3, 4), (3, 1), (1, 1). Deleting the nonbasic arc (1, 4), we are left with the unique basic chain (3, 4), (3, 1), (1, 1). As we already know, these are assigned alternating signs of $+1$ and -1, giving the following representation:

$$\mathbf{a}_{14} = \mathbf{a}_{11} - \mathbf{a}_{31} + \mathbf{a}_{34}$$

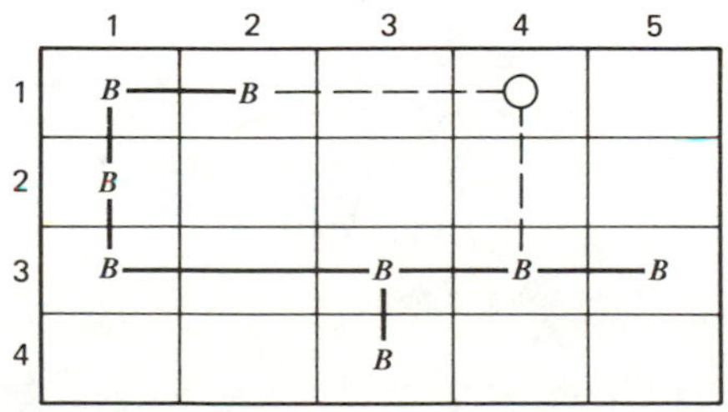

Figure 8.9. Illustration of finding the representation of a nonbasic cell.

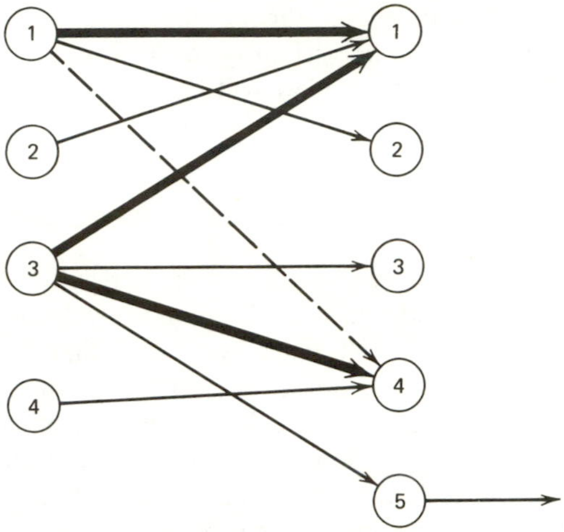

Figure 8.10. **The basis in the graph associated with the basis of Figure 8.9.**

which we may verify by

$$\mathbf{e}_1 + \mathbf{e}_{4+4} = (\mathbf{e}_1 + \mathbf{e}_{4+1}) - (\mathbf{e}_3 + \mathbf{e}_{4+1}) + (\mathbf{e}_3 + \mathbf{e}_{4+4})$$

If we had sought the representation from the transportation tableau, we would first obtain the unique cycle (1, 4), (1, 2), (1, 1), (2, 1), (3, 1), (3, 3), (3, 4). Deleting the nonbasic cell (1, 4), we are left with the unique basic chain. We delete cell (1, 2) since there is no basic cell of the chain in the same column. Similarily we delete cells (2, 1) and (3, 3). We are left with cells (1, 1), (3, 1), and (3, 4) for the representation. This is the same as before.

Ignoring the other basic points, the representation is given below.

As another example, we shall represent the vector $\mathbf{a}_{42}$ associated with the nonbasic cell (4, 2) in terms of the basic variables. Tracing the chain in the transportation tableau between the basic cells (1, 2) and (4, 3), we get

$$C = \{(1, 2), (1, 1), (2, 1), (3, 1), (3, 3), (4, 3)\}$$

for which the required basic cells are (1, 2), (1, 1), (3, 1), (3, 3), and (4, 3). Hence the representation is

$$\mathbf{a}_{42} = \mathbf{a}_{12} - \mathbf{a}_{11} + \mathbf{a}_{31} - \mathbf{a}_{33} + \mathbf{a}_{43}$$

which can easily be verified as correct. Pictorially, the representation is (ignoring other basic points) given below. Although this appears to be two cycles, it is actually one cycle through the basic cells and cell (4, 2), since the cell (3, 2), where the lines cross, is not basic.

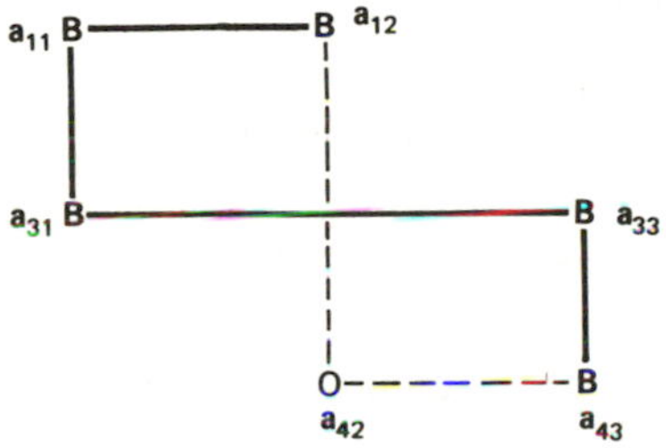

The Role of the Artificial Variable in the Transportation Problem

We note that the representation of a nonbasic vector involves only basic vectors associated with the unique chain through the basic cells. In particular, the artificial vector never becomes involved in any representation, and therefore the artificial variable will always remain zero. This fact will allow us essentially to ignore the artificial variable in the application of the simplex method to transportation problems.

8.4. THE SIMPLEX METHOD FOR TRANSPORTATION PROBLEMS

The general steps in the application of the simplex method to a linear program are as follows.

1. Find a starting basic feasible solution.
2. Compute $z_j - c_j$ for each nonbasic variable. Stop or select the entry column.
3. Determine the exit column.
4. Obtain the new basic feasible solution and repeat step 2.

We shall show how each of these steps can be carried out directly on the transportation tableau.

Finding a Starting Basic Feasible Solution

In Section 8.2 we produced a feasible solution for the transportation problem. However, the solution was not basic. While it would not be a difficult procedure to convert that solution into a basic feasible solution, we shall consider another procedure for obtaining a basic feasible solution. This procedure is called the *northwest corner rule*. During the procedure as a variable x_{ij} is assigned a value, we reduce the corresponding a_i and b_j by that value. Let the reduced values of a_i and b_j be denoted by $\hat{a}_i$ and $\hat{b}_j$ respectively. In particular, to start with, $\hat{a}_i = a_i$ and $\hat{b}_j = b_j$.

Assuming that the total supply equals the total demand, beginning in cell (1, 1) we let

$$x_{11} = \text{Minimum}\{\hat{a}_1, \hat{b}_1\}$$

and replace $\hat{a}_1$ by $\hat{a}_1 - x_{11}$ and $\hat{b}_1$ by $\hat{b}_1 - x_{11}$. Then, if $\hat{a}_1 > \hat{b}_1$, move to cell (1, 2); let

$$x_{12} = \text{Minimum}\{\hat{a}_1, \hat{b}_2\}$$

and replace $\hat{a}_1$ by $\hat{a}_1 - x_{12}$ and $\hat{b}_2$ by $\hat{b}_2 - x_{12}$. However, if $\hat{a}_1 < \hat{b}_1$, move to cell (2, 1); let

$$x_{21} = \text{Minimum}\{\hat{a}_2, \hat{b}_1\}$$

and replace $\hat{a}_2$ by $\hat{a}_2 - x_{21}$ and $\hat{b}_1$ by $\hat{b}_1 - x_{21}$. The case when $\hat{a}_1 = \hat{b}_1$ produces *degeneracy* and will be discussed later. For now we assume that equality never occurs. The process of assigning a variable the minimum of the remaining supply or demand, adjusting both, and moving to the right, or down, one cell continues until all supplies and demands are allocated. Figure 8.11 illustrates how the process might work.

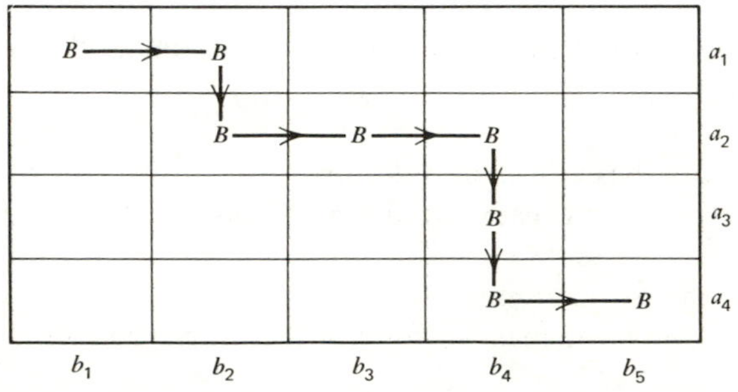

Figure 8.11. **Graphical illustration of how the northwest corner rule might allocate values to the variables.**

The northwest corner rule will (in the absence of degeneracy) produce exactly $m + n - 1$ positive x_{ij}'s. Each time an x_{ij} is assigned some positive value, either a supply or a demand constraint is satisfied. When $m + n - 1$ variables have been assigned positive values, then $m + n - 1$ of the constraints are satisfied. Noting that one of the constraints of the transportation problem is redundant, then all the constraints are met.

The graphical structure of the basic cells is obviously connected and spanning. To demonstrate that the graph is a spanning tree and therefore a basis, it remains only to show that it contains no cycles. Since, at each step, either the row or column index of the variable assigned a positive value is increased by 1, it is not possible to assign a new variable in an earlier row or column a positive value—the only way to produce cycles. Therefore the northwest corner method produces a basic feasible solution.

To illustrate, consider the transportation tableau of Figure 8.12 where the supplies and demands are indicated. We first let $x_{11} = \text{Minimum}\{\hat{a}_1, \hat{b}_1\} = 15$, decrease $\hat{a}_1$ and $\hat{b}_1$ by $x_{11} = 15$, and move to cell (1, 2) since $\hat{a}_1 > \hat{b}_1$. Next, let $x_{12} = \text{Minimum}\{\hat{a}_1, \hat{b}_2\} = \text{Minimum}\{15, 20\} = 15$, decrease $\hat{a}_1$ and $\hat{b}_2$, and move to cell (2, 2) since $\hat{a}_1 < \hat{b}_2$. This process is continued until all supplies and demands are satisfied. Notice that we do have the required number of basic variables, namely $7 = m + n - 1$. The blank cells are nonbasic and the associated variables have zero values.

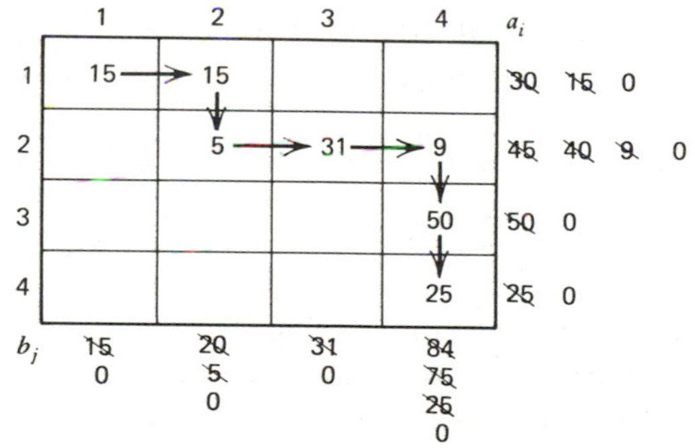

Figure 8.12. **Example of the northwest corner rule.**

Computing $z_{ij} - c_{ij}$ for Each Nonbasic Cell

Given a basic feasible solution, our next task is to determine whether that solution is optimal or to select an entering variable. Now

$$z_{ij} - c_{ij} = \mathbf{c}_B \mathbf{y}_{ij} - c_{ij}$$

We have shown how to obtain the components of $\mathbf{y}_{ij}$, that is, the coefficients of the representation of $\mathbf{a}_{ij}$ in terms of the basic vectors. Since $\mathbf{y}_{ij}$ consists of 1's, −1's, and 0's, then $\mathbf{c}_B\mathbf{y}_{ij}$ is calculated by simply adding and subtracting the costs of some basic variables. To illustrate, consider Figure 8.2. The $z_{ij} - c_{ij}$ value for the nonbasic variable x_{ij} is given by

$$z_{ij} - c_{ij} = (c_{uj} - c_{us} + c_{ls} - c_{lk} + c_{ik}) - c_{ij}$$

Using the data of the example in Section 8.1 and the basis indicated below, we get the following result.

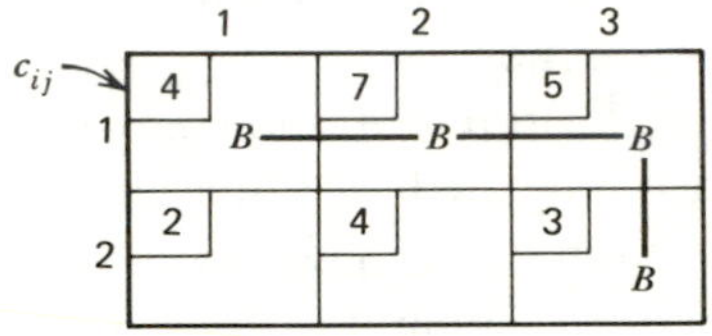

$$z_{21} - c_{21} = 4 - 5 + 3 - 2 = 0$$

$$z_{22} - c_{22} = 7 - 5 + 3 - 4 = 1$$

and x_{22} is a candidate to enter the basis.

The optimality criterion for the transportation problem is given by $z_{ij} - c_{ij} \leqslant 0$ for each nonbasic x_{ij}. A given cell (k, l) is a candidate to enter the basis if $z_{kl} - c_{kl} > 0$.

The foregoing procedure for calculating $z_{ij} - c_{ij}$ utilizes the form

$$z_{ij} - c_{ij} = \mathbf{c}_B\mathbf{B}^{-1}\mathbf{a}_{ij} - c_{ij} = \mathbf{c}_B\mathbf{y}_{ij} - c_{ij}$$

The vector $\mathbf{y}_{ij}$ is determined by constructing the unique cycle through cell (i, j) and some of the basic cells as discussed above. Hence this method is sometimes called the *cycle method*. Note, however, that calculating $z_{ij} - c_{ij}$ can be alternatively performed as follows:

$$z_{ij} - c_{ij} = \mathbf{c}_B\mathbf{B}^{-1}\mathbf{a}_{ij} - c_{ij} = \mathbf{w}\mathbf{a}_{ij} - c_{ij}$$

Let w_i, $i = 1, \ldots, m$ be denoted by u_i and w_{m+j}, $j = 1, \ldots, n$ by v_j. Then the dual vector $\mathbf{w}$ is given by

$$\mathbf{w} = (u_1, \ldots, u_m, v_1, \ldots, v_n)$$

Since $\mathbf{a}_{ij}$ has a 1 in the ith and $(m + j)$th positions, then $\mathbf{w}\mathbf{a}_{ij} = u_i + v_j$. Hence

$z_{ij} - c_{ij} = u_i + v_j - c_{ij}$. This method of calculating $z_{ij} - c_{ij}$ is thus called the *dual variable* or (u_i, v_j) method.

Since the dual vector $\mathbf{w} = \mathbf{c}_B\mathbf{B}^{-1}$, then $\mathbf{w}$ is the solution to the system

$$\mathbf{wB} = \mathbf{c}_B$$

where

$$\mathbf{B} = (\mathbf{a}_{pq}, \ldots, \mathbf{a}_{st}, \mathbf{e}_{m+n})$$

$$\mathbf{c}_B = (c_{pq}, \ldots, c_{st}, c_a)$$

and $\mathbf{a}_{pq}, \ldots, \mathbf{a}_{st}$ are $m + n - 1$ basic columns, $c_{pq}, \ldots, c_{st}$ are their corresponding cost coefficients, $\mathbf{e}_{m+n}$ is the artificial column, and c_a is its cost coefficient. Since we have previously shown that the value of the artificial variable will never vary from zero, the value of c_a does not matter. For convenience, we shall select $c_a = 0$. Since $\mathbf{B}$ is triangular, we have an easy system to solve. The system

$$(u_1, \ldots, u_m, v_1, \ldots, v_n)(\mathbf{a}_{pq}, \ldots, \mathbf{a}_{st}, \mathbf{e}_{m+n}) = (c_{pq}, \ldots, c_{st}, 0)$$

is equivalent to (since $\mathbf{a}_{ij} = \mathbf{e}_i + \mathbf{e}_{m+j}$):

$$\begin{aligned} u_p + v_q &= c_{pq} \\ &\vdots \\ u_s + v_t &= c_{st} \\ v_n &= 0 \end{aligned}$$

The foregoing system has $m + n$ variables and $m + n$ equations. Utilizing the concept of triangularity, we back-substitute the value $v_n = 0$ into each equation where v_n appears and solve for a u-variable. Using this newly found u-variable, we back-substitute to find some v, and so on.

As an illustration, consider the example problem of Section 8.1 with the basis as indicated below. The last dual variable, v_3, receives the value zero from the artificial column.

Use the artificial column		$\Rightarrow$	$v_3 = 0$
Use the basic cell (2, 3):	$u_2 + v_3 = 3$	$\Rightarrow$	$u_2 = 3$
Use the basic cell (1, 3):	$u_1 + v_3 = 5$	$\Rightarrow$	$u_1 = 5$
Use the basic cell (1, 2):	$u_1 + v_2 = 7$	$\Rightarrow$	$v_2 = 2$
Use the basic cell (1, 1):	$u_1 + v_1 = 4$	$\Rightarrow$	$v_1 = -1$

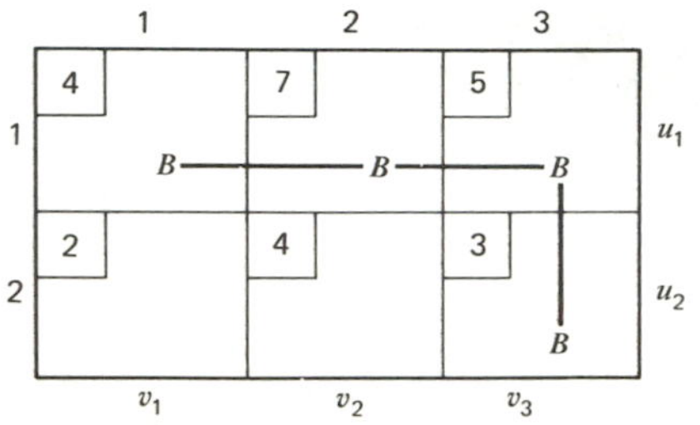

Only the basic cells are used to solve for the dual variables as shown above. Given the **w** vector, we may compute the $z_{ij} - c_{ij}$ for each nonbasic cell in order to determine the entering column. In particular:

$$\begin{aligned} z_{21} - c_{21} &= u_2 + v_1 - c_{21} \\ &= 3 + (-1) - 2 = 0 \\ z_{22} - c_{22} &= u_2 + v_2 - c_{22} \\ &= 3 + 2 - 4 = 1 \end{aligned}$$

and, again, we see that x_{22} is a candidate to enter the basis.

Determination of the Exit Column

Once a column (cell), say (k, l), has been selected to enter the basis, it is an easy matter to determine the exit column. Recall that the coefficients in the basic representation for that column are the negatives of the rates of change of the corresponding basic variables with a unit increase in the nonbasic (entering) variable. Thus if the entry, in column $\mathbf{y}_{kl}$, corresponding to a basic variable is -1, then the basic variable will increase at the same rate as the nonbasic variable x_{kl} increases. If this entry is $+1$, then the basic variable will decrease at the same rate as the nonbasic variable x_{kl} increases.

Let $\hat{x}_{ij}$ be the value of x_{ij} in the current solution and let Δ be the amount by which the nonbasic variable, x_{kl} increases. Since each component of $\mathbf{y}_{kl}$ is either 1, -1, or 0, then the usual minimum ratio test gives

$$\Delta = \text{Minimum } \{\hat{x}_{ij} : \text{basic cell } (i, j) \text{ has a } +1 \text{ in the representation of the nonbasic cell } (k, l)\}$$

Given Δ, we proceed to adjust the values of the variables around the cycle by this amount, according to the sign of the coefficient in the representation. For the example of Section 8.1 with the basis indicated below, as x_{22} enters the basis, we get the following result.

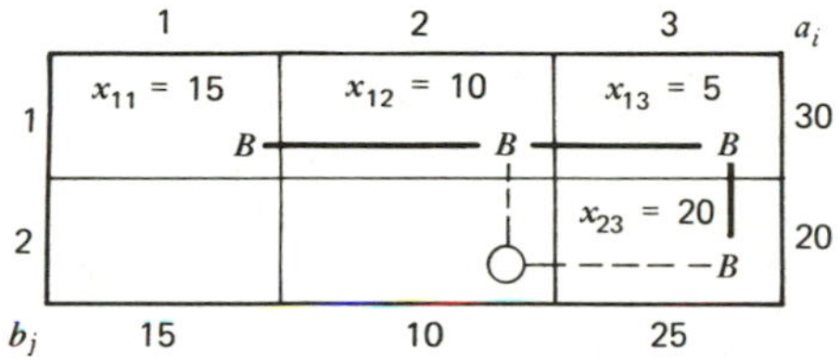

$$\Delta = \text{Minimum}\{x_{12}, x_{23}\} = \text{Minimum}\{10, 20\} = 10$$

The new solution is given by

$$x_{12} = \hat{x}_{12} - \Delta = 10 - 10 = 0 \quad \text{(leaves the basis)}$$

$$x_{13} = \hat{x}_{13} + \Delta = 5 + 10 = 15$$

$$x_{23} = \hat{x}_{23} - \Delta = 20 - 10 = 10$$

$$x_{22} = \Delta \qquad = 10$$

$$x_{11} = 15 \qquad \text{(unchanged)}$$

The new basis is given by the following.

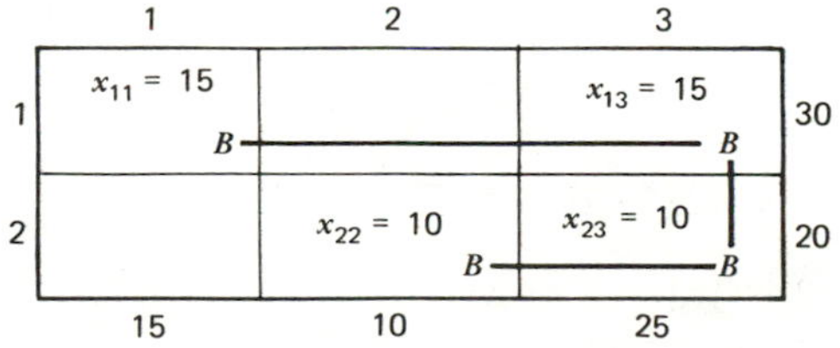

8.5 AN EXAMPLE OF THE TRANSPORTATION ALGORITHM

Consider the transportation problem indicated by the data of Figure 8.13. The number in the upper left hand corner of each cell is the cost associated with the particular variable.

	1	2	3	4	5	6	a_i
1	10	12	13	8	14	19	18
2	15	18	12	16	19	20	22
3	17	16	13	14	10	18	39
4	19	18	20	21	12	13	14
b_j	10	11	13	20	24	15	

Figure 8.13. Example data for a transportation problem.

The starting basic feasible solution produced by the northwest corner method is illustrated in Figure 8.14.

	1	2	3	4	5	6
1	10	8				
2		3	13	6		
3				14	24	1
4						14

Figure 8.14. The northwest corner basic solution.

Beginning with (1, 3), we price out each of the nonbasic cells by the cycle method:

$$z_{13} - c_{13} = (12 - 18 + 12) - 13 = -7$$

$$z_{14} - c_{14} = (12 - 18 + 16) - 8 = 2$$

$$z_{15} - c_{15} = (12 - 18 + 16 - 14 + 10) - 14 = -8$$

$$\vdots$$

$$z_{45} - c_{45} = (10 - 18 + 13) - 12 = -7$$

The current values of the basic variables and the tree are shown in Figure 8.15; the $z_{ij} - c_{ij}$ values for the nonbasic variables are the circled values. Since

	1	2	3	4	5	6	
1	10 B	8 B	$\boxed{-7}$	$\boxed{2}$	$\boxed{-8}$	$\boxed{-5}$	$u_1 = 14$
2	$\boxed{1}$	3 B	13 B	6 B	$\boxed{-7}$	$\boxed{0}$	$u_2 = 20$
3	$\boxed{-3}$	$\boxed{0}$	$\boxed{-3}$	14 B	24 B	1 B	$u_3 = 18$
4	$\boxed{-10}$	$\boxed{-7}$	$\boxed{-15}$	$\boxed{-12}$	$\boxed{-7}$	14 B	$u_4 = 13$
	$v_1 = -4$	$v_2 = -2$	$v_3 = -8$	$v_4 = -4$	$v_5 = -8$	$v_6 = 0$	

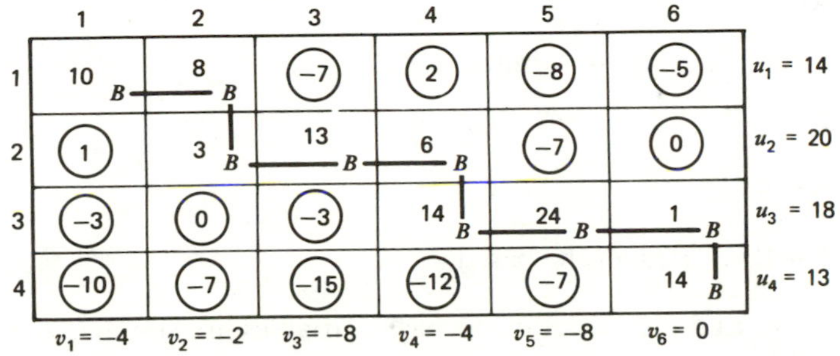

Figure 8.15. The $z_{ij} - c_{ij}$ values for the nonbasic cells.

$z_{ij} - c_{ij} = 0$ for the basic variables, these are not indicated. Note that the $z_{ij} - c_{ij}$ values could have been alternatively calculated as follows. First solve for the dual variables (their values are summarized in Figure 8.15).

artificial column $\Rightarrow v_6 = 0$

basic cell (4, 6): $u_4 + v_6 = 13 \Rightarrow u_4 = 13$

basic cell (3, 6): $u_3 + v_6 = 18 \Rightarrow u_3 = 18$

basic cell (3, 5): $u_3 + v_5 = 10 \Rightarrow v_5 = -8$

basic cell (3, 4): $u_3 + v_4 = 14 \Rightarrow v_4 = -4$

basic cell (2, 4): $u_2 + v_4 = 16 \Rightarrow u_2 = 20$

basic cell (2, 3): $u_2 + v_3 = 12 \Rightarrow v_3 = -8$

basic cell (2, 2): $u_2 + v_2 = 18 \Rightarrow v_2 = -2$

basic cell (1, 2): $u_1 + v_2 = 12 \Rightarrow u_1 = 14$

basic cell (1, 1): $u_1 + v_1 = 10 \Rightarrow v_1 = -4$

Then $z_{ij} - c_{ij} = u_i + v_j - c_{ij}$, for example, $z_{14} - c_{14} = u_1 + v_4 - c_{14} = 14 - 4 - 8 = 2$. Using either alternative for calculating $z_{ij} - c_{ij}$, Figure 8.15 indicates that the maximal $z_{ij} - c_{ij}$ is $z_{14} - c_{14} = 2$. Therefore x_{14} enters the basis. Referring to Figure 8.15, we see that

$$a_{14} = a_{12} - a_{22} + a_{24}$$

and the corresponding cycle is as follows.

From this we find

$x_{14} = \text{Minimum}\{\hat{x}_{12}, \hat{x}_{24}\} = \text{Minimum}\{8, 6\} = 6$

$x_{12} = 8 - 6 = 2$

$x_{22} = 3 + 6 = 9$

$x_{24} = 6 - 6 = 0$ (leaves the basis)

The new basis and the values of the basic variables are summarized in Figure 8.16. Using either the cycle method or the dual variable method, $z_{ij} - c_{ij}$ is calculated for each nonbasic variable.

	1	2	3	4	5	6	
1	10 B	2 B	(−7)	6 B	(−10)	(−7)	$u_1 = 12$
2	(1)	9 B	13 B	(−2)	(−9)	(−2)	$u_2 = 18$
3	(−1)	(2)	(−1)	14 B	24 B	1 B	$u_3 = 18$
4	(−8)	(−5)	(−13)	(−12)	(−7)	14 B	$u_4 = 13$
	$v_1 = -2$	$v_2 = 0$	$v_3 = -6$	$v_4 = -4$	$v_5 = -8$	$v_6 = 0$	

Figure 8.16. Second basic feasible solution.

As indicated by the circled entries in Figure 8.16, cell (3, 2) is the entry cell. The cycle associated with (3, 2) is as follows.

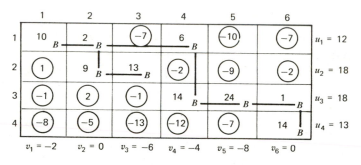

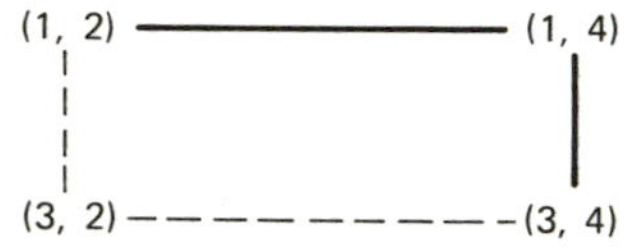

From this we obtain

$x_{32} = \text{Minimum}\{\hat{x}_{12}, \hat{x}_{34}\} = \text{Minimum}\{2, 14\} = 2$

$x_{12} = 2 - 2 = 0$ (leaves the basis)

$x_{14} = 6 + 2 = 8$

$x_{34} = 14 - 2 = 12$

The new basic feasible solution and the new $z_{ij} - c_{ij}$ values are given in Figure 8.17. Examining the $z_{ij} - c_{ij}$ entries, we find that (2, 1) is the entry cell.

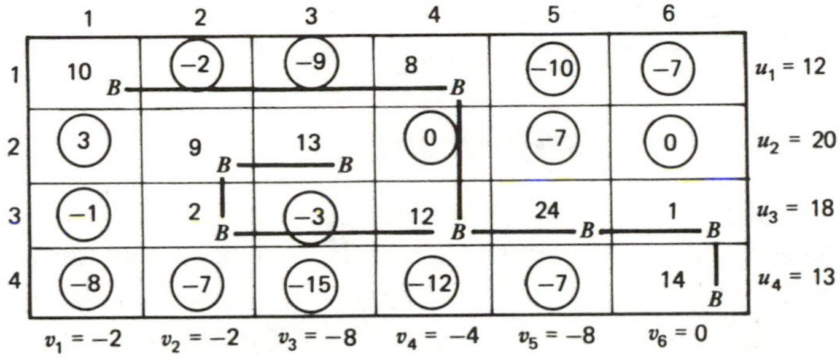

Figure 8.17. **Third basic feasible solution.**

The associated cycle is as follows.

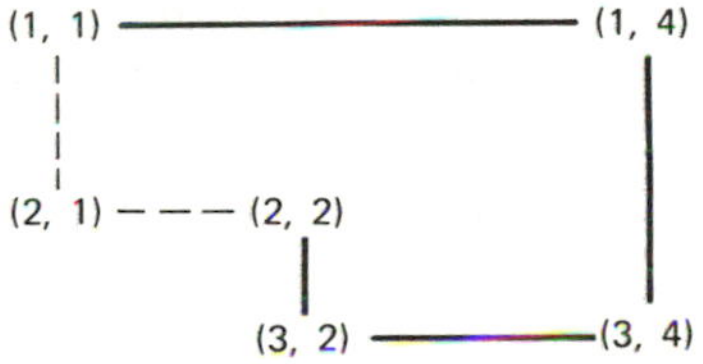

From this we obtain

$x_{21} = \text{Minimum}\{\hat{x}_{11}, \hat{x}_{34}, \hat{x}_{22}\} = \text{Minimum}\{10, 12, 9\} = 9$

$x_{11} = 10 - 9 = 1$

$x_{14} = 8 + 9 = 17$

$x_{34} = 12 - 9 = 3$

$x_{32} = 2 + 9 = 11$

$x_{22} = 9 - 9 = 0$ (leaves the basis)

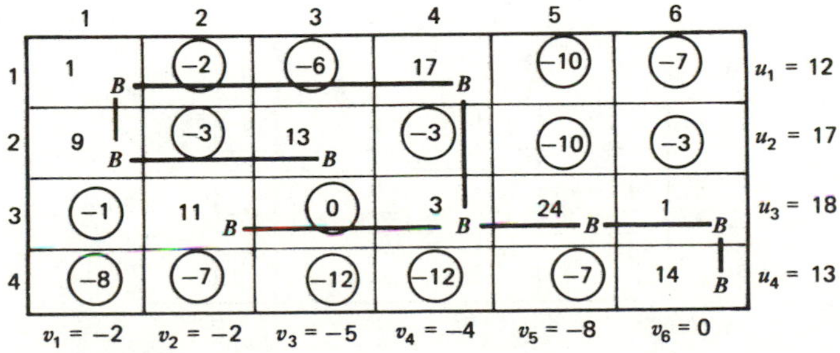

Figure 8.18. **Fourth basic feasible solution.**

Figure 8.18 presents the new basic feasible solution and the new $z_{ij} - c_{ij}$'s. Since $z_{ij} - c_{ij} \leqslant 0$ for each nonbasic variable, the indicated solution is optimal.

8.6 DEGENERACY IN THE TRANSPORTATION PROBLEM

As with any linear programming problem, degeneracy can occur in the transportation problem. Also, analogous to general linear programs, degeneracy and cycling do not represent a serious difficulty in the transportation problem. For completeness we shall examine degeneracy in the transportation problem.

Finding and Maintaining a Basis in the Presence of Degeneracy

In Section 8.4 the northwest corner rule was presented for finding an initial basic feasible solution to the transportation problem. In that section we assumed that we would not reach a point where the reduced supply is equal to the reduced demand. If we relax this assumption, it is still an easy matter to find a starting basis by the northwest corner rule.

Suppose that at some stage in the application of the northwest corner rule we have

$$x_{kl} = \text{Minimum}\{\hat{a}_k, \hat{b}_l\} = \hat{a}_k = \hat{b}_l$$

where either $\hat{a}_k$ or $\hat{b}_l$ was reduced by a previous calculation of $x_{k,\,l-1}$ or $x_{k-1,\,l}$ respectively. Whichever way we go with the northwest corner rule, the next basic variable will be zero and degeneracy occurs. A practical method for obtaining a starting basis is to proceed in either direction, that is, to $(k, l+1)$ or $(k+1, l)$, and assign either $x_{k,\,l+1}$ or $x_{k+1,\,l}$ as a basic variable at zero level. Basic variables at zero level are treated in exactly the same fashion as other basic variables. The northwest corner rule produces a basic feasible solution even in the presence of degeneracy.

Let us illustrate the foregoing concepts with an example. Consider the example of Figure 8.19. Applying the northwest corner rule to this example we

	1	2	3	4	a_i
1	2	3	4	9	20
2	14	12	5	1	30
3	12	15	9	3	40
b_j	10	10	20	50	

Figure 8.19. **An example of degeneracy.**

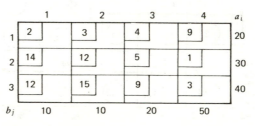

obtain the sequence of calculations:

$$x_{11} = \text{Minimum}\{\hat{a}_1, \hat{b}_1\} = \text{Minimum}\{20, 10\} = 10$$

$$\hat{a}_1 = 20 - 10 = 10, \hat{b}_1 = 10 - 10 = 0$$

$$x_{12} = \text{Minimum}\{\hat{a}_1, \hat{b}_2\} = \text{Minimum}\{10, 10\} = 10$$

$$\hat{a}_1 = 10 - 10 = 0, \hat{b}_2 = 10 - 10 = 0$$

At this point we may move to either (2, 2), or to (1, 3). Suppose that we move to (2, 2).

$$x_{22} = \text{minimum}\{\hat{a}_2, \hat{b}_2\} = \text{Minimum}\{30, 0\} = 0$$

$$\hat{a}_2 = 30 - 0 = 30, \hat{b}_2 = 0 - 0 = 0$$

$$x_{23} = \text{Minimum}\{\hat{a}_2, \hat{b}_3\} = \text{Minimum}\{30, 20\} = 20$$

$$\hat{a}_2 = 30 - 20 = 10, \hat{b}_3 = 20 - 20 = 0$$

$$x_{24} = \text{Minimum}\{\hat{a}_2, \hat{b}_4\} = \text{Minimum}\{10, 50\} = 10$$

$$\hat{a}_2 = 10 - 10 = 0, \hat{b}_4 = 50 - 10 = 40$$

$$x_{34} = \text{Minimum}\{\hat{a}_3, \hat{b}_4\} = \text{Minimum}\{40, 40\} = 40$$

$$\hat{a}_3 = 40 - 40 = 0, \hat{b}_4 = 40 - 40 = 0$$

All other x_{ij}'s are nonbasic and are assigned value zero. The initial basic feasible solution is given in Figure 8.20. As required there are $m + n - 1 = 3 + 4 - 1 = 6$ basic variables forming a connected tree. Note, however, that the basic feasible solution is degenerate since the basic variable $x_{22} = 0$. For each nonbasic variable we calculate $z_{ij} - c_{ij}$ by either the cycle method or the dual variables method. These values are depicted in circles in Figure 8.20 for the

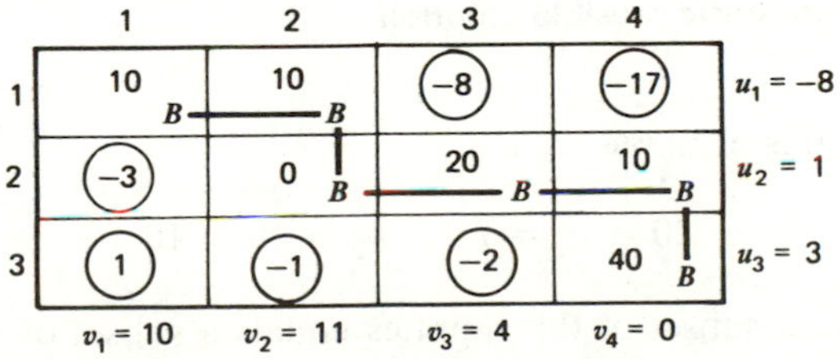

Figure 8.20. **Initial (degenerate) basic feasible solution.**

nonbasic variables. Since $z_{31} - c_{31} = 1$, then x_{31} enters the basis. The corresponding cycle is as follows.

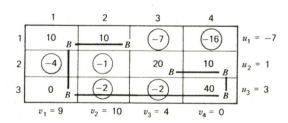

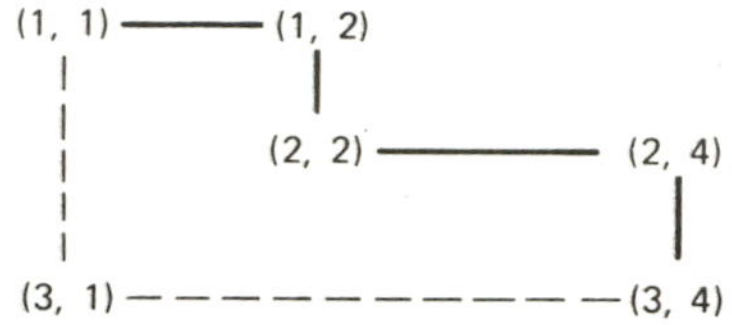

$$x_{31} = \text{Minimum}\{\hat{x}_{11}, \hat{x}_{22}, \hat{x}_{34}\} = \text{Minimum}\{10, 0, 40\} = 0$$

$$x_{11} = 10 - 0 = 10$$

$$x_{21} = 10 + 0 = 10$$

$$x_{22} = 0 - 0 = 0 \qquad \text{(leaves the basis)}$$

$$x_{24} = 10 + 0 = 10$$

$$x_{34} = 40 - 0 = 40$$

Note that x_{22} leaves the basis and x_{31} enters the basis at zero level. We have the same extreme point but a different basis. The new basis and the new $z_{ij} - c_{ij}$'s for the nonbasic variables are shown in Figure 8.21. Since $z_{ij} - c_{ij} \leq 0$ for each nonbasic variable, the current solution is optimal.

	1	2	3	4	
1	10 B	10 B	(−7)	(−16)	$u_1 = -7$
2	(−4)	(−1)	20 B	10 B	$u_2 = 1$
3	0 B	(−2)	(−2)	40 B	$u_3 = 3$
	$v_1 = 9$	$v_2 = 10$	$v_3 = 4$	$v_4 = 0$	

Figure 8.21. **Optimal basic feasible solution.**

Notice that in this example we have

$$20 = a_1 = b_1 + b_2 = 10 + 10$$

or, in other words, a subset of the supplies equals a subset of the demands. We shall show that this is always true when degeneracy occurs in the transportation problem.

A Necessary Condition for Degeneracy in the Transportation Problem

Suppose that at some iteration of the transportation algorithm we obtain a degenerate basic feasible solution, as that shown in Figure 8.22. Deleting one of the degenerate cells separates the connected tree into several components as shown in Figure 8.23. Sum the supply constraints over the variables in one of the components, say, C_1, to obtain

$$\sum_{C_1} x_{ij} = \sum_{C_1} a_i$$

Summing the demand constraints over the variables in the same component, we get

$$\sum_{C_1} x_{ij} = \sum_{C_1} b_j$$

Together, these constraints imply

$$\sum_{C_1} a_i = \sum_{C_1} b_j$$

Thus a necessary condition for the presence of degeneracy is that a proper

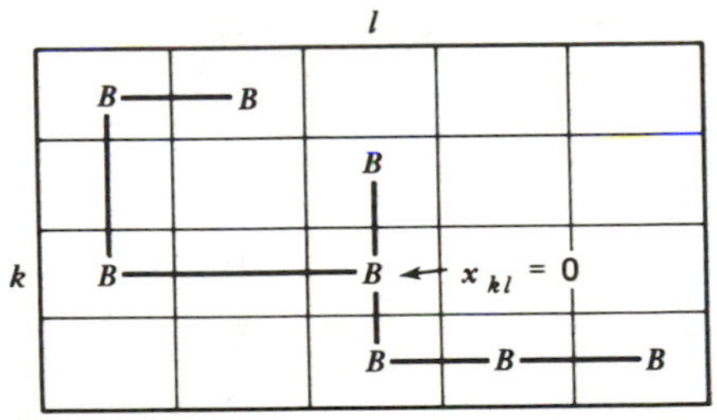

Figure 8.22. A degenerate basis.

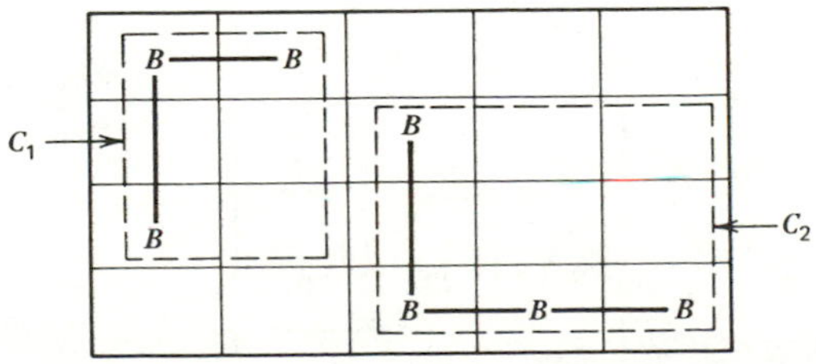

Figure 8.23. Components created by deleting a zero basic cell.

subset of the rows and columns have the total supply equal to the total demand. Therefore degeneracy (and cycling) can be eliminated if no such proper subset exists (see Exercise 8.43).

8.7 THE SIMPLEX TABLEAU ASSOCIATED WITH A TRANSPORTATION TABLEAU

We have all of the information available to construct the simplex tableau associated with a transportation tableau if we so desire. In Section 8.3 a method was described for calculating the updated column vectors $\mathbf{y}_{ij}$. In Section 8.4 two methods were presented for calculating $z_{ij} - c_{ij}$ for a nonbasic variable x_{ij}. This information together with the basic solution provides all the necessary entries in the simplex tableau.

As an example, consider the transportation problem given by the following data.

c_{ij}	1	2	3	a_i
1	10	11	20	20
2	6	9	8	20
b_j	15	15	10	

Figure 8.24 presents the initial transportation tableau and the associated simplex tableau (including the artificial variable). Examining either tableau, we see that x_{21} enters the basis and x_{22} leaves. We ask the reader to verify the entries in the simplex tableau by generating the $z_{ij} - c_{ij}$'s and the $\mathbf{y}_{ij}$'s from the transportation tableau.

	z	x_{11}	x_{12}	x_{13}	x_{21}	x_{22}	x_{23}	x_a	RHS
z	1	0	0	−10	2	0	0	0	375
x_{11}	0	1	0	0	1	0	0	0	15
x_{12}	0	0	1	1	−1	0	0	0	5
x_{22}	0	0	0	−1	1	1	0	0	10
x_{23}	0	0	0	1	0	0	1	0	10
x_a	0	0	0	0	0	0	0	1	0

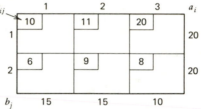

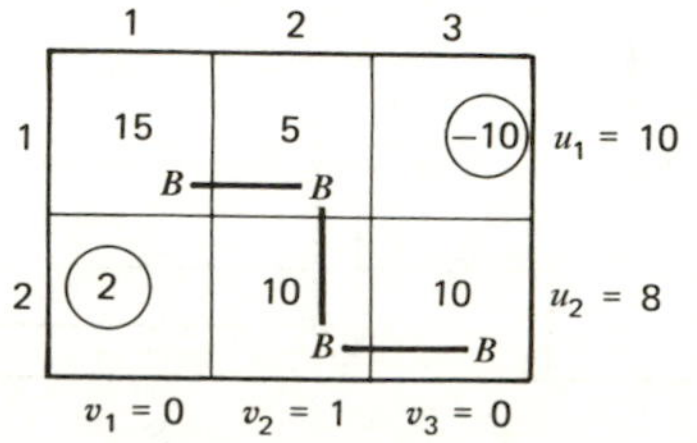

Figure 8.24. An initial transportation tableau and the associated simplex tableau.

8.8 THE ASSIGNMENT PROBLEM

An important special case of the transportation problem is the case where $m = n$ and each $a_i = 1$ and each $b_j = 1$. This special case is called the *assignment problem*. As an example, suppose we have m individuals and m jobs. If individual i is assigned to job j, the cost incurred will be c_{ij}. We wish to find the minimal cost assignment of individuals to jobs. In each basic feasible solution $x_{ij} = 1$ means that individual i is assigned to job j, $x_{ij} = 0$ indicates that individual i is not assigned to job j.

Since the assignment problem is a special case of the transportation problem, we could apply the transportation procedure developed in this chapter. Note, however, as will be discussed in more detail below, that the constraints of the assignment problem admit exactly m positive variables at each basic feasible solution. The number of basic variables is $2m - 1$. Thus if the transportation algorithm is used we would have $m - 1$ basic variables at zero level, leading to a highly degenerate problem. In this section we shall exploit the special structure of the assignment problem to get a more efficient algorithm.

A mathematical model for the assignment problem is given by the following.

$$\begin{aligned}
\text{Minimize} \quad & \sum_{i=1}^{m} \sum_{j=1}^{m} c_{ij} x_{ij} \\
\text{Subject to} \quad & \sum_{j=1}^{m} x_{ij} = 1 \qquad i = 1, \ldots, m \\
& \sum_{i=1}^{m} x_{ij} = 1 \qquad j = 1, \ldots, m \\
& x_{ij} = 0 \text{ or } 1 \qquad i,j = 1, \ldots, m
\end{aligned}$$

In matrix form, the assignment problem becomes as follows.

$$\begin{aligned}
\text{Minimize} \quad & \mathbf{cx} \\
\text{Subject to} \quad & \mathbf{Ax} = \mathbf{1} \\
& x_{ij} = 0 \text{ or } 1 \qquad i, j = 1, \ldots, m
\end{aligned}$$

where $\mathbf{x} = (x_{11}, \ldots, x_{1m}, \ldots, x_{m1}, \ldots, x_{mm})^t$, and $\mathbf{A}$ is a $2m \times m^2$ matrix whose (i, j) column is $\mathbf{a}_{ij} = \mathbf{e}_i + \mathbf{e}_{m+j}$ for $i = 1, 2, \ldots, m$ and $j = 1, 2, \ldots, m$. Thus we see that $\mathbf{A}$ is the same constraint matrix as that for the transportation problem. Applying the total unimodularity property of $\mathbf{A}$, we know that an optimal basic feasible solution to the assignment problem with the constraint $x_{ij} = 0$ or 1 replaced by $x_{ij} \geqslant 0$ will be all integer. Furthermore, as a result of the constraints no x_{ij} can exceed 1. Hence all x_{ij} will be either zero or 1 in an optimal solution to the linear program. This permits us to replace the constraint $x_{ij} = 0$ or 1 by the constraint $x_{ij} \geqslant 0$. Thus we obtain the following.

$$\begin{aligned} \text{Minimize} \quad & \mathbf{cx} \\ \text{Subject to} \quad & \mathbf{Ax} = \mathbf{1} \\ & \mathbf{x} \geqslant \mathbf{0} \end{aligned}$$

The Dual Problem

The dual of the assignment problem with the nonnegativity restrictions replacing the 0 − 1 constraints becomes as follows.

$$\text{Maximize} \quad \sum_{i=1}^{m} u_i + \sum_{j=1}^{m} v_j$$

$$\begin{aligned} \text{Subject to} \quad & u_i + v_i \leqslant c_{ij} & & i, j = 1, \ldots, m \\ & u_i, v_j \quad \text{unrestricted} & & i, j = 1, 2, \ldots, m \end{aligned}$$

The complementary slackness conditions are given by

$$(c_{ij} - u_i - v_j)x_{ij} = 0 \qquad i, j = 1, \ldots, m$$

Thus, if we can find a set of feasible u's, v's, and x's that satisfy complementary slackness, those u's, v's, and x's will be optimal.

A feasible dual solution is given by

$$\hat{u}_i = \underset{1 \leqslant j \leqslant m}{\text{Minimum}} \{c_{ij}\} \qquad i = 1, \ldots, m$$

$$\hat{v}_j = \underset{1 \leqslant i \leqslant m}{\text{Minimum}} \{c_{ij} - \hat{u}_i\} \qquad j = 1, \ldots, m$$

From this we see that $\hat{u}_i$ is the minimum c_{ij} in row i and $\hat{v}_j$ is the minimum $c_{ij} - \hat{u}_i$ in column j.

The Reduced Matrix

Consider a reduced cost coefficient matrix where c_{ij} is replaced by $\hat{c}_{ij} = c_{ij} - u_i - v_j$. In other words, the reduced cost matrix is obtained by first subtracting from each row the minimum in that row; and then on the resulting matrix subtracting from each column the minimum in that column. The reduced matrix will have a zero in every row and column and all of its entries will be nonnegative. The reduced matrix is actually the matrix of dual slack variables (why?).

Suppose that we can find a feasible set of x_{ij}'s such that each x_{ij} with value 1 is associated with a zero cell of the reduced matrix. Then by complementary slackness we conclude that we have the optimal solution. What, then, constitutes a set of feasible x_{ij}'s? Reviewing the constraints of the assignment problem, it is clear that we must have exactly one x_{ij} in each row equal to 1 and exactly one x_{ij} in each column equal to 1. Thus in a feasible solution there will be exactly m of the x_{ij}'s equal to 1, the rest being zero.

Let us illustrate the forgoing ideas with an example. Consider the following cost coefficient matrix for an assignment problem.

	1	2	3	4	ROW MINIMUM
1	3	2	5	4	2
2	0	1	2	3	0
3	4	1	− 1	3	− 1
4	2	5	3	4	2

Subtracting the row minimum from each element in the row, we get the following tableau.

	1	2	3	4	
1	1	0	3	2	
2	0	1	2	3	
3	5	2	0	4	
4	0	3	1	2	
	0	0	0	2	COLUMN MINIMUM

Subtracting the column minimum in the new matrix from each element in the column, we get the reduced matrix as follows.

	1	2	3	4	
1	1	[0]	3	0	
2	[0]	1	2	1	$(= \hat{c}_{ij})$
3	5	2	[0]	2	
4	0	3	1	[0]	

Now if we let $x^*_{12} = x^*_{21} = x^*_{33} = x^*_{44} = 1$ and if we let all other x^*_{ij}'s be zero, then we have a feasible solution with positive x_{ij}'s associated with zero cells of the reduced matrix, thus producing an optimal solution.

It is not always so easy to find an optimal solution. Take, for example, the following cost matrix.

	1	2	3	
1	2	5	7	
2	4	2	1	$(= c_{ij})$
3	2	6	5	

The reduced matrix is given by:

	1	2	3	
1	0	2	5	
2	3	0	0	$(= \hat{c}_{ij})$
3	0	3	3	

Here it is not possible to set three of the x_{ij}'s equal to 1 such that all three positive x_{ij}'s occur in zero cells and no two positive x_{ij}'s occur in the same row or column.

A Partial Solution

Notice that for the reduced matrix above, the maximum number of x_{ij}'s, from among the zero cells, which can be set equal to 1 without any two positive x_{ij}'s occurring in the same row or column is 2. For example, we might let $x_{11} = x_{22} = 1$, $x_{11} = x_{23} = 1$, $x_{31} = x_{22} = 1$, or $x_{31} = x_{23} = 1$. In this case the maximum number of cells with zero $\hat{c}_{ij}$ such that no two cells occupy the same row or column is 2. The corresponding cells are called *independent*. Notice also that if one were to draw a set of lines through the rows and columns to *cover* the zeros so that there is at least one line through each zero, the minimum number of such lines for this matrix is 2, a line through column 1 and a line through row 2.

	1	2	3
1	0	2	5
2	3	0	0
3	0	3	3

We see in this example that the maximum number of independent zero cells and the minimum number of lines required to cover the zeros are equal. This result, which is true in general, is given by the following theorem. We shall not prove this theorem here. (In Chapter 11, Exercise 11.15 asks the reader to show that this theorem is a special case of the maximal flow–minimal cut theorem for networks. At that time we also suggest a method for systematically finding the required number of lines. Also see Exercise 8.33 and Exercise 8.37.)

Theorem 1

The maximum number of independent zero cells in a reduced assignment matrix is equal to the minimum number of lines to cover all zeros in the matrix.

Modifying the Reduced Matrix

Suppose that we have not yet obtained the optimal solution, that is, cannot find a feasible set of positive x_{ij}'s from among the zero cells of the reduced matrix. Consider the covered matrix, the reduced matrix with zeros covered by the fewest number of lines. Let k be the number of lines required. Also let $S_r = \{i_1, i_2, \ldots\}$ be the set of uncovered rows and $S_c = \{j_1, j_2, \ldots\}$ be the set of uncovered columns. Define $\bar{S}_r = M - S_r$ and $\bar{S}_c = M - S_c$ where $M = \{1, 2, \ldots, m\}$. Finally, let p be the number of rows in S_r and q the number of columns in S_c. Then $k = (m - p) + (m - q)$.

Let c_0 be the minimum uncovered element, that is,

$$c_0 = \underset{\substack{i \in S_r \\ j \in S_c}}{\text{Minimum}} \{\hat{c}_{ij}\} > 0$$

It can be easily demonstrated that a new dual feasible solution is given by

$$\begin{aligned}
\bar{u}_i &= \hat{u}_i + c_0 & i &\in S_r \\
\bar{u}_i &= \hat{u}_i & i &\in \bar{S}_r \\
\bar{v}_j &= \hat{v}_j & j &\in S_c \\
\bar{v}_j &= \hat{v}_j - c_0 & j &\in \bar{S}_c
\end{aligned}$$

In the reduced matrix this is equivalent to subtracting c_0 from each uncovered row and adding c_0 to each covered column. Another way to view this is that c_0 is subtracted from each uncovered element and added to each twice-covered element. The new reduced cost coefficient matrix has nonnegative elements and a zero in every row and column (why?).

For the previous 3×3 matrix we have $c_0 = \text{Minimum } \{2, 5, 3, 3\} = 2$ and the new reduced cost matrix is given by

	1	2	3
1	0	[0]	3
2	5	0	[0]
3	[0]	1	1

Notice that now a feasible set of x_{ij}'s exists with positive x_{ij}'s associated with zero cells (zero dual slack variables).

Note that primal feasibility is attained, dual feasibility is maintained (since the entries in the reduced cost matrix are nonnegative), and finally complementary slackness holds (since $x_{ij} = 1$ only if the corresponding dual slack is zero). Thus the Kuhn-Tucker conditions hold and the optimal solution, $x_{12}^* = x_{23}^* = x_{31}^* = 1$ (all other x_{ij}^*'s set equal to 0) is at hand.

Summary of the Assignment Algorithm

The algorithm developed in this section is called the *Hungarian algorithm* and is summarized as follows.

INITIALIZATION STEP

For each row of the cost matrix, subtract the minimum element in the row from each element in the row. For each column of the resulting matrix, subtract the minimum element in the column from each element in the column. The result is a reduced matrix.

MAIN STEP

1. Draw the minimum number of lines through the rows and columns to cover all zeros in the reduced matrix. If the minimum number of lines is m, then an optimal solution is available. Otherwise to go step 2.
2. Select the minimum uncovered element. Subtract this element from each uncovered element and add it to each twice-covered element. Return to step 1.

An Example

Consider the following cost matrix.

	1	2	3	4	5
1	2	3	5	1	4
2	− 1	1	3	6	2
3	− 2	4	3	5	0
4	1	3	4	1	4
5	7	1	2	1	2

The reduced matrix is as follows.

	1	2	3	4	5
1	1	2	3	0	2
2	0	2	3	7	2
3	0	6	4	7	1
4	0	2	2	0	2
5	6	0	0	0	0

Here the minimum number of lines to cover all zeros is 3. The minimum uncovered element is 1. Subtract this from each uncovered element and add it to each twice-covered element.

	1	2	3	4	5
1	1	1	2	0	1
2	0	1	2	7	1
3	0	5	3	7	0
4	0	1	1	0	1
5	7	0	0	1	0

Again, we do not have an optimal solution at hand. The minimum uncovered element is 1. Subtract 1 from each uncovered element and add it to each twice covered element. This leads to the following matrix.

	1	2	3	4	5
1	1	$\boxed{0}$	1	0	1
2	$\boxed{0}$	0	1	7	1
3	0	4	2	7	$\boxed{0}$
4	0	0	0	$\boxed{0}$	1
5	8	0	$\boxed{0}$	2	1

In this matrix an optimal solution is given by $x_{12}^* = x_{21}^* = x_{35}^* = x_{44}^* = x_{53}^* = 1$ and all other x_{ij}^*'s set equal to zero.

In Exercise 8.35 we ask the reader to show that the Hungarian method for assignment problems is precisely the primal-dual algorithm applied to the assignment problem.

Finite Convergence of the Assignment Algorithm

If we could not find a feasible set of x_{ij}'s from among the zero cells of the new matrix, then we would repeat the process of drawing lines and adjusting the matrix. We can only do this a finite number of times before an optimal solution can be found. Clearly, an optimal solution can be found if all reduced costs become zero. To show finiteness we note that the reduced costs are always nonnegative and that

$$
\begin{aligned}
\sum_i \sum_j \hat{c}_{ij} - \sum_i \sum_j \bar{c}_{ij} &= \sum_{(S_r, S_c)} (\hat{c}_{ij} - \bar{c}_{ij}) + \sum_{(S_r, \bar{S}_c)} (\hat{c}_{ij} - \bar{c}_{ij}) \\
&\quad + \sum_{(\bar{S}_r, S_c)} (\hat{c}_{ij} - \bar{c}_{ij}) + \sum_{(\bar{S}_r, \bar{S}_c)} (\hat{c}_{ij} - \bar{c}_{ij}) \\
&= \sum_{(S_r, S_c)} c_0 + \sum_{(S_r, \bar{S}_c)} 0 + \sum_{(\bar{S}_r, S_c)} 0 + \sum_{(\bar{S}_r, \bar{S}_c)} (-c_0) \\
&= pqc_0 - (m - p)(m - q)c_0 \\
&= m(p + q - m)c_0
\end{aligned}
$$

But $p + q$ is the number of uncovered rows and columns, so that

$$p + q - m = (2m - k) - m = m - k$$

where k is the number of covered rows and columns. By Theorem 1, k is also the maximum number of independent zero cells. In particular, $k < m$ because otherwise we would have had an optimal solution at the last iteration. Therefore

$$\sum_i \sum_j (\hat{c}_{ij} - \bar{c}_{ij}) = m(m - k)c_0$$

is a positive integer provided that the original cost matrix consists of integers. Since the entries in the reduced cost matrix are always nonnegative by construction, and since the sum of the entries is reduced by a positive integer at each iteration, the algorithm stops in a finite number of steps. At termination we have an optimal solution since the Kuhn-Tucker conditions hold.

8.9 THE TRANSSHIPMENT PROBLEM

In the transportation problem studied in this chapter we have assumed that each point is either an origin, where goods are available, or a destination, where goods are required. Suppose that, in addition, there are intermediate points where goods are neither available nor required, but where goods can be transshipped through. The problem of finding the shipping pattern with the least cost is called the *transshipment problem* and is illustrated in Figure 8.25.

It is possible to handle transshipment problems by the transportation algorithm. There are several methods of converting a transshipment problem to a transportation format. One such method is as follows. On the transshipment

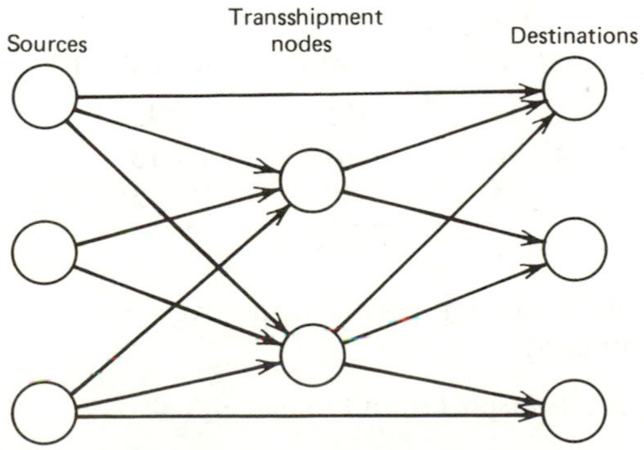

Figure 8.25. **Example of a transshipment problem.**

network solve for the least cost path from each origin i to each destination j (see Chapter 11). We then use this cost as the unit shipping cost in cell (i, j) of a transportation tableau. Another procedure for converting the model into a transportation problem is to add buffer stocks at certain nodes as discussed in Exercise 8.42.

There are other methods of handling transshipment problems besides transforming them to transportation problems. In the next chapter we provide an algorithm for solving the transshipment problem directly, which we call the general *minimal cost network flow algorithm.*

EXERCISES

8.1 Solve the following transportation problem.

Destination

		1	2	3	a_i
Origin	1	3	4	1	2
	2	7	2	5	5
	b_j	1	3	3	

c_{ij} matrix

8.2 On Tuesday the GT Railroad Company will have 4 locomotives at IE Junction, 1 locomotive at Centerville, and 2 locomotives at Wayover City. Student trains each requiring one locomotive will be at A-Station, Fine Place, Goodville, and Somewhere Street. The local map gives the following distances.

	A-STATION	FINE PLACE	GOODVILLE	SOMEWHERE STREET
IE Junction	13	35	42	9
Centerville	6	61	18	30
Wayover City	15	10	5	9

How should they assign power (locomotives) so that the total distance traveled is minimized?

8.3 The following is a transportation tableau.

c_{ij} (boxed, upper left of cell), x_{ij} (lower right of cell)

	1	2	3	4	a_i
1	9 4	8 14	12	13	18
2	10	10	12 24	14	24
3	8 2	9	11 4	12	6
4	10	10	11 7	12 5	12
b_j	6	14	35	5	

a. Is the solution basic?
b. Show that the solution is optimal.
c. Give the original linear programming problem and its dual.
d. Derive the optimal solution to the dual problem.
e. Give the optimal simplex tableau associated with the foregoing transportation tableau.
f. Suppose that c_{43} is increased from 11 to 16. Is the solution still optimal? If not, find the new optimal solution.

8.4 The following tableau depicts a feasible solution to a transportation problem. Is this solution basic? If not, then starting with this solution, construct a basic feasible solution. Compare the costs of the two solutions.

c_{ij} (boxed, upper left of cell), x_{ij} (lower right of cell)

	1	2	3	4	a_i
1	5 15	6	2	4 20	35
2	3	6 10	8 22	2 10	42
3	4 5	1	9 18	10	23
b_j	20	10	40	30	

8.5 Consider the following transportation problem.

	1	2	3	a_i
1	2	5	1	1
2	1	3	4	2
b_j	1	1	1	

c_{ij} matrix

a. Construct a basic feasible solution by the northwest corner method. Show the basic tree.
b. Find the optimal solution.

8.6 Consider Exercise 8.3.
a. Add 10 to each c_{ij}. Applying the cycle method, is the tableau still optimal?
b. Multiply each c_{ij} by 10. Applying the cycle method, is the tableau still optimal?
c. Repeat parts (a) and (b) for a general constant k.
d. What can we conclude from part (c) for transportation problems?

8.7 Consider the following transportation problem.

Destination

	1	2	3	a_i
Origin 1	2	5	1	1
2	1	3	4	2
b_j	1	1	1	

c_{ij} matrix

a. Prove by duality theory that
$(x_{11}, x_{12}, x_{13}, x_{21}, x_{22}, x_{23}) = (0, 3, 0, 1, 0, 4)$
is the optimal solution.
b. Interpret, economically, the dual variables for the solution in part (a) above.

8.8 Consider the following transportation problem.

	1	2	3	4	a_i
1	6	2	−1	0	5
2	4	7	2	5	25
3	3	1	2	1	25
b_j	10	10	20	15	

c_{ij} matrix

a. Give the northwest corner starting basic solution.
b. Find an optimal solution.
c. What is the effect on the optimal solution if c_{11} is changed to -4?

8.9 Consider the data of Exercise 8.8. Apply the method of Section 8.1 to obtain a feasible solution. Convert the feasible solution into a basic feasible solution. Show the basic tree.

8.10 Attempt to find an example of cycling in the transportation problem, or else show that cycling can never occur.

8.11 Devise a method for applying the lexicographic simplex method directly on the transportation tableau.

8.12 Show that if a basis is not connected and is, in fact, several trees, then there are not enough basic variables. (*Hint*. Count the basic variables in Figure 8.5.)

8.13 An airline company can buy gasoline from each of three suppliers. The suppliers have available $2K$, $6K$, and $6K$ gallons respectively. The company needs gasoline at three locations with each location requiring $5K$, $3K$, and $2K$ gallons respectively. The per/K gallon quoted price for gas delivered to each location is as follows.

		Location 1	Location 2	Location 3
Suppliers	1	2	3	1
	2	4	2	5
	3	1	8	9

How can the company buy the gasoline to minimize the total cost?

8.14 Show that if the cost coefficient for the artificial variable is increased by an amount θ, then all of the u_i's and v_j's will "change" by the same amount θ. Thereby, show that the cost coefficient of the artificial variable does not matter in the computation of $z_{ij} - c_{ij}$.

8.15 Show that if we define

$$\begin{aligned} \hat{a}_i &= a_i + \varepsilon & i &= 1, \ldots, m \\ \hat{b}_j &= b_j & j &= 1, \ldots, n-1 \\ \hat{b}_n &= b_n + m\varepsilon \end{aligned}$$

then by a proper choice of ε we can totally avoid degeneracy in the transportation problem.

8.16 An automobile manufacturer has assembly plants located in the Northwest, Midwest, and Southeast. The cars are assembled and sent to major markets in the Southwest, West, East, and Northeast. The appropriate distance matrix, availabilities, and demands are given by the following chart.

	SOUTHWEST	EAST	WEST	NORTHEAST	a_i
NORTHWEST	1000	8000	1800	2000	2,000,000
MIDWEST	400	700	900	1400	1,300,000
SOUTHEAST	800	1200	900	1100	1,600,000
b_j	1,000,000	1,500,000	1,200,000	700,000	

a. Assuming that cost is proportional to distance, find the optimal shipment pattern.

b. Assuming that cost is proportional to the square of distance, find the optimal shipment pattern.

8.17 Solve the following assignment problem by the transportation method.

JOB

PERSON	1	2	3	a_i
1	2	1	0	1
2	1	3	4	1
3	1	2	6	1
b_j	1	1	1	

c_{ij} matrix

8.18 Consider the following problem.

$$\text{Minimize} \quad \sum_{i=1}^{m} \sum_{j=1}^{n} c_{ij} x_{ij}$$

$$\text{Subject to} \quad \sum_{j=1}^{n} x_{ij} = a_i \qquad i = 1, 2, \ldots, m$$

$$\sum_{i=1}^{m} p_{ij} x_{ij} = b_j \qquad j = 1, 2, \ldots, n$$

$$x_{ij} \geqslant 0 \qquad i = 1, \ldots, m$$
$$j = 1, \ldots, n$$

where $p_{ij} > 0$ for all i, j. Extend the transportation algorithm of this chapter to handle the foregoing problem (which is sometimes referred to as the *generalized transportation problem*).

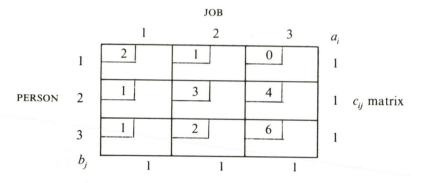

8.19 Formulate the problem of Exercise 1.14 as a generalized transportation model, and use the procedure of Exercise 8.18 above to find the optimal solution.

8.20 Devise a procedure for identifying the unique cycle of basic cells associated with an entering nonbasic cell. The procedure must specify the cells with coefficient 1, -1, and 0 in the representation of the nonbasic cell. Such a procedure must be developed when the transportation algorithm is coded on a digital computer.

8.21 Show how sensitivity analysis of the c_{ij}, a_i, and b_j may be carried out on a transportation tableau.

8.22 Give the dual of a transportation problem. Is it possible to readily specify a feasible solution to the dual problem? If so, starting with this solution, devise a method for applying the primal-dual method directly to a transportation tableau.

8.23 Prove or give a counterexample: For a variable to be basic in a particular row of the linear program for a transportation problem, the variable must have a nonzero entry in that row.

8.24 A company has contracted for five jobs. These jobs can be performed in six of its manufacturing plants. Because of the size of the jobs, it is not feasible to assign more than one job to a particular manufacturing facility. Also, the second job cannot be assigned to the third manufacturing plant. The cost estimates, in thousands of dollars, of performing the jobs in the different manufacturing plants, are summarized below.

	PLANT					
JOB	1	2	3	4	5	6
1	50	55	42	57	48	52
2	66	70	—	68	75	63
3	81	78	72	80	85	78
4	40	42	38	45	46	42
5	62	55	58	60	56	65

a. Formulate the problem of assigning the jobs to the plants so that the total cost is minimized.
b. Solve the problem by the transportation algorithm.
c. Solve the problem by the assignment algorithm.
d. Apply the primal-dual algorithm to this problem. Make all possible simplifications.

8.25 Consider the transportation problem corresponding to the following tableau.

	1	2	3	4	a_i	
1	4	3	6	5	20	
2	7	10	5	6	30	c_{ij} matrix
3	8	9	12	7	50	
b_j	15	35	20	30		

a. Solve the problem by the transportation algorithm.
b. Suppose that c_{24} is replaced by 5. Without resolving the problem, find the new optimal solution.
c. Suppose that c_{12} is replaced by 5. Which of the optimality conditions of the solution in part (a) is violated? Find the new optimal solution.

8.26 Consider the following data for a transportation problem:

	1	2	3	4	a_i	
1	14	56	48	27	71	
2	82	35	21	81	47	c_{ij} matrix
3	99	31	71	63	93	
b_j	71	35	45	60		

a. Indicate a starting basic solution by the northwest corner rule. Give the basic tree.
b. Find the optimal solution.
c. Give the simplex tableau associated with the basic feasible solution in part (a) above.

8.27 The following is *Vogel's approximation method* for obtaining a reasonably good starting basic feasible solution.

Step 0. Begin with all cells unallocated.

Step 1. In the problem at hand, compute for each row and each column the difference between the lowest and next lowest cost cell in the row or column.

Step 2. Among those rows and columns at hand, select the one with maximum difference.

Step 3. Allocate as much as possible to the x_{ij} with the lowest cost cell in the selected row or column. Decrease the corresponding supply and demand. Drop the row or column whose supply or demand is zero.

Step 4. Make any allocations where only one unallocated cell remains in a row or column. After reducing the corresponding supply and demands and dropping the row or column, repeat Step 4 as necessary.

Step 5. Stop if no rows and columns remain. Otherwise return to Step 1 with the reduced problem.

a. Apply Vogel's method to the data of Exercise 8.1.

b. Apply Vogel's method to the data of Exercise 8.8.

c. Apply Vogel's method to the data of Exercise 8.17.

8.28 Show that Vogel's approximation method leads to a basic feasible solution, after including any required zero cells.

8.29 The following is the *matrix minimum method* for obtaining a reasonably good starting feasible solution.

Step 0. Begin with all cells unallocated.

Step 1. Identify the lowest-cost unallocated cell in the matrix and allocate as much as possible to this cell.

Step 2. Reduce the corresponding supply and demand, dropping the one going to zero, and repeat Step 1 until all supplies and demands are allocated.

a. Show that the procedure produces a basic feasible solution.

b. Apply the procedure to Exercises 8.1 and 8.8.

8.30 Can the dual simplex method be applied to transportation problems directly on the transportation tableau?

8.31 Consider the following *capacitated transportation problem*.

$$\text{Minimize} \quad \sum_{i=1}^{m} \sum_{j=1}^{n} c_{ij} x_{ij}$$

$$\text{Subject to} \quad \sum_{j=1}^{n} x_{ij} = a_i \qquad i = 1, 2, \ldots, m$$

$$\sum_{i=1}^{m} x_{ij} = b_j \qquad j = 1, 2, \ldots, n$$

$$0 \leqslant x_{ij} \leqslant u_{ij} \qquad i = 1, \ldots, m, \quad j = 1, \ldots, n$$

Specialize the bounded simplex method of Chapter 5 to solve the foregoing transportation problem. Describe all details: finding a starting basic feasible solution, computing $z_{ij} - c_{ij}$ and the dual variables, the entering cell, and the leaving cell.

8.32 Consider Exercise 1.15.

a. Formulate the problem.

b. Solve the problem by the capacitated transportation algorithm that you developed in Exercise 8.31 above.

c. Suppose that the third manufacturing company lost one of its other contracts so that 700 tons are available to the furniture company. What is the optimal solution?

8.33 Use the theorems of duality to prove that the minimum number of lines to cover all zeros in a reduced assignment matrix equals the maximum number of independent zero cells.
(*Hint*. Consider the problem:

$$\text{Maximize} \quad \sum_i \sum_j x_{ij}$$

$$\text{Subject to} \quad \sum_{\hat{c}_{ij}=0} \mathbf{a}_{ij} x_{ij} \leqslant \mathbf{1}$$

$$x_{ij} \geqslant 0$$

where $\mathbf{a}_{ij} = \mathbf{e}_i + \mathbf{e}_{m+j}$ and the sum is taken only over zero cells of the reduced matrix. Take the dual of this linear program and examine its properties.)

8.34 Show that each new dual solution in the assignment procedure specified in Section 8.8 is feasible.

8.35 Compare the Hungarian method for the assignment problem with the primal-dual method applied to the assignment problem.

8.36 Apply the Hungarian method to the following assignment problem.

	1	2	3	4	5
1	2	6	4	−1	3
2	1	5	2	4	6
3	0	2	5	1	1
4	4	1	3	2	5
5	6	2	4	2	5

Cost matrix

8.37 Describe a procedure suitable for computer coding that will find, directly on the reduced matrix, the maximum number of independent zero cells in the reduced matrix (or equivalently the minimum number of lines to cover all zeros). The reader may wish to study Exercise 11.15.

8.38 Sally, Susan, and Sandra will go on a date with Bob, Bill, and Ben. Sally likes Bill twice as much as Bob and three times as much as Ben. Susan likes Bill three times as much as Bob and five times as much as Ben (Ben is a loser!). Sandra likes Bob about as much as Bill but likes them both about five times as much as Ben. How should the couples pair up so that in the aggregate the girls are as happy as possible? If one girl is willing to stay home, which one should it be? Which boy will lose out? (You guessed it!)

8.39 A carpenter, plumber, and engineer are available to perform certain tasks. Each person can perform only one task in the allotted time. There are four tasks available to be done. The inefficiency matrix for man i assigned to task j is as follows.

	SOLDERING	FRAMING	DRAFTING	WIRING
Carpenter	4	2	5	3
Plumber	1	3	4	2
Engineer	3	3	1	5

Which man should be assigned to which job? (*Hint*. Create a dummy man.) Which job will go unfinished? Now suppose that each man can perform up to two tasks. What should they do?

8.40 Given the optimal reduced matrix for an assignment problem, show how to construct the basic tree for the associated linear program. Demonstrate by the following reduced matrix.

2	3	0	4
0	1	4	0
0	0	2	4
1	0	2	1

(*Hint*. To the zero cells associated with the solution add an additional number of zero cells to create a tree or trees on the matrix. Add the required number of appropriate artificials.)

8.41 Show how one can construct the simplex tableau associated with the optimal assignment matrix.

8.42 Given a transshipment problem, the following procedure is suggested to convert it into a transportation problem. First the nodes are classified into the following mutually exclusive categories.

1. Pure source: a node that only ships.

2. Pure sink: a node that only receives.
3. Transshipment node: a node that may ship and receive.

A transportation tableau is constructed as follows. The origins are the pure sources and the transshipment nodes. The availability at each transshipment node i is replaced by $a_i + B$, where a_i is the maximum of zero and the net out of node i, and B is a buffer stock to be specified later. The destinations are the pure sinks and the transshipment nodes. The requirement at a transshipment node i is $b_i + B$, where b_i is the maximum of 0 and the net into node i. If there is no direct link from node i to node j, then c_{ij} is equal to M, where M is a large positive number. Also $c_{jj} = 0$ for transshipment nodes. Finally, B is a large positive number, say

$$B = \Sigma_i a_i$$

a. Using the foregoing instructions, form the transportation tableau corresponding to the following transshipment problem, where the availability at nodes 1, 2, and 3 are respectively 10, 20, and 15, and the requirement at nodes 5, 6, and 7 are respectively 10, 25, and 10.

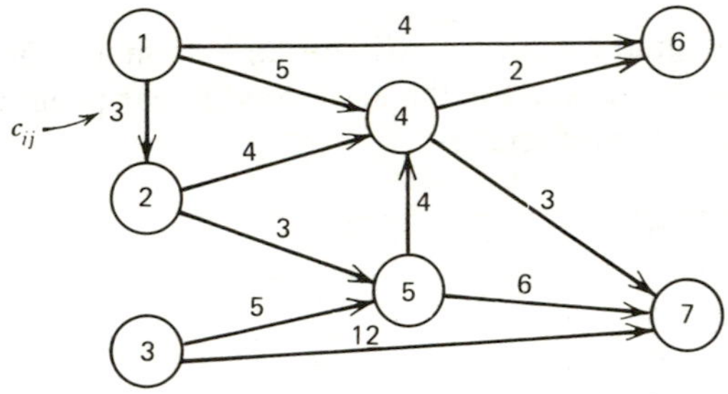

b. Solve the problem using the transportation algorithm. Interpret the solution. What is the interpretation of the buffer B?
c. Show that the procedure outlined in this exercise is valid in general.
d. Convert the problem into a transportation problem using the least cost method discussed in Section 8.9. Apply the transportation algorithm and interpret your solution.

8.43 If it is known in advance that a certain variable will be positive in any optimal solution to a transportation problem, what simplifications can result in the solution method?

NOTES AND REFERENCES

1. Hitchcock [240] is credited with the first formulation and discussion of a transportation model. Dantzig [89] adapted his simplex method to solve transportation problems. Charnes and Cooper [62] developed an intuitive presentation of Dantzig's procedure through what is called the "stepping stone" method. In this chapter we label this the "cycle method."
2. Koopmans [290] was the first to note the relationship between basic solutions in the transportation problem and the tree structure of a graph. Other good discussions are provided by Dantzig [97] and Ellis Johnson [266].
3. The Hungarian method for the assignment problem was developed by Kuhn [294]. His work led to the general primal-dual method for linear programs in the following year.

NINE: MINIMAL COST NETWORK FLOWS

In this chapter we generalize the concepts of the previous chapter, on transportation problems, to the more comprehensive class of network flow problems. Again, we shall find that the class of network flow problems possesses an important special structure that permits the simplification of the (primal) simplex procedure to a point where it may be applied directly on the network without the need of a simplex tableau.

9.1 THE MINIMAL COST NETWORK FLOW PROBLEM

Consider a *directed network* G, consisting of a finite set of *nodes* (points) $N = \{1, 2, \cdots, m\}$ and a set of *directed arcs* (lines) $S = \{(i, j), (k, l), \cdots, (s, t)\}$ joining pairs of nodes in N. Arc (i, j) is said to be *incident* with nodes i and j and is directed from node i to node j. We shall assume that the network has m nodes and n arcs. Figure 9.1 presents a network with 4 nodes and 7 arcs.

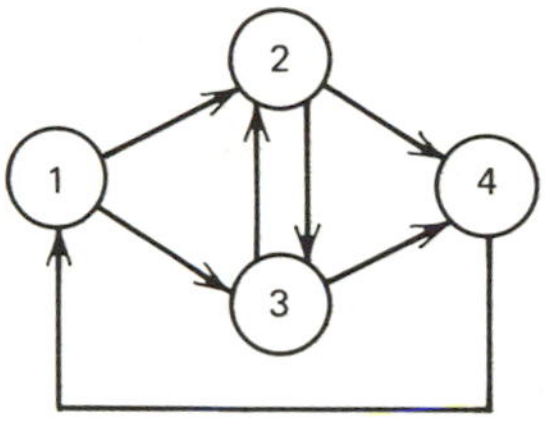

Figure 9.1. **Example of a network.**

With each node i in G we associate a number b_i that is the available supply of an item (if $b_i > 0$) or the required demand for the item (if $b_i < 0$). Nodes with $b_i > 0$ are sometimes called *sources*, and nodes with $b_i < 0$ are sometimes called *sinks*. If $b_i = 0$, then none of the item is available at node i and none is required; in this case node i is sometimes called an *intermediate* (*or transshipment*) node. Associated with each arc (i, j) we let x_{ij} be the amount of flow on the arc (we assume $0 \leqslant x_{ij}$) and c_{ij} be the unit shipping cost along the arc.

We shall assume that the total supply equals the total demand within the network, that is, $\Sigma_{i=1}^m b_i = 0$. If this is not the case, that is, $\Sigma_{i=1}^m b_i > 0$, then add a dummy demand node, $m + 1$, with $b_{m+1} = -\Sigma_{i=1}^m b_i$ and arcs with zero cost from each supply node to the new node.

The minimal cost network flow problem may be stated as follows. Ship the available supply through the network to satisfy demand at minimal cost. Mathematically, this problem becomes (where summations are taken over existing arcs)

$$
\begin{aligned}
\text{Minimize} \quad & \sum_{i=1}^{m} \sum_{j=1}^{m} c_{ij} x_{ij} \\
\text{Subject to} \quad & \sum_{j=1}^{m} x_{ij} - \sum_{k=1}^{m} x_{ki} = b_i \quad i = 1, \ldots, m \qquad (9.1) \\
& x_{ij} \geqslant 0 \quad i, j = 1, \ldots, m
\end{aligned}
$$

Constraints (9.1) are called the *flow conservation* or *Kirchhoff* equations and indicate that the flow may be neither created nor destroyed in the network. In the conservation equations, $\Sigma_{j=1}^m x_{ij}$ represents the total flow out of node i while $\Sigma_{k=1}^m x_{ki}$ indicates the total flow into node i. These equations require that the net flow out of node i, $\Sigma_{j=1}^m x_{ij} - \Sigma_{k=1}^m x_{ki}$, should equal b_i. If $b_i < 0$, then there should be more flow into i than out of i.

The minimal cost flow problem might arise in a logistics network where men and materials are being moved between various points in the world. It may be associated with the movement of locomotives between points in a railroad network to satisfy power for trains at least travel cost. Minimal cost network flow problems occur in the design and analysis of communication systems, oil pipeline systems, tanker scheduling problems, and a variety of other areas.

Clearly, the minimal cost flow problem is a linear program and can be solved in any one of several ways. One way is to apply the ordinary primal simplex algorithm to the problem. What we seek in this chapter is a simplification of the simplex method so that it can be applied directly on the network without the need of a simplex tableau.

The results of this chapter are a direct extension of those in the transportation chapter, as transportation problems are a special case of network flows. To place the constraints of a transportation problem in the form of the constraints specified by Equations (9.1) we multiply each destination constraint of the transportation problem by -1.

Paths, Chains, Circuits, Cycles, and Trees

For clarity of notation we shall provide a brief discussion of four concepts that appear throughout this and later chapters.

A *path* (from node i_o to i_p) is a sequence of arcs $P = \{ (i_o, i_1), (i_1, i_2), \ldots, (i_{p-1}, i_p) \}$ in which the initial node of each arc is the same as the terminal node of the preceding arc in the sequence. Thus each arc in the path is directed "toward" i_p and "away from" i_0. A *chain* is a similar structure to a path except that not all arcs are necessarily directed toward node i_p. Figure 9.2*a* illustrates a

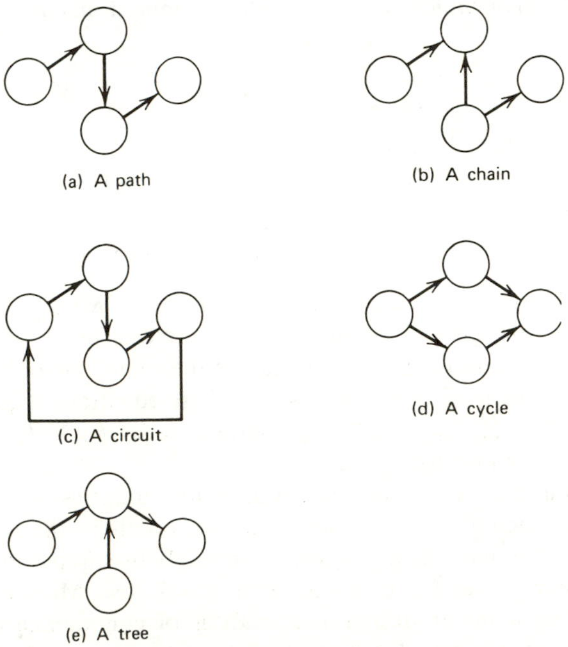

Figure 9.2 Paths, chains, circuits, cycles, and trees.

path while Figure 9.2*b* presents a chain. A *circuit* is a path in which $i_0 = i_p$. Thus a circuit is a closed path. A *cycle* is a closed chain. Figures 9.2*c* and *d* depict circuits and cycles. Every path is a chain but not vice versa. Every circuit is a cycle but not conversely. Throughout this chapter we shall assume that G is *connected*; that is, that a chain exists from every node to every other node in G. A *tree* is a connected graph with no cycles. A *spanning tree* is a tree that includes every node of the graph. Figure 9.2*e* illustrates a spanning tree. Figures 9.2*a* and *b* are also examples of spanning trees; Figures 9.2*c* and *d* are not trees. Additional properties of trees are given in Exercise 9.1.

9.2 PROPERTIES OF THE A MATRIX

Consider the coefficient matrix **A**, associated with the constraint set (9.1). The matrix **A** has one row for each node of the network and one column for each arc. Each column of **A** contains exactly two nonzero coefficients: a "+1" and a "−1." The column associated with arc (i, j) contains a "+1" in row i, a "−1" in row j, and a zero elsewhere. Thus the columns of **A** are given by

$$\mathbf{a}_{ij} = \mathbf{e}_i - \mathbf{e}_j$$

where $\mathbf{e}_i$ and $\mathbf{e}_j$ are unit vectors in E^m, with 1's in the ith and jth positions respectively. The **A** matrix is called the *node-arc incidence matrix* for the graph. The **A** matrix for the network of Figure 9.1 is

$$\begin{array}{cc} & \begin{array}{ccccccc} (1,2) & (1,3) & (2,3) & (2,4) & (3,2) & (3,4) & (4,1) \end{array} \\ \mathbf{A} = & \left[\begin{array}{ccccccc} 1 & 1 & 0 & 0 & 0 & 0 & -1 \\ -1 & 0 & 1 & 1 & -1 & 0 & 0 \\ 0 & -1 & -1 & 0 & 1 & 1 & 0 \\ 0 & 0 & 0 & -1 & 0 & -1 & 1 \end{array}\right] \begin{array}{c} 1 \\ 2 \\ 3 \\ 4 \end{array} \end{array}$$

Rank of the A Matrix

Clearly the **A** matrix does not have full rank since the sum of its rows is the zero vector. To show that **A** has rank $m - 1$ we need only select an $(m - 1) \times (m - 1)$ submatrix from **A** that is nonsingular.

Let T be any spanning tree in the network G. Tree T consists of the m nodes of G together with $m - 1$ arcs of G that do not form a cycle (see Exercise 9.1). Consider the $m \times (m - 1)$ submatrix $\mathbf{A}_T$ of **A** associated with the nodes and arcs in T. Since $m \geqslant 2$, T has at least one *end*, that is, a node k with exactly one arc incident with it. In such a case the kth row of $\mathbf{A}_T$ contains a single nonzero entry. Permute the rows and columns of $\mathbf{A}_T$ so that this nonzero entry is in the

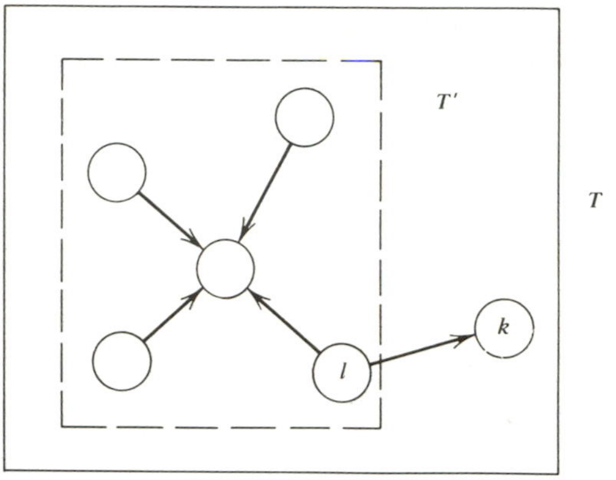

Figure 9.3. **Reduction of T to T'.**

first row and first column. Then $\mathbf{A}_T$ becomes

$$\mathbf{A}_T = \left[\begin{array}{c|c} \pm 1 & \mathbf{0} \\ \hline \mathbf{p} & \mathbf{A}_{T'} \end{array}\right]$$

Delete the first row and column of $\mathbf{A}_T$ and consider the matrix $\mathbf{A}_{T'}$, which is $(m-1) \times (m-2)$. Correspondingly obtain the graph T' from T by removing node k and the incident arc (see Figure 9.3). Here T' is also a tree. It must contain at least one end, say node l. Permuting the rows and columns of $\mathbf{A}_{T'}$ so that the single nonzero entry in row l is in the first row and column, we may write $\mathbf{A}_T$ as

$$\mathbf{A}_T = \left[\begin{array}{c|c|c} \pm 1 & 0 & \mathbf{0} \\ \hline p_1 & \pm 1 & \mathbf{0} \\ \hline \mathbf{p}_2 & \mathbf{q} & \mathbf{A}_{T''} \end{array}\right]$$

We can continue in this manner exactly $m - 1$ times, after which all $m - 1$ columns of $\mathbf{A}_T$ are fixed. Deleting the remaining bottom row of $\mathbf{A}_T$, we have an $(m-1) \times (m-1)$ matrix that is lower triangular with nonzero diagonal elements and therefore nonsingular. Thus the rank of $\mathbf{A}$ is $m - 1$.

If we select columns (1, 3), (2, 3), and (3, 4) from the node–arc incidence matrix for the network of Figure 9.1, we get the following lower triangular matrix after discarding row 4:

$$\begin{array}{c} \\ \mathbf{B} = \end{array} \begin{array}{c} \begin{array}{ccc} (1,3) & (2,3) & (3,4) \end{array} \\ \left[\begin{array}{ccc} 1 & 0 & 0 \\ 0 & 1 & 0 \\ -1 & -1 & 1 \end{array}\right] \begin{array}{c} 1 \\ 2 \\ 3 \end{array} \end{array}$$

The corresponding spanning tree is given as follows.

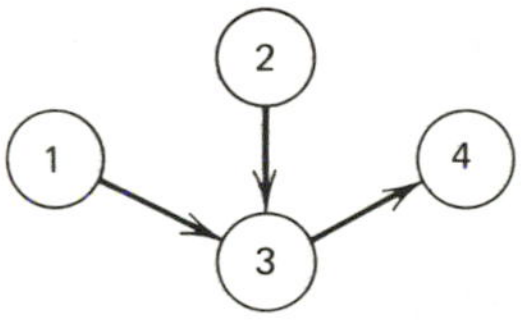

The Artificial Variable

Recall that the simplex method always starts with a full rank constraint matrix. We demonstrated above that the rank of **A** is $m - 1$. Therefore an artificial variable is required so that the rank of the new matrix is m. Introducing an artificial variable corresponding to node m (any other node will do) leads to the constraint matrix $(\mathbf{A}, \mathbf{e}_m)$.

Any basic solution must contain m linearly independent columns, and hence the artificial variable must appear in every basic solution. If we liberalize our definition of an arc, then the new column can be viewed as an arc beginning at node m and terminating in space (see Figure 9.4). This one ended arc is called a *root arc*.

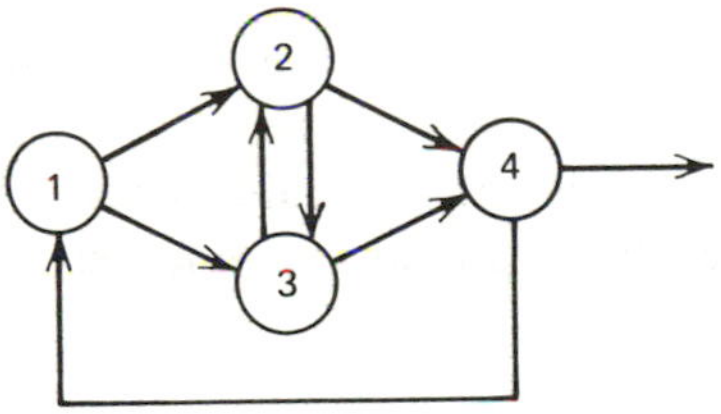

Figure 9.4. **A generalized graph** G.

Characterization of a Basic Matrix

We determined the rank of **A** by examining any submatrix associated with a spanning tree. This also demonstrates that a spanning tree together with a single artificial variable is a basis for the **A** matrix. What we have not yet shown is the converse; that is, any basis is also a spanning tree together with an artificial variable.

Let **B** be an $m \times m$ matrix formed by choosing $m - 1$ linearly independent columns of **A** together with the artificial column so that **B** is a basis matrix. Consider the graph G_B associated with all the nodes in the original graph together with the arcs in **B** (leaving the artificial variable aside). Thus G_B can contain no cycles. By contradiction, suppose that G_B contains a cycle C. (See Figure 9.5.) Select some arc (i, j) in the cycle and assign the cycle an orientation in the direction of that arc. Then, for each column of **B** associated with an arc in the cycle, assign a coefficient of $+1$ if the arc is in the direction of orientation of

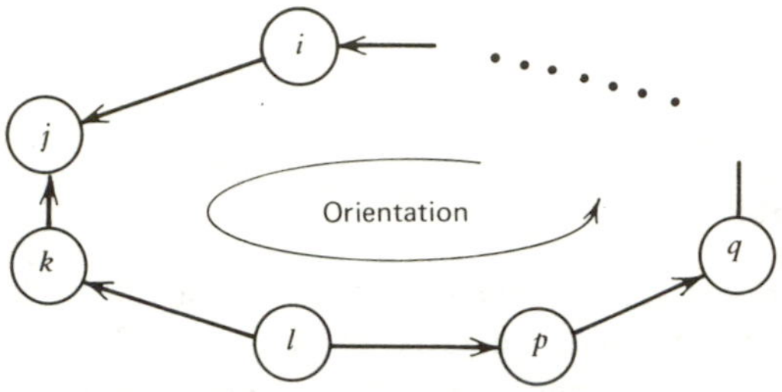

Figure 9.5. **Illustration of a linearly dependent set of columns.**

the cycle and -1 otherwise. Applying these coefficients to the respective columns in **B**, we find that the weighted sum of these columns is the zero vector. In Figure 9.5 we have

$$\mathbf{a}_{ij} - \mathbf{a}_{kj} - \mathbf{a}_{lk} + \mathbf{a}_{lp} + \mathbf{a}_{pq} + \cdots = (\mathbf{e}_i - \mathbf{e}_j) - (\mathbf{e}_k - \mathbf{e}_j) - (\mathbf{e}_l - \mathbf{e}_k) + (\mathbf{e}_l - \mathbf{e}_p) + (\mathbf{e}_p - \mathbf{e}_q) + \cdots = \mathbf{0}$$

Thus they could not have been linearly independent. This contradiction shows that G_B contains no cycles. But G_B contains m nodes and $m - 1$ arcs. Hence G_B is a tree (why?).

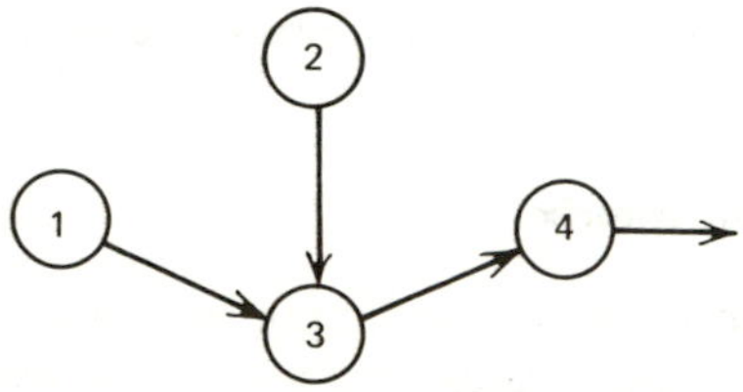

Figure 9.6. **A basis subgraph is a rooted spanning tree.**

To summarize, we have shown that each basis consists of the root together with a spanning tree (see Figure 9.6), and conversely. Thus we have the following theorem.

***Theorem* 1**

A *basis* for the minimal cost network flow problem is characterized by a *rooted spanning tree*.

Triangularity, Integrality and Total Unimodularity

Every basis corresponds to a rooted spanning tree and every rooted spanning tree is (lower) triangular; thus every basis is triangular. From Chapter 8 we see that a triangular basis matrix **B** permits a simple and efficient method for solving for the values of the variables. We shall shortly exploit this fact to solve efficiently for the values of the basic variables directly on the network. Triangularity of the basis matrix also permits efficient, graphic determination of the dual variables.

Recall that **B** is triangular and each of its entries is ± 1 or 0 (its diagonal entries are ± 1). Hence each of the basic variables will take on integral values provided that the b_i's are all integers.

Finally, as with the transportation problem, it is possible to show that the constraint matrix for a network flow problem is totally unimodular. As the proof is similar, we ask the reader to do this in Exercise 9.2. Since **A** is totally unimodular, it follows that $\mathbf{y}_{ij} = \mathbf{B}^{-1}\mathbf{a}_{ij}$ is a vector of zeros or ± 1's. We shall shortly demonstrate this fact constructively.

9.3 REPRESENTATION OF A NONBASIC VECTOR IN TERMS OF THE BASIC VECTORS

Consider the basis subgraph G_B corresponding to a rooted spanning tree; and select any nonbasic arc (r, s). Since G_B is a tree, we know that there is a unique chain between nodes r and s. This chain together with the nonbasic arc (r, s) constitutes a cycle. (See Figure 9.7.) Assigning the cycle an orientation consistent with (r, s), we have

$$\mathbf{a}_{rs} - \mathbf{a}_{rj} + \mathbf{a}_{kj} + \cdots + \mathbf{a}_{sp}$$
$$= (\mathbf{e}_r - \mathbf{e}_s) - (\mathbf{e}_r - \mathbf{e}_j) + (\mathbf{e}_k - \mathbf{e}_j) + \cdots + (\mathbf{e}_s - \mathbf{e}_p) = \mathbf{0}$$

or

$$\mathbf{a}_{rs} = \mathbf{a}_{rj} - \mathbf{a}_{kj} + \cdots - \mathbf{a}_{sp}$$

This development leads to the following simple procedure for representing any nonbasic column in terms of the basic columns. First the unique cycle formed by joining the nonbasic arc to the basis subgraph is determined. The cycle is then given an orientation consistent with the nonbasic variable. A basic column in the cycle along its orientation receives a coefficient of -1 in the

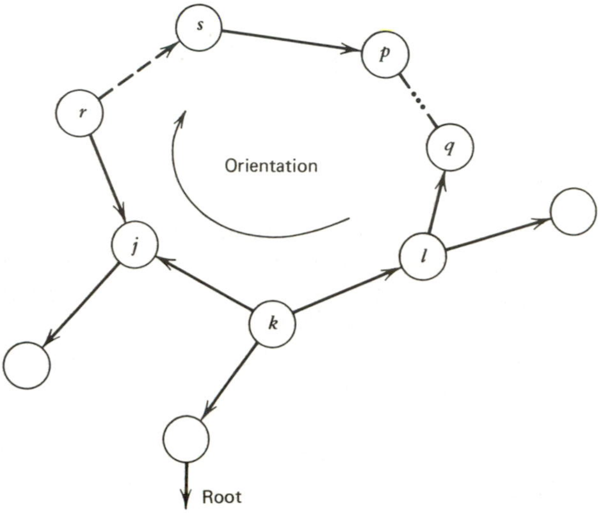

Figure 9.7. **Cycle formed by adding a nonbasic arc to the basis tree.**

representation, and a basic column in the cycle opposite to its orientation receives a coefficient of $+1$ in the representation. Other basic columns receive zero coefficients.

As an example consider the subgraph of Figure 9.6, which is a basis for the network of Figure 9.4. Suppose that we seek the representation of the nonbasic arc (1, 2). Using the foregoing rule, we get

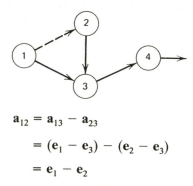

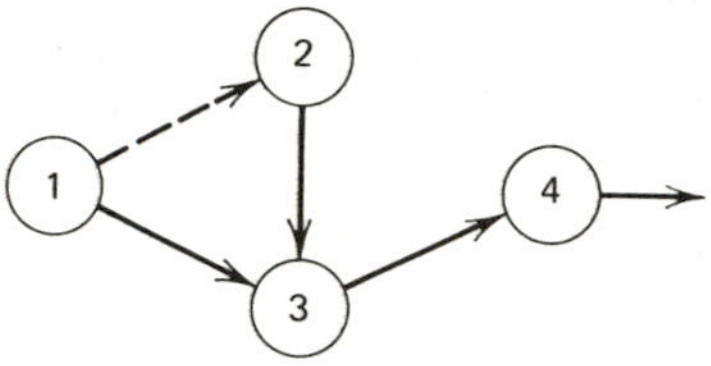

$$\begin{aligned}\mathbf{a}_{12} &= \mathbf{a}_{13} - \mathbf{a}_{23} \\ &= (\mathbf{e}_1 - \mathbf{e}_3) - (\mathbf{e}_2 - \mathbf{e}_3) \\ &= \mathbf{e}_1 - \mathbf{e}_2\end{aligned}$$

Note that the coefficients in the representation of the nonbasic column $\mathbf{a}_{rs}$ in terms of the basic columns give rise to the vector $\mathbf{y}_{rs}$ [that is, the entries in the simplex tableau under the (r, s) column]. Since the artificial column never appears in the representation for any other column and since the artificial variable always remains basic at value zero, we may select any value for the associated cost coefficient, say $c_a = 0$.

9.4 THE SIMPLEX METHOD FOR NETWORK FLOW PROBLEMS

The general steps of the simplex method are as follows. First, find a starting basic feasible solution. Next compute $z_j - c_j$ for each nonbasic variable x_j. If optimality is achieved, stop; otherwise select the entering column. If optimality is not achieved, determine the exit (blocking) column and pivot. The following paragraphs present a discussion of each of these operations applied to network flow problems. For the moment we shall postpone the difficulties associated with identifying a feasible basis (that is, phase I of the simplex method) and assume that one is at hand. To fix ideas we shall apply each of the foregoing steps to the problem presented in Figure 9.8.

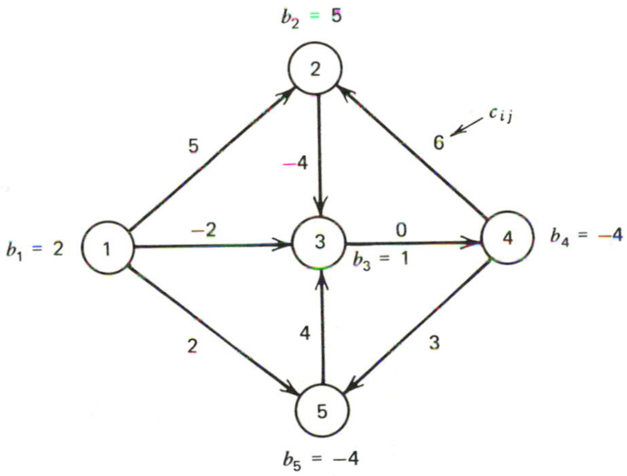

Figure 9.8. **An example network flow problem.**

Computing the Values of the Basic Variables

Adding the artificial arc to node 5, suppose that we select the feasible basis given by the subgraph in Figure 9.9. The basic system of equations $\mathbf{B}\mathbf{x}_B = \mathbf{b}$ to be solved is

$$\begin{bmatrix} 1 & 0 & 0 & 0 & 0 \\ 0 & 1 & 0 & 0 & 0 \\ 0 & -1 & 1 & 0 & 0 \\ 0 & 0 & -1 & 1 & 0 \\ -1 & 0 & 0 & -1 & 1 \end{bmatrix} \begin{bmatrix} x_{15} \\ x_{23} \\ x_{34} \\ x_{45} \\ x_5 \end{bmatrix} = \begin{bmatrix} 2 \\ 5 \\ 1 \\ -4 \\ -4 \end{bmatrix}$$

where x_5 is the artificial variable associated with node 5.

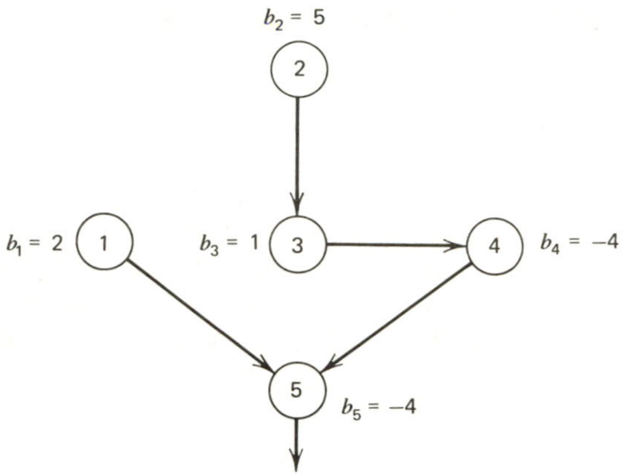

Figure 9.9. **A basic subgraph.**

Taking advantage of the lower triangular structure of the basis matrix, we may iteratively solve for the basic variables. From the top equation, $x_{15} = 2$. From the second equation, $x_{23} = 5$. From the third equation, $x_{34} = 1 + x_{23} = 6$. Next, $x_{45} = -4 + x_{34} = 2$. Finally, $x_5 = -4 + x_{15} + x_{45} = 0$ and thus the basis is feasible. These same computations can be made directly on the graph in Figure 9.9 as follows.

Examining node 1 in Figure 9.9, we see that it is an *end* of the basic tree, that is, a node with only one basic arc incident to it. Hence the corresponding basic equation contains only one variable, and the value of that variable can be readily obtained. In this case arc (1, 5) points out of node 1 and thus x_{15} has a $+1$ in row 1. Thus $x_{15} = b_1$ or $x_{15} = 2$.

Examining node 2, we see that it is an end and hence x_{23} can be computed similiarly:

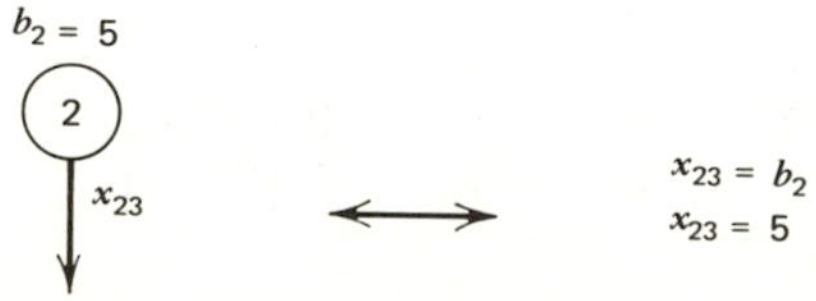

Next, notice that node 3 has all of its incident arc variables assigned values except one. Thus we can use the conservation equation for node 3 to solve for the remaining variable.

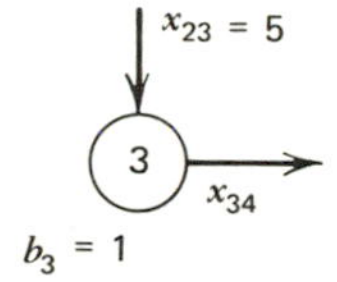

$$x_{34} - x_{23} = 1$$
$$x_{34} = 6$$

We can now solve for x_{45}:

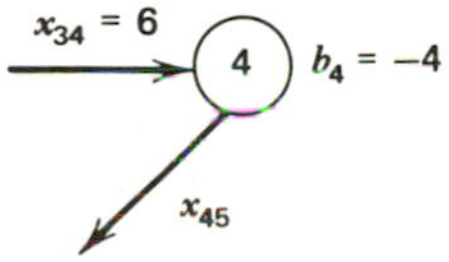

$$x_{45} - x_{34} = -4$$
$$x_{45} = 2$$

Finally we solve for x_5.

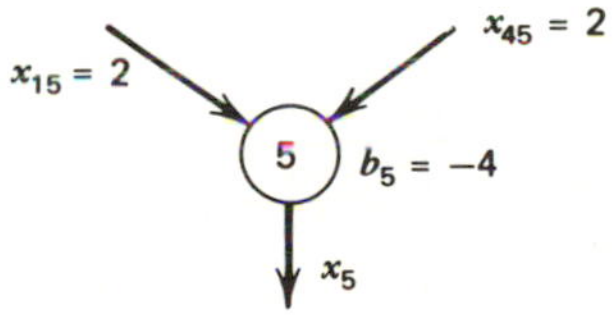

$$x_5 - x_{15} - x_{45} = -4$$
$$x_5 = 0$$

The process of obtaining the basic solution proceeds from the ends of the tree toward the root (see Figure 9.10). As we shall see later, the process of obtaining the dual variables is just reversed.

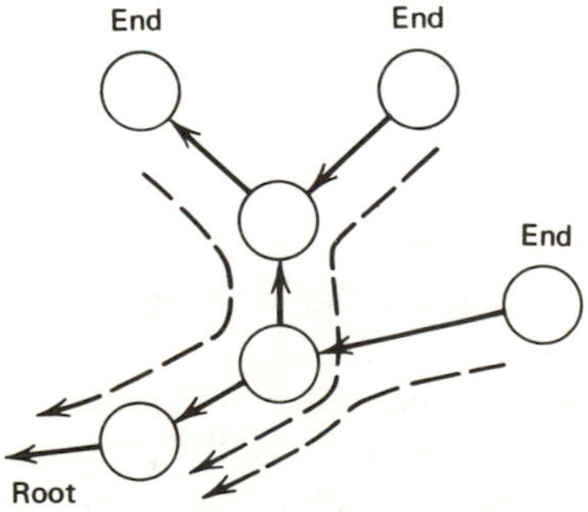

Figure 9.10. **Computing the values of the basic variables.**

Computing $z_{ij} - c_{ij}$

Given a basic subgraph, compute $z_{ij} - c_{ij}$ for each nonbasic variable x_{ij} and either stop or proceed by introducing a nonbasic variable with a positive $z_{ij} - c_{ij}$. As with the transportation problem, we have two methods of making this computation: one involving cycles and one involving the direct computation of the dual variables. We treat the computation of $z_{ij} - c_{ij}$ with cycles first.

Recall that $z_{ij} - c_{ij} = \mathbf{c}_B\mathbf{y}_{ij} - c_{ij}$. In Section 9.3 we discussed how to compute the vector $\mathbf{y}_{ij}$ that represents the nonbasic arc (i, j) in terms of the basic arcs. In particular, to compute z_{ij} we first determine the cycle obtained by joining arc (i, j) to the basic subgraph and then give the cycle an orientation consistent with arc (i, j). Then z_{ij} is the sum of the costs of the basic arcs in the cycle opposite to the orientation minus the sum of the costs of the basic arcs in the cycle along the orientation. From this, $z_{ij} - c_{ij}$ can be computed. As an example, referring to Figure 9.9, we have

$$z_{13} - c_{13} = -c_{34} - c_{45} + c_{15} - c_{13} = 0 - 3 + 2 - (-2) = 1$$

A second method of computing $z_{ij} - c_{ij}$ for a nonbasic arc is to compute the dual vector, $\mathbf{w}$, and determine $z_{ij} - c_{ij}$ through the expression $z_{ij} - c_{ij} = \mathbf{wa}_{ij} - c_{ij}$. In order to compute $\mathbf{w}$, the system $\mathbf{wB} = \mathbf{c}_B$ must be solved. For the basis subgraph of Figure 9.9 we have

$$[w_1, w_2, w_3, w_4, w_5]\begin{bmatrix} 1 & 0 & 0 & 0 & 0 \\ 0 & 1 & 0 & 0 & 0 \\ 0 & -1 & 1 & 0 & 0 \\ 0 & 0 & -1 & 1 & 0 \\ -1 & 0 & 0 & -1 & 1 \end{bmatrix} = [2, -4, 0, 3, 0]$$

Using the last $\mathbf{w}$ equation, the one associated with the root, we get $w_5 = 0$. We may now proceed away from the root in the following fashion,

$$\begin{aligned} w_4 - w_5 &= c_{45} \Rightarrow w_4 = 3 + 0 = 3 \\ w_3 - w_4 &= c_{34} \Rightarrow w_3 = 0 + 3 = 3 \\ w_2 - w_3 &= c_{23} \Rightarrow w_2 = -4 + 3 = -1 \\ w_1 - w_5 &= c_{15} \Rightarrow w_1 = 2 + 0 = 2 \end{aligned}$$

We start with the dual variable for the root node at zero value, then proceed away from the root toward the ends of the tree using the relationship that $w_i - w_j = c_{ij}$ along the basic arcs in the tree.

While the process of computing primal variables consisted of working from the ends of the basis tree inward toward the root (see Figure 9.10), the process of computing dual variables consists of working from the root of the basis tree outward toward the ends (see Figure 9.11).

To compute $z_{ij} - c_{ij}$ for the nonbasic arc (i, j) we apply the definition

$$\begin{aligned} z_{ij} - c_{ij} &= \mathbf{wa}_{ij} - c_{ij} \\ &= \mathbf{w}(\mathbf{e}_i - \mathbf{e}_j) - c_{ij} \\ &= w_i - w_j - c_{ij} \end{aligned}$$

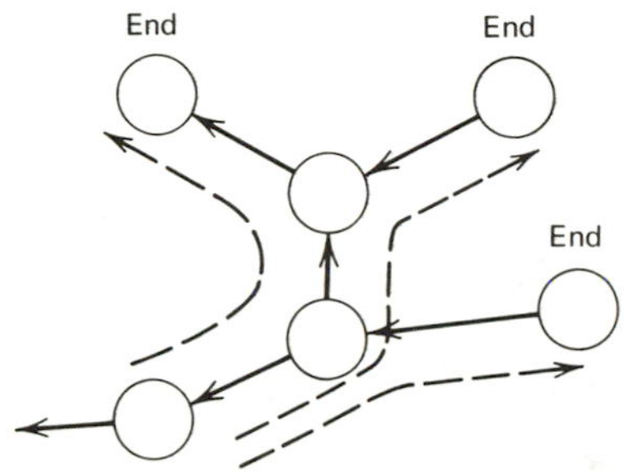

Figure 9.11. **Computing the values of the dual variables.**

Thus the $z_{ij} - c_{ij}$ can be conveniently computed on the network. Notice also that by requiring that $w_i - w_j = c_{ij}$ along basic arcs, we are actually requiring that $z_{ij} - c_{ij} = 0$ for basic variables.

Using the values of the dual variables obtained above, we summarize below the value of $z_{ij} - c_{ij}$ for each nonbasic variable x_{ij}.

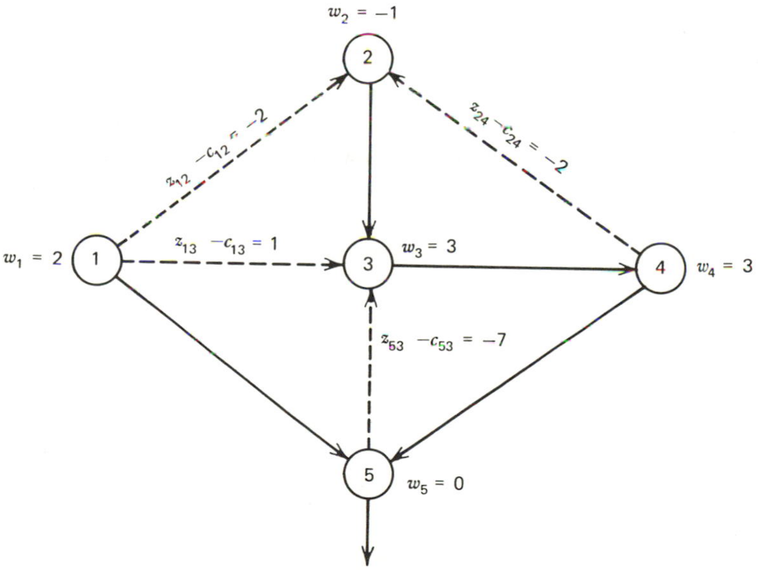

Determining the Exit Column and Pivoting

When we applied the cycle method to compute $z_{ij} - c_{ij}$ for a nonbasic arc, we essentially identified the pivot process. In the foregoing example $z_{13} - c_{13} > 0$ and so x_{13} is a candidate to enter the basis. What we must do is proceed to increase x_{13}, adjust the basic variables to maintain feasibility with respect to the right-hand side, and determine the first basic variable to reach zero. This blocking basic variable is the exit variable and leaves the basis.

Consider the basic tree together with arc (1, 3). If we increase x_{13} by Δ, then to provide balance we must increase x_{34} by Δ, increase x_{45} by Δ, and finally decrease x_{15} by Δ.

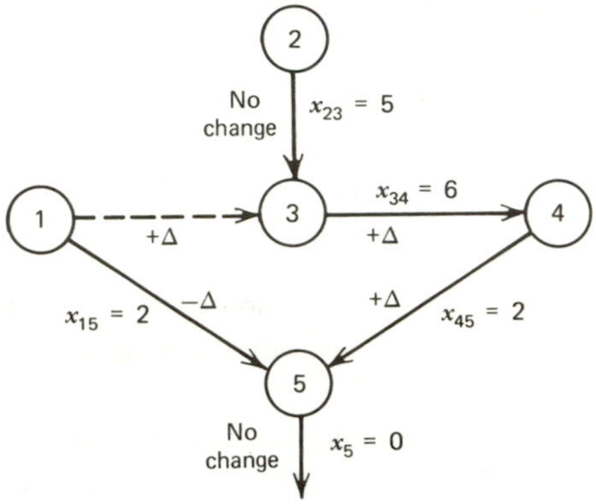

This process of adjustment can be thought of as sending an additional amount of flow Δ around the unique cycle created when the nonbasic arc is added to the basic tree. Sending flow against the direction of an arc corresponds to decreasing flow on the arc.

As x_{13} increases by Δ, the only basic variable to decrease is x_{15} and its new value is $x_{15} = 2 - \Delta$. Thus the critical value of Δ is equal to 2, at which instant x_{15} drops to zero and leaves the basis. All of the other basic variables are adjusted appropriately in value and the new basic solution is given as follows.

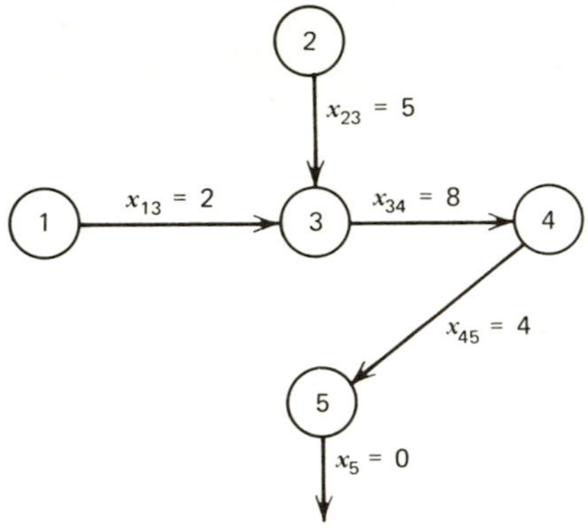

We leave it as an exercise to the reader to show that determining the exiting variable and adjusting the values of the basic variables accordingly as described above is equivalent to performing the usual minimum ratio test and pivoting.

9.5 AN EXAMPLE OF THE NETWORK SIMPLEX METHOD

Consider the network of Figure 9.12. In Figure 9.13 we given the complete solution of this minimal cost network flow problem. The exiting variable is denoted by * in the figure.

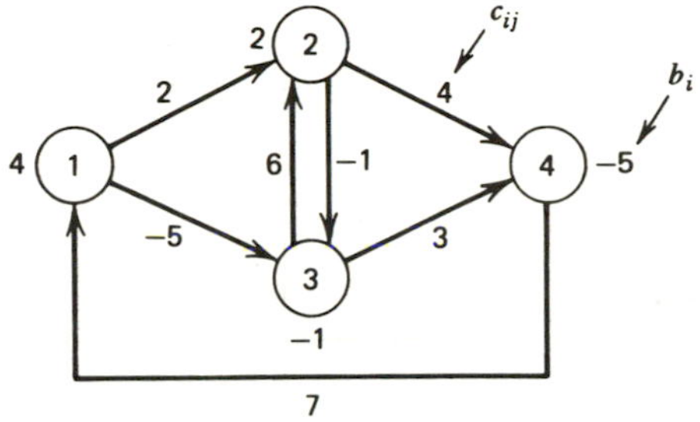

Figure 9.12. **An example network.**

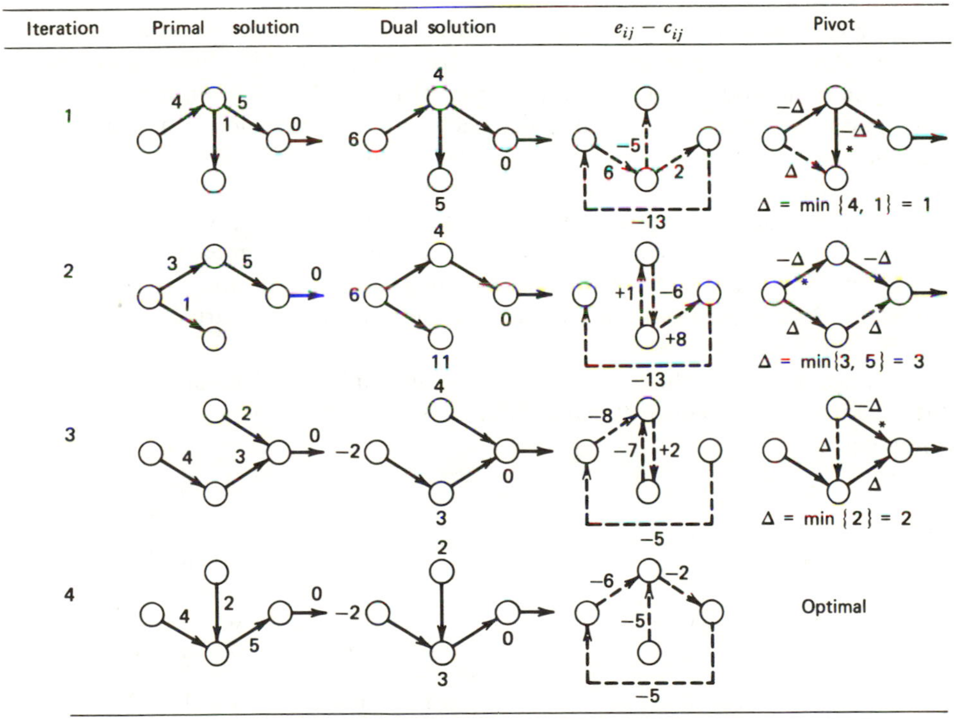

$x_{13}^* = 4$, $x_{23}^* = 2$, $x_{34}^* = 5$, all other $x_{ij}^* = 0$, $z^* = -7$

Figure 9.13. **The solution to Figure 9.12.**

9.6 FINDING AN INITIAL BASIC FEASIBLE SOLUTION

In the previous sections we have assumed that a starting basic feasible solution was at hand. We now give a method of finding one. Consider the **A** matrix without the additional artificial column. Suppose that we add an artificial column for *every* row of **A**, the ith artificial column being $\pm \mathbf{e}_i$ depending on the sign of b_i (that is, $+\mathbf{e}_i$ if $b_i \geq 0$; $-\mathbf{e}_i$ otherwise). Also, let us add a redundant row given by the negative of the sum of the rows of the "extended" **A** matrix.

The problem then becomes

$$
\left[\begin{array}{c|cccc}
\mathbf{A} & \pm 1 & & & \\
 & & \pm 1 & & \\
 & & & \ddots & \\
 & & & & \pm 1 \\
\hline
\mathbf{0} & \mp 1 & \mp 1 & \cdots & \mp 1
\end{array}\right]
\qquad
\left[\begin{array}{c}
\mathbf{b} \\
\hline
0
\end{array}\right]
$$

Since each column of this "new" **A** matrix has exactly one +1 and one −1, we may view it as a node–arc incidence matrix of a graph. This "new" graph has all of the same nodes and arcs as the original graph. In addition, it has a new node and m new arcs—one arc between each original node and the new node. A feasible basis for this new problem is given by the m artificial variables (arcs) plus an additional artificial variable (root) for the new row $(m + 1)$.

Beginning with the artificial basis we may proceed to apply the two-phase method or the big-M method, using appropriate costs in each case, until feasibility is achieved. At that time we may drop all of the artificial arcs (variables) and node $(m + 1)$, and replace these by a single artificial variable (root) at node m.

To illustrate the technique, consider the example problem of Section 9.5. After adding appropriate artificial columns and creating the new row, we get the following.

	x_{12}	x_{13}	x_{23}	x_{24}	x_{32}	x_{34}	x_{41}	x_1	x_2	x_3	x_4	x_5	b
1	1	1	0	0	0	0	−1	1	0	0	0	0	+4
2	−1	0	1	1	−1	0	0	0	1	0	0	0	+2
3	0	−1	−1	0	1	1	0	0	0	−1	0	0	−1
4	0	0	0	−1	0	−1	1	0	0	0	−1	0	−5
5	0	0	0	0	0	0	0	−1	−1	1	1	1	0

Selecting the phase I method, the artificial variables x_1, x_2, x_3, and x_4 will have cost coefficients of 1 while all other variables have zero cost coefficients. This leads to the associated network flow problem of Figure 9.14 where the cost coefficient of the root x_5 is zero. With this feasible basis at hand we proceed to solve the phase I problem, using the procedures developed in this chapter.

9.7 NETWORK FLOWS WITH LOWER AND UPPER BOUNDS

It is simple and straightforward to make the transition from the ordinary simplex method for network flow problems to the lower-upper bound simplex

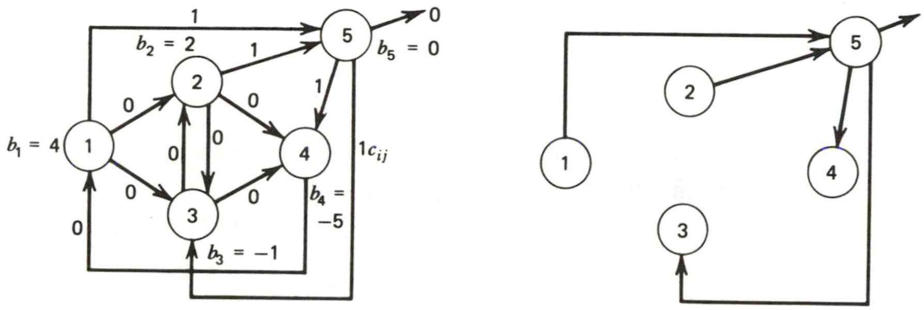

The phase I problem A feasible basis

Figure 9.14. **The phase I network flow problem and a starting feasible basis.**

method for network flow problems. We briefly review the essential changes required to effect such an extension to the lower-upper bound method.

Getting Started

The method of Section 9.6 carries over directly for problems with lower and upper bounds. In this case, since all "real" arcs start out nonbasic during phase I (or at the outset of the big-M method), we set all of these arc flow variables at one or other of their bounds and compute the effect on the b values according to

$$\hat{b}_i = b_i - \sum_j x_{ij} + \sum_k x_{ki}$$

Using the vector $\hat{\mathbf{b}}$, we establish the "direction" (sign) of the artificial columns to be added and begin phase I.

Computing the Values of the Basic Variables

Whether in phase I or phase II, or during the big-M procedure, after adjusting the $\mathbf{b}$ vector to $\hat{\mathbf{b}}$ to reflect the values of the nonbasic variables (arcs), we proceed in the same manner as before to compute the values of the basic flow variables.

Computing the Dual Variables and the $z_{ij} - c_{ij}$'s

Lower and upper bounds have no effect on the computation of the dual variables and on the computation of the $z_{ij} - c_{ij}$'s. Note, however, that in the presence of lower and upper bounds the optimality criteria are

$$x_{ij} = u_{ij} \;\Rightarrow\; z_{ij} - c_{ij} \geqslant 0$$

and

$$x_{ij} = l_{ij} \quad \Rightarrow \quad z_{ij} - c_{ij} \leqslant 0$$

These are easy to check and we can readily determine whether some nonbasic variable x_{ij} should be increased or decreased if optimality is not achieved.

Determining the Exit Column and Pivoting

Once the entering column is selected, it is again an easy task to select the exit column and pivot. We add the entering nonbasic arc, regardless of whether the variable is increasing or decreasing, to the basis tree and determine the unique cycle formed. Then, if the entering variable is increasing, we send an amount Δ around the cycle in the direction of the entering variable. If the entering variable is decreasing, we send an amount Δ around the cycle against the direction of the entering variable. Figure 9.15 illustrates these two possibilities. To compute the critical value of Δ we check those basic variables increasing as well as those decreasing, and the possibility that x_{ij} may reach its other bound. If the last possibility occurs, x_{ij} remains nonbasic (at its other bound) and all basic variables along the cycle are adjusted accordingly. Otherwise, the nonbasic variable enters and some basic variable exists at one or other of its bounds, and all variables along the cycle are adjusted accordingly.

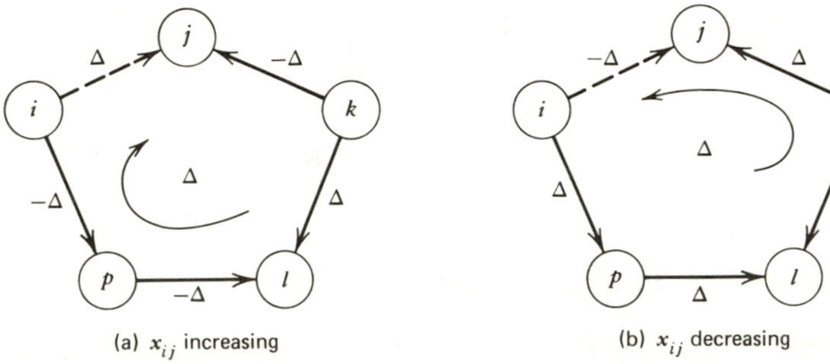

Figure 9.15. Two cases for entering are: (*a*) x_{ij} increasing. (*b*) x_{ij} decreasing.

An Example of the Network Simplex Method in the Presence of Lower and Upper Bounds

Consider the network of Figure 9.16. We present, in Figure 9.17, the complete solution to that network. We have identified a starting basic feasible solution, thus omitting phase I. The notation "$|\overset{k}{\rightarrow}$" represents a nonbasic arc at value k. The exiting variable is noted by *.

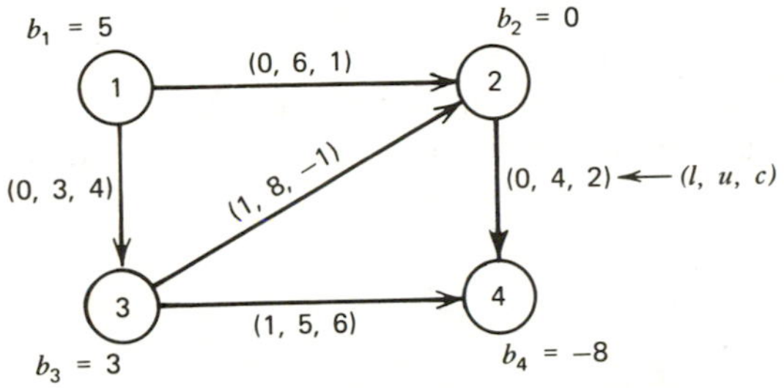

Figure 9.16. **An example network with lower and upper bounds.**

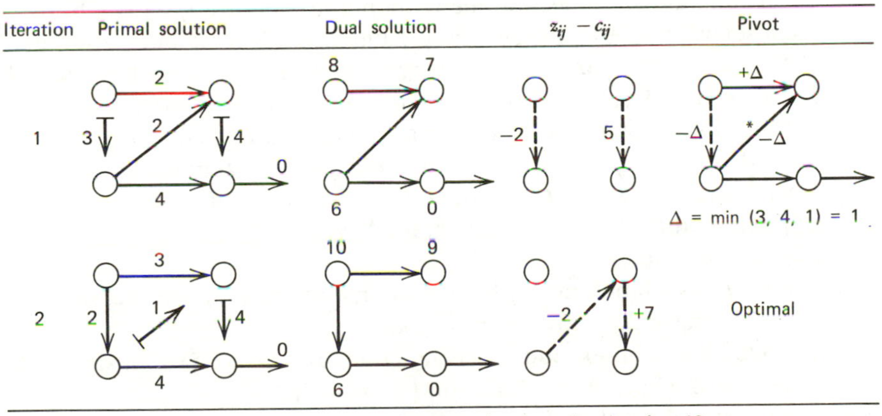

$x_{12}^* = 3,\ x_{13}^* = 2,\ x_{24}^* = 4,\ x_{32}^* = 1,\ x_{34}^* = 4,\ z^* = 42$

Figure 9.17. **The solution to Figure 9.16.**

A Labeling Algorithm for the Network Simplex Method

For either hand or computer calculations there are simple and convenient ways to maintain the information required to solve a minimal cost flow problem with lower and upper bounds by the network simplex method. Suppose that we associate with each node $j \in N$ a label, $L(j) = (\pm i, \Delta_j)$, containing two pieces of information. The second entry, Δ_j, in $L(j)$ indicates the current estimate for the value of the flow change. The first entry, $\pm i$, in $L(j)$ indicates the previous node in the cycle along which flow will be changed. If the first entry in $L(j)$ is $+i$, then flow will be added to arc (i, j); otherwise, if the first entry is $-i$, then flow will be subtracted from arc (j, i). The labeling algorithm becomes the following.

INITIALIZATION STEP

Select a basic feasible solution and set the x_{ij}'s to their required values in the solution. If a basic feasible solution is not readily available, utilize artificial variables.

MAIN STEP

1. (Compute dual variables.) Set $w_m = 0$. If w_i has been computed, w_j has not been computed, and arc (i, j) is a basic arc, then set $w_j = w_i - c_{ij}$. If w_i has been computed, w_j has not been computed, and arc (j, i) is a basic arc, then set $w_j = w_i + c_{ji}$. Repeat step 1 until all w_i's have been computed.
2. (Check optimality or select entering variable.) If each nonbasic variable has $x_{ij} = l_{ij}$ and $z_{ij} - c_{ij} \leqslant 0$ or $x_{ij} = u_{ij}$ and $z_{ij} - c_{ij} \geqslant 0$, stop; the optimal solution is obtained. Otherwise, erase any labels and choose an arc (p, q) such that $z_{pq} - c_{pq} < 0$ and $x_{pq} = u_{pq}$ or $z_{pq} - c_{pq} > 0$ and $x_{pq} = l_{pq}$. In the former case, set $s = p, t = q$, and $L(s) = (-t, \Delta_s)$; in the latter case set $s = q, t = p$, and $L(s) = (+t, \Delta_s)$ where $\Delta_s = u_{pq} - l_{pq}$.
3. (Determine cycle.)
   a. If node i has a label, node j has no label, and arc (i, j) is basic, set $L(j) = (+\mathrm{i}, \Delta_j)$ where $\Delta_j = \text{Minimum}\{\Delta_i, u_{ij} - x_{ij}\}$.
   b. If node i has a label, node j has no label, and arc (j, i) is basic, set $L(j) = (-\mathrm{i}, \Delta_j)$, where $\Delta_j = \text{Minimum}\{\Delta_i, x_{ji} - l_{ji}\}$.
   c. Repeat step 3 until node t is labeled.
4. (Change flow along cycle and determine exit variable.) Let $\Delta = \Delta_t$. If the first entry in $L(t)$ is $+k$, then (i) replace x_{kt} by $x'_{kt} = x_{kt} + \Delta$, and (ii) if $x'_{kt} = u_{kt}$, set $(g, h) = (k, t)$; otherwise, if the first entry in $L(t)$ is $-k$, then (i) replace x_{tk} by $x'_{tk} = x_{tk} - \Delta$, and (ii) if $x'_{tk} = l_{tk}$, set $(g, h) = (t, k)$. Backtrack to node k and repeat this step until node t is reached in the backtrack process.
5. (Update basis.) If $(g, h) = (p, q)$, go to step 2. Otherwise add (p, q) to the basis, remove (g, h) from the basis, and go to step 1.

An Example of the Labeling Algorithm

Since the initialization and computation of dual variables are the same as previously described, we shall illustrate the labeling algorithm in performing the first pivot on the network of Figure 9.16. For the initial basic feasible solution in iteration 1 of Figure 9.17, we have $z_{13} - c_{13} = -2$ and $x_{13} = u_{13}$, and so x_{13} should enter the basis (by decreasing).

The sequence of operations beginning with step 3 of the labeling algorithm is as follows:

1. $s = 1, t = 3, L(1) = (-3, 3)$
2. $L(2) = (+1, 3)$
3. $L(3) = (-2, 1)$
4. $\Delta = 1$
5. $L_1(3) = -2 \Rightarrow x'_{32} = x_{32} - \Delta = 1,\ x'_{32} = l_{32} \Rightarrow (g, h) = (3,2)$
6. $L_1(2) = +1 \Rightarrow x'_{12} = x_{12} + \Delta = 3$
7. $L_1(1) = -3 \Rightarrow x'_{13} = x_{13} - \Delta = 2$
8. $(g, h) = (3, 2) \Rightarrow x_{32}$ leaves

Having completed the pivot, we proceed to compute the new values of the dual variables and locate the next entering variable.

9.8 THE SIMPLEX TABLEAU ASSOCIATED WITH A NETWORK FLOW PROBLEM

In Section 9.3 we showed how to construct the column $\mathbf{y}_{ij}$ for any nonbasic arc (i, j). Elsewhere, we have seen how to obtain the values of the basic variables and the $z_{ij} - c_{ij}$'s. Thus it is possible to construct the entire updated tableau by examining the basis subgraph at the corresponding iteration.

As an example, consider the final (optimal) basis in Figure 9.13 for the network flow problem of Figure 9.12. The simplex tableau for this basis is given in Table 9.1, where x_4 denotes the artificial variable at node 4. To illustrate how a particular nonbasic column is obtained, consider x_{12}. We have already indicated how $z_{12} - c_{12} = -6$ may be computed. To produce the other entries in the column we consider the unique chain formed by adding (1, 2) to the basic subgraph. The unique chain in the basis tree is $C = \{(1, 3), (2, 3)\}$. To reorient this into a path from 1 to 2 we multiply column x_{13} by 1 and x_{23} by -1; thus the

Table 9.1 The Simplex Tableau Associated with the Final Basis in Figure 9.13

	z	x_{12}	x_{13}	x_{23}	x_{24}	x_{32}	x_{34}	x_{41}	x_4	RHS
z	1	-6	0	0	-2	-5	0	-5	0	-7
x_{13}	0	1	1	0	0	0	0	-1	0	4
x_{23}	0	-1	0	1	1	-1	0	0	0	2
x_{34}	0	0	0	0	1	0	1	-1	0	5
x_4	0	0	0	0	0	0	0	0	1	0

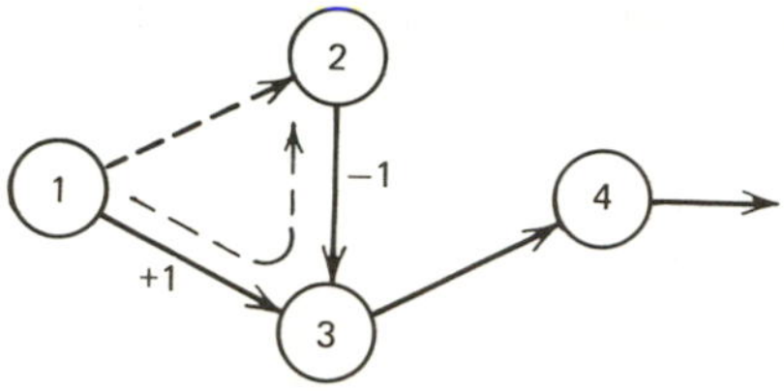

coefficients in the tableau. As a check we see that

$$\mathbf{a}_{13} - \mathbf{a}_{23} = (\mathbf{e}_1 - \mathbf{e}_3) - (\mathbf{e}_2 - \mathbf{e}_3)$$
$$= \mathbf{e}_1 - \mathbf{e}_2 = \mathbf{a}_{12}$$

as required. The other columns of the simplex tableau are obtained in a similiar manner.

EXERCISES

9.1 Let G be a connected graph.
   a. Show that G contains at least one tree.
   b. Let T be a tree in G. If the number of nodes in G is at least 2, show that T has at least two ends.
   c. Show that if an end of the tree and its incident arc are removed, the resulting graph is a tree.
   d. Show that a graph with m nodes, $m - 1$ arcs, and no cycles is a tree.

9.2 Show that the constraint matrix for the minimal cost flow problem is totally unimodular. (*Hint*. Use an analogous reasoning to that for the transportation problem.)

9.3 Can a basis tree ever have two roots (if they were available)?

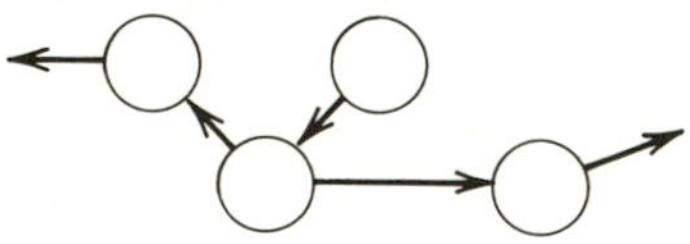

(*Hint*. Show that the two root arcs and the chain between them form a dependent set.)

9.4 Solve the following network flow problem using x_{13}, x_{23}, x_{34} as part of a starting basis.

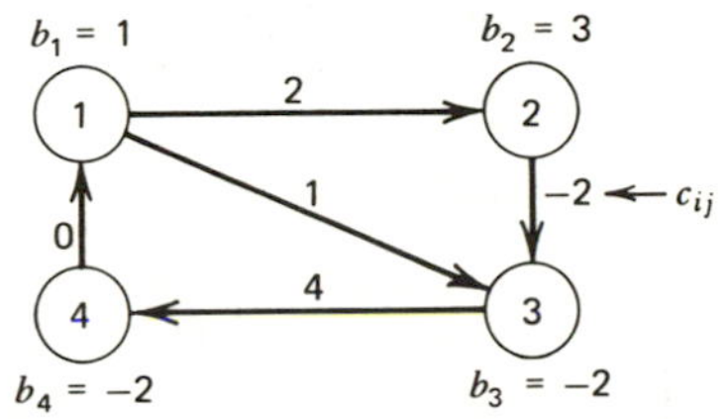

9.5 Apply the two-phase method and the big M method to the network of Exercise 9.4 to get an optimal solution.

9.6 Indicate how the minimal cost flow problem of Section 9.1 can be transformed into a transportation problem. Illustrate by the following network. (*Hint.* Locate the minimal cost path from each supply point to each demand point.)

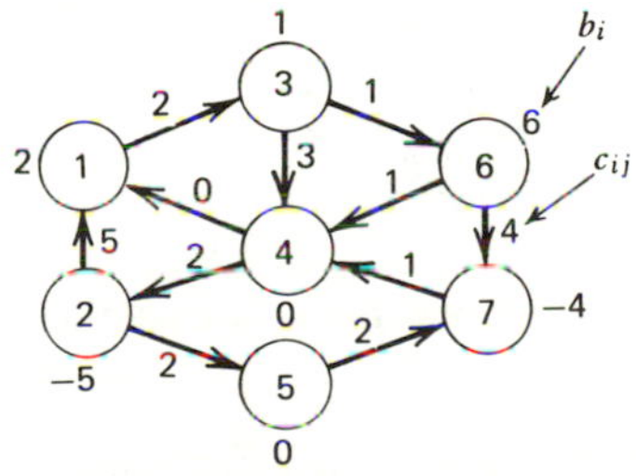

9.7 Suppose, in the flow problem of Section 9.1, that we locate a path from a supply point to a demand point. Putting as much flow as possible on the path, decreasing the corresponding supply and demand, we repeat the whole process until all supplies are allocated (and demands satisfied).

a. Will the feasible solution obtained be basic?
b. If not, how can the solution obtained be made basic?

9.8 Solve the following network flow problem.

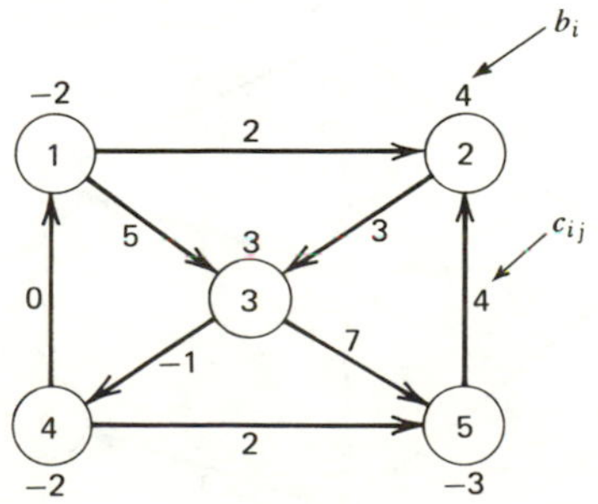

9.9 Solve the following network flow problem by the two-phase method.

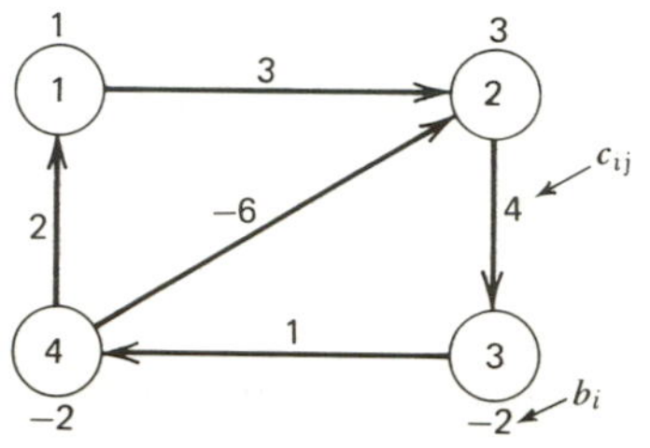

9.10 Solve the following network flow problem.

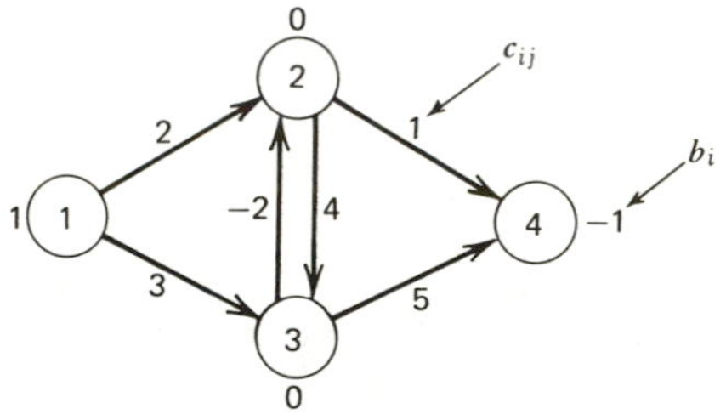

(*Note*. for this choice of b_i's we have found the shortest path node 1 to node 4.)

9.11 Show that if the cost coefficient for the artificial variable (root) is increased by an amount θ, then every dual variable will change by an amount θ. Thus, show that the value of the artificial cost coefficient does not matter in the computation of $z_{ij} - c_{ij}$.

9.12 Suppose that the following figure represents a railroad network. The numbers beside each arc represent the time it takes to traverse the arc. Two locomotives are stationed at point 2 and one locomotive at point 1. Three locomotives are needed at point 6. Find the minimum total time solution to get the power required to point 6.

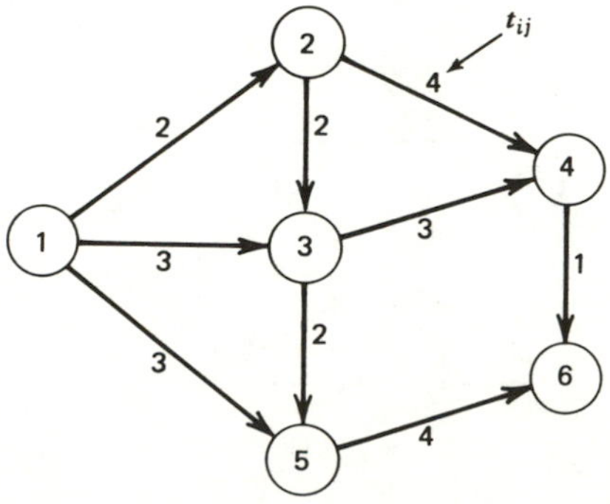

9.13 Solve the following flow problem. (*Note*: $\Sigma b_i \neq 0$.)

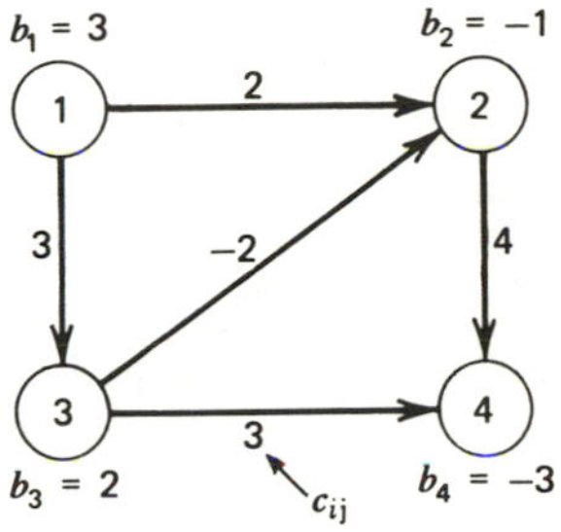

9.14 Consider the following network.
Node 1 has 5 units available.
Node 3 has 2 units available.
Node 2 needs 4 units.
Node 4 needs 1 unit.
a. Set up the linear program for this problem.
b. Solve the problem by the network simplex method.

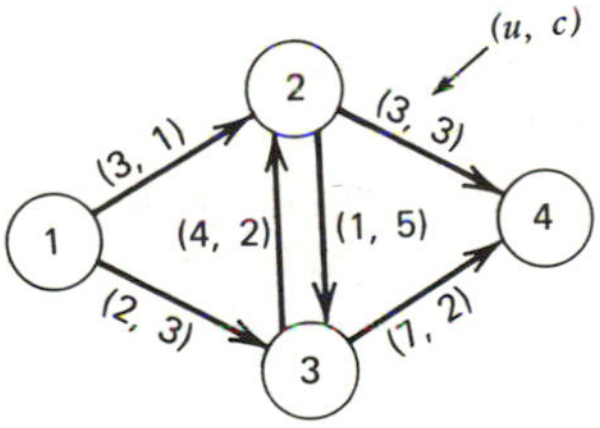

9.15 Solve the following network flow problem.

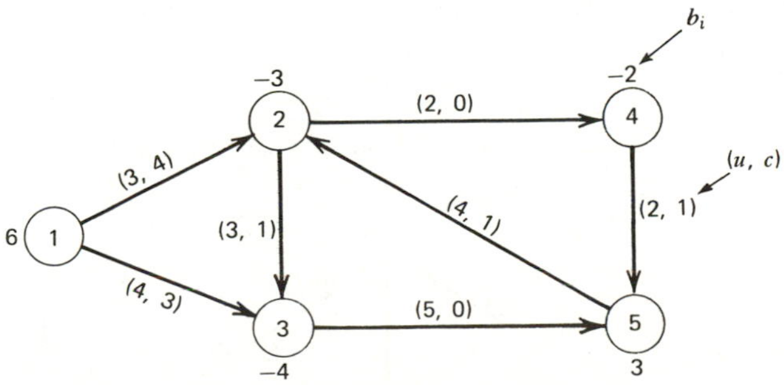

9.16 Starting with x_{12}, x_{24}, and x_{31} as part of a basis where all other x_{ij}'s are nonbasic at their lower bound, solve the following network flow problem.

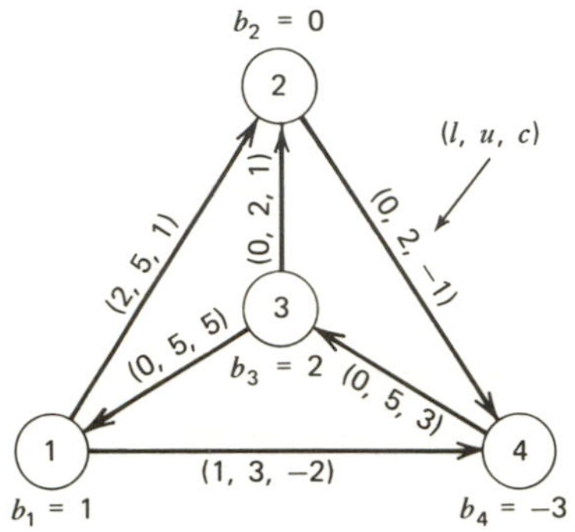

9.17 Apply the two-phase method and the big-M method to Exercise 9.16.

9.18 A company has requirements for a certain type of machine during the next N months of D_i per month, $i = 1, 2, \ldots, N$. Those requirements must be met, although excesses are permitted to any desirable extent. No machines are available at the beginning of month 1. At the beginning of each of the N months, the company may purchase machines. The machines are delivered immediately; that is, there is no lead time. The company may purchase machines that last one month, two months, and so on on up to M months. The number of months a machine is usable is called its service life. The cost of a machine depends on the month in which it is bought and on its service life. In particular, a machine bought in the ith month with service life of k months costs c_{ik} dollars. Naturally a machine that lasts longer costs more, so that $c_{ip} < c_{is}$ for $p < s$.

a. Formulate a mathematical statement for this problem, assuming that the objective is to minimize the sum of the monthly costs. Let x_{ik} be the number of machines bought in month i with service life of k months.

b. Formulate the problem as a flow problem. Use $N = 3$ and $M = 2$. Summarize your work on a network. (*Hint*. Consider elementary row operations to obtain a 1 and -1 in every column.)

9.19 Consider the following network flow problem.
In addition:
Node 1 can produce an unlimited supply of units at no cost.

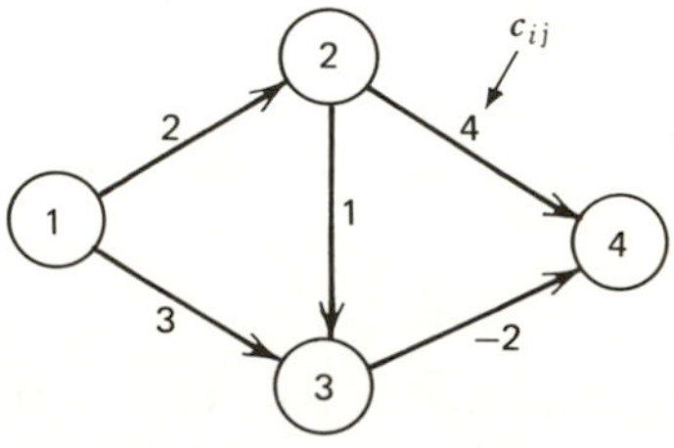

Node 2 can produce up to 3 units at a cost of 4 per unit.
Node 3 needs 2 units.
Node 4 needs 5 units.
Find the optimal shipping policy.

9.20 It is often necessary (particularly in integer programs) to generate a row of the simplex tableau. Indicate how this can easily be done for a network flow problem. Illustrate by generating the row associated with x_{13} in the starting simplex tableau for Exercise 9.4. [*Hint*. Remove some arc (basic in a given row) of the basic tree, thereby separating the nodes into two subsets. Then consider the set of nonbasic arcs going between the two node sets—those in the same direction as the given basic arc and those in the opposite direction.]

9.21 Indicate how the lexicographic simplex method may be applied to the network flow problem.

9.22 Develop in detail two algorithms for solving network flow problems that are based on the dual simplex and the primal-dual methods.

9.23 Solve the following network flow problem.

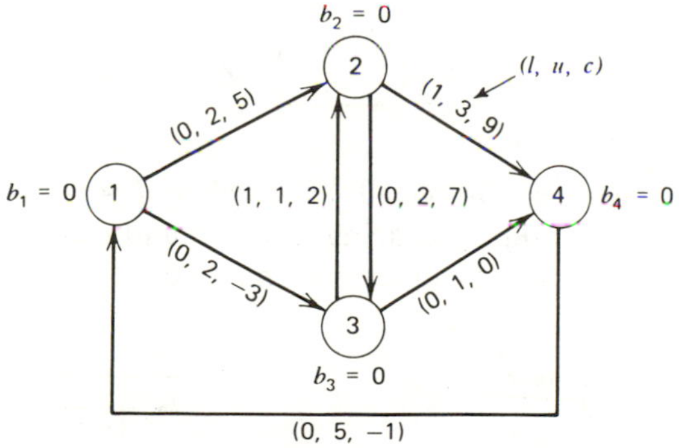

9.24 Indicate how we can generate a row or a column of $\mathbf{B}^{-1}$ associated with a network flow problem. Illustrate by generating the row of $\mathbf{B}^{-1}$ associated with x_{12} in the initial basis for Exercise 9.4.

9.25 According to the rule of Section 9.4, show that the root is the last basic variable to be assigned a value.

9.26 Prove or give a counterexample: In order for a variable to be basic in a given row of the linear program for a network flow problem, the variable must have a nonzero entry (that is, ± 1) in that row of the original constraint matrix; that is, the arc must be incident with the associated node.

9.27 Indicate how we can handle lower and upper capacity constraints on flow through a node i. (*Hint*. Consider splitting node i into two nodes.)

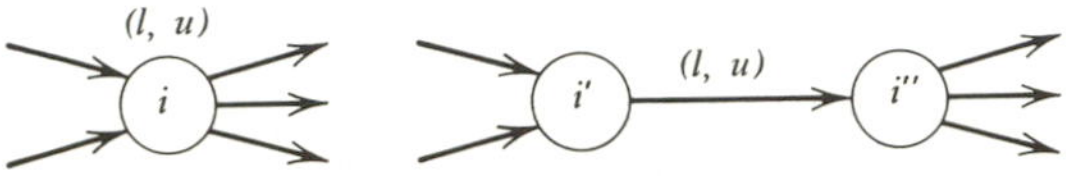

9.28 Consider the following network of cities. Each city must be visited at least once by some salesman. Use network flows to indicate the number of salesmen necessary and their routes to minimize the total distance traveled. Assume that an unlimited number of salesmen are available at no cost to be positioned wherever needed. (*Hint*. Impose a lower bound of 1 on flow through each node.) Can we impose the additional constraint that only one salesman be used? (This would be the classical *traveling salesman problem*.)

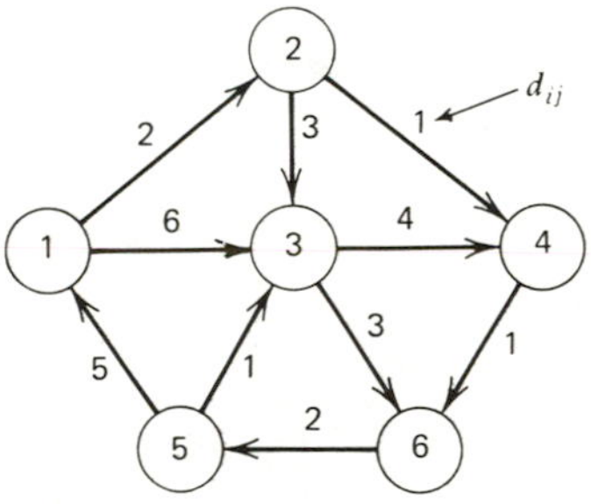

9.29 How can we handle undirected arcs, when $l = 0$, in a network flow problem? (*Hint*. Consider replacing a single undirected arc by two oppositely directed arcs.) What happens when $l > 0$?

9.30 How can we transform a network flow problem with some $l'_{ij} < 0$ into one with all $l \geqslant 0$? [*Hint*. Consider adding two arcs, (i, j) and (j, i) with appropriate lower bounds.] Illustrate by the following network.

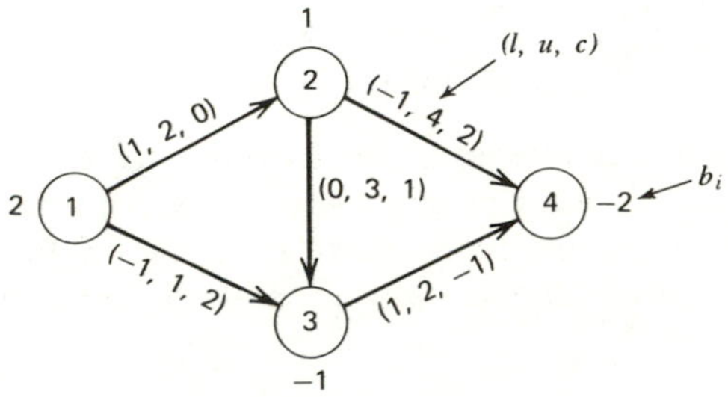

9.31 Show how a network flow problem with $l = 0$, for which the cost function associated with each arc is piecewise-linear and convex, can be solved by the methods of this chapter. (*Hint*. Consider adding parallel arcs—one for each segment—with proper costs and upper bounds.) Can the methods of this chapter be also used for network problems with piecewise-linear and concave functions?

9.32 Show that a network flow problem with nonzero lower bounds can be converted to one with zero lower bounds. (*Hint*. Consider a change of variables: $\hat{x}_{ij} = x_{ij} - l_{ij}$.) Illustrate by converting the following network to one with zero lower bounds.

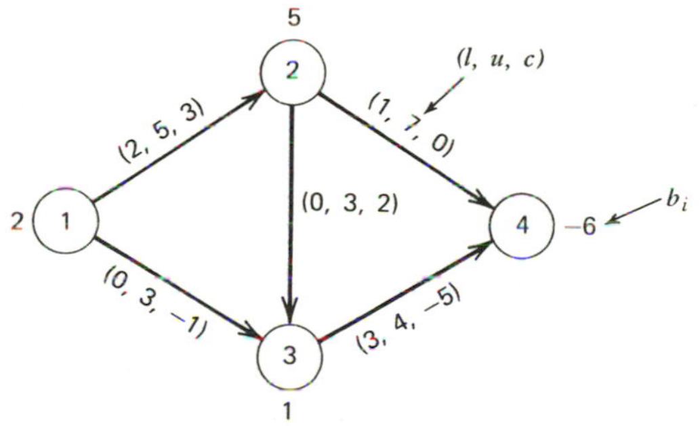

9.33 Show that a network flow problem with zero lower bounds and finite upper bounds can be converted to one with zero lower bounds and no (infinite) upper bounds. [*Hint*. Consider splitting arc (i, j) as shown below.]

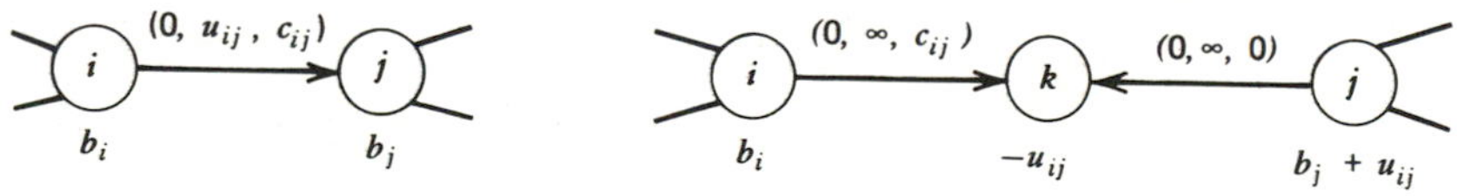

Illustrate by converting the following network to one without upper bounds.

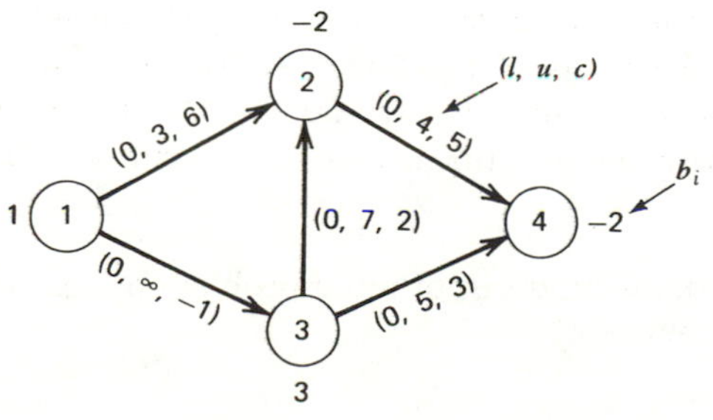

9.34 Prove that when we discard the redundant constraint, a network flow basis is characterized by a *rooted spanning forest* (a collection of trees, each with exactly one root, which spans the node set). Using this characterization, develop the representation of a nonbasic column in terms of the basis. Utilize your results to describe a complete simplex method for network flow problems without the redundant constraint. Apply the method to the network in Exercise 9.4 with the last constraint deleted.

9.35 Show how the results of this chapter can be generalized to handle the *flow with gains* problem where each column of the **A** matrix is of the form $\mathbf{a}_{ij} = \mathbf{e}_i - p_{ij}\mathbf{e}_j$, $p_{ij} > 0$. In particular, show that a basis is a *pseudorooted spanning forest* (a collection of subgraphs, each with either no cycles and exactly one root arc or no root arcs and exactly one cycle, which spans the node set).

9.36 Develop a method to solve a linear program of the form

$$\begin{array}{ll} \text{Minimize} & \mathbf{cx} + c_{n+1}x_{n+1} \\ \text{Subject to} & \mathbf{Ax} + \mathbf{a}_{n+1}x_{n+1} = \mathbf{b} \\ & \mathbf{x} \geq \mathbf{0},\ x_{n+1} \geq 0 \end{array}$$

where **A** is a node-arc incidence matrix. Apply the method to the following problem.

$$\begin{array}{lrcl} \text{Minimize} & 2x_1+3x_2+x_3+3x_4+5x_5+4x_6 & & \\ \text{Subject to} & x_1+x_2 \qquad\qquad\qquad\qquad + x_6 & = & 2 \\ & -x_1 \qquad +x_3+x_4 \qquad -2x_6 & = & 3 \\ & -x_2-x_3 \qquad +x_5+3x_6 & = & -1 \\ & -x_4-x_5-x_6 & = & -4 \\ & x_1,\ x_2,\ x_3,\ x_4,\ x_5,\ x_6 & \geq & 0 \end{array}$$

9.37 Can the results of Exercise 9.36 be generalized to the case where the constraint matrix is of the form (**A**, **D**), where **A** is a node-arc incidence matrix, and **D** is any arbitrary matrix? Apply the method to the previous problem with the additional column x_7 with $\mathbf{a}_7 = (2, -5, 2, 0)^t$ and $c_7 = -3$.

9.38 Solve the following transportation problem by the network simplex method of this chapter.

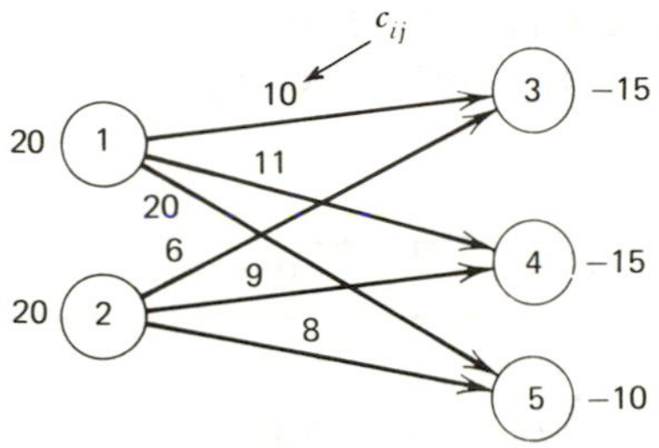

9.39 Consider the following *production-transportation-inventory problem.* A firm produces a single product at two locations in each of two time periods. The unit costs and production limits vary by time period and are given by the following diagram.

PRODUCTION LOCATION	TIME PERIOD 1	TIME PERIOD 2	
1	\$25/6	\$35/2	Unit cost/
2	\$30/10	\$42/9	Production limit

The product will be shipped (instantaneously) to each of two locations to satisfy specified demands over the two periods. These demands are as follows.

CONSUMER LOCATION	TIME PERIOD 1	TIME PERIOD 2
1	3	1
2	5	4

The unit shipping cost varies over time and is given by the following.

PRODUCTION LOCATION	PERIOD 1 CONSUMER LOCATION 1	PERIOD 1 CONSUMER LOCATION 2	PRODUCTION LOCATION	PERIOD 2 CONSUMER LOCATION 1	PERIOD 2 CONSUMER LOCATION 2
1	\$50	\$60	1	\$60	\$80
2	\$40	\$70	2	\$70	\$90

Finally, the product may be held in inventory at the production and consumer locations to satisfy later period needs. The relavent data are given below.

PRODUCTION LOCATION		CONSUMER LOCATION		
1	2	1	2	
\$1/2	\$2/3	\$3/1	\$4/3	Unit cost/ Inventory limit

Set up a network flow problem that can be used to solve the problem of minimizing the total cost to satisfy the demand over the two periods. (*Hint*. Create a separate node that represents each location at each time period. Shipping arcs connect nodes for the same time period; inventory arcs connect nodes in different time periods.)

9.40 Show that the coefficient matrix for the lower-upper bounded network flow problem is totally unimodular. This matrix is of the form

x	$\mathbf{s}_1$	$\mathbf{s}_2$	RHS
A	**0**	**0**	**b**
I	**− I**	**0**	**l**
I	**0**	**I**	**u**

9.41 a. State the dual problem for the lower-upper bounded network flow problem in Section 9.7.
b. Give the values of all of the optimal dual variables for Figure 9.17 at iteration 2.
c. Verify optimality by computing the dual objective function.

9.42 Consider the following network.

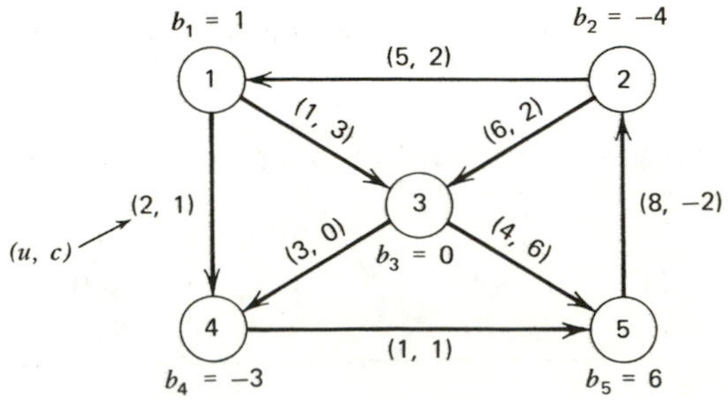

Consider a basis given by x_{21}, x_{23}, x_{34}, x_{52} and the artificial variable x_5 with x_{14} nonbasic at its upper bound and all other variables nonbasic at their lower bounds, that is, zero.

a. Find the basic solution. Is it feasible?
b. Give the simplex tableau associated with the specified basic solution.
c. State the dual program.
d. Find the complementary dual solution. Is it feasible?
e. Regardless of costs, perform one pivot graphically to bring x_{14} into the basis.
f. Is the new basis optimal?

9.43 If it is known in advance that a certain arc will be carrying positive flow in any optimal solution to an uncapacitated network flow problem, what simplifications can be made in the solution method?

9.44 Twenty million barrels of oil must be transported from Dhahran in Saudi Arabia to the ports of Rotterdam, Marseilles, and Naples in Europe. The demands of these ports are respectively 4, 12, and 4 million barrels. The

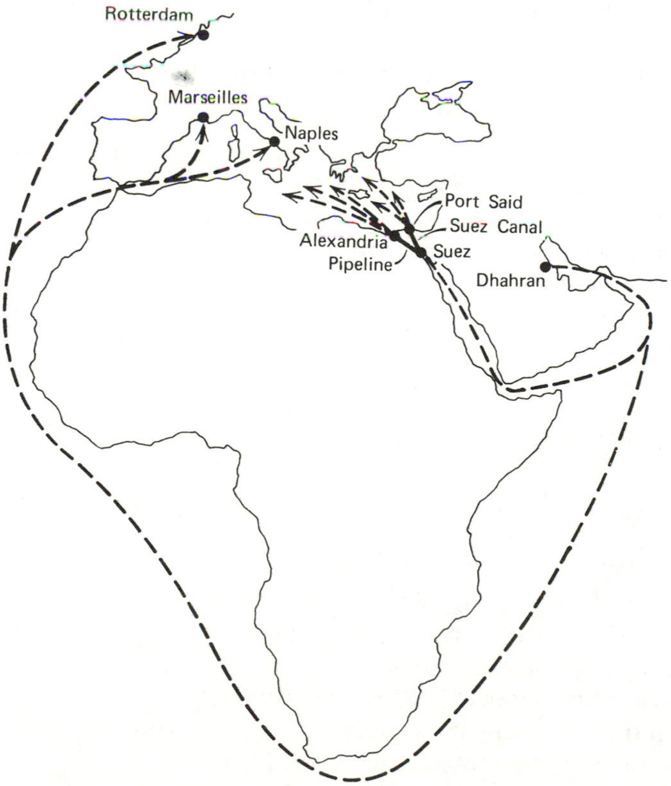

following three alternative routes are possible (see accompanying map).

a. From Dhahran, around Africa to Rotterdam, Marseilles, and Naples. The average transportation and handling cost per barrel is \$1.20, \$1.40, and \$1.40 respectively.

b. From Dhahran to the city of Suez, and then through the Suez Canal to Port Said. From Port Said the oil is shipped to Rotterdam, Marseilles, and Naples. The average transportation and handling cost from Dhahran to the city of Suez is \$0.30, and the additional unit cost of transporting through the canal is \$0.20. Finally, the unit transportation costs from Port Said to Rotterdam, Marseilles, and Naples are respectively \$0.25, \$0.20, and \$0.15.

c. From Dhahran to the city of Suez, and then through the proposed pipeline system from Suez to Alexandria. The average transportation cost per barrel through the pipeline system is \$0.15, and the unit transportation costs from Alexandria to Rotterdam, Marseilles, and Naples are \$0.22, \$0.20, and \$0.15.

Furthermore, 30% of the oil in Dhahran is transported by large tankers that cannot pass through the Suez Canal. Also, the pipeline system form Suez to Alexandria has a capacity of 10 million barrels of oil.

a. Formulate the problem as a general network flow problem.
b. Use the procedures of this chapter to find the optimal shipping pattern.

9.45 The following network represents an electrical power distribution network connecting power generating points with power consuming points. The arcs are undirected; that is, power may flow in either direction. Points 1, 4, 7, and 8 are generation points with generating capacities and unit costs given by the following table.

	GENERATING POINT			
	1	4	7	8
CAPACITY (THOUSANDS OF KILOWATT HOURS)	100	60	80	150
UNIT COST (\$/1000 KILOWATT HOURS)	15.0	13.5	21.0	23.5

Points 2, 5, 6, and 9 are consuming points with demands of 35,000 KWH, 50,000 KWH, 60,000 KWH, and 40,000 KWH respectively. There is no upper bound on the transmission line capacity. The unit cost of transmission on each line segment is \$11.0 per 1000 KWH.

a. Set up the power distribution problem as a network flow problem.
b. Solve the resulting problem.

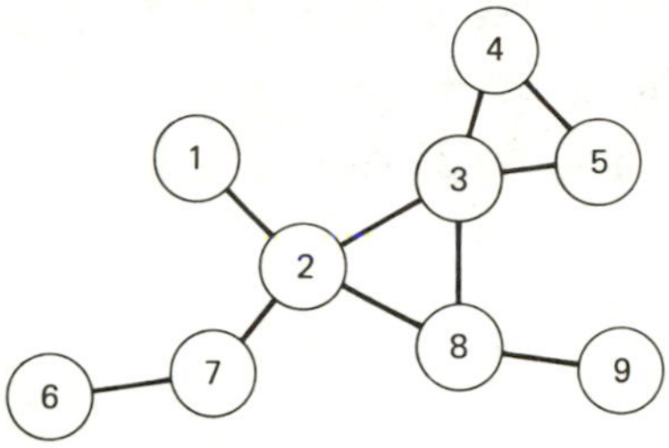

9.46 Formulate Exercise 1.16 as a network flow problem. Solve this problem by the network simplex method.

NOTES AND REFERENCES

1. The simplex method for general network flow problems is a natural extension of the work on the transportation problem and the work of Koopmans [290] relating linear programming bases for transportation problems and trees in a graph.
2. Computational experience with the network simplex algorithm, as reported by Glover, Karney, and Klingman [197], Langley and Kennington [303], and others, indicates that this algorithm compares favorably with other procedures for solving network flow problems.
3. Exercise 9.35 discusses the extension of the models of this chapter to the *flow with gains* network models. Jewell [261] first solved this problem by a primal-dual method. Ellis Johnson [266], Langley [302], and others have since treated the problem via the simplex method.

TEN: THE OUT-OF-KILTER ALGORITHM

In the previous chapter we presented a network simplex method for solving minimal cost network flow problems. In this chapter we present another method for solving minimal cost network flow problems, called the out-of-kilter algorithm. This algorithm is similar to the primal-dual algorithm in that it begins with dual feasibility but not necessarily primal feasibility and iterates between primal and dual problems until optimality is achieved. However, it differs from the primal-dual algorithm (as strictly interpreted) in that the out-of-kilter algorithm does not always maintain complementary slackness. Thus it can be viewed as a generalization of the primal-dual algorithm for network flow problems.

10.1 THE OUT-OF-KILTER FORMULATION OF A MINIMAL COST NETWORK FLOW PROBLEM

For convenience of presentation, the form of the minimal cost flow problem that we shall work with is

$$\begin{aligned}
\text{Minimize} \quad & \sum_{i=1}^{m} \sum_{j=1}^{m} c_{ij} x_{ij} \\
\text{Subject to} \quad & \sum_{j=1}^{m} x_{ij} - \sum_{k=1}^{m} x_{ki} = 0 \qquad i = 1, \ldots, m \qquad (10.1) \\
& x_{ij} \geqslant l_{ij} \qquad i, j = 1, \ldots, m \\
& x_{ij} \leqslant u_{ij} \qquad i, j = 1, \ldots, m
\end{aligned}$$

where it is understood that the sums and inequalities are taken over existing arcs only. We call a *conserving flow* any flow (choice of the x_{ij}'s) satisfying constraints (10.1). A conserving flow that satisfies the remaining constraints $l_{ij} \leqslant x_{ij} \leqslant u_{ij}$ is a *feasible flow* (solution). We shall assume that c_{ij}, l_{ij}, and u_{ij} are integers and that $0 \leqslant l_{ij} \leqslant u_{ij}$.

The foregoing formulation is completely equivalent to the formulation of the minimal cost network flow problem presented in Chapter 9. In Exercise 10.5 we ask the reader to provide the transformation between the two formulations.

Since all right-hand-side values of the flow conservation equations (10.1) are zero, we conclude that the flow in the network does not have a starting point or an ending point, but rather *circulates* continuously throughout the network. Thus all conserving flow in the network will be along circuits (directed cycles).

The Dual of a Network Flow Problem and Its Properties

If we associate a dual variable w_i with each node conservation equation (10.1), a dual variable h_{ij} with the constraint $x_{ij} \leqslant u_{ij}$ (which is treated as $-x_{ij} \geqslant -u_{ij}$ for the purpose of taking the dual), and a dual variable v_{ij} with the constraint $x_{ij} \geqslant l_{ij}$, the dual of the out-of-kilter formulation for the minimal cost network flow problem is given by

$$\begin{aligned}
\text{Maximize} \quad & \sum_{i=1}^{m} \sum_{j=1}^{m} l_{ij} v_{ij} - \sum_{i=1}^{m} \sum_{j=1}^{m} u_{ij} h_{ij} \\
\text{Subject to} \quad & w_i - w_j + v_{ij} - h_{ij} = c_{ij} \qquad i, j = 1, \ldots, m \\
& h_{ij}, v_{ij} \geqslant 0 \qquad i, j = 1, \ldots, m \\
& w_i \quad \text{unrestricted} \qquad i = 1, \ldots, m
\end{aligned}$$

where the summations and the constraints are taken over existing arcs. The dual problem has a very interesting structure. Suppose that we select any set of w_i's (we shall assume throughout the development that the w_i's are integers). Then the dual constraint for arc (i, j) becomes

$$v_{ij} - h_{ij} = c_{ij} - w_i + w_j, \qquad h_{ij} \geqslant 0, \qquad v_{ij} \geqslant 0$$

and can be satisfied by

$$v_{ij} = \text{Maximum } \{0, c_{ij} - w_i + w_j\}$$

$$h_{ij} = \text{Maximum } \{0, -(c_{ij} - w_i + w_j)\}$$

Thus the dual problem always possesses a feasible solution given any set of w_i's. In fact, the choices of v_{ij} and h_{ij} above yield the optimal values of v_{ij} and h_{ij} for a fixed set of w_i's (why?).

The Complementary Slackness Conditions

The complementary slackness conditions for optimality of the out-of-kilter formulation are (review the Kuhn-Tucker optimality conditions) the following:

$$(x_{ij} - l_{ij})v_{ij} = 0 \qquad i, j = 1, 2, \ldots, m \tag{10.2}$$

$$(u_{ij} - x_{ij})h_{ij} = 0 \qquad i, j = 1, 2, \ldots, m \tag{10.3}$$

Define $z_{ij} - c_{ij} \equiv w_i - w_j - c_{ij}$. Then by the definition of v_{ij} and h_{ij} we get

$$v_{ij} = \text{Maximum } \{0, -(z_{ij} - c_{ij})\} \tag{10.4}$$

$$h_{ij} = \text{Maximum } \{0, z_{ij} - c_{ij}\} \tag{10.5}$$

Note that $z_{ij} - c_{ij}$ would be the familiar coefficient of x_{ij} in the objective function row of the lower-upper bounded simplex tableau if we had a basic solution to the primal problem. However, we need not have one here.

Given a set of w_i's we can compute $z_{ij} - c_{ij} = w_i - w_j - c_{ij}$. Noting Equations (10.4) and (10.5), then the complementary slackness conditions (10.2) and (10.3) hold only if

$$z_{ij} - c_{ij} < 0 \Rightarrow v_{ij} > 0 \Rightarrow x_{ij} = l_{ij} \qquad i, j = 1, 2, \ldots, m$$

$$z_{ij} - c_{ij} > 0 \Rightarrow h_{ij} > 0 \Rightarrow x_{ij} = u_{ij} \qquad i, j = 1, 2, \ldots, m$$

We include the obvious additional condition

$$z_{ij} - c_{ij} = 0 \Rightarrow l_{ij} \leqslant x_{ij} \leqslant u_{ij} \quad i, j = 1, \ldots, m$$

Any *conserving flow* that satisfies the three conditions above will be optimal (why?). The problem, then, is to search over values of w_i's and conserving x_{ij}'s until the three conditions above are satisfied.

Consider Figure 10.1*a*. Selecting a set of starting w_i's, say each $w_i = 0$, and a conserving flow, say each $x_{ij} = 0$, we can check for optimality. Figure 10.1*b* gives $z_{ij} - c_{ij}$, x_{ij}, and w_i for the network of Figure 10.1*a*. In Figure 10.1*b* we see that $z_{12} - c_{12} = -2$ and $x_{12} = 0\ (= l_{12})$ and thus arc (1, 2) is said to be *in-kilter,* that is, *well.* On the other hand, $z_{23} - c_{23} = 3$ and $x_{23} = 0\ (< u_{23})$ and hence arc (2, 3) is said to be *out-of-kilter,* that is, *sick.* Thus the name *out-of-kilter*.

To bring arc (2, 3) into kilter we must either increase x_{23} or decrease $z_{23} - c_{23}$ by changing the w_i's. This is exactly what the out-of-kilter algorithm attempts to do. During the primal portion of the out-of-kilter algorithm we shall be changing the x_{ij}'s in an attempt to bring arcs into kilter. During the dual phase we change the w_i's in an attempt to reach an in-kilter state.

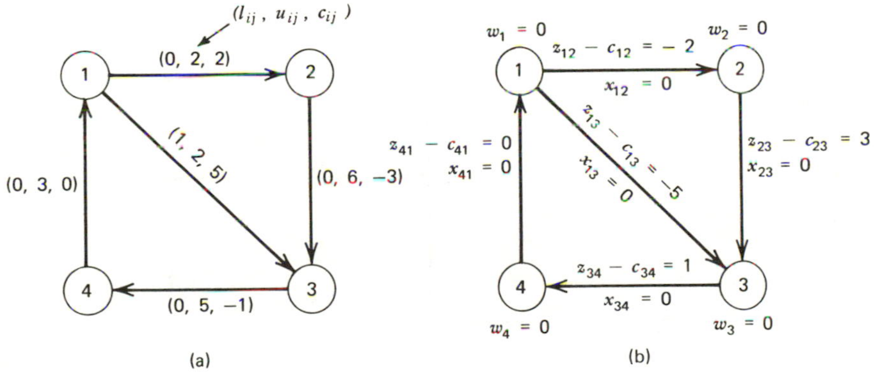

Figure 10.1. An example network: (*a*) The Network. (*b*) w_i, $z_{ij} - c_{ij}$, x_{ij}.

The Kilter States and Kilter Numbers for an Arc

The in-kilter and out-of-kilter states for each arc in a network are given in Figure 10.2. Note that an arc is in kilter if $l_{ij} \leqslant x_{ij} \leqslant u_{ij}$ and conditions (10.2) and (10.3) hold. As we change the flow on arc (i, j), the arc moves up and down a particular column in Figure 10.2 depending on whether x_{ij} is increased or decreased. As we change the w_i's, the arc moves back and forth along a row. Figure 10.2*b* gives a graphical depiction of the kilter states of an arc. Each of the cells in the matrix in Figure 10.2*a* corresponds to a particular subregion in Figure 10.2*b*.

In order to assure that the algorithm will converge, we need some measure of the "distance" from optimality. If we can construct an algorithm that periodically (at finite intervals) reduces the distance from optimality, then the algorithm

	$z_{ij} - c_{ij} < 0$	$z_{ij} - c_{ij} = 0$	$z_{ij} - c_{ij} > 0$
$x_{ij} > u_{ij}$	Out-of-kilter	Out-of-kilter	Out-of-kilter
$x_{ij} = u_{ij}$	Out-of-kilter	In-kilter	In-kilter
$l_{ij} < x_{ij} < u_{ij}$	Out-of-kilter	In-kilter	Out-of-kilter
$x_{ij} = l_{ij}$	In-kilter	In-kilter	Out-of-kilter
$x_{ij} < l_{ij}$	Out-of-kilter	Out-of-kilter	Out-of-kilter

(a)

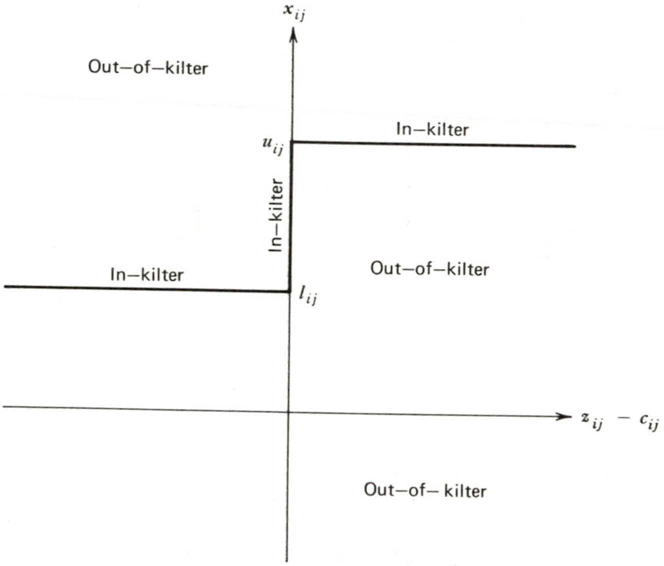

(b)

Figure 10.2. The possible kilter states for an arc.

will eventually converge. (We actually need a slightly stronger argument about the amount of reduction, but as we shall see, reduction is by an integer, so there is no problem with finiteness.)

There are many different measures of distance for the out-of-kilter problem. We present in Figure 10.3 one measure of distance that we call the *kilter number* K_{ij} for an arc (i, j). The kilter number is defined here to be the minimal change

	$z_{ij} - c_{ij} < 0$	$z_{ij} - c_{ij} = 0$	$z_{ij} - c_{ij} > 0$
$x_{ij} > u_{ij}$	$\lvert x_{ij} - l_{ij} \rvert$	$\lvert x_{ij} - u_{ij} \rvert$	$\lvert x_{ij} - u_{ij} \rvert$
$x_{ij} = u_{ij}$	$\lvert x_{ij} - l_{ij} \rvert$	0	0
$l_{ij} < x_{ij} < u_{ij}$	$\lvert x_{ij} - l_{ij} \rvert$	0	$\lvert x_{ij} - u_{ij} \rvert$
$x_{ij} = l_{ij}$	0	0	$\lvert x_{ij} - u_{ij} \rvert$
$x_{ij} < l_{ij}$	$\lvert x_{ij} - l_{ij} \rvert$	$\lvert x_{ij} - l_{ij} \rvert$	$\lvert x_{ij} - u_{ij} \rvert$

(a)

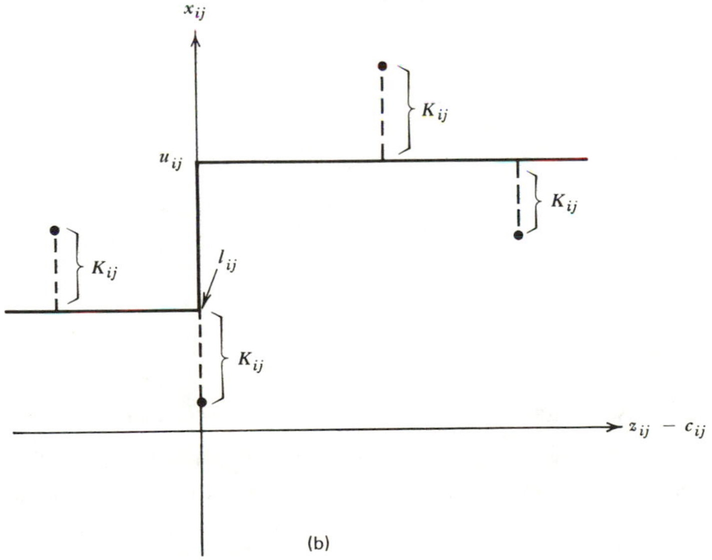

(b)

Figure 10.3. The kilter numbers K_{ij}'s.

of flow on the arc that is needed to bring it into kilter. The kilter number of an arc is illustrated graphically in Figure 10.3*b*. Notice that since all terms involve absolute values, the kilter number for an arc is nonnegative. Also, notice that if the arc is in-kilter, the associated kilter number is zero and if the arc is out-of-kilter, the associated kilter number is strictly positive. Note that if $z_{ij} - c_{ij} < 0$, then arc (i, j) is in-kilter only if the flow is equal to l_{ij} and hence the kilter number $|x_{ij} - l_{ij}|$ indicates how far the current flow x_{ij} is from the ideal case l_{ij}. Similarly, if $z_{ij} - c_{ij} > 0$, then the kilter number $|x_{ij} - u_{ij}|$ gives the distance from the ideal flow of u_{ij}. Finally, if $z_{ij} - c_{ij} = 0$, then the arc is in-kilter if $l_{ij} \leqslant x_{ij} \leqslant u_{ij}$. In particular, if $x_{ij} > u_{ij}$, then the arc is brought in-kilter if the flow decreases by $|x_{ij} - u_{ij}|$, and if $x_{ij} < l_{ij}$, then the arc is brought in-kilter if the flow increases by $|x_{ij} - l_{ij}|$, and hence the entries in Figure 10.3 under the column $z_{ij} - c_{ij} = 0$.

One method of assuring finite convergence of the out-of-kilter algorithm is to show the following.

1. The kilter number of any arc never increases.
2. At finite intervals the kilter number of some arc is reduced (by an integer).

This is exactly what we shall be able to show.

10.2 STRATEGY OF THE OUT-OF-KILTER ALGORITHM

As indicated before, the out-of-kilter algorithm may be viewed as a generalization of the primal-dual algorithm. In this respect the general steps of the algorithm are the following.

1. Begin with a conserving flow, such as each $x_{ij} = 0$, and a feasible solution to the dual, such as each $w_i = 0$, with h_{ij}, v_{ij} as defined in Equations (10.4) and (10.5). Identify the kilter states and compute the kilter numbers.
2. If the network has an out-of-kilter arc, conduct a primal phase of the algorithm. During this phase an out-of-kilter arc is selected and an attempt is made to construct a new conserving flow in such a way that the kilter number of no arc is worsened and that of the selected arc is improved.
3. When it is determined that no such improving flow can be constructed during the primal phase, the algorithm constructs a new dual solution in such a way that no kilter number is worsened and step 2 is repeated.
4. Iterating between steps 2 and 3, the algorithm eventually constructs an optimal solution or determines that no feasible solution exists.

The Primal Phase: Flow Change

During the primal phase the out-of-kilter algorithm attempts to decrease the kilter number on an out-of-kilter arc by changing the conserving flows in such a way that the kilter number on any other arc is not worsened. Examining Figure 10.3, we see that the flows must be changed in such a way that the corresponding kilter states move closer to the in-kilter states. For example, for the out-of-kilter state $x_{ij} > u_{ij}$ and $z_{ij} - c_{ij} < 0$, we can decrease x_{ij} by as much as $|x_{ij} - l_{ij}|$ before the arc comes into kilter. If we decrease x_{ij} beyond this, the arc will pass the in-kilter state (we do not want this to happen). Also, we do not permit any increase in this x_{ij}. A similar analysis of the other kilter states produces the results in Figure 10.4*a*.

Several cells in Figure 10.4*a* deserve special attention. The out-of-kilter state $x_{ij} > u_{ij}$ and $z_{ij} - c_{ij} = 0$ indicates that the flow can be decreased by as much as

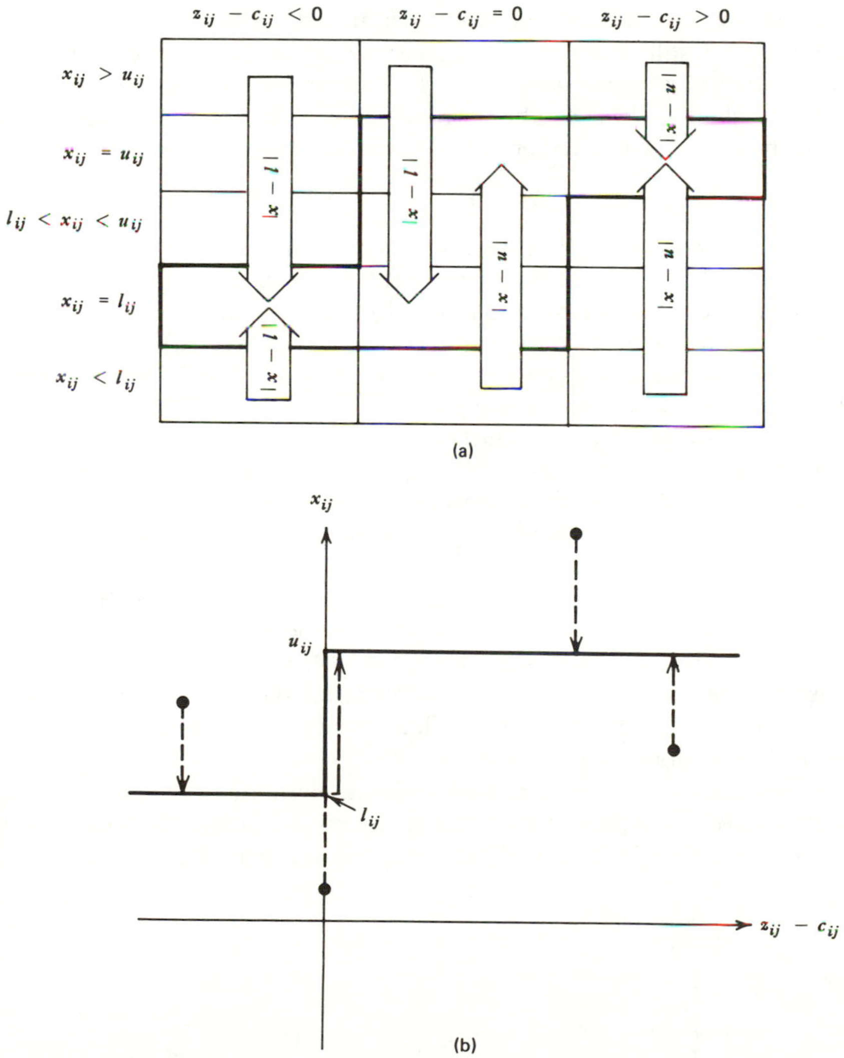

Figure 10.4. **Permitted flow change directions and amounts.**

$|x_{ij} - l_{ij}|$. Referring to Figure 10.3, we see that we really only need to decrease the particular x_{ij} by $|x_{ij} - u_{ij}|$, a smaller amount, to reach an in-kilter state. However, as can be seen in Figure 10.3, we may continue to decrease x_{ij} by an amount up to $|x_{ij} - l_{ij}|$ and the arc will still remain in-kilter. It is often desirable to do this to aid other arcs in reaching in-kilter states. Also, an arc in the in-kilter state $l_{ij} < x_{ij} < u_{ij}$ and $z_{ij} - c_{ij} = 0$ may have its flow either increased or decreased, while still maintaining its in-kilter status. Again we may illustrate the permitted flow changes graphically. This is done in Figure 10.4*b*.

Now that we have determined how much an individual flow on an arc may change, we must still determine what combination of flows we can change in order to maintain a *conserving flow*. If $\bar{\mathbf{x}}$ is the vector of (current) conserving flows, then Equation (10.1) can be rewritten as $\mathbf{A}\bar{\mathbf{x}} = \mathbf{0}$, where $\mathbf{A}$ is the node-arc incidence matrix. If $\boldsymbol{\Delta}$ is a vector of flow changes, then we must have

$$\mathbf{A}(\bar{\mathbf{x}} + \boldsymbol{\Delta}) = \mathbf{0} \qquad \text{or} \qquad \mathbf{A}\boldsymbol{\Delta} = \mathbf{0}$$

If $\mathbf{A}\boldsymbol{\Delta} = \mathbf{0}$ for a nonzero $\boldsymbol{\Delta}$, then the columns of $\mathbf{A}$ corresponding to the nonzero components of $\boldsymbol{\Delta}$ must be linearly dependent. Since $\mathbf{A}$ is a node-arc incidence matrix, then each column of $\mathbf{A}$ has exactly one $+1$ and one -1, and the nonzero components of $\boldsymbol{\Delta}$ must correspond to a (undirected) cycle or set of cycles (why?). Hence flows must be changed along a cycle or set of cycles in order to maintain the conservation equations.

Given an out-of-kilter arc, we must construct a cycle containing that arc. This cycle must have the property that when assigned an orientation and when flow is added, no arc has its kilter number worsened. A convenient method for doing this is to construct a new network G' from the original network according to the information in Figure 10.4. First, every node of the original network is in the new network. Next, if an arc (i, j) is in the original network and the flow may be increased, then arc (i, j) becomes part of the new network with the appropriate permitted flow change as indicated in Figure 10.4. Finally, if an arc (i, j) is in the original network and the flow can be decreased, then arc (j, i) becomes part of the new network with the permitted flow change as indicated for arc (i, j) in Figure 10.4. Arcs in the original network with $l_{ij} < x_{ij} < u_{ij}$ and $z_{ij} - c_{ij} = 0$ will produce two arcs, (i, j) and (j, i), with differing permitted flow changes in the new network. Arcs not permitted to change in flow are omitted entirely from G'.

Given the example indicated in Figure 10.1, a new network G' is constructed

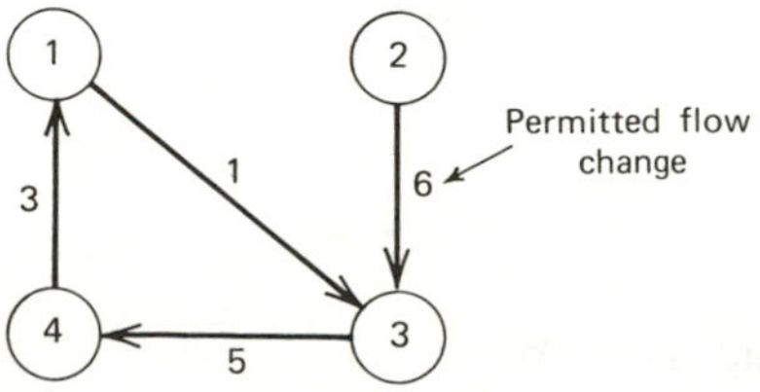

Figure 10.5. **The modified network G' for Figure 10.1.**

by the foregoing rules and is presented in Figure 10.5. To illustrate, consider arc (1, 3) in Figure 10.1. Note that $x_{13} < l_{13}$ and $z_{13} - c_{13} < 0$. From Figure 10.4 the flow on (1, 3) can increase to $l_{13} = 1$. This results in arc (1, 3) in Figure 10.5 with permitted flow of 1.

Once the new network G' is constructed and an out-of-kilter arc (p, q) is selected, we find a circuit (directed cycle) containing that arc in G'. This circuit in G' corresponds to a cycle in G. The flow in the cycle in G is changed according to the orientation provided by the circuit in G'. The amount of change is specified by the smallest permitted flow change of any arc that is a member of the circuit in G'. If no circuit containing the selected out-of-kilter arc exists in G', then we must proceed to the dual phase of the algorithm.

As an illustration of the primal phase, consider the modified network G' of Figure 10.5. We select an out-of-kilter arc, say (1, 3). From Figure 10.5 we see that a circuit exists containing arc (1, 3), namely $C = \{(1, 3), (3, 4), (4, 1)\}$. Hence we can change the flow around the associated cycle in G, increasing flows on arcs with the orientation of the circuit in G' and decreasing flows on arcs against the orientation of the circuit in G', and obtain an improved (in the kilter number sense) solution. The amount of permitted change in flow is $\Delta = \text{Minimum } \{1, 5, 3\} = 1$. The new solution and associated modified network is given in Figure 10.6*a*. Arcs (2, 3) and (3, 4) are still out-of-kilter in G. Selecting one of the associated arcs in G' (see Figure 10.6*b*), say (2, 3), we attempt to find a circuit in G' containing the selected arc. Because no such circuit exists, we must pass to the dual phase of the out-of-kilter algorithm.

It is convenient (but not necessary) for the various proofs of convergence to work on the same out-of-kilter arc (p, q) until it comes in-kilter. We shall assume throughout our discussion of the algorithm that this is done.

The Dual Phase: Dual Variable Change

When it is no longer possible to construct a circuit in G' containing a specific out-of-kilter arc, then we must change the $z_{ij} - c_{ij}$'s in such a way that no kilter

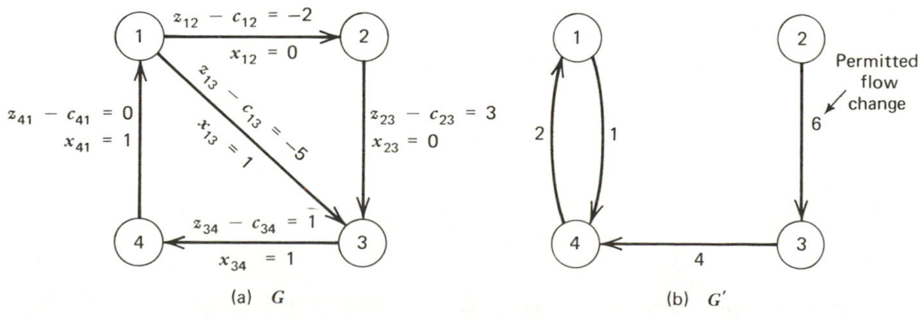

Figure 10.6. The new solution for the network of Figure 10.1.

number is worsened and (it is hoped) new arcs are introduced into G' that would allow us to find a circuit containing the out-of-kilter arc under consideration.

Since $z_{ij} - c_{ij} = w_i - w_j - c_{ij}$, we must change the w_i's in order to change the $z_{ij} - c_{ij}$'s. Let (p, q) be an out-of-kilter arc and let X be the set of nodes in G' that can be reached from node q along some *path* in G'. Let $\bar{X} \equiv N - X$, where $N = \{1, \ldots, m\}$. Note that neither X nor $\bar{X}$ is empty since $q \in X$ and $p \in \bar{X}$ when we pass to the dual phase. For $(p, q) = (2, 3)$ in Figure 10.6 we have $X = \{3, 4, 1\}$ and $\bar{X} = \{2\}$. In Figure 10.7 we illustrate the sets X and $\bar{X}$.

We would like to change the w_i's in such a way that no kilter number is worsened and the set X gets larger periodically. If another node comes into X at finite intervals, then eventually p will come into X and a circuit is created in G'. We have implicitly assumed that X will not get smaller. To ensure that this will not happen we should change the w_i's in such a way that all arcs with both ends in X are retained in the modified graph.

Consider $z_{ij} - c_{ij} = w_i - w_j - c_{ij}$. If w_i and w_j are changed by the same amount, then $z_{ij} - c_{ij}$ remains unchanged. Thus we can ensure that the set X will contain at least all of the same nodes after a dual variable change if we change all of the w_i's in X by the same amount θ. Suppose that we leave the w_i's in $\bar{X}$ unchanged. Then the only arcs that will be affected will be arcs from X to $\bar{X}$ and from $\bar{X}$ to X. Specifically, if $\theta > 0$ and we change the w_i's according to

$$w_i = \begin{cases} w_i + \theta & i \in X \\ w_i & i \in \bar{X} \end{cases}$$

then

$$(z_{ij} - c_{ij})' = z_{ij} - c_{ij} \quad \text{if} \quad i \in X, j \in X$$
$$\text{or} \quad i \in \bar{X}, j \in \bar{X}$$

Now, if $i \in X$ and $j \in \bar{X}$, we get

$$\begin{aligned}(z_{ij} - c_{ij})' &= (w_i + \theta) - w_j - c_{ij} \\ &= (z_{ij} - c_{ij}) + \theta\end{aligned}$$

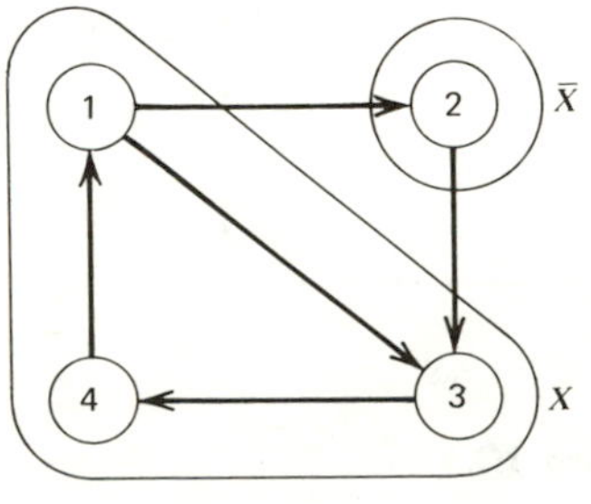

Figure 10.7. X AND $\bar{X}$ for $(p, q) = (2, 3)$ in Figure 10.6.

Also, for $i \in \bar{X}$ and $j \in X$ we get

$$(z_{ij} - c_{ij})' = w_i - (w_j + \theta) - c_{ij}$$
$$= (z_{ij} - c_{ij}) - \theta$$

Thus arcs from X to $\bar{X}$ will have their $z_{ij} - c_{ij}$'s increased by θ and those arcs from $\bar{X}$ to X will have their $z_{ij} - c_{ij}$'s decreased by θ. We must determine θ so that the kilter number of no arc is worsened and the kilter state of some arc is changed. First we must identify the arcs that can be in the set $(X, \bar{X})$ and in the set $(\bar{X}, X)$. (The notation (X, Y) represents the set $S = \{(x, y) : x \in X, y \in Y\}$).

Examining Figure 10.4, we see that the set $(X, \bar{X})$ cannot contain an arc associated with the kilter state $x_{ij} < l_{ij}$ and $z_{ij} - c_{ij} < 0$, since such an arc (i, j) in G would become an arc in G' with the result that if i can be reached (along a path) from q, then j can be reached from q and thus $j \in X$ (a contradiction). Examining the remaining kilter states, we find that the only candidates for membership in $(X, \bar{X})$ are those identified in Figure 10.8. Recall that arcs from X to $\bar{X}$ in G have their $z_{ij} - c_{ij}$'s increased. Thus these arcs change kilter states in a left-to-right fashion as indicated in Figure 10.8*a*. Examining an arc from X to $\bar{X}$ in G that is in the kilter state $x_{ij} > u_{ij}$ and $z_{ij} - c_{ij} < 0$, we see from Figure 10.3 that as θ increases, K_{ij} decreases from $K_{ij} = |x_{ij} - l_{ij}|$ to $K_{ij} = |x_{ij} - u_{ij}|$ and thereafter remains constant. Thus, for such an arc, we can increase θ as much as we like and the arc's kilter number will never increase. Hence such an arc gives rise to an upper limit on θ of ∞ as indicated in Figure 10.8*a*. Any arc from X to $\bar{X}$ in G that is in the kilter state $x_{ij} = u_{ij}$ and $z_{ij} - c_{ij} < 0$ will have its kilter number first decrease and then remain unchanged as θ increases (why?). Thus, again ∞ is an upper limit on the permitted change in θ for such an arc to ensure that no kilter number will worsen. However, examining an arc from X to $\bar{X}$ in G in the kilter state $l_{ij} < x_{ij} < u_{ij}$ and $z_{ij} - c_{ij} < 0$, we see that the associated kilter number K_{ij} first decreases (to zero), then starts to increase. In order to eliminate the potential increase in K_{ij} for the arc we must place a limit of $|z_{ij} - c_{ij}|$ on θ. Similarly, we must place a limit of $|z_{ij} - c_{ij}|$ on θ for arcs in the state $x_{ij} = l_{ij}$ and $z_{ij} - c_{ij} < 0$. This analysis justifies the entries in Figure 10.8. Each of the permissible kilter states for arcs in $(X, \bar{X})$ is graphically portrayed in Figure 10.8*b*.

A similar analysis of arcs from $\bar{X}$ to X in G gives rise to the information in Figure 10.9.

Insofar as worsening of kilter numbers is concerned, Figures 10.8 and 10.9 indicate that we need only compute θ based on arcs from X to $\bar{X}$ with $x_{ij} < u_{ij}$ and arcs from $\bar{X}$ to X with $x_{ij} > l_{ij}$. However, if we proceed to define a method of computing θ based only on these considerations, difficulties would arise in interpreting the meaning of the value $\theta = \infty$. Matters are greatly simplified if,

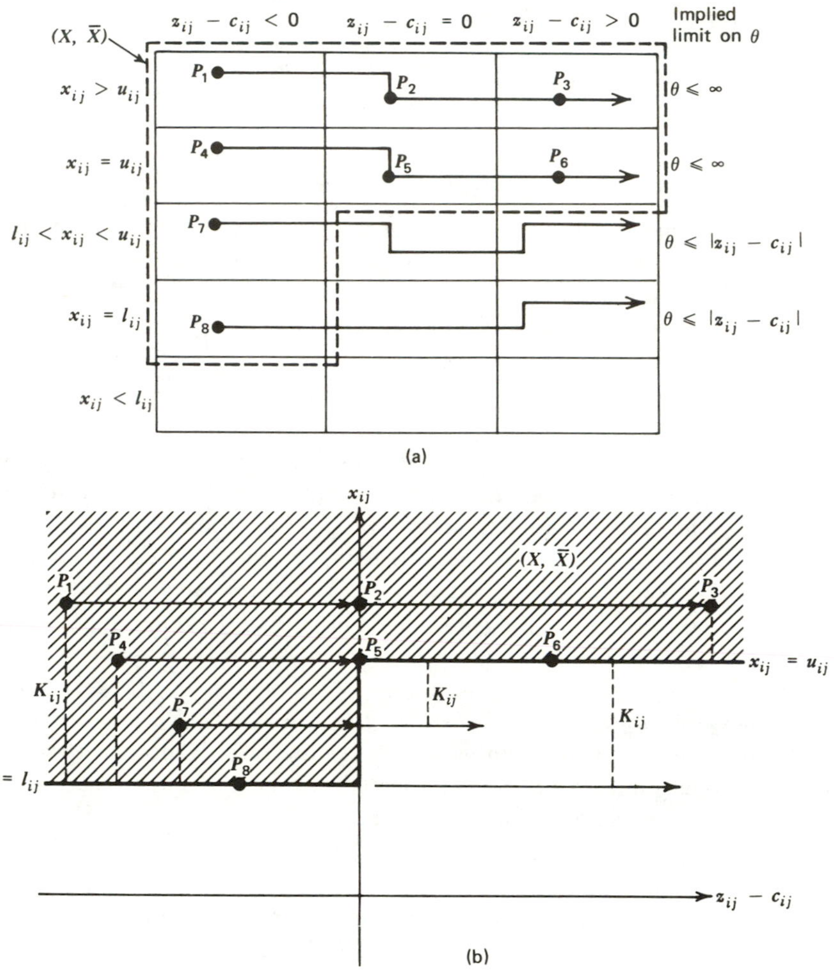

Figure 10.8. **Possible kilter states for arcs from X to $\bar{X}$ in G and limits on θ.**

instead of strict inequalities on flow (that is, $x_{ij} < u_{ij}$ and $x_{ij} > l_{ij}$), we admit weak inequalities on flow (that is, $x_{ij} \leqslant u_{ij}$ and $x_{ij} \geqslant l_{ij}$). The reason for this deviation from intuition will become apparent when we proceed to establish convergence of the algorithm.

The previous discussion concerning limits on θ based on kilter number considerations and on (yet to be established) convergence properties leads to the following formal procedure for computing θ.

In G define S_1 and S_2 by

$$S_1 \equiv \left\{(i, j) : i \in X, j \in \bar{X}, z_{ij} - c_{ij} < 0, x_{ij} \leqslant u_{ij}\right\}$$

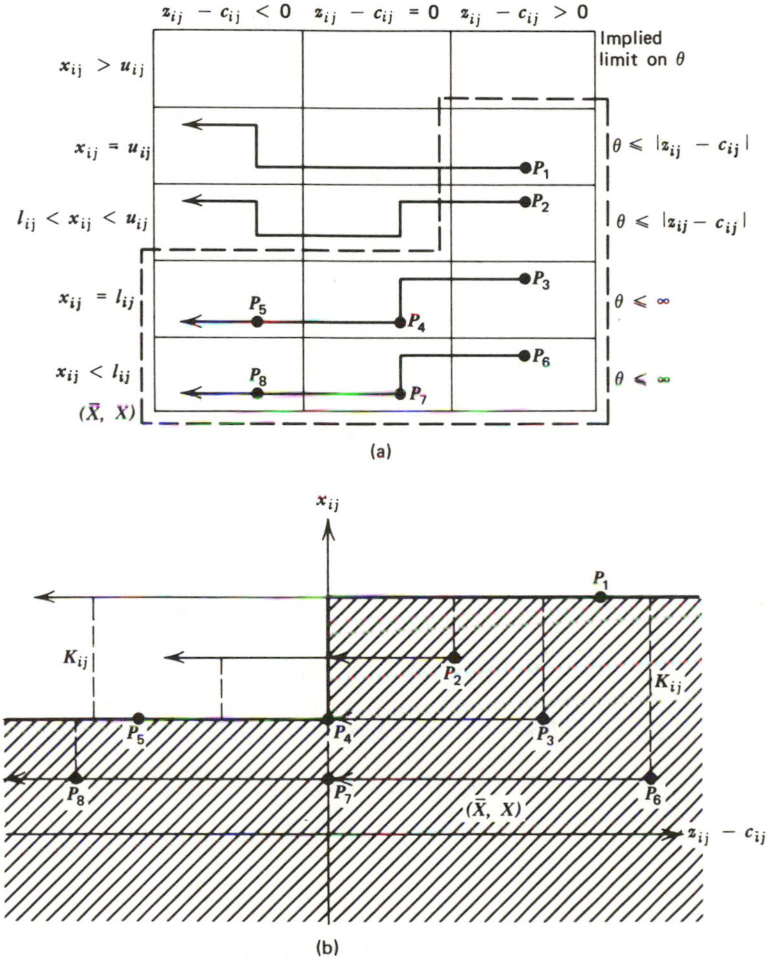

Figure 10.9. **Possible kilter states for arcs from $\overline{X}$ to X in G and limits on θ.**

and

$$S_2 \equiv \left\{(i,j) : i \in \overline{X}, j \in X, z_{ij} - c_{ij} > 0, x_{ij} \geqslant l_{ij}\right\}$$

Let

$$\theta_1 = \underset{(i,j)\in S_1}{\text{Minimum}} \left\{|z_{ij} - c_{ij}|\right\}$$

$$\theta_2 = \underset{(i,j)\in S_2}{\text{Minimum}} \left\{|z_{ij} - c_{ij}|\right\}$$

$$\theta = \text{Minimum}\{\theta_1, \theta_2\}$$

where $\theta_i \equiv \infty$ if S_i is empty. Thus θ is strictly positive (why?). Also, θ is either a positive integer or ∞ (why?). These two possibilities are briefly discussed below.

Case 1: $0 < \theta < \infty$.

In this case we make the appropriate changes in w_i, (that is, $w_i' = w_i + \theta$ if $i \in X$ and $w_i' = w_i$ if $i \in \bar{X}$) and pass to the primal phase of the algorithm.

Case 2: $\theta = \infty$.

In this case the primal problem has no feasible solution. (We shall show this shortly.)

This completes the specification of the dual phase of the out-of-kilter algorithm and provides the foundation of the overall out-of-kilter algorithm.

As an illustration, consider the example of Figure 10.1 with the current solution specified by Figures 10.6 and 10.7. Here

$$S_1 = \{(1, 2)\}, \qquad \theta_1 = |-2| = 2$$

$$S_2 = \{(2, 3)\}, \qquad \theta_2 = |3| = 3$$

$$\theta = \text{Minimum}\{2, 3\} = 2$$

This gives rise to the following change in dual variables:

$$\begin{aligned} w_1' &= w_1 + \theta = 2 \\ w_2' &= w_2 \qquad = 0 \\ w_3' &= w_3 + \theta = 2 \\ w_4' &= w_4 + \theta = 2 \end{aligned}$$

The x_{ij}'s and new values of $z_{ij} - c_{ij}$'s are given in Figure 10.10*a*. Passing to the primal phase of the out-of-kilter algorithm, we see that G' in Figure 10.10*b* contains a circuit involving arc (2, 3) so we may change flows. The remaining iterations are not shown.

There is really no need to work directly with the dual variables themselves since we may transform the $z_{ij} - c_{ij}$'s directly by

$$(z_{ij} - c_{ij})' = \begin{cases} (z_{ij} - c_{ij}) & \text{if } i \in X, j \in X \text{ or } i \in \bar{X}, j \in \bar{X} \\ (z_{ij} - c_{ij}) + \theta & \text{if } i \in X, j \in \bar{X} \\ (z_{ij} - c_{ij}) - \theta & \text{if } i \in \bar{X}, j \in X \end{cases}$$

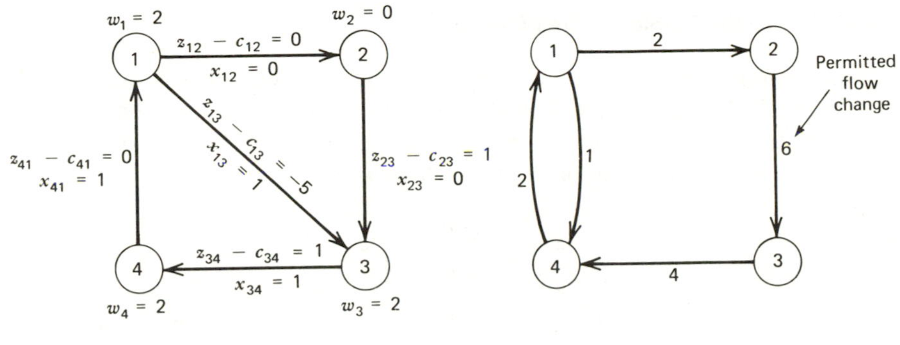

(a) G (b) G'

Figure 10.10. The new solution obtained from Figure 10.6 after the first dual variable change.

In Exercise 10.20 we ask the reader to show how the dual variables can be recovered anytime we need them. Note that $(z_{ij} - c_{ij})'$ is an integer (why?).

As an example of infeasibility, consider the example network of Figure 10.11*a*. Selecting a set of x_{ij}'s and w_i's, we find in Figure 10.11*b* that arc (2, 1) is out-of-kilter. Setting up G' in Figure 10.11*c*, we find no circuit containing the arc (2, 1). In this case $X = \{1\}$ and $\bar{X} = \{2\}$. Here $S_1 = \Phi$ (the empty set) and $S_2 = \Phi$ and thus $\theta = \infty$. It is clear by examining u_{12} and l_{21} that no feasible solution exists.

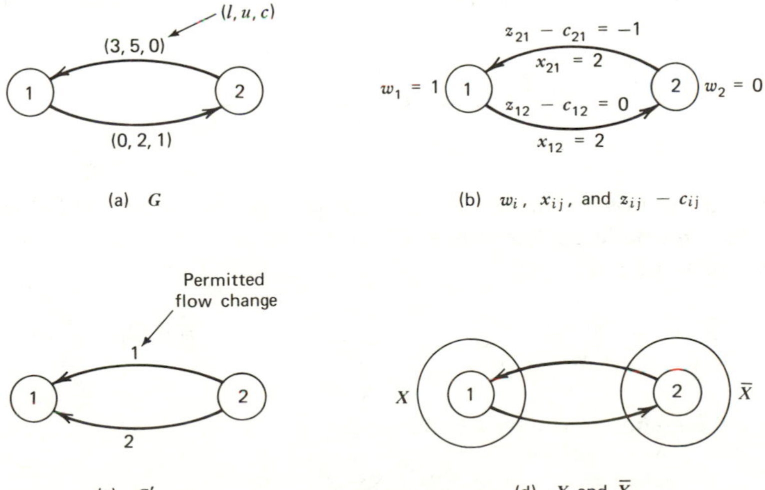

(a) G (b) w_i, x_{ij}, and $z_{ij} - c_{ij}$

(c) G' (d) X and $\bar{X}$

Figure 10.11. An example of an infeasible network.

Infeasibility of the Problem When $\theta = \infty$

Suppose during some application of the dual phase of the out-of-kilter algorithm that we reach the case where $\theta = \infty$. When this occurs, we must have $S_1 = S_2 = \Phi$. Since $S_1 = \Phi$, then by reviewing the definition of S_1 we conclude that $i \in X$ and $j \in \bar{X}$ imply one of the following cases:

1. $z_{ij} - c_{ij} < 0$ and $x_{ij} > u_{ij}$
2. $z_{ij} - c_{ij} = 0$
3. $z_{ij} - c_{ij} > 0$

From Figure 10.8 and since $i \in X$ and $j \in \bar{X}$, possibility (2) or (3) above can hold only if $x_{ij} \geqslant u_{ij}$. Hence $S_1 = \Phi$ holds only if $x_{ij} \geqslant u_{ij}$ for $i \in X$ and $j \in \bar{X}$. Similarly, $S_2 = \Phi$ holds only if $i \in \bar{X}$ and $j \in X$ implies that $x_{ij} \leqslant l_{ij}$. Hence $S_1 = S_2 = \Phi$ implies

$$x_{ij} \geqslant u_{ij} \quad \text{if} \quad i \in X, j \in \bar{X} \tag{10.6}$$

and

$$x_{ij} \leqslant l_{ij} \quad \text{if} \quad i \in \bar{X}, j \in X \tag{10.7}$$

In particular, consider the out-of-kilter arc (p, q) in G'. If (p, q) is in G, then by inequality (10.7) $x_{pq} \leqslant l_{pq}$. Suppose that $x_{pq} = l_{pq}$. Since (p, q) is out-of-kilter, then $z_{pq} - c_{pq} > 0$, violating the assumption that $S_2 = \Phi$. Thus, $x_{pq} < l_{pq}$. If, on the other hand, (q, p) is in G then by a similar argument we may show that $x_{qp} > u_{qp}$. Thus, at least one of the inequalities (10.6) or (10.7) is strict. Summing these two inequalities we get

$$\sum_{\substack{i \in X \\ j \in \bar{X}}} x_{ij} - \sum_{\substack{i \in \bar{X} \\ j \in X}} x_{ij} > \sum_{\substack{i \in X \\ j \in \bar{X}}} u_{ij} - \sum_{\substack{i \in \bar{X} \\ j \in X}} l_{ij} \tag{10.8}$$

Since the current flow given by the x_{ij}'s is conserving, then equation (10.1) holds. Noting that the node set consists of $X \cup \bar{X}$ and $X \cap \bar{X} = \Phi$, then equation (10.1) can be written as

$$\sum_{j \in X} x_{ij} + \sum_{j \in \bar{X}} x_{ij} - \sum_{j \in X} x_{ji} - \sum_{j \in \bar{X}} x_{ji} = 0 \qquad i = 1, 2, \ldots, m$$

Summing these equations over $i \in X$, we get

$$\sum_{\substack{i \in X \\ j \in X}} x_{ij} + \sum_{\substack{i \in X \\ j \in \bar{X}}} x_{ij} - \sum_{\substack{j \in X \\ i \in X}} x_{ji} - \sum_{\substack{i \in X \\ j \in \bar{X}}} x_{ji} = 0$$

Noting that

$$\sum_{\substack{i \in X \\ j \in X}} x_{ij} = \sum_{\substack{j \in X \\ i \in X}} x_{ji}$$

and that

$$\sum_{\substack{i \in X \\ j \in \bar{X}}} x_{ji} = \sum_{\substack{i \in \bar{X} \\ j \in X}} x_{ij}$$

the foregoing equation reduces to

$$\sum_{\substack{i \in X \\ j \in \bar{X}}} x_{ij} - \sum_{\substack{i \in \bar{X} \\ j \in X}} x_{ij} = 0 \tag{10.9}$$

Substituting in inequality (10.8), we get

$$0 > \sum_{\substack{i \in X \\ j \in \bar{X}}} u_{ij} - \sum_{\substack{i \in \bar{X} \\ j \in X}} l_{ij} \tag{10.10}$$

Suppose by contradiction that there is a feasible flow represented by $\hat{x}_{ij}$ for $i, j = 1, 2, \ldots, m$. Therefore $u_{ij} \geqslant \hat{x}_{ij}$ and $-l_{ij} \geqslant -\hat{x}_{ij}$ and so inequality (10.10) gives

$$0 > \sum_{\substack{i \in X \\ j \in \bar{X}}} u_{ij} - \sum_{\substack{i \in \bar{X} \\ j \in X}} l_{ij} \geqslant \sum_{\substack{i \in X \\ j \in \bar{X}}} \hat{x}_{ij} - \sum_{\substack{i \in \bar{X} \\ j \in X}} \hat{x}_{ij} \tag{10.11}$$

But since the $\hat{x}_{ij}$'s represent a feasible flow, they must be conserving. In a fashion similar to Equation (10.9) it is clear that the right-hand side of inequality (10.11) is equal to zero. Therefore inequality (10.11) implies that $0 > 0$, which is impossible. This contradiction shows that if $\theta = \infty$, there could be no feasible flow.

Note that if we had defined S_1 and S_2 by strict inequalities on x (namely, $x_{ij} < u_{ij}$ and $x_{ij} > l_{ij}$ respectively), we could not have produced the strict inequality needed in (10.8).

Convergence of the Out-of Kilter Algorithm

For the purpose of the following finite convergence argument we make the assumption that the vectors **l**, **u**, and **c** are integer valued.

In developing a finite convergence argument for the out-of-kilter algorithm, there are several properties of the algorithm that should be noted. First, every time a circuit is constructed in G' containing an out-of-kilter arc, the kilter number of that arc and of the total network is reduced by an integer (why?). We can construct only a finite number of circuits containing out-of-kilter arcs before an optimal solution is obtained (why?). Second, after each dual variable change, the kilter state of each arc in G that has both ends in X remains unchanged. Hence, if (p, q) is not in kilter, then after a dual variable change, each node in X before the change is in X after the change. Two possibilities exist. One possibility is that a new node k may be brought into X by virtue of an arc being added in G' from some node in X to node k. Each time this occurs the set X grows by at least one node. This can occur at most a finite number of times before node p becomes a member of X and a circuit is created containing (p, q). Thus, if the algorithm is not finite, it must be the case that an infinite number of dual variable changes take place without the set X increasing or $\theta = \infty$. We shall show that this cannot occur.

Suppose that after a dual variable change no new node becomes a member of X; that is, X does not increase. Then upon passing to the next dual phase we have the same sets X and $\bar{X}$ and the same x_{ij}'s. In addition, each arc from X to $\bar{X}$ has had its $z_{ij} - c_{ij}$ increased and each arc from $\bar{X}$ to X has had its $z_{ij} - c_{ij}$ decreased. Thus, after the dual variable change, the new sets S_1' and S_2' satisfy

$$S_1' \subset S_1 \quad \text{and} \quad S_2' \subset S_2$$

(why?). Further, by the choice of the (finite) value of θ, at least one arc has been dropped from either S_1 or S_2. Thus at least one of the foregoing inclusions is proper. Now, S_1 and S_2 may decrease at most a finite number of times before $S_1 \cup S_2 = \Phi$ and $\theta = \infty$, in which case the algorithm stops.

This completes a finiteness argument for the out-of-kilter algorithm. We now summarize the algorithm and present an example.

10.3 SUMMARY OF THE OUT-OF-KILTER ALGORITHM

The complete algorithm consists of three phases: the initiation phase, the primal phase, and the dual phase.

Initiation Phase

Begin with a conserving (integer) flow, say each $x_{ij} = 0$, and an initial set of (integer) dual variables, say each $w_i = 0$. Compute $z_{ij} - c_{ij} = w_i - w_j - c_{ij}$.

Primal Phase

Determine the kilter state and the kilter number for each arc. If all arcs are in kilter, stop; the optimal solution is obtained. Otherwise, select or continue with a previously selected out-of-kilter arc (p, q). From the network G construct a new network G' according to Figure 10.4. For each arc (i, j) in G that is in one of the kilter states that permit a flow increase, place an arc (i, j) in G' with a permitted flow increase as indicated in Figure 10.4. For any arc (i, j) in G that is in one of the kilter states that permit a flow decrease, place an arc (j, i) in G' with the permitted flow indicated in Figure 10.4. For those arcs in G that are members of states that permit no flow change, place no arc in G'. In G', attempt to construct a circuit containing the out-of-kilter arc (p, q). Finding such a circuit is called *breakthrough*. If such a circuit is available, determine a flow change Δ equal to the minimum of the permitted flow changes on arcs of the circuit. Change the flow on each arc of the associated cycle in G by the amount Δ using the orientation specified by the circuit as the direction of increase. In particular, let $x'_{ij} = x_{ij} + \Delta$ if (i, j) was a member of the circuit in G'; let $x'_{ij} = x_{ij} - \Delta$ if (j, i) was a member of the circuit in G'; let $x'_{ij} = x_{ij}$ otherwise. Repeat the primal phase. If no circuit containing arc (p, q) is available in G', pass to the dual phase. Finding no such circuit is called *nonbreakthrough*.

Dual Phase

Determine the set of nodes X which can be reached from node q along a path in G'. Let $\bar{X} = N - X$. In G, define S_1 and S_2 by

$$S_1 = \left\{(i, j) : i \in X, j \in \bar{X}, z_{ij} - c_{ij} < 0, x_{ij} \leqslant u_{ij}\right\}$$

$$S_2 = \left\{(i, j) : i \in \bar{X}, j \in X, z_{ij} - c_{ij} > 0, x_{ij} \geqslant l_{ij}\right\}$$

Let

$$\theta = \underset{(i,j) \in S_1 \cup S_2}{\text{Minimum}} \left\{|z_{ij} - c_{ij}|, \infty\right\}$$

If $\theta = \infty$, stop; no feasible solution exists. Otherwise, change the w_i's and the $z_{ij} - c_{ij}$'s according to

$$w'_i = \begin{cases} w_i + \theta & \text{if} \quad i \in X \\ w_i & \text{if} \quad i \in \bar{X} \end{cases}$$

$$(z_{ij} - c_{ij})' = \begin{cases} (z_{ij} - c_{ij}) & \text{if} \quad (i, j) \in (X, X) \cup (\bar{X}, \bar{X}) \\ (z_{ij} - c_{ij}) + \theta & \text{if} \quad (i, j) \in (X, \bar{X}) \\ (z_{ij} - c_{ij}) - \theta & \text{if} \quad (i, j) \in (\bar{X}, X) \end{cases}$$

and pass to the primal phase.

10.4 AN EXAMPLE OF THE OUT-OF-KILTER ALGORITHM

Consider the network given in Figure 10.12. Initiating the out-of-kilter algorithm with each $x_{ij} = 0$ and each $w_i = 0$, we get the sequence of primal and dual phases given in Figure 10.13.

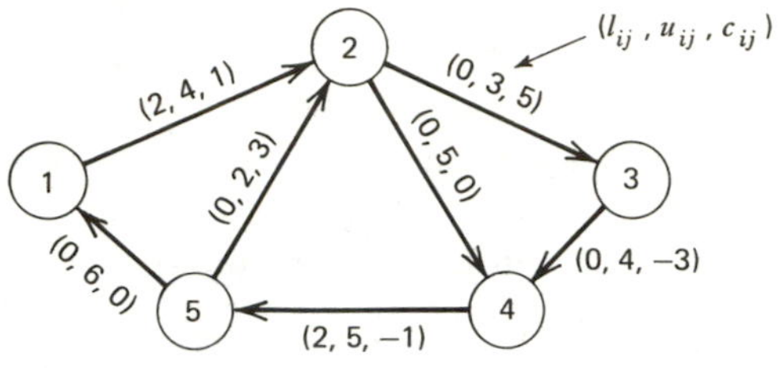

Figure 10.12. **An example network.**

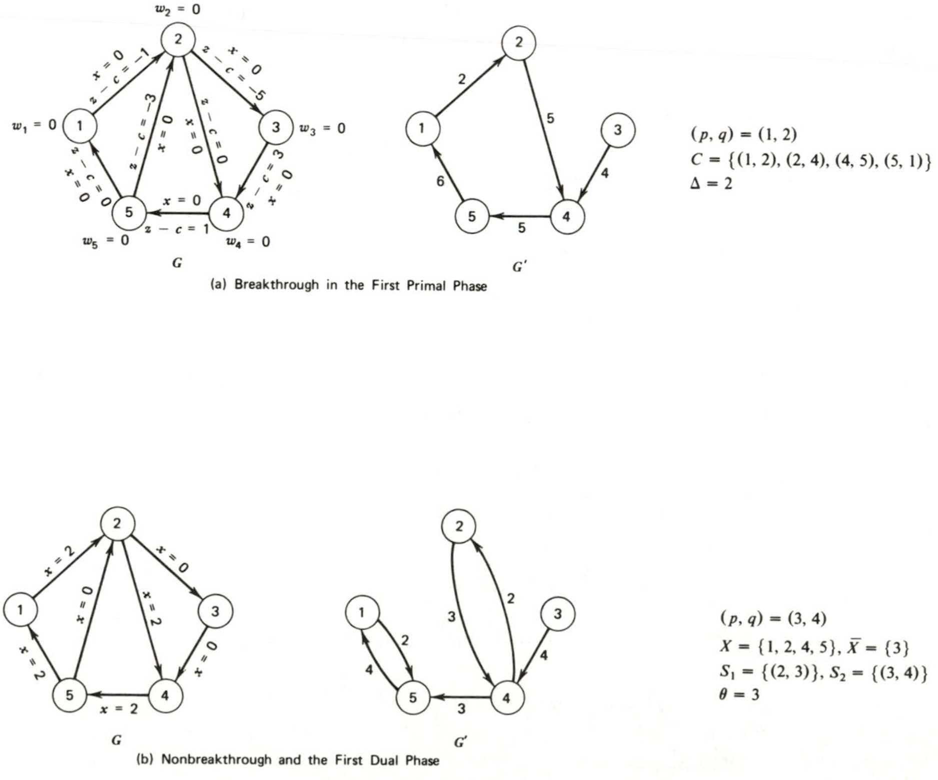

Figure 10.13. **The out-of-kilter solution for Figure 10.12.**

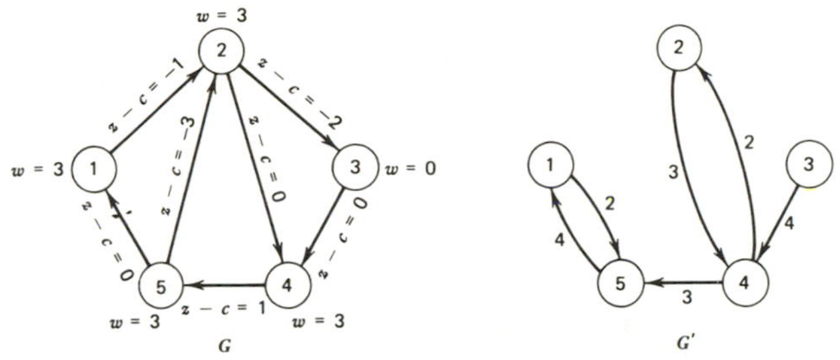

(c) Nonbreakthrough and the Second Dual Phase

(3, 4) is in - kilter

$(p, q) = (4, 5)$

$X = \{1, 5\}, \bar{X} = \{2, 3, 4\}$

$S_1 = \{(1, 2), (5, 2)\}$

$S_2 = \{(4, 5)\}$

$\theta = 1$

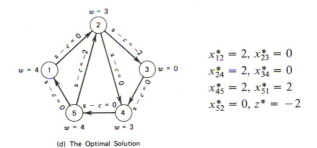

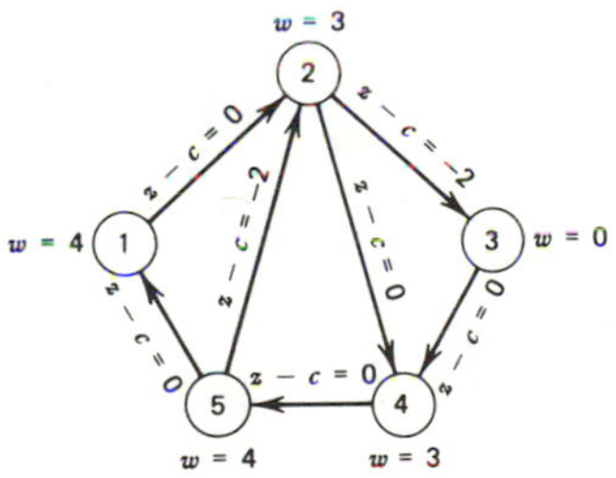

(d) The Optimal Solution

$x_{12}^* = 2, x_{23}^* = 0$

$x_{24}^* = 2, x_{34}^* = 0$

$x_{45}^* = 2, x_{51}^* = 2$

$x_{52}^* = 0, z^* = -2$

Figure 10.13. (Cont.)

A Labeling Procedure for the Out-of-Kilter Algorithm

Either for hand or computer calculations there are simple and convenient ways to maintain the information required to solve a minimal cost flow problem by the out-of-kilter algorithm. Suppose that we associate with each node j a label $L(j) = (\pm i, \Delta_j)$. A label (i, Δ_j) indicates that the flow on arc (i, j) could be increased by an amount Δ_j without worsening the kilter number of any arc. A label $(-i, \Delta_j)$ indicates that the flow on arc (j, i) could be decreased by an amount Δ_j without worsening the kilter number of any arc. Note that Δ_j represents the current estimate of the amount of flow change that can take place along some cycle containing an out-of-kilter arc and either arc (i, j) or (j, i) in such a way that the kilter number of no arc is increased. The labeling algorithm becomes as follows.

INITIALIZATION STEP

Select a conserving flow, for example, each $x_{ij} = 0$, and a set of dual variables, such as each $w_i = 0$.

MAIN STEP

1. If all arcs are in kilter according to Figure 10.2, stop; the optimal solution is obtained. Otherwise, select (or continue with a previously selected) out-of-kilter arc, say (p, q). Erase all labels. If (p, q) is in one of the states where a flow increase, Δ_{pq}, is required according to Figure 10.4, then set $s = q$, $t = p$ and $L(s) = (+t, \Delta_{pq})$. Otherwise, if (p, q) is in one of the states where a flow decrease, Δ_{pq}, is required according to Figure 10.4, then set $s = p$, $t = q$ and $L(s) = (-t, \Delta_{pq})$.
2. If node i has a label, node j has no label, and flow may be increased by an amount Δ_{ij} along arc (i, j) according to Figure 10.4, then assign node j the label $L(j) = (+i, \Delta_j)$ where $\Delta_j = \text{Minimum } \{\Delta_i, \Delta_{ij}\}$. If node i has a label, node j has no label, and flow may be decreased by an amount Δ_{ji} along arc (j, i) according to Figure 10.4, then give node j the label $L(j) = (-i, \Delta_j)$ where $\Delta_j = \text{Minimum}\{\Delta_i, \Delta_{ji}\}$. Repeat step 2 until either node t is labeled or until no more nodes can be labeled. If node t is labeled, go to step 3 (*breakthrough* has occurred); otherwise, go to step 4 (*nonbreakthrough* has occurred).
3. Let $\Delta = \Delta_t$. Change flow along the identified cycle as follows. Begin at node t. If the first entry in $L(t)$ is $+k$, then add Δ to x_{kt}. Otherwise, if the first entry in $L(t)$ is $-k$, then subtract Δ from x_{tk}. Backtrack to node k and repeat the process until node t is reached again in the backtrack process. Go to step 1.
4. Let X be the set of labeled nodes and let $\bar{X} = N - X$. Define $S_1 = \{(i, j) : i \in X,\ j \in \bar{X},\ z_{ij} - c_{ij} < 0,\ x_{ij} \leqslant u_{ij}\}$ and $S_2 = \{(i, j) : i \in \bar{X}, j \in X, z_{ij} - c_{ij} > 0, x_{ij} \geqslant l_{ij}\}$. Let $\theta = \text{Minimum}\{|z_{ij} - c_{ij}|, \infty : (i, j) \in S_1 \cup S_2\}$. If $\theta = \infty$, stop; no feasible solution exists. Otherwise, let

$$w_i' = \begin{cases} w_i + \theta & \text{if } i \in X \\ w_i & \text{if } i \in \bar{X} \end{cases}$$

and

$$(z_{ij} - c_{ij})' = \begin{cases} (z_{ij} - c_{ij}) & \text{if } (i, j) \in (X, X) \cup (\bar{X}, \bar{X}) \\ (z_{ij} - c_{ij}) + \theta & \text{if } (i, j) \in (X, \bar{X}) \\ (z_{ij} - c_{ij}) - \theta & \text{if } (i, j) \in (\bar{X}, X) \end{cases}$$

Go to Step 1.

An Example of the Labeling Algorithm

We shall illustrate the labeling method for the out-of-kilter algorithm by performing the first two iterations represented in Figure 10.13 *a* and *b*. From Figure 10.13*a* we find that arc (1, 2) is an out-of-kilter arc whose flow must be increased.

The sequence of operations of the labeling algorithm are as follows:

1. $(p, q) = (1, 2)$, $s = 2$, $t = 1$, $L(2) = (+1, 2)$
2. $L(4) = (+2, 2)$
3. $L(5) = (+4, 2)$
4. $L(1) = (+5, 2)$
5. Breakthrough: $\Delta = 2$
6. $L_1(1) = +5 \Rightarrow x'_{51} = x_{51} + \Delta = 2$
7. $L_1(5) = +4 \Rightarrow x'_{45} = x_{45} + \Delta = 2$
8. $L_1(4) = +2 \Rightarrow x'_{24} = x_{24} + \Delta = 2$
9. $L_1(2) = +1 \Rightarrow x'_{12} = x_{12} + \Delta = 2$
10. Erase all labels, $(p, q) = (3, 4)$, $s = 4$, $t = 3$, $L(4) = (+3, 4)$
11. $L(5) = (+4, 3)$
12. $L(1) = (+5, 3)$
13. $L(2) = (-4, 2)$
14. Nonbreakthrough: $X = \{1, 2, 4, 5\}$, $\bar{X} = \{3\}$, $\theta = 3$
15. $w_1 = w_2 = w_4 = w_5 = 3$, $w_3 = 0$

Since arc (3, 4) is now in-kilter, we select another out-of-kilter arc, erase all labels, and continue.

EXERCISES

10.1 Solve the following problem by the out-of-kilter algorithm.

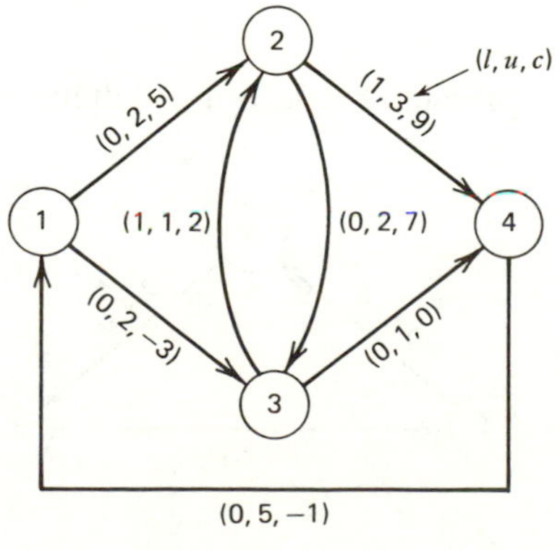

10.2 Consider the following network flow problem.

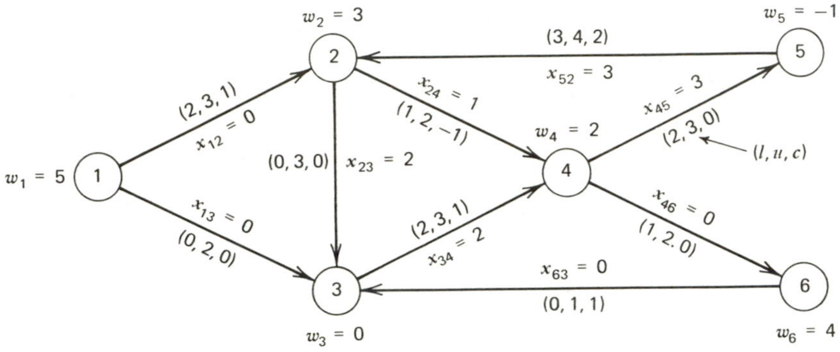

a. Give the kilter state of each arc.
b. Solve the problem by the out-of-kilter algorithm.

10.3 Solve the following problem by the out-of-kilter algorithm.

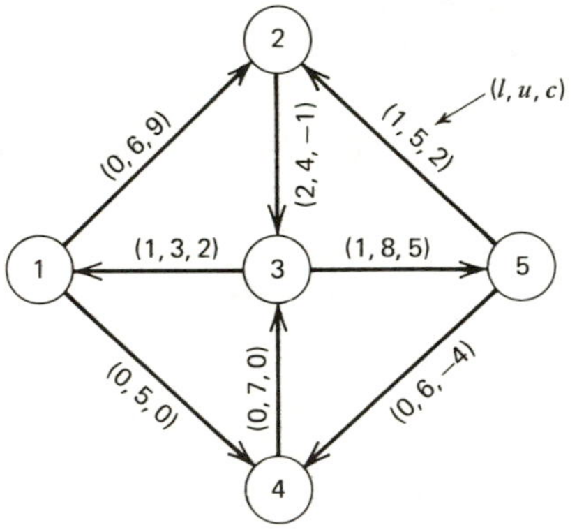

10.4 Solve the following problem by the out-of-kilter method.

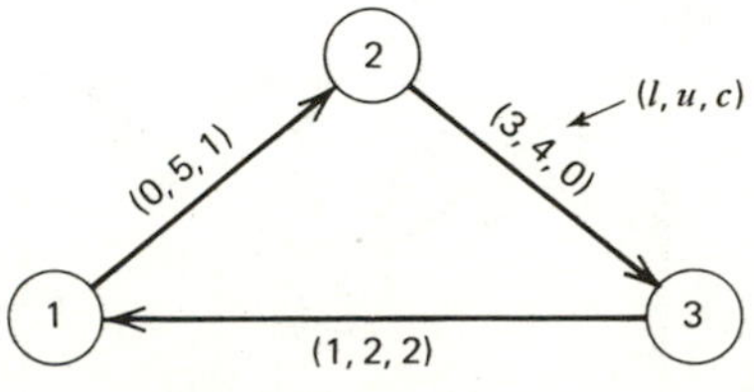

10.5 Show by manipulating the constraint equations mathematically that any minimal cost network flow problem can be transformed into the out-of-kilter form by adding an additional node and at most m additional arcs.

10.6 How can alternate optimal solutions be detected in the out-of-kilter algorithm?

10.7 Solve the following problem by the out-of-kilter algorithm.

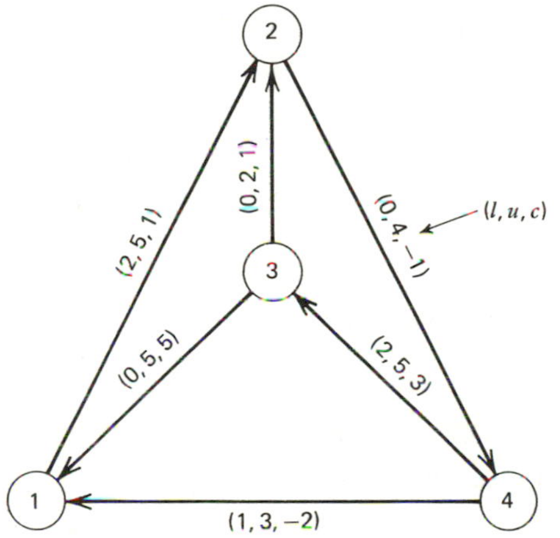

10.8 Consider the following network flow problem.

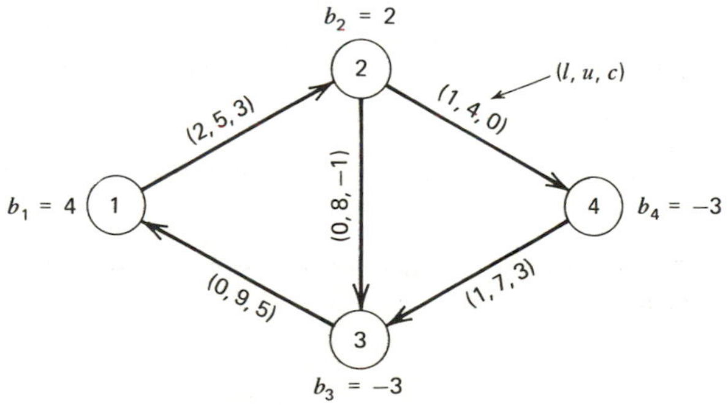

a. Solve the problem by the network simplex method of Chapter 9.
b. Transform the problem into a circulation form and solve it by the out-of-kilter algorithm.

10.9 Show that after each dual phase we can replace each new w_i by $w_i - w_k$, where k is some arbitrary node, and the results are the same. In a computer, we might do this to force one dual variable, such as w_k, to remain zero and keep all of the dual variables from getting too large.

10.10 Show explicitly how the dual objective value changes after each dual phase of the out-of-kilter algorithm. Does it always increase?

10.11 Consider the following network flow problem with flows as indicated.

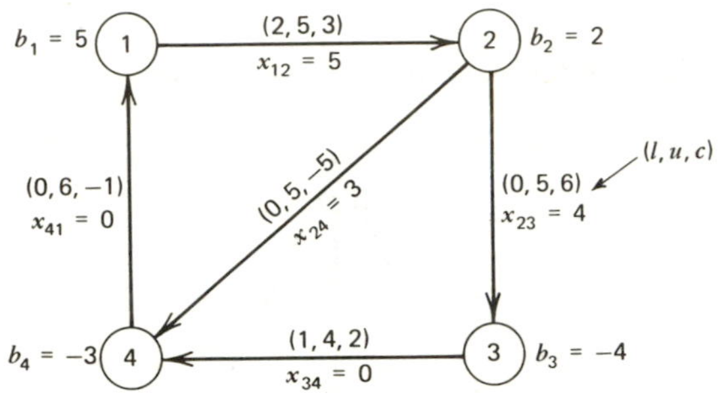

a. Ignoring the fact that the b_i's are not zero, apply the out-of-kilter algorithm directly to the foregoing network with the starting x_{ij}'s as given.
b. Solve by the network simplex method of Chapter 9.
c. Are the solutions in (a) and (b) the same? Discuss!

10.12 Consider the general minimal cost network flow problem: Minimize $\mathbf{cx}$ subject to $\mathbf{Ax} = \mathbf{b}$, $\mathbf{l} \leqslant \mathbf{x} \leqslant \mathbf{u}$ where $\mathbf{A}$ is a node-arc incidence matrix. Define a "conserving flow" to be any $\mathbf{x}$ satisfying $\mathbf{Ax} = \mathbf{b}$ (the conservation equations). Show that without transforming the network to the out-of-kilter form, the out-of-kilter algorithm can be applied directly on the original network with a starting "conserving flow" to obtain the optimal solution.

10.13 Interpret the dual of the out-of-kilter formulation for a minimal cost flow problem.

10.14 Considering the out-of-kilter problem, show that a feasible solution exists if and only if for every choice of X and $\overline{X} = N - X$ we have

$$\sum_{\substack{i \in X \\ j \in \overline{X}}} l_{ij} \leqslant \sum_{\substack{i \in \overline{X} \\ j \in X}} u_{ij}$$

10.15 How can one handle $l_{ij} < 0$ in the out-of-kilter algorithm?

10.16 Is there any difficulty with the out-of-kilter algorithm when $l_{ij} = u_{ij}$ for some (i, j)? Carefully work through the development of the out-of-kilter algorithm for this case!

10.17 Is there any problem with degeneracy in the out-of-kilter algorithm?

10.18 Suppose that we have a feasible solution to the out-of-kilter problem. Assuming that the selected arc remains out-of-kilter, is it possible for no new node to come into X after a dual variable change? Discuss!

10.19 Show that the dual solution given in Section 10.1 is optimal for a fixed set of w_i's.

10.20 Suppose that we work only with the $z_{ij} - c_{ij}$'s after the initial dual solution, and never bother to change the w_i's. Show how the w_i's can be recovered anytime we want them. (*Hint*. The w_i's are not unique. Set any one $w_i = 0$.)

10.21 In the out-of-kilter algorithm, show that if no cycle exists in the subset of arcs in G with $x_{ij} \neq l_{ij}$ and $x_{ij} \neq u_{ij}$, then the current solution corresponds to a basic solution of the associated linear program. Indicate how the out-of-kilter algorithm can be initiated with a basic solution if one is not readily available. Illustrate by the following network with the indicated conserving flow. (*Hint*. Start with a conserving flow. If a cycle exists among arcs where $x_{ij} \neq l_{ij}$ and $x_{ij} \neq u_{ij}$, consider modifying the flow around the cycle.)

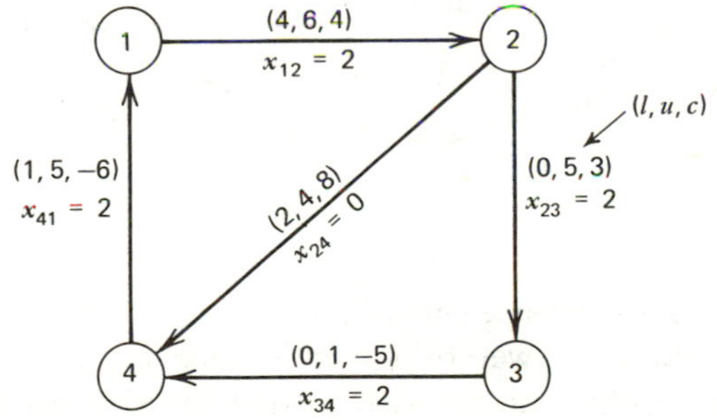

10.22 Using the results of Exercise 10.21, if the out-of-kilter algorithm is initiated with a basic solution to the linear program, show how a basic solution can be maintained thereafter. (*Hint*. Let $E = \{(i, j) : x_{ij} \neq l_{ij}$ and $x_{ij} \neq u_{ij}\}$. Start with only appropriate arcs associated with E as members of G'. If a circuit exists, change flows. Otherwise, after developing X, add an appropriate arc to G', which is not a member of E, and which enlarges X; then work with E as much as possible again. If no circuit still exists in G', add another arc that is not in E but does enlarge X. Continue as often as necessary. If no such arc not in E exists that enlarges X, then pass to the dual phase. This is an example of *block pivoting*.)

10.23 Extend the out-of-kilter algorithm to handle noninteger values of c_{ij}, l_{ij}, and u_{ij} directly.

10.24 If during the primal phase we permit some kilter numbers to increase as long as the sum of all the kilter numbers decreases, will the out-of-kilter algorithm work? How could this be made operational?

10.25 Suppose that we are given a network with m nodes and n arcs, all lower bounds equal to zero, positive upper bounds, and no costs involved. Show how the out-of-kilter algorithm can be used to find the maximum amount of flow from node 1 to node m. (*Hint*. Consider adding an arc from node m to node 1 with $l_{m1} = 0$, $u_{m1} = \infty$, $c_{m1} = -1$ with all other c_{ij}'s set at zero.)

10.26 Find the maximum flow in the following network from node 1 to node 4 using the out-of-kilter algorithm. (*Hint*. Refer to Exercise 10.25.)

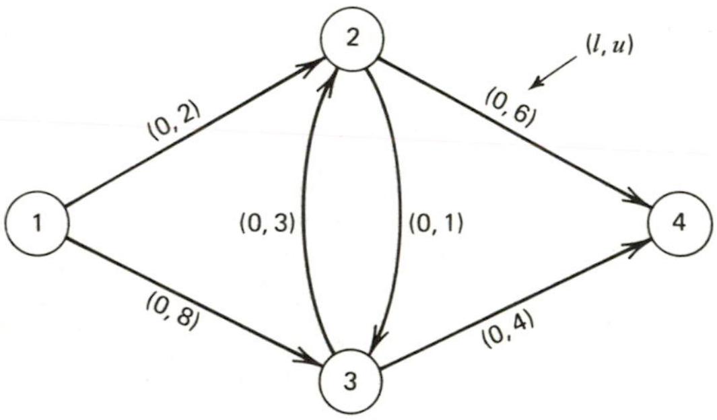

10.27 Suppose that we are given a network of m nodes and n arcs with a cost c_{ij} for each arc and no lower or upper bounds involved. Show how the out-of-kilter algorithm can be used to find the shortest (least) cost path from node 1 to node m. (*Hint*. Consider adding an arc from node m to node 1 with $l_{m1} = u_{m1} = 1$, and $c_{m1} = 0$. Set the lower and upper bounds of all other arcs at 0 and 1 respectively.)

10.28 Consider a general linear program of the form: Minimize $\mathbf{cx}$ subject to $\mathbf{Ax} = \mathbf{b}$, $\mathbf{l} \leqslant \mathbf{x} \leqslant \mathbf{u}$. Suppose that we begin with a solution $\mathbf{x}$ that satisfies $\mathbf{Ax} = \mathbf{b}$. Develop primal and dual phases of a linear programming algorithm, based on the out-of-kilter algorithm, for solving this general linear program.

10.29 Coal is being hauled out of Kentucky bound for locations in the Southeast, Southwest, Midwest, Northwest, and Northeast. The network of routes is given below.

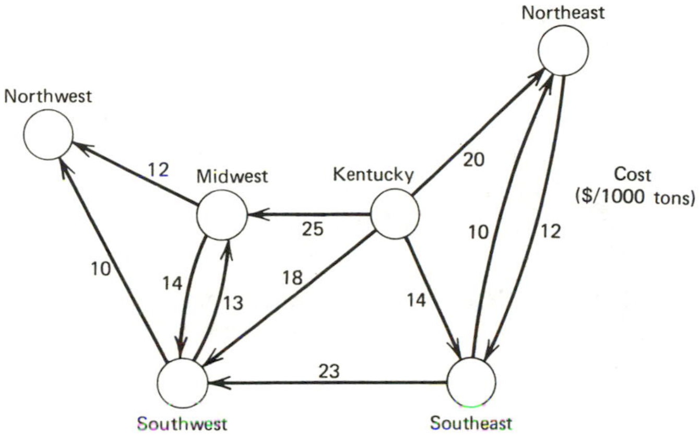

The demands are given by the following chart.

LOCATION	DEMAND (1000'S OF TONS)
Southeast	5
Southwest	3
Northwest	10
Midwest	8
Northeast	15

Kentucky has a supply of 65,000 tons per week. Except for nonnegativity there is an upper limit on flow of 17,000 tons on each arc. Ignoring the return route for coal cars, use the out-of-kilter algorithm to find the least cost distribution system for coal.

10.30 Show how a transportation problem and an assignment problem can be solved by the out-of-kilter algorithm.

10.31 The "Plenty of Water" Company wishes to deliver water for the purpose of irrigation to three oases: the sin oasis, the devil's oasis, and the pleasure oasis. The company has two stations A and B in the vicinity of these oases. Because of other commitments, at most 600 kilo-tons and 200 kilo-tons can be delivered by the two stations to the oases. Station A is connected with the sin oasis by a 10 kilometer pipeline system and with the devil's oasis by a 15 kilometer pipeline system. Similarly station B is connected with the pleasure oasis by a 15 kilometer pipeline system and with the devil's oasis by a 5 kilometer pipeline system. Furthermore, the pleasure oasis and the devil's oasis are connected by a road allowing the transportation of water by trucks. Suppose that the sin oasis, the devil's oasis, and the pleasure oasis require 200, 350, and 150 kilo-tons of water. Further suppose that the transportation cost from station A is $0.01 per kilo-ton per kilometer, and the transportation cost from station B is $0.012 per kilo-ton per kilometer. Finally suppose that the transportation cost between the pleasure oasis and the devil's oasis is $0.15 per kilo-ton.

a. Formulate the problem so that the out-of-kilter algorithm can be used.
b. Solve the problem by the out-of-kilter algorithm.
c. Suppose that a road is built joining the sin oasis and the devil's oasis with a shipping cost of $0.10 per kilo-ton. Would this affect the optimal solution? If so, find the new optimal.

10.32 Suppose that the air freight charge per ton between locations is given by the following table (except where no direct air freight service is available).

LOCATION	1	2	3	4	5	6	7
1	—	12	27	—	45	35	15
2	12	—	10	25	32	—	22
3	27	10	—	28	50	28	10
4	—	25	28	—	16	20	32
5	45	32	50	16	—	26	35
6	36	—	28	20	26	—	20
7	15	22	10	32	35	20	—

A certain corporation must ship a certain perishable commodity from locations 1, 2, and 3 to locations 4, 5, 6, and 7. A total of 30, 50, and 20 tons of this commodity are to be sent from locations 1, 2, and 3 respectively. A total of 15, 30, 25, and 30 tons are to be sent to locations 4, 5, 6, and 7 respectively. Shipments can be sent through intermediate locations at a cost equal to the sum of the costs for each of the legs of the journey. The problem is to determine the shipping plan that minimizes the total freight cost. Formulate the problem and solve it by the out-of-kilter algorithm.

10.33 A manufacturer must produce a certain product in sufficient quantity to meet contracted sales in the next four months. The production facilities available for this product are limited, but by different amounts in the respective months. The unit cost of production also varies according to the facilities and personnel available. The product can be produced in one month and then held for sale in a later month, but at an estimated storage cost of $1 per unit per month. No storage cost is incurred for goods sold in the same month in which they are produced. There is presently no inventory of this product, and none is desired at the end of the four months. Pertinent data are given below.

MONTH	CONTRACTED SALES	MAXIMUM PRODUCTION	UNIT COST OF PRODUCTION	UNIT STORAGE COST PER MONTH
1	20	40	14	1
2	30	50	16	1
3	50	30	15	1
4	40	50	17	1

Formulate the production problem as a network problem and solve it by the out-of-kilter algorithm.

10.34 Water is to be transported through a network of pipelines from the big dam to the low valley for irrigation. The network is shown below where arcs represent pipelines and the number on each arc represents the maximum permitted rate of water flow in kilo-tons per hour. It is desired to determine the maximum rate of flow from the big dam to the low valley.

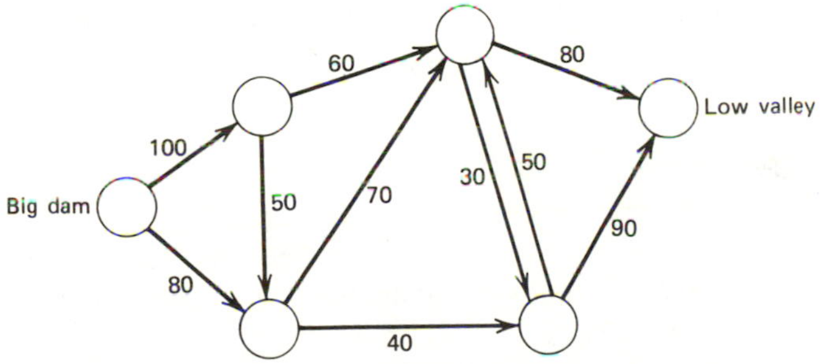

a. Formulate the problem so that it can be solved by the out-of-kilter algorithm.
b. Solve the problem by the out-of-kilter algorithm.
c. Through the use of a more powerful pumping system the maximum rate of flow on any arc can be increased by a maximum of 10 kilo-tons of water per hour. If the rate is to be increased on only one pipeline, which one would you recommend and why?

10.35 Let c_{ij} be the length associated with arc (i, j) in a given network. It is desired to find the path with shortest distance and that with the maximum distance between any two given nodes. Formulate the two problems so that the out-of-kilter algorithm can be used. Make all possible simplifications in the application of the out-of-kilter algorithm for these two problems. (Hint. See Exercise 10.27.)

10.36 An assembly consists of three parts A, B, and C. These parts go through the following operations in order: forging, drilling, grinding, painting, and assembling. The duration of these operations in days is summarized below.

	DURATION OF OPERATION			
PART	FORGING	DRILLING	GRINDING	PAINTING
A	1.0	0.5	1.0	0.3
B	2.0	0.3	0.5	0.3
C	3.0	1.0	—	0.4

Upon painting, parts A and B are assembled in two days and then A, B, and C are assembled in one day. It is desired to find the least time required for the assembly (this problem is called the *critical path problem*).

a. Formulate the problem as a network problem.
b. Solve the problem by any method you wish.
c. Solve the problem by the out-of-kilter algorithm.
d. Solve the problem by the simplified procedure you obtained in Exercise 10.35 above.
e. Because of the shortage of forging machines, suppose that at most two parts can go through forging at any particular time. What is the effect of this restriction on the total processing duration?

NOTES AND REFERENCES

1. Fulkerson [167] developed the out-of-kilter algorithm for network flow problems. For a slightly different development of the out-of-kilter algorithm, see Ford and Fulkerson [158].
2. The presentation of the out-of-kilter algorithm in this chapter follows that of Clasen [78], especially the division of states according to values of flows x_{ij} and reduced costs $z_{ij} - c_{ij}$.
3. The spirit of the out-of-kilter algorithm can be extended to a procedure for general linear programs. This has been done by Jewell [264]. The corresponding steps in the general case require the solution to linear programs instead of finding cycles or changing dual variables in a simple way.

ELEVEN: MAXIMAL FLOW, SHORTEST PATH, AND MULTICOMMODITY FLOW PROBLEMS

Two special and important network flow problems are the maximal flow problem and the shortest path problem. Both of these problems can be solved by either the network simplex method of Chapter 9 or the out-of-kilter algorithm of Chapter 10. However, their frequent occurrence in practice and the specialized, more efficient procedures for handling these two problems provide a strong case for considering them separately.

We also include in this chapter an introduction to the class of network flow problems called multicommodity network flows. In Chapters 8, 9, and 10 we have considered network flow problems in which it was not necessary to distinguish among the units flowing in the network. There was essentially a single commodity or type of unit. There are network flow problems in which different types of units must be treated. In these instances supplies and demands are by commodity type, and the distinction among the commodities must be maintained. We shall examine this multicommodity flow problem, consider the difficulty in dealing with it, and present a decomposition-based procedure for solving it.

11.1 THE MAXIMAL FLOW PROBLEM

Consider a network with m nodes and n arcs through which a single commodity will flow. We associate with each arc (i, j) a lower bound on flow of $l_{ij} = 0$ and an upper bound on flow of u_{ij}. We shall assume throughout the development that the u_{ij}'s (*arc capacities*) are integers. There are no costs involved in the maximal flow problem. In such a network, we wish to find the maximum amount of flow from node 1 to node m.

Let f represent the amount of flow in the network from node 1 to node m. Then the maximal flow problem may be stated as follows:

$$\begin{aligned} &\text{Maximize} \quad f \\ &\text{Subject to} \quad \sum_{j=1}^{m} x_{ij} - \sum_{k=1}^{m} x_{ki} = \begin{cases} f & \text{if} \quad i = 1 \\ 0 & \text{if} \quad i \neq 1 \text{ or } m \\ -f & \text{if} \quad i = m \end{cases} \\ &\qquad\qquad\qquad x_{ij} \leq u_{ij} \qquad i, j = 1, 2, \ldots, m \\ &\qquad\qquad\qquad x_{ij} \geq 0 \qquad i, j = 1, 2, \ldots, m \end{aligned}$$

where the sums and inequalities are taken over existing arcs in the network. This is called the *node-arc* formulation for the maximal flow problem since the constraint matrix is a node-arc incidence matrix. (See Exercise 11.17 for another formulation.) Noting that f is a variable and denoting the node-arc incidence matrix by $\mathbf{A}$, we can write the maximal flow problem in matrix form as

$$\begin{aligned} &\text{Maximize} \quad f \\ &\text{Subject to} \quad (\mathbf{e}_m - \mathbf{e}_1)f + \mathbf{A}\mathbf{x} = \mathbf{0} \\ &\qquad\qquad\qquad \mathbf{x} \leq \mathbf{u} \\ &\qquad\qquad\qquad \mathbf{x} \geq \mathbf{0} \end{aligned}$$

Since the activity vector for f is $(\mathbf{e}_m - \mathbf{e}_1)$, the difference of two unit vectors, we may view f as a flow variable on an arc from node m to node 1. This provides the direct formulation of the maximal flow problem in out-of-kilter form (with zero right-hand-side values for the flow conservations). Recalling that the out-of-kilter problem dealt with minimization, we assign a coefficient of zero to every flow variable except $x_{m1} \equiv f$, which receives -1.

Arc $(m, 1)$ is sometimes called the *return arc*. Figure 11.1 presents an example of the maximal flow problem and its equivalent out-of-kilter network flow problem. In Figure 11.1 the lower bound $l_{m1} = l_{41} = 0$ is derived from the fact that all $x_{ij} = 0$, $x_{m1} = 0$ is a feasible solution to the maximal flow problem. Thus the maximal value of x_{m1} will never be less than zero.

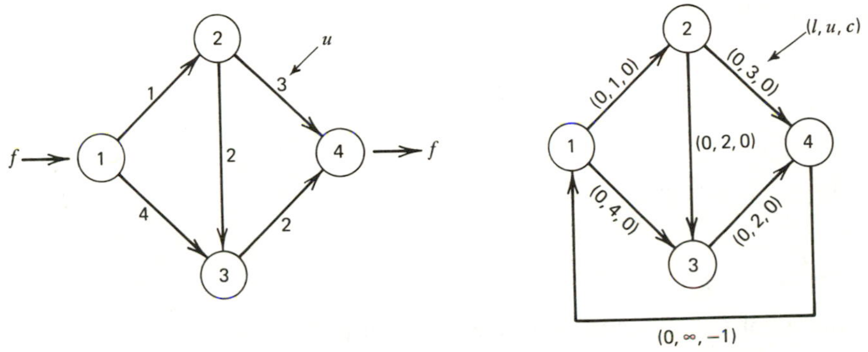

(a) (b)

Figure 11.1. An example of a maximal flow problem: (*a*) **Maximal flow problem.** (*b*) **Out-of-kilter equivalent problem.**

Before continuing with the development of the maximal flow problem, we introduce the useful and important concept of cut-sets.

Cut-Set (Separating Node m from Node 1)

Let X be any set of nodes in the network such that X contains node 1 but not node m. Let $\bar{X} = N - X$. Then $(X, \bar{X}) \equiv \{(i, j) : i \in X, j \in \bar{X}\}$ is called a *cut-set* separating node m from node 1.

Capacity of a Cut-Set

Let $(X, \bar{X})$ be any cut-set in a network G. Then $u(X, \bar{X}) \equiv \sum_{(i,j)\in(X,\bar{X})} u_{ij}$ is called the capacity of the cut-set. In Figure 11.1a there are several cut-sets separating node 4 from node 1 in G. They are

$$\begin{array}{llll}
X = \{1\}, & \bar{X} = \{2, 3, 4\}, & (X, \bar{X}) = \{(1, 2), (1, 3)\}, & u(X, \bar{X}) = 5 \\
X = \{1, 2\}, & \bar{X} = \{3, 4\}, & (X, \bar{X}) = \{(1, 3), (2, 3), (2, 4)\}, & u(X, \bar{X}) = 9 \\
X = \{1, 3\}, & \bar{X} = \{2, 4\}, & (X, \bar{X}) = \{(1, 2), (3, 4)\}, & u(X, \bar{X}) = 3 \\
X = \{1, 2, 3\}, & \bar{X} = \{4\}, & (X, \bar{X}) = \{(2, 4), (3, 4)\}, & u(X, \bar{X}) = 5
\end{array}$$

Let $(X, \bar{X})$ be any cut-set separating node m from node 1 in G. Summing the flow conservation equations of the maximal flow problem over nodes in X, flow variables with both ends in X cancel and we get

$$\sum_{\substack{i\in X \\ j\in \bar{X}}} x_{ij} - \sum_{\substack{i\in \bar{X} \\ j\in X}} x_{ij} = f \tag{11.1}$$

Using $x_{ij} \geqslant 0$ and $x_{ij} \leqslant u_{ij}$, we get

$$\sum_{\substack{i \in X \\ j \in \bar{X}}} u_{ij} - 0 \geqslant f \tag{11.2}$$

This leads to the following.

Lemma 1

The value f of any (feasible) flow is less than or equal to the capacity $u(X, \bar{X})$ of any cut-set (separating node m from node 1).

The Dual of the Maximal Flow Problem

Consider the dual of the maximal flow problem:

$$\begin{aligned} \text{Minimize} \quad & \sum_{i=1}^{m} \sum_{j=1}^{m} u_{ij} h_{ij} \\ \text{Subject to} \quad & w_m - w_1 = 1 \\ & w_i - w_j + h_{ij} \geqslant 0 \qquad i, j = 1, 2, \ldots, m \\ & h_{ij} \geqslant 0 \qquad i, j = 1, 2, \ldots, m \end{aligned}$$

where $\mathbf{w}$ corresponds to the conservation equations and $\mathbf{h}$ corresponds to $\mathbf{x} \leqslant \mathbf{u}$. Note that the first dual constraint above is associated with the flow f whose column is $\mathbf{e}_m - \mathbf{e}_1$. A typical column $\mathbf{a}_{ij}$ of the node-arc incidence matrix $\mathbf{A}$ has $+1$ at the ith position and -1 at the jth position, which leads to the dual constraints $w_i - w_j + h_{ij} \geqslant 0$.

Let $(X, \bar{X})$ be any cut-set and consider the dual problem above. If we let

$$w_i = \begin{cases} 0 & \text{if} \quad i \in X \\ 1 & \text{if} \quad i \in \bar{X} \end{cases}$$

$$h_{ij} = \begin{cases} 1 & \text{if} \quad (i, j) \in (X, \bar{X}) \\ 0 & \text{otherwise} \end{cases}$$

then this particular choice of $\mathbf{w}$ and $\mathbf{h}$ provides a feasible solution to the dual problem (why?), whose dual objective is equal to the capacity of the cut-set.

Thus, Lemma 1 above also follows from the duality theorem, which states that any feasible solution to a minimization problem has an objective value greater than or equal to that of the associated maximization problem. As the reader may suspect, we shall show that the capacity of the minimal cut-set (the one with minimal capacity) is equal to the value of the maximal flow. We shall prove this constructively.

An Algorithm for the Maximal Flow Problem

From Lemma 1 if we are able to find a flow and a cut-set such that $u(X, \bar{X}) = f$, we have the maximal flow (and the minimal cut-set). We shall do this constructively.

Suppose that we start with any feasible (integer) flow in G, say each $x_{ij} = 0$. From G construct G' as follows.

1. If arc (i, j) is in G and $x_{ij} < u_{ij}$, then place arc (i, j) in G' with permitted flow change value $\Delta_{ij} = u_{ij} - x_{ij}$.
2. If arc (i, j) is in G and $x_{ij} > 0$, then place arc (j, i) in G' with permitted flow change value $\Delta_{ji} = x_{ij}$.

Now, in G' two possibilities exist:

Case 1

A *path* P exists, in G', from node 1 to node m.

Case 2

No path exists, in G', from node 1 to node m.

In case 1 we may construct a new feasible flow with greater objective value (that is, flow out of node 1). Let Δ be equal to the minimum permitted flow on the path P from node 1 to node m in G', that is, $\Delta = \text{Minimum } \{\Delta_{ij} : (i, j) \text{ is in the path}\}$. Now Δ is a positive integer (why?). Consider the associated chain P' (undirected path) in G. Construct a new flow as follows. Add Δ to flows on arcs of the associated chain in G with the direction of the path in G', subtract Δ from flows on arcs of the associated chain in G against the direction of the path in G', and leave all other arc flows unchanged. The new flow is feasible (why?). The value of the new flow is $f' = f + \Delta$ (why?).

Assuming that the capacities are finite, case 1 can occur only a finite number of times before case 2 occurs (why?). When case 2 occurs, let X be the set of nodes in G' that can be reached along some path in G' from node 1. Let $\bar{X} = N - X$ and note that node m belongs to $\bar{X}$ (why?). Consider the arcs in G between X and $\bar{X}$. First, every arc (i, j) in G from X to $\bar{X}$ must have $x_{ij} = u_{ij}$; otherwise there would be an arc (i, j) in G' and j would be a member of X (a contradiction). Second, every arc (i, j) in G from $\overline{X}$ to X must have $x_{ij} = 0$; otherwise there would be an arc (j, i) in G' and i would be a member of X (a contradiction). Substituting in Equation (11.1), we get

$$\sum_{\substack{i \in X \\ j \in \overline{X}}} u_{ij} = f \quad \text{or} \quad u(X, \overline{X}) = f.$$

Thus we must have the maximal flow by noting inequality (11.2). Hence we have constructively proved the following.

***Theorem* 1** (Maximal Flow–Minimal Cut Theorem)

The value of the maximal flow in G is equal to the capacity of the minimal cut-set in G.

Summary of the Maximal Flow Algorithm

The constructive proof of the maximal flow–minimal cut theorem leads to the following maximal flow algorithm.

INITIALIZATION STEP

Select a set of feasible (integer) flows, say each $x_{ij} = 0$.

MAIN STEP

From G construct G' as follows.

1. All of the nodes in G are in G'.
2. If $x_{ij} < u_{ij}$ in G, place (i, j) in G' with the permitted flow change on (i, j) of $\Delta_{ij} = u_{ij} - x_{ij}$.
3. If $x_{ij} > 0$ in G, place (j, i) in G' with the permitted flow change on (j, i) of $\Delta_{ji} = x_{ij}$.

Note that arc (i, j) in G will give rise to two arcs in G' if $0 < x_{ij} < u_{ij}$. Attempt to locate a path P in G' from node 1 to node m. If no such path exists, stop; the optimal solution is obtained. Otherwise, let Δ be the minimum permitted flow change on P in G'. Add Δ to flows on arcs of the associated chain in G with the direction of the path in G', subtract Δ from flows on arcs of the associated chain in G against the direction of the path in G', and leave all other arc flows unchanged. Repeat the main step.

Locating a path is called *breakthrough*, whereas finding no path is called *nonbreakthrough*.

An Example of the Maximal Flow Problem

Consider the network of Figure 11.1*a*. Figure 11.2 presents the complete solution to the maximal flow problem for this network.

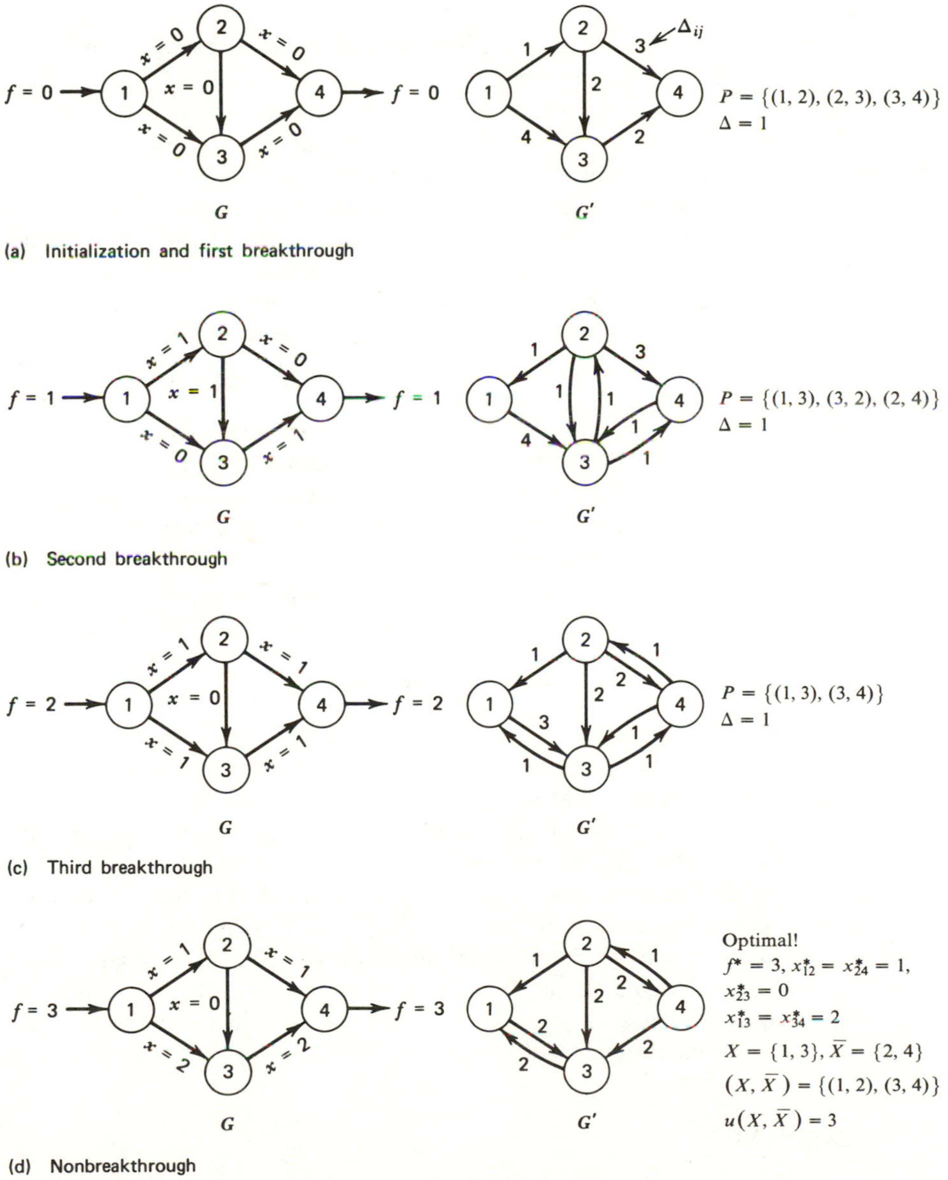

Figure 11.2. **The solution for the network of Figure 11.1*a*.**

Basic Solutions in the Maximal Flow Algorithm

In Chapter 9 we characterized basic solutions to a network flow problem. Recall that a basic solution to a network flow problem consists of a set of nonbasic variables at one of their lower or upper bounds plus a set of variables that form a rooted spanning tree. Thus we may conclude that if the set $E = \{(i, j) : 0 < x_{ij} < u_{ij}\}$ does not contain a cycle, then we have a basic feasible solution at each iteration of the maximal flow algorithm (why?).

To identify a basis after each flow change in the maximal flow algorithm we take all of the variables in the set E above plus an additional number of variables at one of their bounds to form a spanning tree. This set together with the artificial variable (located at node m) forms a rooted spanning tree (and the nonbasic variables are at their bounds). Note that since f, the flow in the network, is a variable, it must be taken onto the left-hand side of the constraint system and becomes an arc from node m to node 1 (as discussed previously).

In Figure 11.3 we present bases corresponding to the solutions at each iteration of the example in Figure 11.2. Notice that the bases in Figure 11.3*b* and *d* are unique. Also, notice that in Figure 11.3*c* no basis is possible that corresponds to the maximal flow algorithm solution at that point. In Exercise 11.9 we suggest a procedure for finding paths in G' in such a way that we shall always have a basic solution available. Finally, notice that the bases presented in Figure 11.3*a* and *b* are not *adjacent*. To obtain the basis in Figure 11.3*b* we have replaced two basic variables in the basis of Figure 11.3*a* by two nonbasic variables. This is an example of *block pivoting* discussed in Chapter 3.

A Labeling Algorithm for the Maximal Flow Problem

Either for hand or computer calculations, there are simple and convenient ways to maintain the information required to solve a maximal flow problem. We shall present a labeling algorithm that does not require the creation of the network G'. Suppose that we associate with each node j a label $L(j) = (\pm i, \Delta_j)$ containing two pieces of information. The second entry, Δ_j, in $L(j)$ indicates the amount of flow that can be sent to node j from node 1 through the current network with given flows without violating the capacity constraints $0 \leqslant x_{ij} \leqslant u_{ij}$. The first entry, $\pm i$, in $L(j)$ indicates the previous node in the chain along which flow can be changed. If the first entry in $L(j)$ is $+i$, then flow will be added to arc (i, j); otherwise, if the first entry is $-i$, then flow will be subtracted from arc (j, i). The labeling algorithm becomes as follows.

INITIALIZATION STEP

Set $x_{ij} = 0$ for $i, j = 1, \ldots, m$.

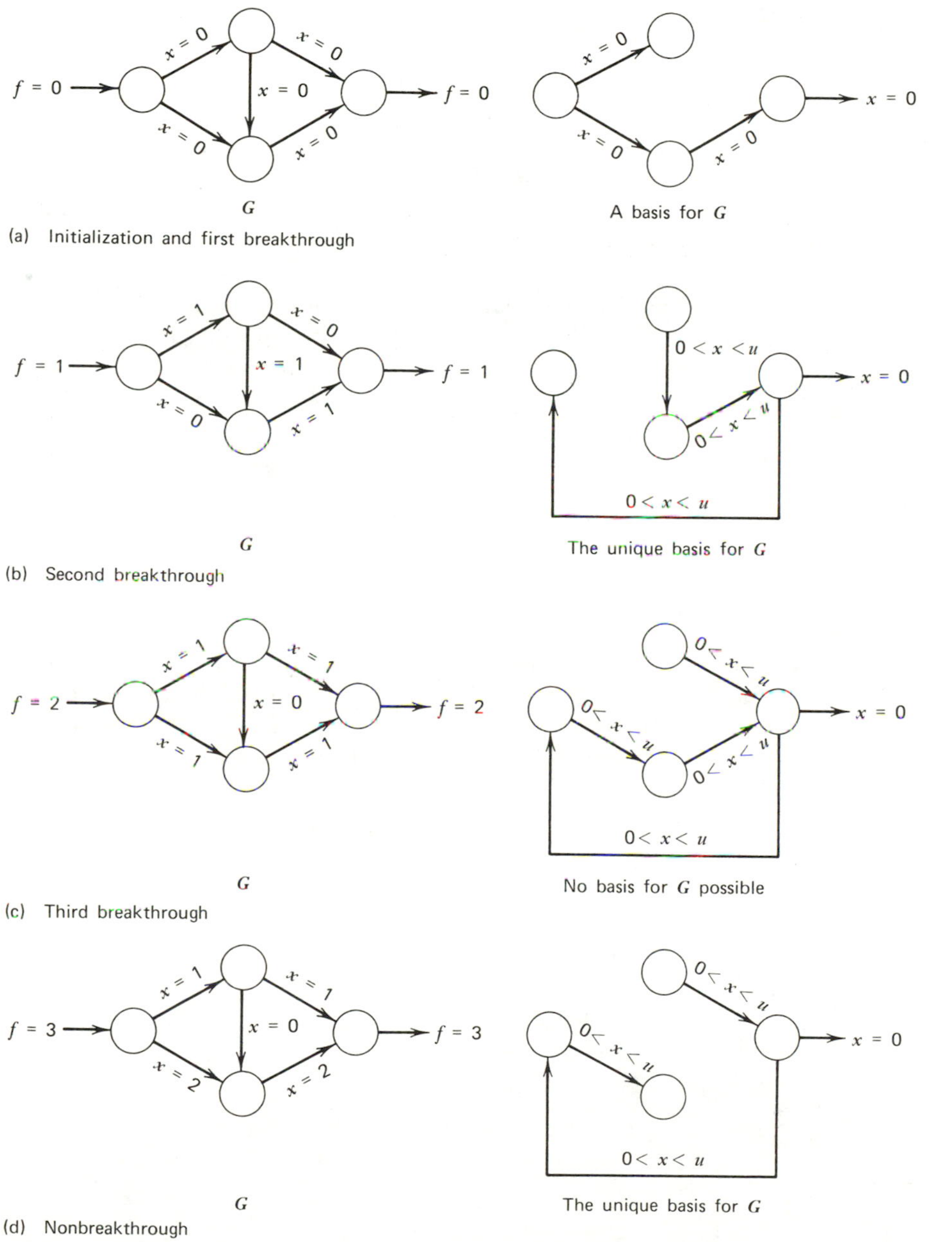

(a) Initialization and first breakthrough

(b) Second breakthrough

(c) Third breakthrough

(d) Nonbreakthrough

Figure 11.3. **Comparison between solutions in the maximal flow algorithm and bases (when possible) for Figure 11.2.**

MAIN STEP

1. Erase any labels and set $L(1) = (-, \infty)$.
2. If node i has a label, node j has no label, and $x_{ij} < u_{ij}$, then set $L(j) = (+i, \Delta_j)$, where $\Delta_j = \text{Minimum}\{\Delta_i, u_{ij} - x_{ij}\}$. If node i has a label, node j has no label, and $x_{ji} > 0$, then set $L(j) = (-i, \Delta_j)$, where $\Delta_j = \text{Minimum}\{\Delta_i, x_{ji}\}$. Repeat step 2 until either node m is labeled or until no more nodes can be labeled. Go to step 3.
3. If node m is not labeled, stop; the optimal solution is obtained. Otherwise, if node m is labeled, then change flows in the network as follows. Set $\Delta = \Delta_m$. Begin at node m and consider the first entry of $L(m)$. If the first entry is $+k$, then add Δ to x_{km}. If the first entry of $L(m)$ is $-k$, then subtract Δ from x_{mk}. Backtrack to node k and repeat the process until node 1 is reached. Return to step 1.

When the algorithm stops, let X be the set of labeled nodes and $\bar{X} = N - X$. The set $(X, \bar{X})$ is the minimal cut-set.

An Example of the Labeling Algorithm

Suppose that we apply the labeling algorithm to the maximal flow problem of Figure 11.1a to produce the first two iterations of the maximal flow algorithms represented in Figure 11.2a and Figure 11.2b. We begin with each $x_{ij} = 0$.

The sequence of labeling operations are as follows:

1. $L(1) = (-, \infty)$
2. $L(2) = (+1, 1)$
3. $L(3) = (+2, 1)$
4. $L(4) = (+3, 1)$
5. Breakthrough: $\Delta = 1$
6. $L_1(4) = +3 \Rightarrow x_{34} = 0 + \Delta = 1$
7. $L_1(3) = +2 \Rightarrow x_{23} = 0 + \Delta = 1$
8. $L_1(2) = +1 \Rightarrow x_{12} = 0 + \Delta = 1$
9. Erase all labels, $L(1) = (-, \infty)$
10. $L(3) = (+1, 4)$
11. $L(2) = (-3, 1)$
12. $L(4) = (+2, 1)$
13. Breakthrough: $\Delta = 1$
14. $L_1(4) = +2 \Rightarrow x_{24} = 0 + \Delta = 1$
15. $L_1(2) = -3 \Rightarrow x_{23} = 1 - \Delta = 0$
16. $L_1(3) = +1 \Rightarrow x_{13} = 0 + \Delta = 1$

Having completed the change of flows, we erase all labels and continue. The foregoing sequence of labels is not unique. We selected it because it illustrated the method completely.

11.2 THE SHORTEST PATH PROBLEM

Suppose that we are given a network G with m nodes and n arcs and a cost c_{ij} associated with each arc (i, j) in G. The shortest path problem is: Find the shortest (least costly) path from node 1 to node m in G. The cost of the path is the sum of the costs on the arcs in the path.

The Mathematical Formulation of a Shortest Path Problem

We may think of the shortest path problem in a network flow context if we set up a network in which we wish to send a single unit of flow from node 1 to node m at minimal cost. Thus $b_1 = 1$, $b_m = -1$, and $b_i = 0$ for $i \neq 1$ or m. The mathematical formulation becomes:

$$\text{Minimize} \quad \sum_{i=1}^{m} \sum_{j=1}^{m} c_{ij} x_{ij}$$

$$\text{Subject to} \quad \sum_{j=1}^{m} x_{ij} - \sum_{k=1}^{m} x_{ki} = \begin{cases} 1 & \text{if} \quad i = 1 \\ 0 & \text{if} \quad i \neq 1 \text{ or } m \\ -1 & \text{if} \quad i = m \end{cases}$$

$$x_{ij} = 0 \text{ or } 1 \qquad i, j = 1, 2, \ldots, m$$

where the sums and the $0 - 1$ requirements are taken over existing arcs in G. The constraints $x_{ij} = 0$ or 1 indicate that each arc is either in the path or not.

Ignoring the $0 - 1$ constraints, we again find the familiar flow conservation equations. From Chapter 9 we know that the node-arc incidence matrix associated with the flow conservation equations is totally unimodular. Thus if we replace $x_{ij} = 0$ or 1 by $x_{ij} \geqslant 0$, and if an optimal solution exists, then the simplex method would still obtain an optimal integer solution wherein the value of each variable is zero or one. (This can be shown by selecting any basis and applying Cramer's rule.) Thus we may solve the integer program as a linear program. That formulation becomes as follows.

$$\text{Minimize} \quad \sum_{i=1}^{m} \sum_{j=1}^{m} c_{ij} x_{ij}$$

$$\text{Subject to} \quad \sum_{j=1}^{m} x_{ij} - \sum_{k=1}^{m} x_{ki} = \begin{cases} 1 & \text{if} \quad i = 1 \\ 0 & \text{if} \quad i \neq 1 \text{ or } m \\ -1 & \text{if} \quad i = m \end{cases}$$

$$x_{ij} \geqslant 0 \qquad i, j = 1, 2, \ldots, m$$

Since the shortest path problem is a minimal cost network flow problem, we can solve it by one of the methods described in Chapter 9 or 10. However, we shall soon see that more efficient methods exist for this problem.

Consider the dual of the shortest path problem:

$$\begin{array}{lll} \text{Maximize} & w_1 - w_m & \\ \text{Subject to} & w_i - w_j \leqslant c_{ij} & i, j = 1, 2, \ldots, m \\ & w_i \quad \text{unrestricted} & i = 1, 2, \ldots, m \end{array}$$

It will be more convenient to make the substitution $w_i' \equiv -w_i$. As we shall shortly see, $w_i' - w_1'$ is the shortest distance from node 1 to node i at optimality. Hence we can get the shortest distance from node 1 to all nodes of the network.

A Shortest Path Procedure When All Costs Are Nonegative

Consider the case when all $c_{ij} \geqslant 0$. In this case a very simple and efficient procedure exists for finding the shortest path (from node 1 to node m).

INITIALIZATION STEP

Set $w_1' = 0$ and let $X = \{1\}$.

MAIN STEP

Let $\bar{X} = N - X$ and consider the arcs in the set $(X, \bar{X}) = \{(i, j) : i \in X, j \in \bar{X}\}$. Let

$$w_p' + c_{pq} = \underset{(i,j)\in(X,\bar{X})}{\text{Minimum}} \{w_i' + c_{ij}\}$$

Set $w_q' = w_p' + c_{pq}$ and place node q in X. Repeat the main step exactly $m - 1$ times (including the first time) and then stop; the optimal solution is at hand.

Validation of the Algorithm

We now prove that the algorithm produces an optimal solution. Assume, inductively, that each w_i' for $i \in X$ represents the cost of the shortest path from node 1 to node i. This is certainly true for $i = 1$ (why?). Consider the algorithm at some point when a new node q is about to be added to X. Suppose that

$$w_p' + c_{pq} = \underset{(i,j)\in(X,\bar{X})}{\text{Minimum}} \{w_i' + c_{ij}\} \tag{11.3}$$

We shall show that the shortest path from node 1 to node q has length $w'_q = w'_p + c_{pq}$ and can be constructed iteratively as the shortest path from node 1 to node p plus the arc (p, q). Let P be any path from node 1 to node q. It suffices to show that the length of P is $\geq w'_q$. Since node 1 is in X and node q is currently in $\bar{X}$, then P must contain an arc (i, j) where $i \in X$ and $j \in \bar{X}$ (i and j could be p and q respectively). The length of the path P is thus equal to the sum of the following.

1. The length from node 1 to node i.
2. The length of arc (i, j), that is, c_{ij}.
3. The length from j to q.

By the induction hypothesis the length from node 1 to node i is greater than or equal to w'_i. Since the costs of all arcs are assumed nonnegative, then the length in part 3 above is ≥ 0. Therefore the length of P is greater than or equal to $w'_i + c_{ij}$. In view of Equation (11.3) and since $w'_q = w'_p + c_{pq}$, it is clear that the length of P is $\geq w'_q$. This completes the induction argument and the algorithm is verified.

An Example of the Shortest Path Problem with Nonnegative Costs

Consider the network of Figure 11.4. It is desired to find the shortest path from node 1 to all other nodes. Figure 11.5 presents the complete solution for this example. The darkened arcs are those used in the selection of the node to be added to X at each iteration. These arcs can be used to trace the shortest path from node 1 to any given node i (how?). As the reader may suspect, it is no accident that darkened arcs form a tree!

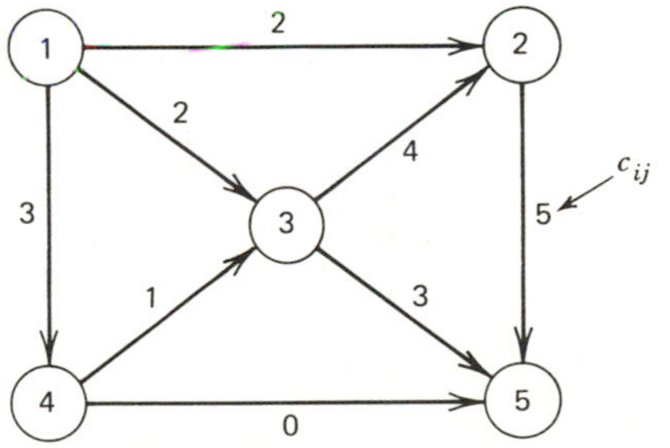

Figure 11.4. **An example of a shortest path problem.**

A Shortest Path Procedure for Arbitrary Costs

The shortest path algorithm described earlier in this section does not generalize to the case when the costs are allowed to be negative. Figure 11.6 illustrates this where the previous algorithm would select node 3 to enter X with $w'_3 = w'_1 + c_{13}$

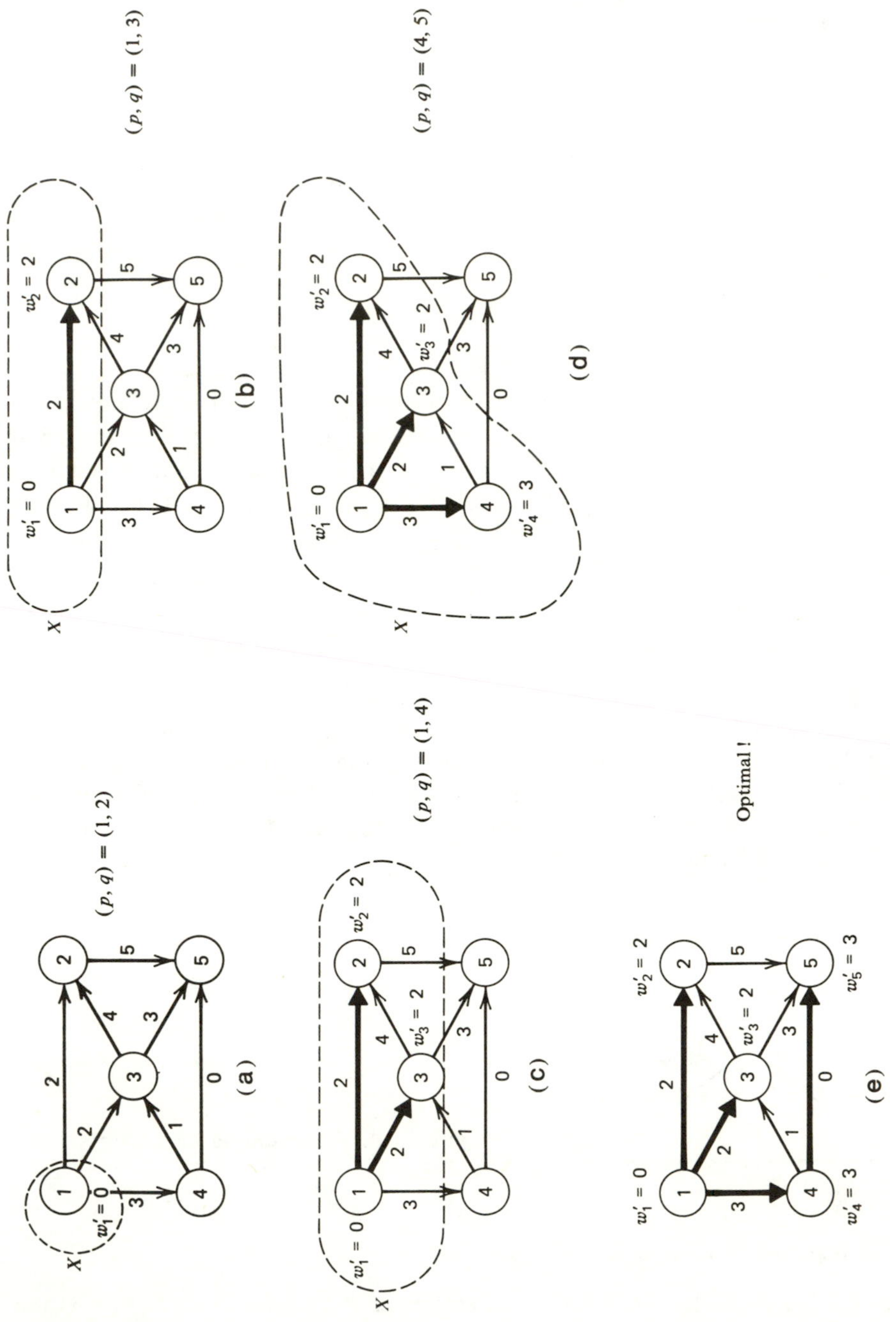

Figure 11.5. Solution of the example of Figure 11.4.

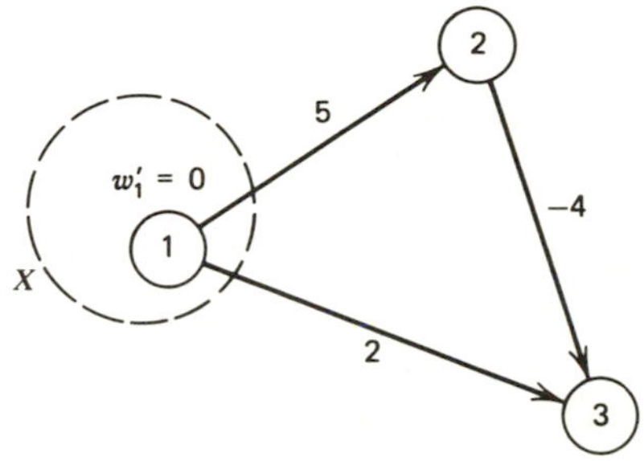

Figure 11.6. **An example where the nonnegative cost algorithm will not work.**

= 2 as the value of the shortest path from node 1 to node 3. However, it would be better to first travel to node 2, incurring a higher cost, and then go on to node 3 for a saving.

There still is a fast and efficient method for the shortest path problem with negative costs. We shall assume, however, that the sum of the costs on arcs of any circuit in G is nonnegative. Without this assumption, a "traveler" would proceed directly to the circuit in G and traverse it an infinite number of times with his cost decreasing after each time around the circuit.

The algorithm works with the dual of the shortest path problem. Recall that the dual problem with the substitution $w_i' = -w_i$ for $i = 1, 2, \ldots, m$ is given by the following.

$$\begin{aligned} &\text{Maximize} \quad w_m' - w_1' \\ &\text{Subject to} \quad w_j' - w_i' \leqslant c_{ij} \qquad i, j = 1, 2, \ldots, m \\ &\qquad\qquad\qquad w_i' \text{ unrestricted} \quad i = 1, 2, \ldots, m \end{aligned}$$

Since the objective and the constraints involve only differences in variables we may set one variable to any value, say $w_1' = 0$ (why?).

In the algorithm for negative costs we shall begin with a choice of $\mathbf{w}'$ that is "best" with respect to the dual objective but which may violate one or more of the dual constraints. We shall show that by iteratively modifying $\mathbf{w}'$ to satisfy the constraints, one at a time, we shall be able to terminate in a finite number of steps with the optimal solution.

The algorithm proceeds as follows (we shall use the convention that $\infty + a = \infty$, and $a - \infty = -\infty$, where $-\infty < a < \infty$).

INITIALIZATION STEP

Set $w_1' = 0$, $w_i' = \infty \; i \neq 1$.

MAIN STEP

If $w'_j \leqslant w'_i + c_{ij}$ for $i, j = 1, \ldots, m$, stop; the optimal solution is at hand. Otherwise, select (p, q) such that $w'_q > w'_p + c_{pq}$ and set $w'_q = w'_p + c_{pq}$. Repeat the main step as often as necessary.

To identify the arcs in the shortest path we begin at node m. If $w'_m = \infty$, then there is no path at all from node 1 to node m (see Exercise 11.25). If $w'_m < \infty$, then there must be a node k such that $w'_m - w'_k = c_{km}$ (why?). Arc (k, m) is an arc in the shortest path. Back up to node k and repeat the argument (except that $w'_k \neq \infty$) until node 1 is reached.

An Example of the Shortest Path Algorithm for Negative Costs

Consider the network of Figure 11.7 where we wish to find the shortest path from node 1 to node 4. In Figure 11.8 we present the complete solution of the example by the previous algorithm. There is no required order in which the arcs must be considered for the algorithm to converge. When the algorithm stops, the result is that $w'_4 - w'_1 = -8$ is the length of the shortest path. Arcs along the shortest path have $w'_j - w'_i = c_{ij}$.

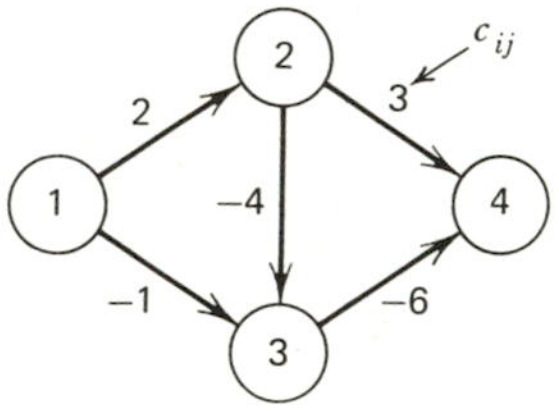

Figure 11.7. **A shortest path example with negative costs.**

Verification of the Algorithm for Negative Costs

In this section it will be necessary to distinguish between simple and nonsimple paths. A *nonsimple* path is a path that contains one or more circuits. We shall first show that when $w'_i < \infty$, it represents the cost of some path (not necessarily simple) from node 1 to node i. We note that in computing the cost of a (not necessarily simple) path, we must count the cost of an arc as many times as the arc appears in the path. Thus for a nonsimple path $P = \{(1, 3), (3, 4), (4, 5), (5, 3), (3, 4), (4, 6)\}$ the associated cost would be $c_{13} + c_{34} + c_{45} + c_{53} + c_{34} + c_{46}$. As we shall see, nonsimple paths play a part in the algorithm only when there is a negative circuit in G.

We shall demonstrate that w'_i ($< \infty$) represents the cost of a path from node 1 to node i by induction on the number of iterations of the shortest path procedure (an iteration being a change in any one w'_i).

Now, since we start with $w'_1 = 0$ and $w'_i = \infty$ for $i \neq 1$, the result is true at the first iteration. This is true because $w'_1 = 0$ represents the cost of the (empty)

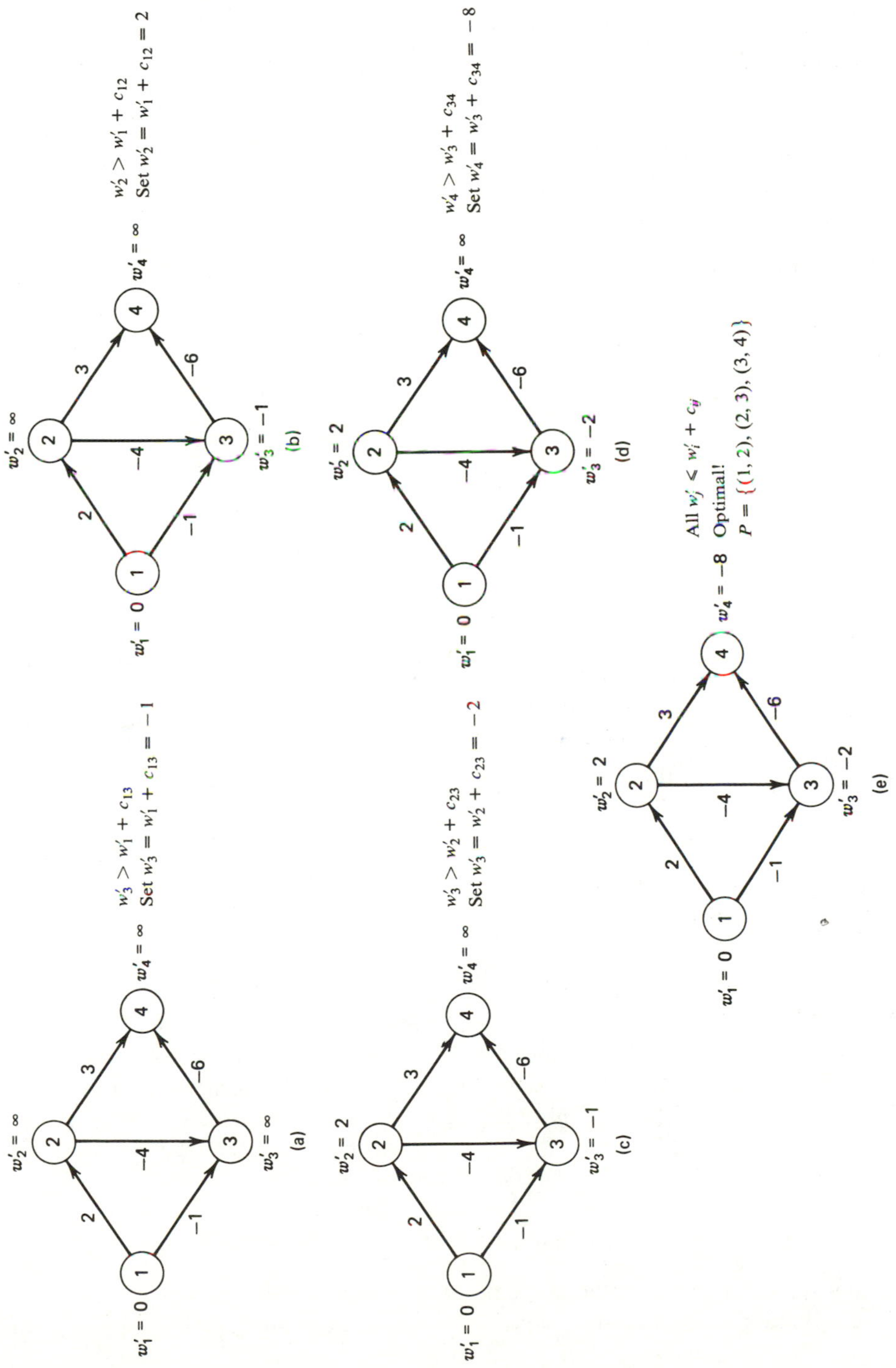

Figure 11.8. The solution for the network of Figure 11.7.

path, which contains no arcs, from node 1 to node 1.

Assume that the result is true at iteration t; that is, that w_i' ($< \infty$) represents the cost of some (not necessarily simple) path from node 1 to node i. Consider iteration $t + 1$. Either for every i and j, $w_i' + c_{ij} \geqslant w_j'$ and we stop, or else for some i and j, $w_i' + c_{ij} < w_j'$ in which case we set $w_j' = w_i' + c_{ij}$. In the former case we stop without changing any w_i', and so the result is true. In the latter case, by assumption there exists a path P_i from node 1 to node i at iteration t with cost w_i'. Consider the path $P_j = P_i \cup \{(i, j)\}$. This path has cost $w_i' + c_{ij} = w_j'$ and the result is true at iteration $t + 1$. Thus we have the following.

Theorem 2

If $w_k' < \infty$, then there exists a (not necessarily simple) path P from node 1 to node k along which $\Sigma_{(i\,j) \in P} c_{ij} = w_k'$.

Corollary 1

$$w_k' \geqslant \underset{P_k}{\text{Minimum}} \left(\sum_{(i,j) \in P_k} c_{ij} \right)$$

where P_k is any (not necessarily simple) path from node 1 to node k.

If there are no negative circuits, then the cost of any nonsimple path is greater than or equal to the cost of the corresponding simple path after deleting the circuits (why?). Hence, if there are no negative circuits, w_i' is bounded from below by the cost of the shortest simple path and thus by a finite integer (why?). Finally, since $\Sigma_i w_i'$ decreases by a positive integer at each iteration (why?), the shortest path algorithm will stop in a finite number of steps if there are no negative circuits.

We also have at hand a way to determine whether the network contains a negative circuit reachable from node 1. If no negative circuits exist, then $c_0 = \Sigma_{c_{ij}<0}\, c_{ij}$ is a lower bound on w_i' (why?) Thus, if any w_i' falls below c_0 during the algorithm, a negative circuit must exist and we stop the shortest path algorithm.

We have shown that the algorithm terminates in a finite number of steps. If at termination $w_m' = \infty$, there is no path from node 1 to node m. If $w_m' < \infty$, then there is a node l such that $w_m' - w_l' = c_{lm}$. Similarly there is a node k such that $w_l' - w_k' = c_{kl}$. Continuing in this fashion, node 1 is eventually reached (why?). This backtracking procedure defines the shortest path from node 1 to node m. To show that this is the case, note that if we let $x_{ij} = 1$ for arcs on the path and $x_{ij} = 0$ otherwise, and if we let the dual vector $\mathbf{w}$ be $-\mathbf{w}'$, then primal feasibility, dual feasibility, and complementary slackness hold.

A Labeling Algorithm for the Shortest Path Problem

Either for hand or computer calculations there are simple and convenient ways to maintain the information required to solve a shortest path problem with arbitrary costs. Suppose that we associate with each node j a label $L(j) = (i, w_j')$ containing two pieces of information. The second entry, w_j', in $L(j)$ indicates the cost (length) of the current "best" path from node 1 to node j. The first entry, i, in $L(j)$ gives the node just prior to node j in the path. Let $c_0 = \sum_{c_{ij}<0} c_{ij}$. The labeling algorithm becomes as follows.

INITIALIZATION STEP

Set $L(1) = (-, 0)$ and $L(i) = (-, \infty)$ for $i = 2, \ldots, m$.

MAIN STEP

If $w_j' \leqslant w_i' + c_{ij}$ for $i, j = 1, \ldots, m$, stop; the optimal solution is obtained. Otherwise, select (p, q) such that $w_q' > w_p' + c_{pq}$ and set $L(q) = (p, w_q' = w_p' + c_{pq})$. If $w_q' < c_0$, stop; there is a negative circuit in G. Otherwise, repeat the main step.

To identify the arcs of the shortest path, begin at node m. If the second label in $L(m)$, w_m', is ∞, then there is no path from node 1 to node m in G. Otherwise, the first entry in $L(m)$, say k, gives the previous node in the shortest path. Backtrack to node k and repeat the process until node 1 is reached.

An Example of the Labeling Algorithm

Suppose that we use the labeling algorithm to solve the shortest path problem of Figure 11.7. First, $c_0 = -1 - 4 - 6 = -11$.

The sequence of operations of the labeling algorithms are as follows.

1. $L(1) = (-, 0)$, $L(2) = (-, \infty)$, $L(3) = (-, \infty)$, $L(4) = (-, \infty)$
2. $L(3) = (1, -1)$
3. $L(2) = (1, 2)$
4. $L(3) = (2, -2)$
5. $L(4) = (3, -8)$
6. Optimal: $L_2(4) = -8 \; (< \infty) \Rightarrow$ a shortest path P exists with length of -8
7. $L_1(4) = 3 \Rightarrow (3, 4)$ is in P
8. $L_1(3) = 2 \Rightarrow (2, 3)$ is in P
9. $L_1(2) = 1 \Rightarrow (1, 2)$ is in P

Thus the shortest path is $P = \{(1, 2), (2, 3), (3, 4)\}$.

Identifying Negative Circuits With the Shortest Path Algorithm

We have already indicated that if at some point in the shortest path algorithm any $w_k' < c_0$ then a negative circuit reachable from node 1 exists in G. To find such a negative circuit, begin at node k and apply the following procedure.

INITIALIZATION STEP

Let $p = k$.

MAIN STEP

Let $L_1(p)$ be the first entry in $L(p)$.

1. If $L_1(p) > 0$ let $l = L_1(p)$ and replace $L_1(p)$ by $-L_1(p)$. Set $p = l$ and repeat the main step.
2. If $L_1(p) < 0$ then stop, a negative circuit has been found.

The original node k may not be part of the circuit. In this case it is necessary to discard the path, from node p to node k, from the sequence of nodes and arcs obtained by the above procedure.

11.3 MULTICOMMODITY FLOWS

In all of the flow problems we have considered to this point, it has not been necessary to distinguish among the units flowing in the network. This class of network flow problems is called single-commodity flow problems. There is also a class of network flow problems called multicommodity flow problems in which it is necessary to distinguish among the flows in the network.

The most natural example of multicommodity flows occurs in rush hour traffic in any metropolitan city. If the area is divided into zones, then there are a number of people in zone i who must travel to work in zone j. There are also a number of people who must travel from zone j to work in zone i. Where people are located corresponds to supply ($b > 0$) and where they wish to go corresponds to demand ($b < 0$). If we treat the problem as a single-commodity flow problem, a minimal cost flow procedure (network simplex or out-of-kilter) would use the supply of people in a given zone to satisfy the demand in the same zone. This is an unacceptable solution. In this problem and ones like it, we must distinguish between the different types of flow and be careful to retain their identity and flow pattern throughout the optimization procedure. That is, we must essentially have a different flow vector and set of conservation equations for each commodity.

As we shall see, multicommodity flow problems do not enjoy the same special properties as single-commodity flow problems. As an example, consider the network of Figure 11.9. Suppose that there are three commodities that flow through the network. The source for commodity 1 is node 1, and the sink for commodity 1 is node 3. That is, commodity 1 must originate only at node 1 and terminate only at node 3. Similarly, let the source and sink for commodity 2 be nodes 2 and 1 respectively. Finally, the source and sink for commodity 3 are

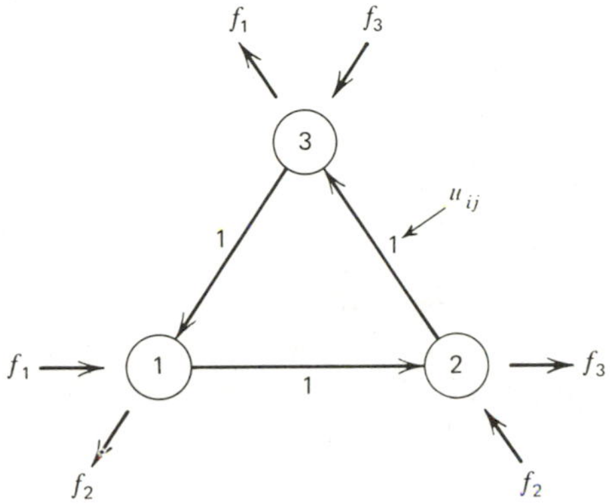

Figure 11.9. **A three-commodity maximal flow problem.**

nodes 3 and 2 respectively. With the restriction that the sum of all commodities flowing on an arc should not exceed the arc capacity $u_{ij} = 1$, what is the maximal sum of commodity flows, $f_1 + f_2 + f_3$, possible in the network?

Finding the maximal flow for the three-commodity problem of Figure 11.9 is relatively simple since there is only one path that each commodity can take on its way from its source to its sink. The paths for commodity 1, 2, and 3 respectively are

$$P_1 = \{(1, 2), (2, 3)\}$$

$$P_2 = \{(2, 3), (3, 1)\}$$

$$P_3 = \{(3, 1), (1, 2)\}.$$

If we place a single unit of flow on any one of the paths, then the other paths are completely blocked (that is, must have zero flow) and thus the total flow would be 1. However, there is a better solution available if we do not require integer flows. Suppose that we place $\frac{1}{2}$ unit of flow of commodity 1 on P_1, $\frac{1}{2}$ unit of flow of commodity 2 on P_2, and $\frac{1}{2}$ unit of flow of commodity 3 on P_3. In this case none of the arc capacities are violated and the total flow of all commodities is $\frac{3}{2}$. From this we see that multicommodity flow problems do not necessarily provide integer flows.

Even though multicommodity flow problems do not have as "nice" a structure as single-commodity flow problems, they still are linear programs (if we ignore integrality of the variables). As we shall soon see, multicommodity flow problems do have some special structure that permits the application of decomposition techniques.

The Multicommodity Minimal Cost Flow Problem

Suppose that we are given a network G with m nodes and n arcs in which there will flow t different commodities. Let $\mathbf{u}_i$ represent the vector of upper limits on flow for commodity i in the arcs of the network. Then u_{ipq} is the upper limit on flow of commodity i in arc (p, q). Also, let $\mathbf{u}$ represent the vector of upper limits on the sum of all commodities flowing in the arcs of the network. Then u_{pq} is the upper limit on the sum of all commodity flows in arc (p, q). Let $\mathbf{c}_i$ represent the vector of arc costs in the network for commodity i. Then c_{ipq} is the unit cost of commodity i on arc (p, q). Finally, let $\mathbf{b}_i$ represent the vector of supplies (or demands) of commodity i in the network. Then b_{iq} is the supply (if $b_{iq} > 0$) or demand (if $b_{iq} < 0$) of commodity i at node q.

The linear programming formulation for the multicommodity minimal cost flow problem is as follows:

$$\text{Minimize} \quad \sum_{i=1}^{t} \mathbf{c}_i\mathbf{x}_i$$

$$\text{Subject to} \quad \sum_{i=1}^{t} \mathbf{x}_i \leqslant \mathbf{u}$$

$$\mathbf{A}\mathbf{x}_i = \mathbf{b}_i \qquad i = 1, \ldots, t$$

$$\mathbf{0} \leqslant \mathbf{x}_i \leqslant \mathbf{u}_i \qquad i = 1, \ldots, t$$

where $\mathbf{x}_i$ is the vector of flows of commodity i in the network and $\mathbf{A}$ is the node-arc incidence matrix of the graph. The foregoing formulation is called the *node-arc* formulation for the multicommodity flow problem since it uses the node-arc incidence matrix.

The multicommodity minimal cost flow problem possesses the block diagonal structure discussed in Section 7.5 of the decomposition chapter. Thus we may apply the block diagonal decomposition technique to the foregoing problem. The multicommodity minimal cost flow problem has $(t + 1)n$ variables and $n + mt$ constraints (including the slack variables for the coupling constraints and ignoring the nonnegativity and upper bound constraints $\mathbf{0} \leqslant \mathbf{x}_i \leqslant \mathbf{u}_i$). Thus, even for moderate-sized problems, the constraint matrix will be large. For example, suppose that we have a problem with 100 nodes, 250 arcs, and 10 commodities. The problem will have 2750 variables and 1250 constraints.

Consider the application of the decomposition algorithm to the minimal cost multicommodity flow problem. Let $X_i = \{\mathbf{x}_i : \mathbf{A}\mathbf{x}_i = \mathbf{b}_i, \mathbf{0} \leqslant \mathbf{x}_i \leqslant \mathbf{u}_i\}$. Assume that each component of $\mathbf{u}_i$ is finite so that X_i is bounded (See Exercise 11.47 for a relaxation of this assumption). Then any $\mathbf{x}_i$ can be expressed as a convex combination of the extreme points of X_i as follows:

$$\mathbf{x}_i = \sum_{j=1}^{k_i} \lambda_{ij}\mathbf{x}_{ij}$$

where

$$\sum_{j=1}^{k_i} \lambda_{ij} = 1$$

$$\lambda_{ij} \geqslant 0 \qquad j = 1, \ldots, k_i$$

and $\mathbf{x}_{i1}, \mathbf{x}_{i2}, \ldots, \mathbf{x}_{ik_i}$ are the extreme points of X_i. Substituting for $\mathbf{x}_i$ in the multicommodity minimal cost flow problem and denoting the vector of slacks by $\mathbf{s}$, we get the following.

$$\text{Minimize} \quad \sum_{i=1}^{t} \sum_{j=1}^{k_i} (\mathbf{c}_i\mathbf{x}_{ij})\lambda_{ij}$$

$$\text{Subject to} \quad \sum_{i=1}^{t} \sum_{j=1}^{k_i} \mathbf{x}_{ij}\lambda_{ij} + \mathbf{s} = \mathbf{u}$$

$$\sum_{j=1}^{k_i} \lambda_{ij} = 1 \qquad i = 1, \ldots, t$$

$$\lambda_{ij} \geqslant 0 \qquad \begin{array}{l} j = 1, \ldots, k_i \\ i = 1, \ldots, t \end{array}$$

$$\mathbf{s} \geqslant \mathbf{0}$$

Suppose that we have a basic feasible solution to the multicommodity minimal cost flow problem in terms of the λ_{ij}'s and let $(\mathbf{w}, \boldsymbol{\alpha})$ be the vector of dual variables corresponding to the basic feasible solution ($\mathbf{w}$ has n components and $\boldsymbol{\alpha}$ has t components). Then dual feasibility is given by the following two conditions:

(i) $w_{pq} \leqslant 0$ corresponding to each s_{pq}, and
(ii) $\mathbf{w}\mathbf{x}_{ij} + \alpha_i - \mathbf{c}_i\mathbf{x}_{ij} \leqslant 0$ corresponding to each λ_{ij}.

If any of these conditions is violated, the corresponding variable (s_{pq} or λ_{ij}) is a candidate to enter the master basis. Here s_{pq} is a candidate to enter the basis if $w_{pq} > 0$. For a given commodity i, a nonbasic variable among the λ_{ij}'s could enter the basis if the optimal objective of the following subproblem is positive (why?).

$$\begin{aligned} \text{Maximize} \quad & (\mathbf{w} - \mathbf{c}_i)\mathbf{x}_i + \alpha_i \\ \text{Subject to} \quad & \mathbf{A}\mathbf{x}_i = \mathbf{b}_i \\ & \mathbf{0} \leqslant \mathbf{x}_i \leqslant \mathbf{u}_i \end{aligned}$$

But, since **A** is a node-arc incidence matrix, this is simply a single-commodity flow problem. Thus it may be solved by one of the efficient techniques for solving single-commodity network flow problems (network simplex or out-of-kilter).

Summary of the Decomposition Algorithm Applied to the Multicommodity Minimal Cost Flow Problem

We now specialize the decomposition algorithm of Chapter 7 to the multicommodity minimal cost flow problem.

INITIALIZATION STEP

Begin with a basic feasible solution to the master problem. Store $\mathbf{B}^{-1}$, $\bar{\mathbf{b}} = \mathbf{B}^{-1}\binom{\mathbf{u}}{\mathbf{1}}$, and $(\mathbf{w}, \boldsymbol{\alpha}) = \hat{\mathbf{c}}_B\mathbf{B}^{-1}$, where $\hat{c}_{ij} = \mathbf{c}_i\mathbf{x}_{ij}$ (the two-phase or the big-M method may be required).

MAIN STEP

1. Let $(\mathbf{w}, \boldsymbol{\alpha})$ be the vector of dual variables corresponding to the current basic feasible solution to the master problem. If any $w_{pq} > 0$, then the corresponding s_{pq} is a candidate to enter the master basis. If $w_{pq} \leqslant 0$ for each arc, consider the following ith subproblem.

$$\begin{aligned} \text{Maximize} \quad & (\mathbf{w} - \mathbf{c}_i)\mathbf{x}_i + \alpha_i \\ \text{Subject to} \quad & \mathbf{A}\mathbf{x}_i = \mathbf{b}_i \\ & \mathbf{0} \leqslant \mathbf{x}_i \leqslant \mathbf{u}_i \end{aligned}$$

   This is a single-commodity flow problem. If the solution $\mathbf{x}_{ik}$ to this problem has $z_{ik} - c_{ik} = (\mathbf{w} - \mathbf{c}_i)\mathbf{x}_{ik} + \alpha_i > 0$, then λ_{ik} is a candidate to enter the master basis.
2. If there is no candidate to enter the master basis, then stop; the optimal solution is at hand. Otherwise, select a candidate variable, update its

column accordingly, $\mathbf{B}^{-1}\begin{pmatrix}\mathbf{e}_{pq}\\ \mathbf{0}\end{pmatrix}$ for s_{pq} and $\mathbf{B}^{-1}\begin{pmatrix}\mathbf{x}_{ik}\\ \mathbf{e}_i\end{pmatrix}$ for λ_{ik}, and pivot. [Note that $\mathbf{e}_{pq}$ is a unit vector with the 1 in the row associated with arc (p, q).] This updates the basis inverse, the dual variables, and the right-hand side. Return to step 1.

An Example of the Multicommodity Minimal Cost Flow Algorithm

Consider the two-commodity minimal cost flow problem whose data are given in Figure 11.10.

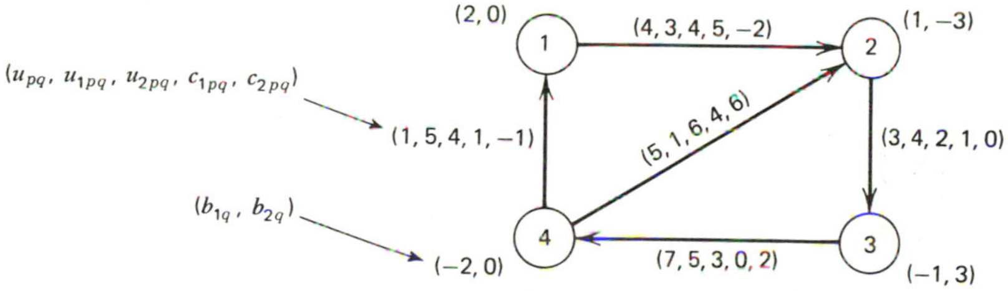

Figure 11.10. A two-commodity minimal cost flow problem.

The constraint matrix and the right-hand side are displayed in Figure 11.11 (the lower and upper bound constraints $\mathbf{0} \leqslant \mathbf{x}_1 \leqslant \mathbf{u}_1$ and $\mathbf{0} \leqslant \mathbf{x}_2 \leqslant \mathbf{u}_2$ are not displayed). Notice the structure of the coupling constraints and the special structured block diagonal constraints. Also note that x_1 and x_2 represent the artificial variables for the two commodities.

Initialization

To avoid the two-phase or the big-M methods, suppose that we begin with the following feasible solutions.

$$\mathbf{x}_{11} = \begin{bmatrix} x_{112}\\ x_{123}\\ x_{134}\\ x_{141}\\ x_{142}\end{bmatrix} = \begin{bmatrix}2\\3\\2\\0\\0\end{bmatrix} \quad \text{and} \quad \mathbf{x}_{21} = \begin{bmatrix} x_{212}\\ x_{223}\\ x_{234}\\ x_{241}\\ x_{242}\end{bmatrix} = \begin{bmatrix}0\\0\\3\\0\\3\end{bmatrix}$$

Note that the master basis (in the space of the slack variables and the λ_{ij}'s) consists of all the slacks, λ_{11} and λ_{21}. The basis and its inverse are

	FIRST COMMODITY VARIABLES						SECOND COMMODITY VARIABLES						SLACK VARIABLES					
	x_{112}	x_{123}	x_{134}	x_{141}	x_{142}	x_1	x_{212}	x_{223}	x_{234}	x_{241}	x_{242}	x_2	s_{12}	s_{23}	s_{34}	s_{41}	s_{42}	RHS
	−5	−1	0	−1	−4	0	2	0	−2	1	−6	0	0	0	0	0	0	0
Coupling Constraints	1	0	0	0	0	0	1	0	0	0	0	0	1	0	0	0	0	4
	0	1	0	0	0	0	0	1	0	0	0	0	0	1	0	0	0	3
	0	0	1	0	0	0	0	0	1	0	0	0	0	0	1	0	0	7
	0	0	0	1	0	0	0	0	0	1	0	0	0	0	0	1	0	1
	0	0	0	0	1	0	0	0	0	0	1	0	0	0	0	0	1	5
Node-arc incidence matrix for subproblem 1	1	0	0	−1	0	0	0	0	0	0	0	0	0	0	0	0	0	2
	−1	1	0	0	−1	0	0	0	0	0	0	0	0	0	0	0	0	1
	0	−1	1	0	0	0	0	0	0	0	0	0	0	0	0	0	0	−1
	0	0	−1	1	1	1	0	0	0	0	0	0	0	0	0	0	0	−2
Node-arc incidence matrix for subproblem 2	0	0	0	0	0	0	1	0	0	−1	0	0	0	0	0	0	0	0
	0	0	0	0	0	0	−1	1	0	0	−1	0	0	0	0	0	0	−3
	0	0	0	0	0	0	0	−1	1	0	0	0	0	0	0	0	0	3
	0	0	0	0	0	0	0	0	−1	1	1	1	0	0	0	0	0	0

Figure 11.11. The constraint matrix for the two-commodity problem of Figure 11.10.

$$\mathbf{B} = \begin{array}{c} \begin{matrix} s_{12} & s_{23} & s_{34} & s_{41} & s_{42} & \lambda_{11} & \lambda_{21} \end{matrix} \\ \begin{bmatrix} 1 & 0 & 0 & 0 & 0 & 2 & 0 \\ 0 & 1 & 0 & 0 & 0 & 3 & 0 \\ 0 & 0 & 1 & 0 & 0 & 2 & 3 \\ 0 & 0 & 0 & 1 & 0 & 0 & 0 \\ 0 & 0 & 0 & 0 & 1 & 0 & 3 \\ 0 & 0 & 0 & 0 & 0 & 1 & 0 \\ 0 & 0 & 0 & 0 & 0 & 0 & 1 \end{bmatrix} \end{array} \qquad \mathbf{B}^{-1} = \begin{bmatrix} 1 & 0 & 0 & 0 & 0 & -2 & 0 \\ 0 & 1 & 0 & 0 & 0 & -3 & 0 \\ 0 & 0 & 1 & 0 & 0 & -2 & -3 \\ 0 & 0 & 0 & 1 & 0 & 0 & 0 \\ 0 & 0 & 0 & 0 & 1 & 0 & -3 \\ 0 & 0 & 0 & 0 & 0 & 1 & 0 \\ 0 & 0 & 0 & 0 & 0 & 0 & 1 \end{bmatrix}$$

Here $\mathbf{c}_1\mathbf{x}_{11} = 13$ and $\mathbf{c}_2\mathbf{x}_{21} = 24$. Denoting $\begin{bmatrix} \mathbf{u} \\ \mathbf{1} \end{bmatrix}$ by $\hat{\mathbf{b}}$, we have

$$(\mathbf{w}, \boldsymbol{\alpha}) = \hat{\mathbf{c}}_B\mathbf{B}^{-1} = (0, 0, 0, 0, 0, 13, 24)\mathbf{B}^{-1} = (0, 0, 0, 0, 0, 13, 24)$$

$$\mathbf{x}_B = \mathbf{B}^{-1}\hat{\mathbf{b}} = \mathbf{B}^{-1}\begin{bmatrix} \mathbf{u} \\ 1 \\ 1 \end{bmatrix} = \mathbf{B}^{-1}\begin{bmatrix} 4 \\ 3 \\ 7 \\ 1 \\ 5 \\ 1 \\ 1 \end{bmatrix} = \begin{bmatrix} 2 \\ 0 \\ 2 \\ 1 \\ 2 \\ 1 \\ 1 \end{bmatrix}$$

$$z = \hat{\mathbf{c}}_B\mathbf{B}^{-1}\hat{\mathbf{b}} = 37$$

Setting up the revised simplex array

$(\mathbf{w}, \boldsymbol{\alpha})$	$\hat{\mathbf{c}}_B\mathbf{B}^{-1}\hat{\mathbf{b}}$
$\mathbf{B}^{-1}$	$\mathbf{B}^{-1}\hat{\mathbf{b}}$

for the master problem, we get the following.

	w_{12}	w_{23}	w_{34}	w_{41}	w_{42}	α_1	α_2	RHS
z	0	0	0	0	0	13	24	37
s_{12}	1	0	0	0	0	−2	0	2
s_{23}	0	1	0	0	0	−3	0	0
s_{34}	0	0	1	0	0	−2	−3	2
s_{41}	0	0	0	1	0	0	0	1
s_{42}	0	0	0	0	1	0	−3	2
λ_{11}	0	0	0	0	0	1	0	1
λ_{12}	0	0	0	0	0	0	1	1

Iteration 1

First, all $w_{pq} \leqslant 0$. Next we check whether a candidate from either subproblem (or commodity) is eligible to enter the master basis.

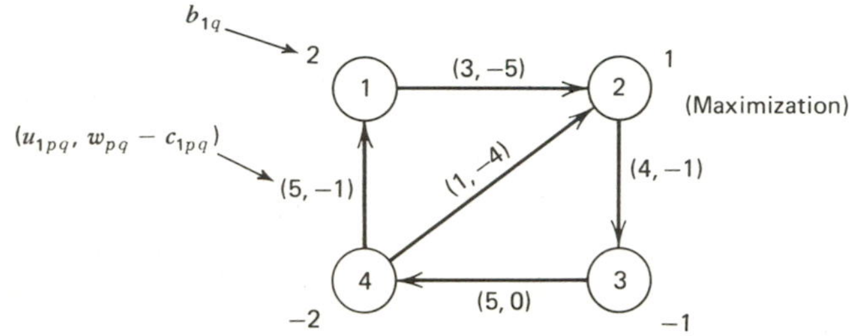

Figure 11.12. **Subproblem 1 at the first iteration.**

SUBPROBLEM 1

$\mathbf{w} - \mathbf{c}_1 = \mathbf{0} - \mathbf{c}_1 = (-5, -1, 0, -1, -4)$

Subproblem 1 is the single-commodity flow problem defined in Figure 11.12. The optimal (maximal cost) solution is $\mathbf{x}_{12} = (2, 3, 2, 0, 0)^t$ and the value of the subproblem 1 objective is

$z_{12} - c_{12} = (\mathbf{w} - \mathbf{c}_1)\mathbf{x}_{12} + \alpha_1 = -13 + 13 = 0$

Thus there is no candidate from subproblem 1.

SUBPROBLEM 2

$\mathbf{w} - \mathbf{c}_2 = \mathbf{0} - \mathbf{c}_2 = (2, 0, -2, 1, -6)$

Subproblem 2 is the single-commodity flow problem defined in Figure 11.13. The optimal (maximal cost) solution is $\mathbf{x}_{22} = (3, 0, 3, 3, 0)^t$ and

$z_{22} - c_{22} = (\mathbf{w} - \mathbf{c}_2)\mathbf{x}_{22} + \alpha_2 = 3 + 24 = 27$

Thus λ_{22} is a candidate to enter the basis. The updated column for λ_{22} (exclusive

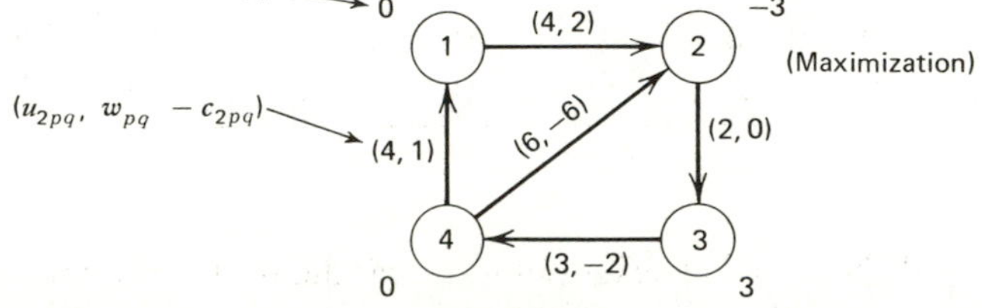

Figure 11.13. **Subproblem 2 at the first iteration.**

of $z_{22} - c_{22}$) is

$$\mathbf{B}^{-1}\begin{bmatrix}\mathbf{x}_{22}\\ 0\\ 1\end{bmatrix} = (3, 0, 0, 3, -3, 0, 1)^t$$

The pivoting process is as follows.

	w_{12}	w_{23}	w_{34}	w_{41}	w_{42}	α_1	α_2	RHS	λ_{22}
z	0	0	0	0	0	13	24	37	27
s_{12}	1	0	0	0	0	−2	0	2	3
s_{23}	0	1	0	0	0	−3	0	0	0
s_{34}	0	0	1	0	0	−2	−3	2	0
s_{41}	0	0	0	1	0	0	0	1	(3)
s_{42}	0	0	0	0	1	0	−3	2	−3
λ_{11}	0	0	0	0	0	1	0	1	0
λ_{21}	0	0	0	0	0	0	1	1	1

	w_{12}	w_{23}	w_{34}	w_{41}	w_{42}	α_1	α_2	RHS
z	0	0	0	−9	0	13	24	28
s_{12}	1	0	0	−1	0	−2	0	1
s_{23}	0	1	0	0	0	−3	0	0
s_{34}	0	0	1	0	0	−2	−3	2
λ_{22}	0	0	0	$\frac{1}{3}$	0	0	0	$\frac{1}{3}$
s_{42}	0	0	0	1	1	0	−3	3
λ_{11}	0	0	0	0	0	1	0	1
λ_{21}	0	0	0	$-\frac{1}{3}$	0	0	1	$\frac{2}{3}$

Iteration 2

Again all $w_{pq} \leqslant 0$ (so no s_{pq} is a candidate to enter the master basis).

SUBPROBLEM 1

$(\mathbf{w} - \mathbf{c}_1) = (-5, -1, 0, -10, -4)$

Subproblem 1 is the single-commodity flow problem defined in Figure 11.14. The optimal solution is $\mathbf{x}_{13} = (2, 3, 2, 0, 0)^t$ with

$z_{13} - c_{13} = (\mathbf{w} - \mathbf{c}_1)\mathbf{x}_{13} + \alpha_1 = -13 + 13 = 0$

Thus there is no candidate from subproblem 1.

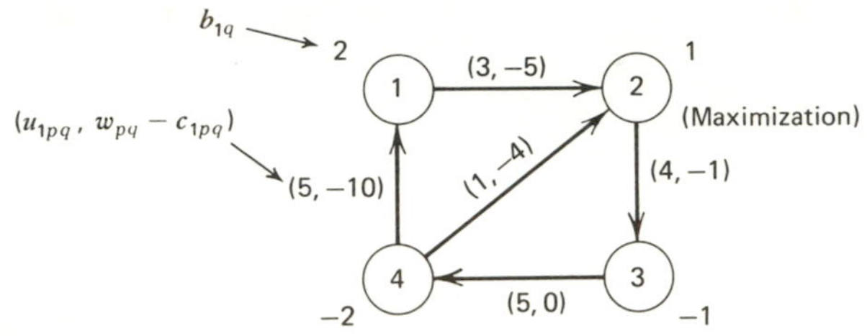

Figure 11.14. **Subproblem 1 at the second iteration.**

SUBPROBLEM 2

$(\mathbf{w} - \mathbf{c}_2) = (2, 0, -2, -8, -6)$

Subproblem 2 is the single-commodity flow problem defined in Figure 11.15. An optimal solution is $\mathbf{x}_{23} = (3, 0, 3, 3, 0)^t$ with

$$z_{23} - c_{23} = (\mathbf{w} - \mathbf{c}_2)\mathbf{x}_{23} + \alpha_2 = -24 + 24 = 0$$

Thus there is no candidate from subproblem 2.

Therefore we already have the optimal solution as follows:

$$z^* = 28$$

$$\mathbf{x}_1^* = \lambda_{11}\mathbf{x}_{11} = (2, 3, 2, 0, 0)^t$$

$$\begin{aligned}\mathbf{x}_2^* &= \lambda_{21}\mathbf{x}_{21} + \lambda_{22}\mathbf{x}_{22} \\ &= \tfrac{2}{3}(0, 0, 3, 0, 3)^t + \tfrac{1}{3}(3, 0, 3, 3, 0)^t \\ &= (1, 0, 3, 1, 2)^t\end{aligned}$$

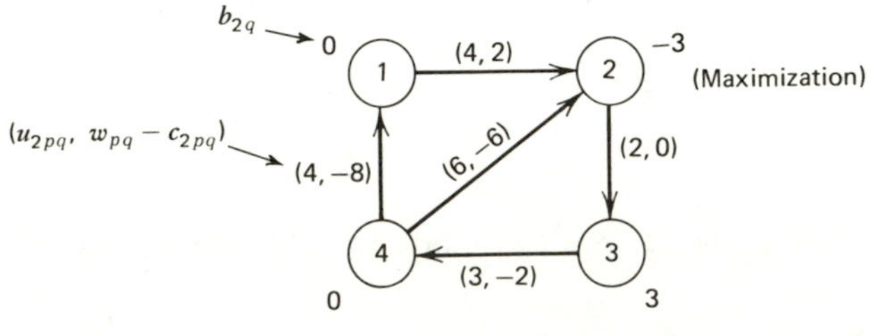

Figure 11.15. **Subproblem 2 at the second iteration.**

11.4 CHARACTERIZATION OF A BASIS FOR THE MULTICOMMODITY MINIMAL COST FLOW PROBLEM

Suppose that we proceed to apply the simplex method directly to the multicommodity minimal cost flow problem. We first note from Chapter 9 that the system

$\mathbf{Ax}_i = \mathbf{b}_i$ has rank $m - 1$ so that it is necessary to add an artificial variable for each commodity. Adding this artificial column vector, the overall constraint matrix is given by

$$\begin{array}{c} \begin{array}{ccccc} \mathbf{x}_1 & \mathbf{x}_2 & \cdots & \mathbf{x}_t & \mathbf{x} \end{array} \\ \left[\begin{array}{ccccc} \bar{\mathbf{I}} & \bar{\mathbf{I}} & \cdots & \bar{\mathbf{I}} & \mathbf{I} \\ \bar{\mathbf{A}} & \mathbf{0} & \cdots & \mathbf{0} & \mathbf{0} \\ \mathbf{0} & \bar{\mathbf{A}} & \cdots & \mathbf{0} & \mathbf{0} \\ & & \cdots & & \\ \mathbf{0} & \mathbf{0} & \cdots & \bar{\mathbf{A}} & \mathbf{0} \end{array}\right] \end{array} \begin{array}{l} \\ \} \; n \text{ rows} \\ \} \; m \text{ rows} \\ \} \; m \text{ rows} \\ \vdots \\ \} \; m \text{ rows} \end{array}$$

where $\bar{\mathbf{A}} = [\mathbf{A}, \mathbf{e}_m]$ and $\bar{\mathbf{I}} = [\mathbf{I}, \mathbf{0}]$. Selecting a basis submatrix from this matrix, we get

$$\hat{\mathbf{B}} = \left[\begin{array}{ccccc} \mathbf{E}_1 & \mathbf{E}_2 & \cdots & \mathbf{E}_t & \mathbf{E} \\ \bar{\mathbf{A}}_1 & \mathbf{0} & \cdots & \mathbf{0} & \mathbf{0} \\ \mathbf{0} & \bar{\mathbf{A}}_2 & \cdots & \mathbf{0} & \mathbf{0} \\ \vdots & \vdots & & \vdots & \vdots \\ \mathbf{0} & \mathbf{0} & \cdots & \bar{\mathbf{A}}_t & \mathbf{0} \end{array}\right]$$

where $\mathbf{E}_i$ and $\mathbf{E}$ are matrices formed by taking selected columns of $\bar{\mathbf{I}}$ and $\mathbf{I}$ respectively. The row location of the 1 for a particular column of $\mathbf{E}_i$ identifies the arc used in $\bar{\mathbf{A}}_i$.

Since $\hat{\mathbf{B}}$ is a basis matrix, each $\bar{\mathbf{A}}_i$ must contain a submatrix that spans E^m. Therefore each $\bar{\mathbf{A}}_i$ contains an $m \times m$ basis (why?). Partition $\bar{\mathbf{A}}_i$ into $[\mathbf{B}_i|\mathbf{D}_i]$ where $\mathbf{B}_i$ is a basis matrix for $\mathbf{Ax}_i = \mathbf{b}_i$. Note that $\mathbf{B}_i$ must contain the artificial column (why?). From Chapter 9, since $\mathbf{B}_i$ is a basis for a set of single-commodity flow conservation constraints, $\mathbf{B}_i$ must correspond to a *rooted spanning tree* in G with the artificial variable as the root. Similarly partition $\mathbf{E}_i$ into $[\mathbf{E}'_i|\mathbf{E}''_i]$. Substituting into $\hat{\mathbf{B}}$ and rearranging the columns, we get

$$\hat{\mathbf{B}} = \left[\begin{array}{cccccccccc} \mathbf{E}'_1 & \mathbf{E}'_2 & \cdots & \mathbf{E}'_t & \mathbf{E}''_1 & \mathbf{E}''_2 & \cdots & \mathbf{E}''_t & \mathbf{E} \\ \mathbf{B}_1 & \mathbf{0} & \cdots & \mathbf{0} & \mathbf{D}_1 & \mathbf{0} & \cdots & \mathbf{0} & \mathbf{0} \\ \mathbf{0} & \mathbf{B}_2 & \cdots & \mathbf{0} & \mathbf{0} & \mathbf{D}_2 & \cdots & \mathbf{0} & \mathbf{0} \\ \vdots & \vdots & & \vdots & \vdots & \vdots & & \vdots & \vdots \\ \mathbf{0} & \mathbf{0} & \cdots & \mathbf{B}_t & \mathbf{0} & \mathbf{0} & \cdots & \mathbf{D}_t & \mathbf{0} \end{array}\right]$$

In other words, the basis matrix $\hat{\mathbf{B}}$ has the following general structure:

$$\hat{\mathbf{B}} = \begin{bmatrix} \mathbf{E}' & \mathbf{E}'' & \mathbf{E} \\ \mathbf{B} & \mathbf{D} & \mathbf{0} \end{bmatrix}$$

Denoting the right-hand side $\begin{bmatrix} \mathbf{u} \\ \mathbf{b} \end{bmatrix}$ by $\hat{\mathbf{b}}$ (where $\mathbf{b}$ is a column vector consisting of $\mathbf{b}_1, \mathbf{b}_2, \ldots,$ and $\mathbf{b}_t$), the basic system $\hat{\mathbf{B}}\mathbf{x}_B = \hat{\mathbf{b}}$ reduces to

$$\begin{bmatrix} \mathbf{E}' & \mathbf{E}'' & \mathbf{E} \\ \mathbf{B} & \mathbf{D} & \mathbf{0} \end{bmatrix} \begin{bmatrix} \mathbf{x}_F \\ \mathbf{x}_D \\ \mathbf{s}_B \end{bmatrix} = \begin{bmatrix} \mathbf{u} \\ \mathbf{b} \end{bmatrix}$$

where $\mathbf{x}_B$ is decomposed into $\begin{bmatrix} \mathbf{x}_F \\ \mathbf{x}_D \\ \mathbf{s}_B \end{bmatrix}$. This system is not easy to solve. However, by utilizing the following change of variables, the system can be solved, as we shall outline shortly:

$$\begin{bmatrix} \mathbf{x}_F \\ \mathbf{x}_D \\ \mathbf{s}_B \end{bmatrix} = \begin{bmatrix} \mathbf{I} & -\mathbf{B}^{-1}\mathbf{D} & \mathbf{0} \\ \mathbf{0} & \mathbf{I} & \mathbf{0} \\ \mathbf{0} & \mathbf{0} & \mathbf{I} \end{bmatrix} \begin{bmatrix} \mathbf{x}'_F \\ \mathbf{x}'_D \\ \mathbf{s}'_B \end{bmatrix} \tag{11.4}$$

This is a nonsingular transformation and thus we have an equivalent system to work with. On substituting for $\mathbf{x}_B$ in $\hat{\mathbf{B}}\mathbf{x}_B = \hat{\mathbf{b}}$, we get

$$\begin{bmatrix} \mathbf{E}' & \mathbf{E}'' & \mathbf{E} \\ \mathbf{B} & \mathbf{D} & \mathbf{0} \end{bmatrix} \begin{bmatrix} \mathbf{I} & -\mathbf{B}^{-1}\mathbf{D} & \mathbf{0} \\ \mathbf{0} & \mathbf{I} & \mathbf{0} \\ \mathbf{0} & \mathbf{0} & \mathbf{I} \end{bmatrix} \begin{bmatrix} \mathbf{x}'_F \\ \mathbf{x}'_D \\ \mathbf{s}'_B \end{bmatrix} = \begin{bmatrix} \mathbf{u} \\ \mathbf{b} \end{bmatrix}$$

$$\begin{bmatrix} \mathbf{E}' & \mathbf{E}'' - \mathbf{E}'\mathbf{B}^{-1}\mathbf{D} & \mathbf{E} \\ \mathbf{B} & \mathbf{0} & \mathbf{0} \end{bmatrix} \begin{bmatrix} \mathbf{x}'_F \\ \mathbf{x}'_D \\ \mathbf{s}'_B \end{bmatrix} = \begin{bmatrix} \mathbf{u} \\ \mathbf{b} \end{bmatrix}$$

Now the second set of equations $\mathbf{B}\mathbf{x}'_F = \mathbf{b}$ is easy to solve since it corresponds to a series of rooted spanning trees, one for each commodity.

Consider the first set of equations in the transformed system after having solved for $\mathbf{x}'_F$:

$$[\mathbf{E}'' - \mathbf{E}'\mathbf{B}^{-1}\mathbf{D}, \mathbf{E}] \begin{bmatrix} \mathbf{x}'_D \\ \mathbf{s}'_B \end{bmatrix} = \mathbf{u} - \mathbf{E}'\mathbf{x}'_F$$

The solution is

$$\begin{bmatrix} \mathbf{x}'_D \\ \mathbf{s}'_B \end{bmatrix} = [\mathbf{E}'' - \mathbf{E}'\mathbf{B}^{-1}\mathbf{D}, \mathbf{E}]^{-1}(\mathbf{u} - \mathbf{E}'\mathbf{x}'_F)$$

and requires the inversion of the matrix $[\mathbf{E}'' - \mathbf{E}'\mathbf{B}^{-1}\mathbf{D}, \mathbf{E}]$. While this matrix is not as easy to invert as $\mathbf{B}$, it is easy to form and understand. Here $\mathbf{E}$ is a matrix of unit columns corresponding to the slack variables in the basis. Now let us investigate the matrix $\mathbf{E}'' - \mathbf{E}'\mathbf{B}^{-1}\mathbf{D}$.

A typical column of $\mathbf{E}'' - \mathbf{E}'\mathbf{B}^{-1}\mathbf{D}$ corresponding to commodity k is given by $\mathbf{e}_{ij} - \mathbf{E}'_k\mathbf{B}_k^{-1}\mathbf{a}_{ij}$, where $\mathbf{e}_{ij}$ is a unit vector in E^n with a 1 in the row corresponding to arc (i, j) and $\mathbf{a}_{ij}$ is a vector in E^m with a 1 in row i and a -1 in row j. From Chapter 9 recall that $\mathbf{y}_{ij} = \mathbf{B}_k^{-1}\mathbf{a}_{ij}$ corresponds to a chain from node i to node j through the basis tree. Note that the coefficients of $\mathbf{y}_{ij}$ actually reorient the chain into a path. Then $-\mathbf{B}_k^{-1}\mathbf{a}_{ij}$ corresponds to a chain from j to i in the rooted spanning tree of commodity k. Each coefficient in $-\mathbf{B}_k^{-1}\mathbf{a}_{ij}$ corresponds to a basic variable in $\mathbf{B}_k$. Now $\mathbf{E}'_k$ is an $n \times m$ matrix with its columns being unit vectors in E^n identifying the basic variables in $\mathbf{B}_k$. Thus $-\mathbf{E}'_k\mathbf{B}_k^{-1}\mathbf{a}_{ij}$ simply expands the m vector $-\mathbf{B}_k^{-1}\mathbf{a}_{ij}$ to an n vector by assigning zero coefficients corresponding to all nonbasic arcs of commodity k. Hence $-\mathbf{E}'_k\mathbf{B}_k^{-1}\mathbf{a}_{ij}$ is an n vector corresponding to the chain from node j to node i in the rooted spanning tree of commodity k. Finally $\mathbf{e}_{ij} - \mathbf{E}'_k\mathbf{B}_k^{-1}\mathbf{a}_{ij}$ corresponds to the unique cycle formed when the arc (i, j) is added to the basis tree (and the basis arcs are properly oriented). Thus, knowing $\mathbf{B}$, it is easy to form $\mathbf{E}'' - \mathbf{E}'\mathbf{B}^{-1}\mathbf{D}$.

The important conclusion is the following.

***Theorem* 3**

A transformed basis matrix for the multicommodity minimal cost flow problem corresponds to a rooted spanning tree for each commodity plus a set of cycles and slacks.

Once $[\mathbf{E}'' - \mathbf{E}'\mathbf{B}^{-1}\mathbf{D}, \mathbf{E}]$ is formed as described above, we can solve for $\mathbf{x}'_D$ and $\mathbf{s}'_B$. With the vector

$$\begin{bmatrix} \mathbf{x}'_F \\ \mathbf{x}'_D \\ \mathbf{s}'_B \end{bmatrix}$$

now known, we can solve for the basic variables $\mathbf{x}_F$, $\mathbf{x}_D$, and $\mathbf{s}_B$ from Equation (11.4). In Exercise 11.55 we ask the reader to develop a systematic procedure for computing the dual variables, updating the column of the entering variable, and the basis inverse. This, coupled with the foregoing procedure for computing the basic variables, represents a direct application of the simplex method for solving multicommodity flow problems.

An Example of a Basis Matrix for the Multicommodity Minimal Cost Flow Problem

Consider the multicommodity minimal cost flow problem of Figure 11.10 without the upper bound constraints on the individual commodities. Recall that the constraint matrix is shown in Figure 11.11.

Suppose that we select the basis submatrix (the reader is asked to verify that this is a basis submatrix) indicated in Figure 11.16.

x_{112}	x_{123}	x_{134}	x_{141}	x_1	x_{212}	x_{234}	x_{241}	x_{242}	x_2	s_{12}	s_{34}	s_{42}
1	0	0	0	0	1	0	0	0	0	1	0	0
0	1	0	0	0	0	0	0	0	0	0	0	0
0	0	1	0	0	0	1	0	0	0	0	1	0
0	0	0	1	0	0	0	1	0	0	0	0	0
0	0	0	0	0	0	0	0	1	0	0	0	1
1	0	0	− 1	0	0	0	0	0	0	0	0	0
− 1	1	0	0	0	0	0	0	0	0	0	0	0
0	− 1	1	0	0	0	0	0	0	0	0	0	0
0	0	− 1	1	1	0	0	0	0	0	0	0	0
0	0	0	0	0	1	0	− 1	0	0	0	0	0
0	0	0	0	0	− 1	0	0	− 1	0	0	0	0
0	0	0	0	0	0	1	0	0	0	0	0	0
0	0	0	0	0	0	− 1	1	1	1	0	0	0

Figure 11.16. A basis submatrix.

Applying the transformation of Equation (11.4), we get the matrix

$$\begin{bmatrix} \mathbf{E}_1' & \mathbf{E}_2' & \mathbf{E}_1'' - \mathbf{E}_1'\mathbf{B}_1^{-1}\mathbf{D}_1 & \mathbf{E}_2'' - \mathbf{E}_2'\mathbf{B}_2^{-1}\mathbf{D}_2 & \mathbf{E} \\ \mathbf{B}_1 & \mathbf{0} & \mathbf{0} & \mathbf{0} & \mathbf{0} \\ \mathbf{0} & \mathbf{B}_2 & \mathbf{0} & \mathbf{0} & \mathbf{0} \end{bmatrix} \tag{11.5}$$

Here $\mathbf{B}_1$ consists of x_{112}, x_{123}, x_{134}, and x_1 whereas $\mathbf{B}_2$ consists of x_{234}, x_{241}, x_{242}, and x_2. These two rooted spanning trees are illustrated in Figure 11.17 below. In addition $\mathbf{D}_1$ and $\mathbf{D}_2$ are represented by x_{141} and x_{212} and correspond to the cycles of Figure 11.17.

Examining Figure 11.17, we see that

$$\mathbf{E}_1'' - \mathbf{E}_1'\mathbf{B}_1^{-1}\mathbf{D}_1 = \begin{bmatrix} 0-(-1) \\ 0-(-1) \\ 0-(-1) \\ 1-0 \\ 0-0 \end{bmatrix} \begin{matrix} (1,2) \\ (2,3) \\ (3,4) \\ (4,1) \\ (4,2) \end{matrix} \qquad \mathbf{E}_2'' - \mathbf{E}_2'\mathbf{B}_2^{-1}\mathbf{D}_2 = \begin{bmatrix} 1-0 \\ 0-0 \\ 0-0 \\ 0-(-1) \\ 0-(+1) \end{bmatrix} \begin{matrix} (1,2) \\ (2,3) \\ (3,4) \\ (4,1) \\ (4,2) \end{matrix}$$

Substituting this information into the transformed basis submatrix of matrix (11.5), we get the basis representation shown in Figure 11.18.

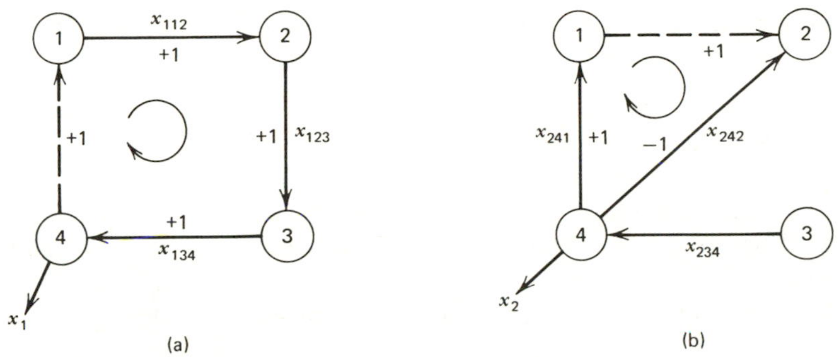

Figure 11.17. **Graphical illustration of the basis matrix: (*a*) Commodity 1 subbasis. (*b*) Commodity 2 subbasis.**

x'_{112}	x'_{123}	x'_{134}	x'_1	x'_{234}	x'_{241}	x'_{242}	x'_2	x'_{141}	x'_{212}	s_{12}	s_{34}	s_{42}
1	0	0	0	0	0	0	0	1	1	1	0	0
0	1	0	0	0	0	0	0	1	0	0	0	0
0	0	1	0	1	0	0	0	1	0	0	1	0
0	0	0	0	0	1	0	0	1	1	0	0	0
0	0	0	0	0	0	1	0	0	−1	0	0	1
1	0	0	0	0	0	0	0	0	0	0	0	0
−1	1	0	0	0	0	0	0	0	0	0	0	0
0	−1	1	0	0	0	0	0	0	0	0	0	0
0	0	−1	1	0	0	0	0	0	0	0	0	0
0	0	0	0	0	−1	0	0	0	0	0	0	0
0	0	0	0	0	0	−1	0	0	0	0	0	0
0	0	0	0	1	0	0	0	0	0	0	0	0
0	0	0	0	−1	1	1	1	0	0	0	0	0

Figure 11.18. **The transformed basis submatrix.**

EXERCISES

11.1 Find the maximal flow from node 1 to node 7 in the following network.

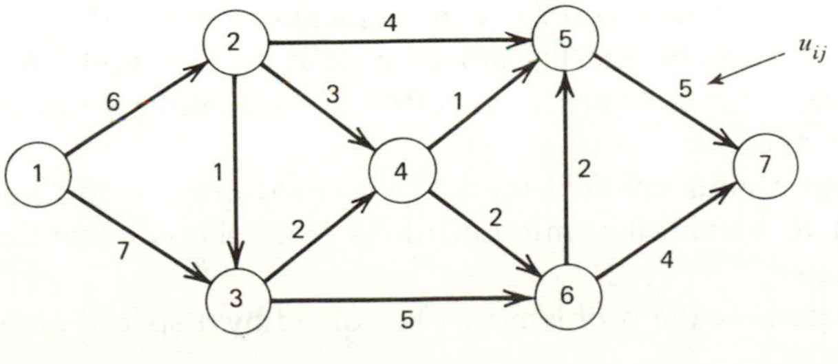

11.2 Discuss the economic meaning of the dual variables in the maximal flow problem. Consider both the w_i's and the h_{ij}'s.

11.3 Consider the production process shown below indicating the various paths that a product can take on its way to assembly through a plant. The number in each box represents the upper limit on items per hour that can be processed at the station.

a. What is the maximal number of parts per hour that the plant can handle?

b. Which operations should you try to improve?

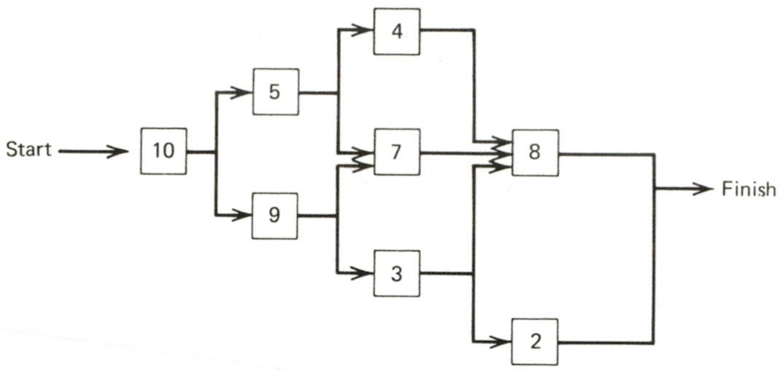

11.4 Two paths are said to be *arc disjoint* if they contain no common arcs. Prove that the maximal number of arc disjoint paths from node 1 to node m in a network is equal to the minimal number of arcs that must be deleted in order to separate node 1 from node m.

11.5 Find the maximal flow from node 1 to node 8 in the following network.

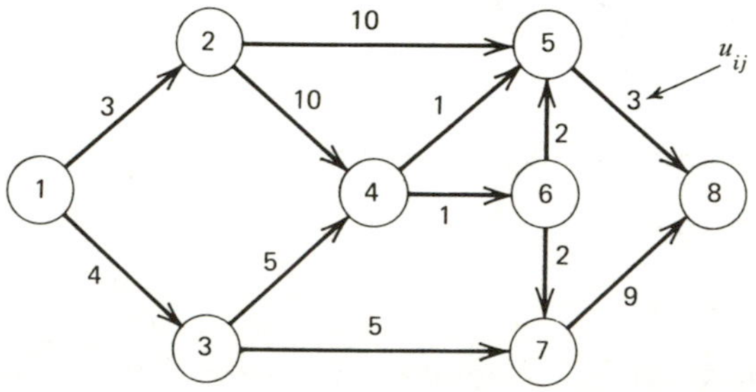

11.6 In a command and control communications network a commander is located at one node and his subordinate at another node. Associated with each link in the network is an effort u_{ij} required to eliminate that link from the network.

a. Present a mathematical model that could be used to find the minimal effort to block all communications from the commander to his subordinate.

b. Indicate how the problem can be solved by a special algorithm.

11.7 Indicate how lower bounds on flow can be handled in the maximal flow algorithm. (*Hint*. Apply a phase I procedure as follows. From G construct G' by: (1) All nodes in G are in G'. (2) In addition G' contains two new nodes $m+1$ and $m+2$. (3) All arcs in G are in G'. (4) $l'_{ij} = 0$, $u'_{ij} = u_{ij} - l_{ij}$. (5) If $l_{ij} > 0$, then place arc $(i, m+2)$ in G' with $u'_{i,m+2} = l_{ij}$ and $l'_{i,m+2} = 0$; place arc $(m+1, j)$ in G' with $u'_{m+1,j} = l_{ij}$ and $l'_{m+1,j} = 0$. (6) Solve for the maximal flow from node $m+1$ to node $m+2$ in G'. (7) If at optimality all arcs out of node $m+1$ are *saturated* (that is, $x'_{ij} = u'_{ij}$), then a feasible flow exists in G and $x_{ij} = x'_{ij} + l_{ij}$; otherwise, no feasible solution exists.)

11.8 Consider the following procedure for reducing the size of a network while finding the maximal flow from node 1 to node m.

Step 1. Remove all arcs entering node 1 (the source) and leaving node m (the sink).

Step 2. Discard any node that has no arcs incident with it.

Step 3. Discard any node, except node 1, that only has arcs leaving it. Also discard these arcs.

Step 4. Discard any node, except node m, that only has arcs entering it. Also discard these arcs.

Repeat steps 2, 3, and 4 until no change results. If node 1 or m is discarded, stop; the maximum flow is zero. Otherwise use the maximum flow algorithm.

a. Show that the maximum flow in the resulting network is the same as in the original network.
b. Show that there is a path from node 1 to every node in the resulting network (assuming that node 1 has not been discarded).
c. Can one state and prove a similar result concerning node m?

11.9 Show that we have a basic feasible solution to the maximal flow problem if there exist no cycles in the set $E = \{(i, j) : 0 < x_{ij} < u_{ij}\}$. In this case show how the basic variables and the simplex tableau can be obtained at any iteration of the maximal flow algorithm. Furthermore, show that the following procedure will maintain basic feasible solutions in the maximal flow algorithm.

Step 1. At each iteration, begin with E as defined above.

Step 2. Try to find a path from node 1 to node m in G' associated only with arcs in E.

Step 3. If no path is available from step 2, then add one arc in G' not associated with arcs in E to E if it permits the labeling of a new node. With this new arc in E, return to step 2.

11.10 What simplifications would result if the out-of-kilter algorithm is used to solve the maximal flow problem? Give all details.

11.11 What simplifications would result if the network simplex method (of Chapter 9) is used to solve the maximal flow problem? Give all details.

11.12 Develop a dual simplex method for the maximal flow problem.

11.13 How can node capacities be handled in the maximal flow algorithm?

11.14 Modify the maximal flow algorithm to handle undirected arcs.

11.15 Consider the problem of finding the minimum number of lines to cover all zeros in the assignment algorithm (refer to Section 8.8). Show that the maximal flow algorithm can be used to provide the result. (*Hint*. Given the reduced assignment matrix, construct a maximal flow network G as follows. Let nodes $1, \ldots, n$ represent the n rows of the assignment matrix and nodes $n + 1, \ldots, 2n$ represent the columns of the assignment matrix. If the ijth entry in the reduced matrix is zero, draw an arc from node i to node $n + j$ with $u_{i,\, n+j} = 1$. Add two additional nodes $2n + 1$ and $2n + 2$. Add an arc $(2n + 1, i)$ with $u_{2n+1,\, i} = 1$ for $i = 1, \ldots, n$ and an arc $(n + i, 2n + 2)$ with $u_{n+i,\, 2n+2} = 1$ for $i = 1, \ldots, n$. Solve for the maximal flow from node $2n + 1$ to node $2n + 2$ in G. The value of the maximal flow is equal to the minimum number of lines to cover all zeros in the reduced matrix. To find which lines to use, consider the sets X and $\bar{X}$ when the maximal flow algorithm stops. If $(2n + 1, i)$ is in $(X, \bar{X})$, draw a line through row i. If $(n + i, 2n + 2)$ is in $(X, \bar{X})$, draw a line through column i. It still must be shown that this procedure works.)

11.16 Apply the procedure of the previous problem to find the minimum number of lines to cover all zeros in the following reduced assignment matrix.

	1	2	3	4
1	0	0	0	1
2	2	5	3	0
3	1	4	2	0
4	2	0	4	1

11.17 In this chapter we have provided the node-arc formulation for the maximal flow problem. Consider an *arc-path* formulation as follows. Let $j = 1, 2, \ldots, t$ be an enumeration of all of the paths from node 1 to node m in the network. Number the arcs from 1 to n and let

$$p_{ij} = \begin{cases} 1 & \text{if arc } i \text{ is in path } j \\ 0 & \text{otherwise} \end{cases}$$

The arc-path formulation for the maximal flow problem is

$$\text{Maximize} \quad \sum_{j=1}^{t} x_j$$

$$\text{Subject to} \quad \sum_{j=1}^{t} p_{ij}x_j \leqslant u_i \qquad i = 1, \ldots, n$$

$$x_j \geqslant 0 \qquad j = 1, 2, \ldots, t$$

where x_j represents the flow on path j.

a. Give the complete arc-path formulation for the maximal flow problem of Figure 11.1.

b. Solve the linear program of part (a).

11.18 Consider the arc-path formulation for the maximal flow problem as given in Exercise 11.17. Suppose that we do not enumerate any paths to begin with but, rather, decide to apply the revised simplex method with all slack variables in the starting basic feasible solution. At any iteration of the revised simplex method let **w** be the dual vector.

a. What is the simplex entry criterion for (i) a slack variable and (ii) a path variable?

b. Show that there is an easy method to test the simplex entry criterion for path variables using the shortest path algorithm.

c. If you always first enter slacks until no more slacks are eligible to enter, show that you may use the shortest path algorithm for nonnegative costs to test the entry criterion for path variables.

d. Describe the complete steps of the revised simplex method thus obtained.

e. Apply the revised simplex method developed in this exercise to the maximal flow problem in Figure 11.1.

11.19 a. Give the dual of the arc-path formulation for the maximal flow problem as stated in Exercise 11.17.

b. If we add the restriction that the dual variables must be zero or 1, what interpretation can you give the dual problem?

c. Interpret the dual solution obtained in part (e) of Exercise 11.18.

11.20 Is the constraint matrix for the arc-path formulation of a maximal flow problem always unimodular? Prove or give a counterexample.

11.21 Find the shortest path from node 1 to all nodes of the following network.

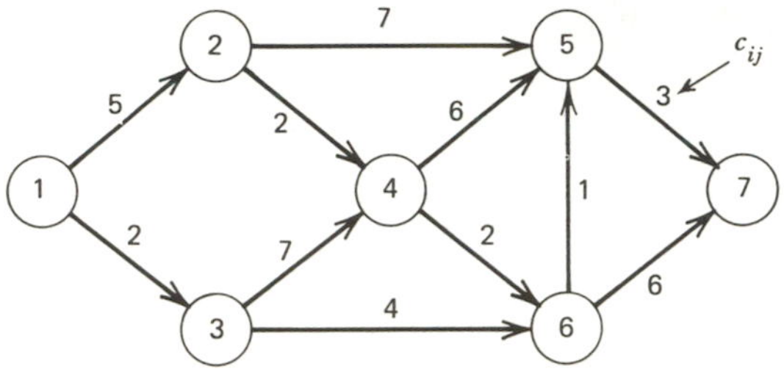

11.22 Find the shortest path from node 1 to node 7 in the following network.

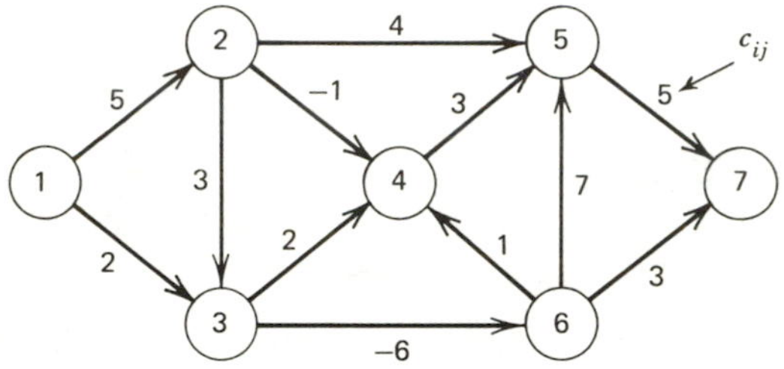

11.23 a. Show how the shortest path algorithm can be used to find the longest path from node 1 to node m in a network.

b. When finding the longest path, what assumption must be made?

c. Use the results of part (a) to devise direct algorithms for the longest path problem.

11.24 Show that at optimality of the shortest path problem $w_i - w_m$ represents a lower bound on the cost of the shortest path from node i to node m for each i.

11.25 In the shortest path algorithm for negative costs, show that at optimality there always exists a path from node 1 to node m along which $w'_j = w'_i + c_{ij}$ provided that $w'_m < \infty$. Also, show that if $w'_m = \infty$, then no path exists from node 1 to node m.

11.26 Find both the shortest path and the longest path from node 1 to node 6 in the following network.

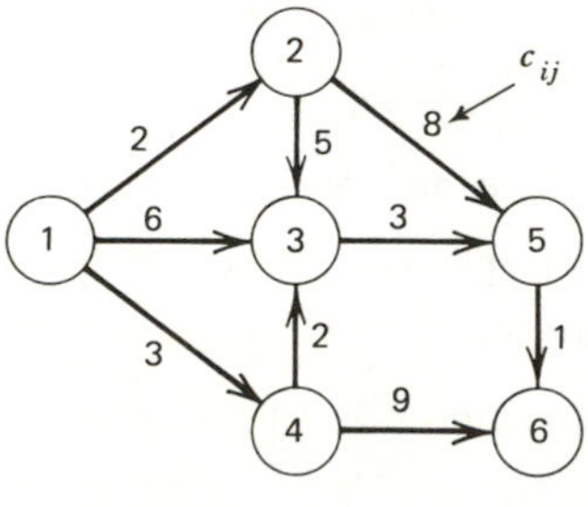

11.27 Find the shortest path from node 1 to every other node in the following network.

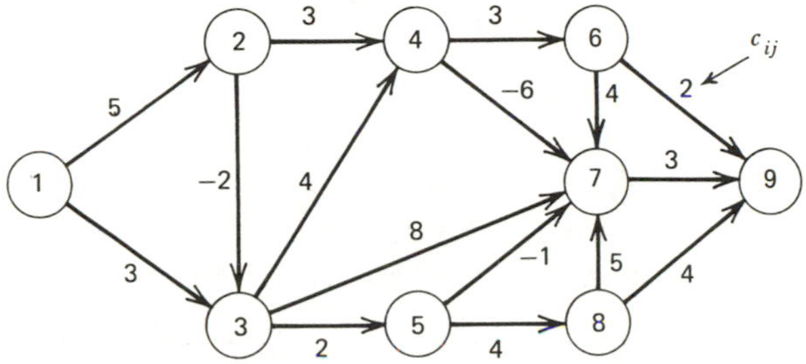

11.28 Find the shortest path from every node to node 6 in the following network. (*Hint.* Apply the shortest path algorithm with nonnegative costs in reverse.)

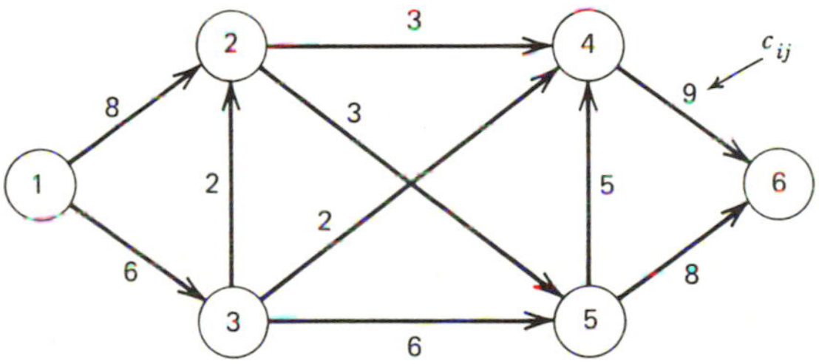

11.29 Modify the shortest path algorithm for negative costs to find the shortest path from every node to node m.

11.30 Find the shortest path from every node to node 7 in the following network.

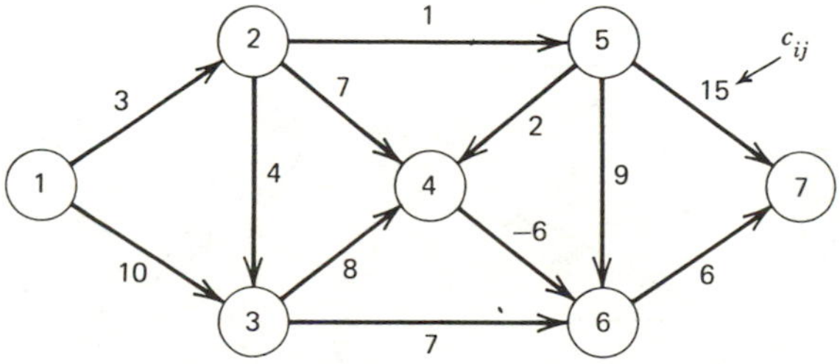

11.31 a. Modify the shortest path algorithm for nonnegative costs to handle an undirected network.

b. Apply the procedure of part (a) to find the shortest path from node 1 to node 5 in the following network.

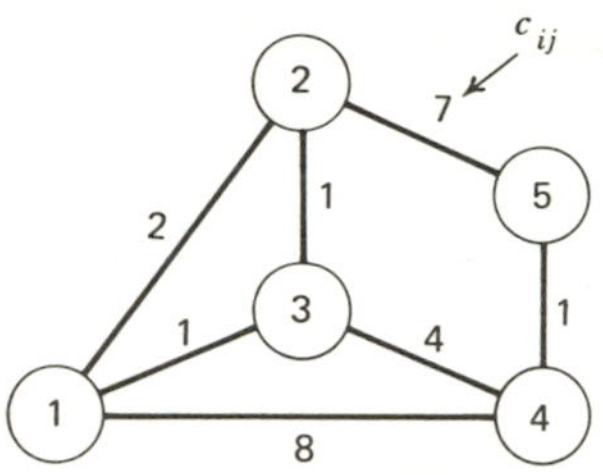

11.32 Can the shortest path algorithm for negative costs be modified to handle undirected arcs? (*Hint.* Consider one of the arcs with negative cost.)

11.33 a. Apply the shortest path procedure to find the shortest path from node 1 to node 5 in the following network.
b. What is the difficulty in part (a)?
c. Solve the problem by the network simplex method of Chapter 9. Compare with the result in (a).

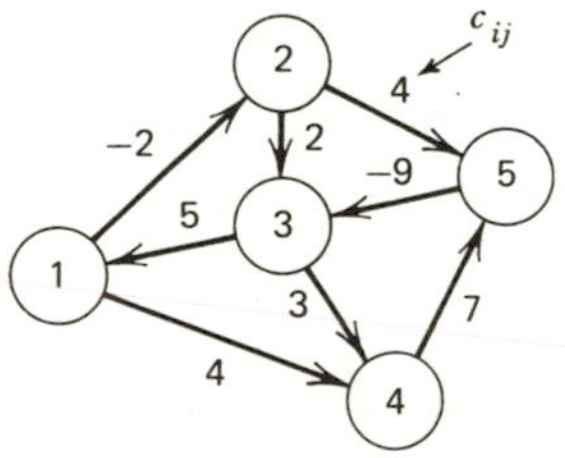

11.34 Bob, Ed, and Stu are in a car pool. They each live at the points 1, 2, and 7 respectively in the following network. They agree to meet at point 10 every morning at a certain time and proceed from there to their work in a single car. The numbers on the arcs represent the travel times in minutes.
a. What is the fastest route for each man to the meeting point?
b. How early (counting back from the meeting time) should each man leave?

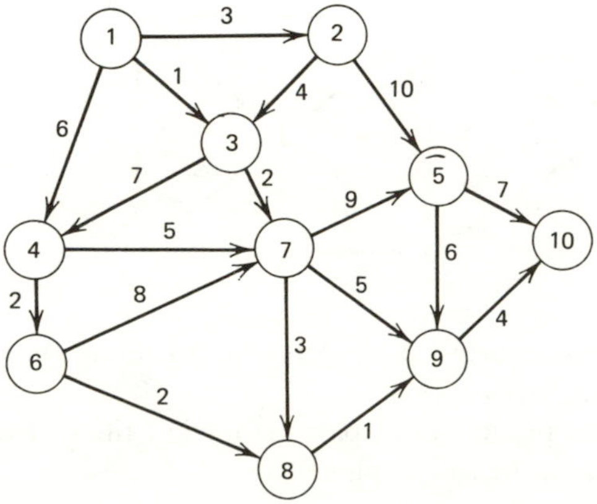

11.35 A single machine is needed to perform a specified function for the next four years, after which the function and machine will no longer be needed. The purchase price of a machine varies over the next four years according to the following table.

Year	Now	One year from now	Two years from now	Three years from now
Purchase price	$25,000	$33,000	$38,000	$47,000

The salvage value of a machine depends only on its length of service and is given by the following table.

Length of Service	1 year	2 years	3 years	4 years
Salvage value	$10,000	$6,000	$3,000	$1,000

The annual operating cost varies with length of service, as follows.

Length of Service	New	1 year	2 years	3 years
Annual operating cost	$3,000	$5,000	$8,000	$12,000

What is the optimal policy of purchasing, operating, and salvaging machines over the next four years if management wishes to minimize the total cost?

11.36 In any project there are usually a set of activities that constitute the project, a completion time for each activity, and a set of precedence relationships specifying which activities must be completed before a given activity can start. *Project management* is concerned with the scheduling and control of activities in such a way that the project can be completed as soon as possible after its start. The *critical path* is that sequence of activities that limits the early completion time of the project. (It is generally activities on the critical path that project managers watch closely.)

Consider the following activities with indicated completion times and precedence relationships.

ACTIVITY	COMPLETION TIME (DAYS)	PREDECESSORS
A	3	—
B	4	—
C	2	A
D	6	A, B
E	8	B, C
F	5	D, E
G	3	B, C, F

Find the critical path for this project and the associated project time. [*Hint*. Draw an arc with its own beginning and ending nodes for each activity with the arc cost equal to the completion time of the activity. If activity Q must precede activity R, then draw an arc from the ending node of activity Q to the beginning node of activity R with zero cost on the arc. Provide a starting node to proceed all activities and a finishing node to succeed all activities. In the network thus obtained the longest path (why not shortest path?) will be the critical path.]

11.37 a. How can the shortest path algorithm be used to obtain a starting (not necessarily feasible) solution when the out-of-kilter algorithm is applied to a minimal cost network flow problem with nonzero right-hand side values?

b. Is there any advantage to doing this?

11.38 Since Dijkstra's shortest path algorithm for nonnegative costs is extremely efficient, it would be highly desirable to be able to convert a network with negative costs to an equivalent network with nonnegative costs. Consider the following procedure for accomplishing this in a network G with m nodes.

INITIALIZATION STEP

Let $t = 1$ (t is the iteration counter)

MAIN STEP

1. Let $i = 1$
2. Let $\bar{c}_i = \underset{j}{\text{minimum}}\ c_{ij}$. If $\bar{c}_i < 0$, replace c_{ij} by $c_{ij} - \bar{c}_i$ for all j and replace c_{ki} by $c_{ki} + \bar{c}_i$ for all k.
3. If $i < m$ replace i by $i + 1$ and go to step 2. Otherwise go to step 4.
4. If all $c_{ij} \geqslant 0$, stop, the equivalent network is obtained. Otherwise, if

$t < m + 1$ replace t by $t + 1$ and go to step 1; if $t = m + 1$, stop, there is a negative circuit in G.

Upon completing the above procedure if all $c_{ij} \geqslant 0$ we may apply Dijkstra's shortest path algorithm to the equivalent network. Note that while the proper path will be found its length must be adjusted.

a. Show that the method works.
b. Apply the method to the networks in exercises 11.22 and 11.33.
c. Show that if $c_{ij} < 0$ at iteration $m + 1$ then there is a negative circuit in G which includes node j. Is it possible to develop a labeling procedure to find the negative circuit?

11.39 Consider a network with upper bounds and costs (all lower bounds are zero). Suppose that we wish to find, among all maximal flows, from node 1 to node m, that maximum flow that minimizes the total cost. This is sometimes called the *minimal cost–maximal flow* problem.

a. Give a linear programming formulation for the minimal cost– maximal flow problem from node 1 to node m in a network.
b. Show how the out-of-kilter method can be used to solve this problem.
c. Apply parts (a) and (b) to the following network to obtain the minimal cost–maximal flow from node 1 to node 4.

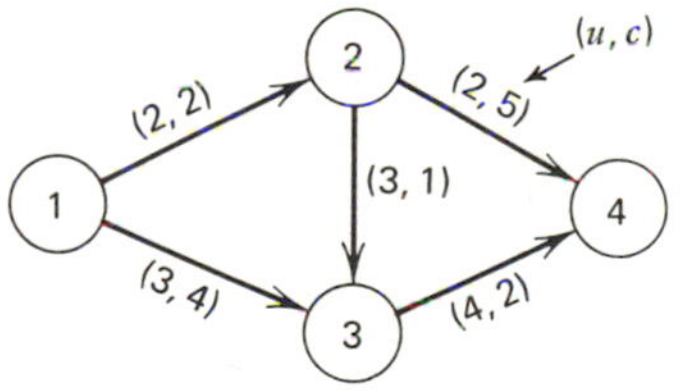

11.40 Consider the following procedure, due to Busacker and Gowen, for finding the minimal cost–maximal flow from node 1 to node m in a network with nonnegative costs and all $l_{ij}' = 0$.

INITIALIZATION STEP

Let all $x_{ij} = 0$.

SHORTEST PATH STEP

From G construct G' as follows. All nodes in G are in G'. If $x_{ij} < u_{ij}$ in G, place (i, j) in G' with $\Delta_{ij} = u_{ij} - x_{ij}$ and $c'_{ij} = c_{ij}$. If $x_{ij} > 0$ in G, place (j, i) in G' with $\Delta_{ji} = x_{ij}$ and $c'_{ji} = - c_{ij}$. Find the shortest path from node

1 to node m in G'. If no path exists, stop; the optimal solution is at hand. Otherwise pass to the flow change step.

FLOW CHANGE STEP

Let $\Delta = \text{Minimum } \{\Delta_{ij} : (i, j) \text{ is in the shortest path}\}$. Adjust flows along the associated chain in G by Δ, increasing flows on arcs with the orientation of the path and decreasing flows on arcs against the orientation of the path. Pass to the shortest path step.

a. Apply the algorithm to the example network of the previous problem.
b. Prove that the algorithm converges to the optimal solution in a finite number of steps. It is necessary to show that (i) negative circuits never occur in G', (ii) after a finite number of flow changes no path from node 1 to node m will exist, and (iii) on termination the optimal solution is obtained. (*Hint*. Consider the flow in the network as a parameter and show that after each flow change we have the minimal cost solution for that amount of flow.)
c. What difficulties would occur when we admit negative costs?

11.41 Consider the following algorithm, due to Klein, for finding the minimal cost-maximal flow from node 1 to node m in a network with arbitrary costs and all $l_{ij} = 0$.

INITIALIZATION STEP

Find the maximal flow from node 1 to node m in G.

NEGATIVE CIRCUIT STEP

From G construct G' as follows. All nodes in G are in G'. If $x_{ij} < u_{ij}$ in G, place (i, j) in G' with $\Delta_{ij} = u_{ij} - x_{ij}$ and $c'_{ij} = c_{ij}$. If $x_{ij} > 0$ in G, then place (j, i) in G' with $\Delta_{ji} = x_{ij}$ and $c'_{ji} = -c_{ij}$. Use the shortest path algorithm or the method of Exercise 11.38 to find a negative circuit in G'. If no negative circuit exists, stop; the optimal solution is at hand. Otherwise, pass to the flow change step.

FLOW CHANGE STEP

Let $\Delta = \text{Minimum } \{\Delta_{ij} : (i, j) \text{ is in the negative circuit}\}$. Adjust flows along the associated cycle in G by Δ, increasing flows on arcs with the orientation of the circuit and decreasing flows on arcs against the orientation of the circuit. Pass to the negative circuit step.

a. Apply the algorithm to the network of Exercise 11.39.
b. Prove that the algorithm converges to the optimal solution in a finite number of steps.

11.42 a. Give the linear programming formulation for the two-commodity maximal flow problem shown below (with no individual commodity upper bounds).
b. Find the two commodity maximal flow in the network.

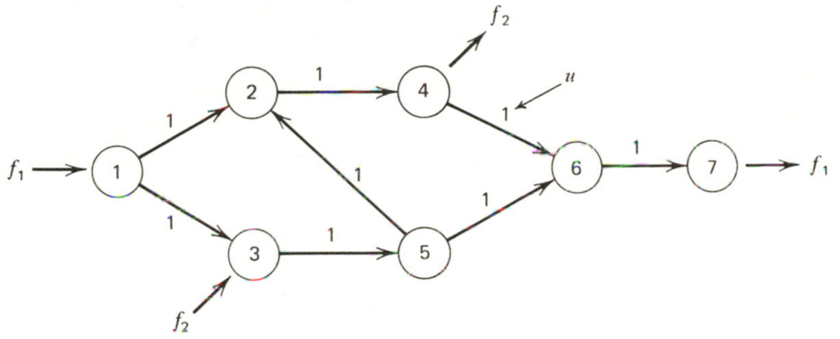

11.43 How can lower bounds be handled in the multicommodity minimal cost flow problem?

11.44 Apply the decomposition algorithm to the following three commodity minimal cost flow problem.

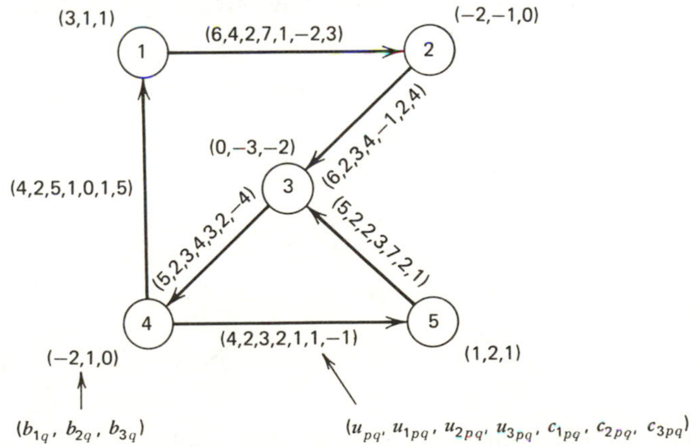

11.45 Given the optimal solution obtained from the decomposition procedure for the minimal cost multicommodity flow problem, indicate how the dual variables for the individual commodity constraints ($\mathbf{Ax}_i = \mathbf{b}_i$ and $\mathbf{x}_i \leqslant \mathbf{u}_i$)

can be recovered. Apply the procedure to the example problem in Section 11.3.

11.46 Discuss the economic meaning of the dual variables for the various constraints in the multicommodity minimal cost flow problem (namely, $\Sigma_i \mathbf{x}_i \leqslant \mathbf{u}$, $\mathbf{A}\mathbf{x}_i = \mathbf{b}_i$, and $\mathbf{x}_i \leqslant \mathbf{u}_i$).

11.47 Modify the decomposition algorithm for the minimal cost multicommodity flow problem when the set $X_i = \{\mathbf{x}_i : \mathbf{A}\mathbf{x}_i = \mathbf{b}_i, \mathbf{0} \leqslant \mathbf{x}_i \leqslant \mathbf{u}_i\}$ is not bounded. (This is only possible when for some i at least one component of $\mathbf{u}_i$ is ∞; that is, there is no upper bound on some arc.)

11.48 Discuss the difficulties, if any, in developing an algorithm for the multicommodity minimal cost flow problem that begins with the minimal cost flow for each commodity and proceeds to adjust these flows to satisfy the common upper bounds.

11.49 Give a node-arc formulation for the multicommodity maximal flow problem. Develop a decomposition procedure for this formulation and discuss the nature of the ith subproblem when $\mathbf{x}_i \leqslant \mathbf{u}_i$ is present and when it is absent.

11.50 Develop an arc-path formulation for the multicommodity maximal flow problem without the presence of the constraints $\mathbf{x}_i \leqslant \mathbf{u}_i$ for $i = 1, \ldots, t$. Develop a decomposition procedure for this formulation. (*Hint*. Consider the formulation given in Exercise 11.17.)

11.51 How can undirected arcs be handled in the multicommodity maximal flow problem? Illustrate on the following three-commodity network.

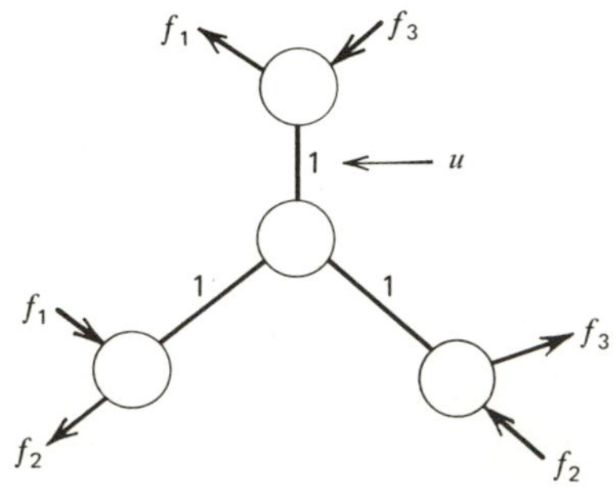

11.52 Consider the multicommodity maximal flow problem without the individual capacity constraints $\mathbf{x}_i \leqslant \mathbf{u}_i$ for $i = 1, \ldots, t$. A *disconnecting set* is a generalization of a cut-set for the single-commodity flow problem. A multicommodity disconnecting set is a set of arcs that "disconnects" (cuts all paths between) the source and sink for every commodity. A *multicommodity minimal disconnecting set* is the one for which the sum of (common) arc capacities is minimal.

a. Give a mathematical formulation for the minimal disconnecting set problem. (*Hint.* Take the dual of the arc-path formulation for the maximal flow problem and require the dual variables to be zero or 1. Give the interpretation of this dual problem.)
b. Show that the capacity of the multicommodity minimal disconnecting set is greater than or equal to the value of the multicommodity maximal flow. (*Hint*. Apply duality theorems to the formulation in part (a).)
c. Give a minimal disconnecting set for the network of Figure 11.9 and to the network of Exercise 11.42.
d. Compare the capacity of the minimal disconnecting set and the value of the maximal flow for both problems of Part (c) above.

11.53 Show that a multicommodity minimal disconnecting set is the union of single-commodity cut-sets. Is a multicommodity minimal disconnecting set necessarily the union of single-commodity *minimal* cut-sets?

11.54 Consider a metropolitan city with the area divided into four zones and a highway network connecting the zones. Let the following matrix, called the *origin-destination* matrix, specify the travel requirements from each (row) zone to every other (column) zone.

	1	2	3	4	5
1	0	10	7	8	5
2	2	0	3	4	4
3	6	2	0	1	5
4	2	4	7	0	5
5	1	1	3	4	0

Travel times and arc (upper) capacities are given by the following.

ARC	(1, 2)	(2, 4)	(2, 5)	(3, 2)	(4, 3)	(5, 1)	(5, 4)
Travel time (min)	15	35	15	20	10	15	10
Capacity	43	38	37	27	35	20	10

Find the minimal time traffic assignment in the network.

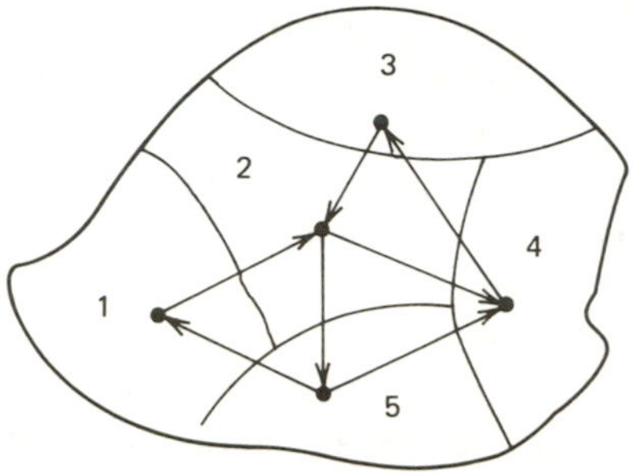

11.55 In the multicommodity minimal cost flow problem suppose that we have the capability of inverting the matrix $[\mathbf{E}'' - \mathbf{E}'\mathbf{B}^{-1}\mathbf{D}, \mathbf{E}]$. Show how the original primal and dual variables can be obtained. Use this information to develop a simplex procedure for solving the multicommodity flow problem directly on the graph. Give all details and illustrate by the problem of Figure 11.10.

11.56 Resolve the multicommodity minimal cost flow problem of Figure 11.10 with $u_{234} = 4$ instead of 3. Is it reasonable to expect this solution in practice?

NOTES AND REFERENCES

1. Ford and Fulkerson [152] first developed the maximal flow algorithm for networks. Dantzig and Fulkerson [105] provided a proof of the maximal flow–minimal cut theorem.
2. Algorithms for the shortest path problem have been developed by many individuals including Bellman [30], Dantzig [95], Dijkstra [122], and Floyd [151]. A particularly good comparison of shortest path procedures is given in Dreyfus [127].
3. Ford and Fulkerson [157] first proposed a column generation procedure for the multicommodity maximal flow problem. This was the forerunner to the Dantzig-Wolfe decomposition procedure for general linear programs. Hartman and Lasdon [233] proposed a procedure based on the simplex method for solving multicommodity flow problems.
4. The procedure of Exercise 11.38 was developed by Bazaraa and Langley [23] and is based on ideas presented by Nemhauser [352].

APPENDIX: PROOF OF THE REPRESENTATION THEOREM

In this appendix we provide the proof of the main representation theorem of polyhedral sets in terms of their extreme points and extreme directions. This theorem was presented in Section 2.6 without proof. The casual reader should skip this appendix since the proof involves some relatively advanced material.

The proof of the main theorem relies on the following two lemmas. The first lemma shows that there is a hyperplane that separates a closed convex set and a point outside the set, and the second lemma shows that the number of extreme points and extreme directions is finite.

Lemma 1

Let S be a closed† convex set in E^n and $\mathbf{x} \notin S$. Then there is a nonzero vector $\mathbf{c}$ in E^n and an $\epsilon > 0$ such that $\mathbf{cx} \geqslant \epsilon + \mathbf{cy}$ for each $\mathbf{y} \in S$.

†The statement that S is closed means that every converging sequence in S has its limit in S.

Proof

Let $\mathbf{y}_0$ be the closest point to $\mathbf{x}$ in S (since S is closed, such a point exists and is unique; see Figure A.1). Note that $\|\mathbf{x} - \mathbf{y}_0\| > 0$ since $\mathbf{x} \notin S$. We first show that $(\mathbf{x} - \mathbf{y}_0)(\mathbf{y} - \mathbf{y}_0) \leqslant 0$ for all $\mathbf{y} \in S$. Let $\mathbf{y} \in S$. By convexity of S, $\lambda\mathbf{y} + (1 - \lambda)\mathbf{y}_0 \in S$ for all $\lambda \in (0, 1)$. Since $\mathbf{y}_0$ is the closest point in S to $\mathbf{x}$, then

$$\begin{aligned}\|\mathbf{x} - \mathbf{y}_0\|^2 &\leqslant \|\mathbf{x} - \lambda\mathbf{y} - (1 - \lambda)\mathbf{y}_0\|^2 \\ &= \|(\mathbf{x} - \mathbf{y}_0) + \lambda(\mathbf{y}_0 - \mathbf{y})\|^2 \\ &= \|\mathbf{x} - \mathbf{y}_0\|^2 + 2\lambda(\mathbf{x} - \mathbf{y}_0)(\mathbf{y}_0 - \mathbf{y}) + \lambda^2\|\mathbf{y}_0 - \mathbf{y}\|^2\end{aligned}$$

This inequality implies that

$$0 \leqslant 2\lambda(\mathbf{x} - \mathbf{y}_0)(\mathbf{y}_0 - \mathbf{y}) + \lambda^2\|\mathbf{y}_0 - \mathbf{y}\|^2$$

Dividing by λ and letting λ go to zero, it follows that $(\mathbf{x} - \mathbf{y}_0)(\mathbf{y}_0 - \mathbf{y}) \geqslant 0$. Let $\mathbf{c} = \mathbf{x} - \mathbf{y}_0$ and note that $\mathbf{c} \neq \mathbf{0}$.

Now for any $\mathbf{y} \in S$ we have

$$\begin{aligned}0 &\leqslant (\mathbf{x} - \mathbf{y}_0)(\mathbf{y}_0 - \mathbf{y}) \\ &= (\mathbf{x} - \mathbf{y}_0)(\mathbf{x} - \mathbf{x} + \mathbf{y}_0 - \mathbf{y}) \\ &= (\mathbf{x} - \mathbf{y}_0)(\mathbf{x} - \mathbf{y}) + (\mathbf{x} - \mathbf{y}_0)(\mathbf{y}_0 - \mathbf{x}) \\ &= \mathbf{cx} - \mathbf{cy} - \|\mathbf{x} - \mathbf{y}_0\|^2\end{aligned}$$

Therefore $\mathbf{cx} \geqslant \mathbf{cy} + \|\mathbf{x} - \mathbf{y}_0\|^2$. Letting $\epsilon = \|\mathbf{x} - \mathbf{y}_0\|^2 > 0$, the result follows.

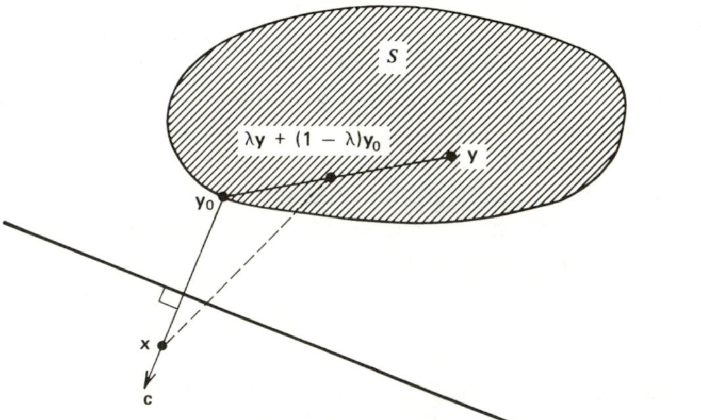

Figure A.1. **Separation of a closed convex set and a point.**

Lemma 2

Consider the (polyhedral) set $X = \{\mathbf{x} : \mathbf{A}\mathbf{x} = \mathbf{b},\ \mathbf{x} \geqslant \mathbf{0}\}$ where $\mathbf{A}$ is an $m \times n$ matrix. Then X has a finite number (perhaps none) of extreme points and extreme directions.

Proof

Without loss of generality suppose that rank $(\mathbf{A}) = m$. From Section 3.2, $\mathbf{x}$ is an extreme point if and only if $\mathbf{x}$ is a basic feasible solution. The number of basic feasible solutions is bounded above by

$$\binom{n}{m} = \frac{n!}{m!\,(n-m)!}$$

and hence the number of extreme points is finite. Now let $\mathbf{d}$ be an extreme direction of X. Note that $\mathbf{A}\mathbf{d} = \mathbf{0}$ and $\mathbf{d} \geqslant \mathbf{0}$. Possibly after rearranging the components of $\mathbf{d}$, suppose that

$$\mathbf{d}^t = (d_1, d_2, \ldots, d_k, 0, 0, \ldots, 0, d_n)$$

where $d_j > 0$ for $j = 1, 2, \ldots, k$ and $j = n$. We first show that the columns $\mathbf{a}_1, \mathbf{a}_2, \ldots, \mathbf{a}_k$ are linearly independent. By contradiction, suppose that they were linearly dependent. Then there would exist scalars $\lambda_1, \lambda_2, \ldots, \lambda_k$ not all zero such that $\Sigma_{j=1}^{k}\lambda_j\mathbf{a}_j = \mathbf{0}$. Since $d_1, d_2, \ldots, d_k > 0$, there exists an $\alpha > 0$ such that $d_j - \alpha\lambda_j > 0$ and $d_j + \alpha\lambda_j > 0$ for $j = 1, 2, \ldots, k$. Construct the following two vectors.

$$\mathbf{d}_1 = \begin{bmatrix} d_1 - \alpha\lambda_1 \\ d_2 - \alpha\lambda_2 \\ \vdots \\ d_k - \alpha\lambda_k \\ 0 \\ 0 \\ \vdots \\ 0 \\ d_n \end{bmatrix} \qquad \mathbf{d}_2 = \begin{bmatrix} d_1 + \alpha\lambda_1 \\ d_2 + \alpha\lambda_2 \\ \vdots \\ d_k + \alpha\lambda_k \\ 0 \\ 0 \\ \vdots \\ 0 \\ d_n \end{bmatrix}$$

Note by the choice of α that $\mathbf{d}_1$ and $\mathbf{d}_2 \geqslant \mathbf{0}$. Furthermore $\mathbf{A}\mathbf{d}_1 = \Sigma_{j=1}^{k}\mathbf{a}_j(d_j - \alpha\lambda_j) + \mathbf{a}_n d_n = \mathbf{A}\mathbf{d} - \alpha\Sigma_{j=1}^{k}\lambda_j\mathbf{a}_j = \mathbf{0} - \mathbf{0} = \mathbf{0}$. Similarly $\mathbf{A}\mathbf{d}_2 = \mathbf{0}$, and hence both $\mathbf{d}_1$ and $\mathbf{d}_2$ are directions of X. In addition $\mathbf{d}_1$ is not a multiple of $\mathbf{d}_2$ and so $\mathbf{d}_1$ and $\mathbf{d}_2$ are distinct (or nonequivalent) directions of X. But $\mathbf{d} = 1/2\ \mathbf{d}_1 + 1/2\ \mathbf{d}_2$, contradicting our assumption that $\mathbf{d}$ is an extreme direction. Therefore $\mathbf{a}_1, \mathbf{a}_2, \ldots, \mathbf{a}_k$ are linearly independent. Since rank $(\mathbf{A}) = m$, then $k \leqslant m$ and there must exist $m - k$ columns from the vectors $\mathbf{a}_{k+1}, \mathbf{a}_{k+2}, \ldots, \mathbf{a}_{n-1}$, which together with $\mathbf{a}_1, \ldots, \mathbf{a}_k$ form a linearly independent set of vectors (why?). To simplify the notation, suppose that these vectors are $\mathbf{a}_{k+1}, \ldots, \mathbf{a}_m$. Denote $\mathbf{a}_1, \mathbf{a}_2, \ldots, \mathbf{a}_m$ by $\mathbf{B}$ and note that $\mathbf{B}$ is an $m \times m$ invertible matrix. Since $\mathbf{Ad} = \mathbf{0}$ we get

$$\mathbf{0} = \mathbf{Ad} = \sum_{j=1}^{m} \mathbf{a}_j d_j + \mathbf{a}_n d_n = \mathbf{B}\hat{\mathbf{d}} + \mathbf{a}_n d_n$$

where $\hat{\mathbf{d}}$ is the first m components of $\mathbf{d}$. Multiplying the last equation by $\mathbf{B}^{-1}$, we get $\hat{\mathbf{d}} = -\mathbf{B}^{-1}\mathbf{a}_n d_n$. Therefore the extreme direction is given by

$$\mathbf{d} = \begin{bmatrix} \hat{\mathbf{d}} \\ 0 \\ 0 \\ \vdots \\ 0 \\ d_n \end{bmatrix} = \begin{bmatrix} -\mathbf{B}^{-1}\mathbf{a}_n d_n \\ 0 \\ 0 \\ \vdots \\ 0 \\ d_n \end{bmatrix} = d_n \begin{bmatrix} -\mathbf{B}^{-1}\mathbf{a}_n \\ 0 \\ 0 \\ \vdots \\ 0 \\ 1 \end{bmatrix}$$

We have shown that every extreme direction must be a vector of the foregoing form. Since there is only a finite number of ways of choosing an $m \times m$ invertible matrix from $\mathbf{A}$, and for each such choice there are $n - m$ ways of choosing a column from the remaining $n - m$ columns, there is only a finite number of extreme directions [at most $n!/m!(n - m - 1)!$] and the proof is complete.

We are now ready to state and prove the main representation theorem. Lemma 2 above is not explicitly used to prove the theorem. However, Lemma 2 is needed since in the statement of the theorem we are implicitly assuming that the number of extreme points and extreme directions is finite. Existence of extreme points was proved by construction in Section 3.2. Existence of extreme directions (if X is unbounded) follows as a corollary of the main theorem.

THE REPRESENTATION THEOREM

Consider the nonempty (polyhedral) set $X = \{\mathbf{x} : \mathbf{Ax} = \mathbf{b}, \mathbf{x} \geqslant \mathbf{0}\}$ where $\mathbf{A}$ is an $m \times n$ matrix. Let the extreme points be $\mathbf{x}_1, \mathbf{x}_2, \ldots, \mathbf{x}_k$ and the extreme directions be $\mathbf{d}_1, \mathbf{d}_2, \ldots, \mathbf{d}_l$ (finiteness of the number of extreme points and extreme directions follows from Lemma 2 above). Then $\mathbf{x} \in X$ if and only if it can be represented as a convex combination of the extreme points plus a nonnegative linear combination of the extreme directions, that is,

$$\mathbf{x} = \sum_{j=1}^{k} \lambda_j \mathbf{x}_j + \sum_{j=1}^{l} \mu_j \mathbf{d}_j$$

$$\sum_{j=1}^{k} \lambda_j = 1$$

$$\lambda_j \geqslant 0 \qquad j = 1, 2, \ldots, k$$

$$\mu_j \geqslant 0 \qquad j = 1, 2, \ldots, l$$

Proof

Note that any $\mathbf{x}$ with the foregoing representation belongs to X (why?). We now show the converse. Suppose that rank $(\mathbf{A})$ = rank $(\mathbf{A}, \mathbf{b}) = m$ because otherwise all redundant constraints can be thrown away. Now suppose that $\mathbf{x} \in X$ and suppose by contradiction that $\mathbf{x}$ cannot be represented as a convex combination of the extreme points plus a nonnegative linear combination of the extreme directions, and consider the following set:

$$S = \left\{ \sum_{j=1}^{k} \lambda_j \mathbf{x}_j + \sum_{j=1}^{l} \mu_j \mathbf{d}_j : \sum_{j=1}^{k} \lambda_j = 1, \lambda_j \geqslant 0 \text{ all } j, \mu_j \geqslant 0 \text{ all } j \right\}$$

Since X is not empty, then it has at least one extreme point (see Section 3.2). Therefore S is not empty. Furthermore, S is a closed convex set and $\mathbf{x} \notin S$ by assumption. By Lemma 1 there exists a nonzero vector $\mathbf{c}$ in E^n and an $\epsilon > 0$ such that

$$\mathbf{cx} \geqslant \mathbf{c}\left(\sum_{j=1}^{k} \lambda_j \mathbf{x}_j + \sum_{j=1}^{l} \mu_j \mathbf{d}_j \right) + \epsilon \tag{A.1}$$

for any set of nonnegative λ_j's and nonnegative μ_j's satisfying $\Sigma_{j=1}^{k} \lambda_j = 1$. Since μ_j can be made arbitrarily large, then by inequality (A.1) above we must have

$$\mathbf{cd}_j \leqslant 0 \qquad \text{for} \quad j = 1, 2, \ldots, l \tag{A.2}$$

Now consider the extreme point defined by

$$\mathbf{cx}_p = \underset{1 \leqslant j \leqslant k}{\text{Maximum}} \ \mathbf{cx}_j \tag{A.3}$$

Letting $\lambda_p = 1$, $\lambda_j = 0$ for $j \neq p$, and $\mu_j = 0$ for all j, inequality (A.1) gives

$$\mathbf{cx} \geqslant \mathbf{cx}_p + \epsilon \tag{A.4}$$

Since $\mathbf{x}_p$ is an extreme point, then $\mathbf{x}_p$ can be represented as $\begin{pmatrix} \mathbf{B}^{-1}\mathbf{b} \\ \mathbf{0} \end{pmatrix}$ where $\mathbf{B}$ is an $m \times m$ invertible submatrix of $\mathbf{A}$ and $\mathbf{B}^{-1}\mathbf{b} \geqslant \mathbf{0}$ (see Section 3.2). Without loss of generality, suppose that $\mathbf{B}^{-1}\mathbf{b} > \mathbf{0}$ (see Exercise 4.44). Recall that $\mathbf{x}$ was assumed to be in X, that is, $\mathbf{Ax} = \mathbf{b}$ and $\mathbf{x} \geqslant \mathbf{0}$. Decomposing $\mathbf{x}$ into $\begin{pmatrix} \mathbf{x}_B \\ \mathbf{x}_N \end{pmatrix}$, we get $\mathbf{b} = \mathbf{Ax} = \mathbf{Bx}_B + \mathbf{Nx}_N$, which implies that $\mathbf{x}_B = \mathbf{B}^{-1}\mathbf{b} - \mathbf{B}^{-1}\mathbf{Nx}_N$. Decomposing $\mathbf{c}$ into $\mathbf{c}_B$ and $\mathbf{c}_N$, we thus have

$$\mathbf{cx} = \mathbf{c}_B\mathbf{x}_B + \mathbf{c}_N\mathbf{x}_N = \mathbf{c}_B\mathbf{B}^{-1}\mathbf{b} + (\mathbf{c}_N - \mathbf{c}_B\mathbf{B}^{-1}\mathbf{N})\mathbf{x}_N \qquad \text{and}$$

$$\mathbf{cx}_p = \mathbf{c}_B\mathbf{B}^{-1}\mathbf{b}$$

Substituting in inequality (A.4), we get

$$(\mathbf{c}_N - \mathbf{c}_B\mathbf{B}^{-1}\mathbf{N})\mathbf{x}_N \geqslant \epsilon > 0$$

Since each component of $\mathbf{x}_N$ is nonnegative, the foregoing inequality holds only if there is a component j of the vector $\mathbf{c}_N - \mathbf{c}_B\mathbf{B}^{-1}\mathbf{N}$ that is positive, and for which $x_j > 0$. Denoting $\mathbf{B}^{-1}\mathbf{a}_j$ by $\mathbf{y}_j$, we thus have $x_j > 0$ and

$$c_j - \mathbf{c}_B\mathbf{y}_j > 0 \tag{A.5}$$

We first show that $\mathbf{y}_j \nleqslant \mathbf{0}$. Suppose by contradiction that $\mathbf{y}_j \leqslant \mathbf{0}$. This implies that $\mathbf{d} = (-\mathbf{y}_j^t, 0, 0, \ldots, 1, \ldots, 0, 0)^t$, where the 1 appears at the jth position, is a direction of X (by noting that $\mathbf{d} \neq \mathbf{0}$, $\mathbf{d} \geqslant \mathbf{0}$ and $\mathbf{Ad} = \mathbf{0}$). Furthermore, by the definition of $\mathbf{d}$ and noting inequality (A.5), we have $\mathbf{cd} = -\mathbf{c}_B\mathbf{y}_j + c_j > 0$, which violates (A.2) (why?). Therefore $\mathbf{y}_j \nleqslant \mathbf{0}$.

Now construct the point $\hat{\mathbf{x}} = \mathbf{x}_p + \lambda\mathbf{d}$ where

$$\lambda = \underset{1 \leqslant i \leqslant m}{\text{Minimum}} \left\{ \frac{\bar{b}_i}{y_{ij}} : y_{ij} > 0 \right\} = \frac{\bar{b}_r}{y_{rj}} > 0$$

and $\bar{\mathbf{b}} = \mathbf{B}^{-1}\mathbf{b}$. By this choice of λ, each component of $\hat{\mathbf{x}}$ is $\geqslant \mathbf{0}$. Furthermore

$$\mathbf{A}\hat{\mathbf{x}} = \mathbf{A}(\mathbf{x}_p + \lambda\mathbf{d}) = \mathbf{Ax}_p + \lambda\mathbf{Ad} = \mathbf{b} + \mathbf{0} = \mathbf{b}$$

Therefore $\hat{\mathbf{x}}$ is feasible. In addition $\hat{\mathbf{x}}$ has at most m positive components since $\hat{x}_r = 0$. Consider the submatrix $\hat{\mathbf{B}}$ of $\mathbf{A}$ where $\mathbf{a}_j$ replaces $\mathbf{a}_r$ in $\mathbf{B}$. Since $y_{rj} \neq 0$, then the columns of $\hat{\mathbf{B}}$ are linearly independent. This shows that $\hat{\mathbf{x}}$ is indeed an extreme point of X (see the characterization of extreme points in Section 3.2). Also note that

$$\mathbf{c}\hat{\mathbf{x}} = \mathbf{cx}_p + \lambda\mathbf{cd}$$

Since $\lambda > 0$ and $\mathbf{cd} > 0$, then $\mathbf{c}\hat{\mathbf{x}} > \mathbf{cx}_p$, which violates Equation (A.3) and the fact that $\hat{\mathbf{x}}$ is an extreme point. This contradiction shows that $\mathbf{x}$ can be represented as a convex combination of the extreme points plus a nonnegative linear combination of the extreme directions, and the proof is complete.

Corollary (Existence of Extreme Directions)

A nonempty polyhedral set X of the form $\{\mathbf{x} : \mathbf{Ax} = \mathbf{b}, \mathbf{x} \geqslant \mathbf{0}\}$ is unbounded if and only if it has at least one extreme direction.

Proof

If X has an extreme direction, then it is obviously unbounded. Conversely, suppose that X is unbounded. By contradiction, suppose that X has no extreme directions. By the theorem

$$X = \left\{ \sum_{j=1}^{k} \lambda_j \mathbf{x}_j : \sum_{j=1}^{k} \lambda_j = 1, \quad \lambda_j \geqslant 0 \qquad j = 1, 2, \ldots, k \right\}$$

Let $\mathbf{x} \in X$. Then $\mathbf{x} = \sum_{j=1}^{k} \lambda_j \mathbf{x}_j$. By the Schwartz inequality and since $0 \leqslant \lambda_j \leqslant 1$ for each j, we have

$$\|\mathbf{x}\| = \left\| \sum_{j=1}^{k} \lambda_j \mathbf{x}_j \right\| \leqslant \sum_{j=1}^{k} \lambda_j \|\mathbf{x}_j\| \leqslant \sum_{j=1}^{k} \|\mathbf{x}_j\|$$

Denoting $\sum_{j=1}^{k} \|\mathbf{x}_j\|$ by ϵ, we have shown that $\|\mathbf{x}\| \leqslant \epsilon$ for each $\mathbf{x} \in X$, contradicting the assumption that X is unbounded. Therefore X has at least one extreme direction, and the proof is complete.

BIBLIOGRAPHY

1. Abadie, J. On the Decomposition Principle, ORC Report 63-20, Operations Research Center, University of California, Berkeley, Calif., August 1963.
2. Abadie, J., and A. C. Williams. "Dual and Parametric Methods in Decomposition," in R. Graves and P. Wolfe (eds.), *Recent Advances in Mathematical Programming*, McGraw-Hill Book Co., New York, 1963.
3. Advani, Suresh. "A Linear Programming Approach to Air-Cleaner Design," *Operations Research*, **22** (2), pp. 295–297, March–April 1974.
4. Aggarwal, S. P. "A Simplex Technique for a Particular Convex Programming Problem," *Canadian Operational Research Society Journal*, **4** (2), pp. 82–88, July 1966.
5. Arinal, J. C. "Two Algorithms for Hitchcock Problem," *Revue Française de Recherche Opérationnelle* (France), **33**, pp. 359–374, 1964.
6. Aronofsky, J. S., and A. C. Williams. "The Use of Linear Programming and Mathematical Models in Underground Oil Production," *Management Science*, **8** (4), pp. 394–407, July 1962.
7. Arrow, K. J., L. Hurwicz, and H. Uzawa (eds.). *Studies in Linear and Nonlinear Programming*, Stanford University Press, Stanford, Calif., 1958.

8. Asher, D. T. "A Linear Programming Model for the Allocation of *R* and *D* Efforts," *IEEE Transactions on Engineering Management*, *EM*-9 (4), pp. 154–157, December 1962.
9. Au, Tung, and Thomas E. Stelson. *Introduction to Systems Engineering, Deterministic Models*, Addison-Wesley, Reading, Mass. 1969.
10. Azpeitia, A. G., and Dickinson, D. J. "A Decision Rule in the Simplex Method that Avoids Cycling," *Numerische Mathematik*, **6**, pp. 329–331, 1964.
11. Balas, E. Solution of Large-Scale Transportation Problems Through Aggregation, working paper, Institute of Mathematics, Rumanian Academy of Sciences, Bucharest, November 1963.
12. Balas, E. "The Dual Method for the Generalized Transportation Problem," *Management Science*, **12**, pp. 555–568, 1966.
13. Balas, E. "An Infeasibility Pricing Decomposition Method for Linear Programs," *Operations Research*, **14**, pp. 847–873, 1966.
14. Balas, E., and P. L. Ivanescu. "On the Generalized Transportation Problem," *Management Science*, **11**, pp. 188–203, 1964.
15. Balinski, M. L. "Integer Programming: Methods, Uses, Computation," *Management Science*, **12** (3), pp. 253–313, November 1965.
16. Balinski, M. L., and R. E. Gomory. "A Mutual Primal-Dual Simplex Method," in R. L. Graves and P. Wolfe (eds.), *Recent Advances in Mathematical Programming*, McGraw-Hill Book Co., New York, 1963.
17. Balinski, M. L., and R. E. Gomory. "A Primal Method for the Assignment and Transportation Problems," *Management Science*, **10** (3), pp. 578–593, April 1964.
18. Barnes, J. W., and R. M. Crisp. "Linear Programming: A Survey of General Purpose Algorithms," *AIIE Transactions*, **7** (3), September 1975.
19. Barnett, S. "Stability of the Solution to a Linear Programming Problem," *Operational Research Quarterly*, **13** (3), pp. 219–228, September 1962.
20. Barnett, S. "A Simple Class of Parametric Linear Programming Problems," *Operations Research*, **16**, (6), pp. 1160–1165, November–December 1968.
21. Battersby, Albert. *Network Analysis for Planning and Scheduling*, St. Martin's Press, New York, 1964.
22. Bazaraa, M. S. An Infeasibility Pricing Algorithm for the Multicommodity Minimum Cost Flow Problem, unpublished paper, School of Industrial and Systems Engineering, Georgia Institute of Technology, Atlanta, Georgia, 1973.
23. Bazaraa, M. S., and R. W. Langley. "A Dual Shortest Path Algorithm," *SIAM Applied Mathematics*, **26** (3), pp. 496–501, May 1974.
24. Bazaraa, M. S., and J. J. Goode. A New Decomposition Technique, unpublished paper, School of Industrial and Systems Engineering, Georgia Institute of Technology, Atlanta, Georgia, 1975.
25. Bazaraa, M. S., and C. M. Shetty. *Nonlinear Programming*, to be published, John Wiley & Sons, New York, 1977.

26. Bazaraa, M. S., and C. M. Shetty. *Foundations of Optimization*, Springer Verlag, New York, 1976.
27. Beale, E. M. L. "Cycling in the Dual Simplex Algorithm," *Naval Research Logistics Quarterly*, **2** (4), pp. 269–276, December 1955.
28. Beale, E. M. L. "An Algorithm for Solving the Transportation Problem When the Shipping Cost Over Each Route is Convex," *Naval Research Logistics Quarterly*, **6** (1), pp. 43–56, March 1959.
29. Bell, Earl J. Primal-Dual Decomposition Programming, U.S.G.R.&D.R. Order AD-625 365 from CFSTI, Operations Research Center, University of California, Berkeley, Calif., August 1965.
30. Bellman, Richard. "On a Routing Problem," *Quarterly Applied Mathematics*, **16** (1), pp. 87–90, April 1958.
31. Bellmore, M., H. J. Greenberg, and J. J. Jarvis. "Multicommodity Disconnecting Sets," *Management Science*, **16**, pp. B427–B433, 1970.
32. Bellmore, M., and R. Vemuganti. "On Multicommodity Maximal Dynamic Flows," *Operations Research*, **21**, pp. 10–21, 1973.
33. Benders, J. F. "Partitioning Procedures for Solving Mixed Variables Programming Problems," *Numerische Methamatik*, **4**, pp. 238–252, 1962.
34. Ben-Israel, A., and A. Charnes. "An Explicit Solution of a Special Class of Linear Programming Problems," *Operations Research*, **16** (6), pp. 1160–1175, November–December 1968.
35. Ben-Israel, Adi, and Philip D. Robers. "A Decomposition Method for Interval Linear Programming," *Management Science*, **16**, (5), pp. 374–387, January 1970.
36. Bennett, John M., and David R. Green. "An Approach to Some Structured Linear Programming Problems," *Operations Research*, **17** (4), pp. 749–750, July–August 1969.
37. Berge, C., and A. Ghouila-Houri. *Programming, Games and Transportation Networks*, John Wiley & Sons, New York, 1965.
38. Bialy, H. "An Elementary Method for Treating the Case of Degeneracy in Linear Programming," *Unternehmensforschung* (Germany), **10** (2), pp. 118–123, 1966.
39. Bitran, G. R., and A. G. Novaes. "Linear Programming with a Fractional Objective Function," *Operations Research*, **21** (1), pp. 22–29, January–February 1973.
40. Boldyreff, A. W. "Determination of the Maximal Steady State Flow of Traffic Through a Railroad Network," *Operations Research*, **3** (4), pp. 443–465, November 1955.
41. Boulding, K. E., and W. A. Spivey. *Linear Programming and the Theory of the Firm*, The Macmillan Company, New York, 1960.
42. Bouška, Jiří, and Martin Cerný. "Decomposition Methods in Linear Programming," *Ekonomicko-matematicky obzor* (Czechoslovakia), **1** (4), pp. 337–369, 1965.
43. Bowman, E. H. "Production Scheduling by the Transportation Method of Linear Programming," *Operations Research*, 4 (1), 1956.

44. Bowman, E. H. "Assembly-Line Balancing by Linear Programming," *Operations Research*, **8**, pp. 385–389, 1960.
45. Boyer, D. D. A Modified Simplex Algorithm for Solving the Multicommodity Maximum Flow Problem, Technical Memorandum TM-14930, The George Washington University, Washington, D.C., 1968.
46. Bradley, G. H. "Survey of Deterministic Networks," *AIIE Transactions*, **7** (3), September 1975.
47. Bramucci, Franco. "The Inversion Algorithm in Linear Programming," *Metra* (France), **6** (2), pp. 357–381, June 1967.
48. Briggs, F. E. A. "A Dual Labelling Method for the Hitchcock Problem," *Operations Research*, **10** (4), pp. 507–517, July–August 1962.
49. Brown, G. W., and T. C. Koopmans. "Computational Suggestions for Maximizing a Linear Function Subject to Linear Inequalities," in T. C. Koopmans (ed.), *Activity Analysis of Production and Allocation*, John Wiley & Sons, New York, pp. 377–380, 1951.
50. Busacker, R. G., and P. J. Gowen. A Procedure for Determining a Family of Minimal-Cost Network Flow Patterns, ORO Technical Report 15, Operations Research Office, John Hopkins University, 1961.
51 Busacker, R. G., and T. L. Saaty. *Finite Graphs and Networks: An Introduction with Applications*, McGraw-Hill Book Co., New York, 1965.
52. Cabot, A. Victor, Richard L. Francis, and Michael A. Stary. "A Network Flow Solution to a Rectilinear Distance Facility Location Problem," *AIIE Transactions*, **2** (2), pp. 132–141, June 1970.
53. Cahn, A. S. "The Warehouse Problem," *Bulletin of American Mathematical Society*, **54**, p. 1073, 1948.
54. Calman, Robert F. *Linear Programming and Cash Management, CASH ALPHA*, The M.I.T. Press, Cambridge, Mass., 1968.
55. Camion, P. "Characterization of Totally Unimodular Matrices," *Proceedings of American Mathematical Society*, **16**, pp. 1068–1073, 1965.
56. Catchpole, A. R. "The Application of Linear Programming to Integrated Supply Problems in the Oil Industry," *Operational Research Quarterly*, **13** (2), pp. 161–169, June 1962.
57. Cederbaum, I. "Matrices All of Whose Elements and Subdeterminants are 1, -1, or 0," *Journal of Mathematics and Physics*, **36**, pp. 351–361, 1958.
58. Chadda, S. S. "A Decomposition Principle for Fractional Programming," *Opsearch*, **4** (3), pp. 123–132, 1967.
59. Chandrascharan, R. "Total Unimodularity of Matrices," *SIAM Journal of Applied Mathematics*, **17**, pp. 1032–1034, 1968.
60. Chandy, K. M., and Tachen Lo. "The Capacitated Minimum Spanning Tree," *Networks*, **3** (2), pp. 173–181, 1973.
61. Charnes, A. "Optimality and Degeneracy in Linear Programming," *Econometrica*, **20** (2), pp. 160–170, 1952.
62. Charnes, A., and W. W. Cooper. "The Stepping Stone Method of Explaining Linear Programming Calculations in Transportation Problems," *Management Science*, **1** (1), 1954.

63. Charnes, A., and W. W. Cooper. *Management Models and Industrial Applications of Linear Programming*, John Wiley & Sons, New York, 1961.
64. Charnes, A., and W. W. Cooper. "Programming with Linear Fractional Functionals," *Naval Research Logistics Quarterly*, **9**, pp. 181–186, 1962.
65. Charnes, A., and W. W. Cooper. "On Some Works of Kantorovich, Koopmans, and Others," *Management Science*, **8** (3), pp. 246–263, April 1962.
66. Charnes, A., W. W. Cooper, and A. Henderson. *An Introduction to Linear Programming*, John Wiley & Sons, New York, 1953.
67. Charnes, A., W. W. Cooper, and R. Mellon. "Blending Aviation Gasolines: A Study in Programming Interdependent Activities in an Integrated Oil Company," *Econometrica*, **20**, (2), 1952.
68. Charnes, A., W. W. Cooper, and G. L. Thompson. "Some Properties of Redundant Constraints and Extraneous Variables in Direct and Dual Linear Programming Problems," *Operations Research*, **10**, pp. 711–723, 1962.
69. Charnes, A., and C. E. Lemke. A Modified Simplex Method for Control of Round-Off Error in Linear Programming, unpublished paper, Carnegie Institute of Technology, 1952.
70. Charnes, A., and C. E. Lemke. "Minimization of Nonlinear Separable Convex Functionals," *Naval Research Logistics Quarterly*, **1**, (4), pp. 301–312, December 1954.
71. Charnes, A., and C. E. Lemke. Computational Theory of Linear Programming, I: The Bounded Variables Problem, ONR Research Memorandum 10, Graduate School of Industrial Administration, Carnegie Institute of Technology, Pittsburgh, Pennsylvania, January 1954.
72. Charnes, A., and W. M. Raike. "One Pass Algorithms for Some Generalized Network Problems," *Operations Research*, **14**, pp. 914–924, 1966.
73. Chester, Louie B. Analysis of the Effect of Variance on Linear Programming Problems, U.S.G.R.&D.R. Order AD-611 273 from Clearinghouse, Air Force Institute of Technology, Wright-Patterson AFB, Washington, D.C., August 1964.
74. Chien, R. T. "Synthesis of a Communication Net," *IBM Journal of Research and Development*, **4**, 1960.
75. Chisman, James A. "Using Linear Programming to Determine Time Standards," *Journal of Industrial Engineering*, **17** (4), pp. 189–191, April 1966.
76. Chung, An-min. *Linear Programming*, Charles E. Merrill Books, Columbus, Ohio, 1963.
77. Clark, Charles E. "The Optimum Allocation of Resources Among the Activities of a Network," *Journal of Industrial Engineering*, **12** (1), pp. 11–17, January–February 1961.
78. Clasen, R. J. The Numerical Solution of Network Problems Using Out-of-Kilter Algorithm, RAND Report RM-5456 PR, March 1968.

79. Clements, R. A. "Linear Programming for Multiple Feed Formulation," *NZOR* (New Zealand), **2** (2), pp. 100–107, July 1974.
80. Cline, R. E. "Representations for the Generalized Inverse of Matrices Partitioned as **A** = [**U**, **V**]," in R. Graves and P. Wolfe (eds.), *Recent Advances in Mathematical Programming*, McGraw-Hill Book Co., New York, 1963.
81. Commoner, F. G. "A Sufficient Condition for a Matrix to be Totally Unimodular," *Networks*, **3**, pp. 351–365, 1973.
82. Courtillot, M. "On Varying All the Parameters in a Linear Programming Problem and Sequential Solution of a Linear Programming Problem," *Operations Research*, **10** (4), 1962.
83. Craven, B. D. "A Generalization of the Transportation Method of Linear Programming," *Operational Research Quarterly*, **14** (2), pp. 157–166, June 1963.
84. Curtis, Floyd H. "Linear Programming the Management of a Forest Property," *Journal of Forestry*, **61** (9), pp. 611–616, September 1962.
85. Dantzig, G. B. Programming in a Linear Structure, Comptroller, United States Air Force, Washington, D.C., February 1948.
86. Dantzig, G. B. "Programming of Interdependent Activities, II, Mathematical Model," in T. C. Koopmans (ed.), *Activity Analysis of Production and Allocation*, John Wiley & Sons, New York, 1951, also published in *Econometrica*, **17** (3–4), pp. 200–211, July–October 1949.
87. Dantzig, G. B. "Maximization of a Linear Function of Variables Subject to Linear Inequalities," in T. C. Koopmans (ed.), *Activity Analysis of Production and Allocation*, John Wiley & Sons, New York, pp. 339–347, 1951.
88. Dantzig, G. B. "A Proof of the Equivalence of the Programming Problem and the Game Problem," in T. C. Koopmans (ed.), *Activity Analysis of Production and Allocation*, John Wiley & Sons, New York, pp. 359–373, 1951.
89. Dantzig, G. B. "Application of the Simplex Method to a Transportation Problem," in T. C. Koopmans (ed.), *Activity Analysis of Production and Allocation*, John Wiley & Sons, New York, pp. 359–373, 1951.
90. Dantzig, G. B. Computational Algorithm of the Revised Simplex Method, RAND Report RM-1266, The Rand Corporation, Santa Monica, Calif. 1953.
91. Dantzig, G. B. Notes on Linear Programming, Part VII. The Dual Simplex Algorithm, RAND Report RM-1270, The Rand Corporation, Santa Monica, Calif., July 1954.
92. Dantzig, G. B. Notes on Linear Programming. Part XI, Composite Simplex-Dual Simplex Algorithm-I, Research Memorandum RM-1274, The Rand Corporation, Santa Monica, Calif., April 1954.
93. Dantzig, G. B. Notes on Linear Programming: Parts VIII, IX, X-Upper Bounds, Secondary Constraints, and Block Triangularity in Linear Pro-

gramming, Research Memorandum RM-1367, The Rand Corporation, Santa Monica, Calif., October 1954, also published in *Econometrica*, **23** (2), pp. 174–183, April 1955.

94. Dantzig, G. B. "Discrete Variable Extremum Problems," *Operations Research*, **5** (2), pp. 266–277, April 1957.
95. Dantzig, G. B. On the Shortest Route Through a Network, RAND Report P-1345, The Rand Corporation, Santa Monica, Calif., April 1958. Also, *Management Science*, **6** (2), pp. 187–190, 1960.
96. Dantzig, G. B. On the Significance of Solving Linear Programming Problems with Some Integer Variables, RAND Report P-1486, The Rand Corporation, Santa Monica, Calif., September 1958.
97. Dantzig, G. B. *Linear Programming and Extensions*, Princeton University Press, Princeton, N.J., 1963.
98. Dantzig, G. B. "Compact Basis Triangularization for the Simplex Method," in R. Graves and P. Wolfe (eds.), *Recent Advances in Mathematical Programming*, McGraw-Hill Book Co., New York, 1963.
99. Dantzig, G. B. Optimization in Operations Research: Some Examples, U.S.G.R.&D.R. Order AD-618 748, Operations Research Center, University of California, Berkeley, Calif., April 1965.
100. Dantzig, G. B. All Shortest Routes in a Graph, Technical Report 66-3, Operations Research House, Stanford University, Stanford, Calif., November 1966.
101. Dantzig, G. B., W. D. Blanttner, and M. R. Rao. All Shortest Routes from a Fixed Origin in a Graph, Technical Report 66-2, Operations Research House, Stanford University, Stanford, Calif., November 1966.
102. Dantzig, G. B., L. R. Ford, and D. R. Fulkerson. "A Primal-Dual Algorithm for Linear Programs," in H. W. Kuhn and A. W. Tucker (eds.), *Linear Inequalities and Related Systems*, Annals of Mathematics Study No. 38, Princeton University Press, Princeton, N.J., pp. 171–181, 1956.
103. Dantzig, G. B., and D. R. Fulkerson. "Minimizing the Number of Tankers to Meet a Fixed Schedule," *Naval Research Logistics Quarterly*, **1** (3), pp. 217–222, September 1954.
104. Dantzig, G. B., and D. R. Fulkerson. "Computation of Maximal Flows in Networks," *Naval Research Logistics Quarterly*, **2** (4), 1955.
105. Dantzig, G. B., and D. R. Fulkerson. "On the Max-Flow Min-Cut Theorem of Networks," in H. W. Kuhn and A. W. Tucker (eds.), *Linear Inequalities and Related Systems*, Annals of Mathematics Study No. 38, Princeton University Press, Princeton, N.J., pp. 215–221, 1956.
106. Dantzig, G. B., D. R. Fulkerson, and S. M. Johnson. "Solution of a Large-Scale Traveling-Salesman Problem," *Operations Research*, **2** (4), pp. 393–410, November 1954.
107. Dantzig, G. B., D. R. Fulkerson, and S. M. Johnson. On a Linear Programming-Combinatorial Approach to the Traveling Salesman Prob-

lem, RAND Report P-1281, The Rand Corporation, Santa Monica, Calif., April 1958.

108. Dantzig, G. B., R. Harvey, and R. McKnight. Updating the Product Form of the Inverse for the Revised Simplex Method, ORC Report 64-33, Operations Research Center, University of California, Berkeley, Calif., December 1964.

109. Dantzig, G. B., and D. L. Johnson. "Maximum Payloads Per Unit Time Delivered Through an Air Network," *Operations Research*, **12** (2), March–April 1964.

110. Dantzig, G. B., S. Johnson, and W. White. A Linear Programming Approach to the Chemical Equilibrium Problem, RAND Report P-1060, The Rand Corporation, Santa Monica, Calif., April 1958.

111. Dantzig, G. B., and A. Orden. Notes on Linear Programming: Part II-Duality Theorems, Research Memorandum RM-1265, The Rand Corporation, Santa Monica, Calif., October 1953.

112. Dantzig, G. B., A. Orden, and P. Wolfe. "The Generalized Simplex Method for Minimizing a Linear Form Under Linear Inequality Restraints," *Pacific Journal of Mathematics*, **5** (2), pp. 183–195, June 1955.

113. Dantzig, G. B., and W. Orchard-Hays. "The Product Form for the Inverse in the Simplex Method," *Mathematical Tables and Aids to Computation*, **8** (46), pp. 64–67, 1954.

114. Dantzig, G. B., and W. Orchard-Hays. Notes on Linear Programming: Part V—Alternate Algorithm for the Revised Simplex Method Using Product Form for the Inverse, Research Memorandum RM-1268, The Rand Corporation, Santa Monica, Calif., November 1953.

115. Dantzig, G. B., and R. M. Van Slyke. Generalized Upper Bounded Techniques for Linear Programming I, II, ORC reports 64-17 (1964), 64-18 (1965), Operations Research Center, University of California, Berkeley, Calif.

116. Dantzig, G. B., and P. Wolfe. "Decomposition Principle for Linear Programs," *Operations Research*, **8** (1), pp. 101–111, January–February 1960.

117. Dantzig, G. B., and P. Wolfe. "The Decomposition Algorithm for Linear Programs," *Econometrica*, **29** (4), pp. 767–778, October 1961.

118. Dantzig, G. B., and P. Wolfe. Linear Programming in a Markov Chain. Notes on Linear Programming and Extensions. Part 59, Research Memorandum RM-2957-PR, The Rand Corporation, Santa Monica, Calif., April 1962.

119. Denardo, Eric V. "On Linear Programming in a Markov Decision Problem," *Management Science*, **16** (5), pp. 281–288, January 1970.

120. Dennis, J. B. *Mathematical Programming and Electrical Networks*, John Wiley & Sons, New York, 1959.

121. Dent, J. B., and H. Casey. *Linear Programming and Animal Nutrition*, J. B. Lippincott Co., Philadelphia, Pa., 1968.

122. Dijkstra, E. W. "A Note on Two Problems in Connection with Graphs," *Numerical Mathematics*, **1**, pp. 269–271, 1959.
123. Doig, A. G. "The Minimum Number of Basic Feasible Solutions to a Transportation Problem," *Operational Research Quarterly*, **14** (4), 1963.
124. Doig, A. G., and A. H. Land. "An Automatic Method of Solving Discrete Programming Problems," *Econometrica*, **28**, pp. 497–520, 1960.
125. Dorfman, R. *Application of Linear Programming to the Theory of the Firm*, University of California Press, Berkeley, Calif., 1951.
126. Dorfman, R., P. A. Samuelson, and R. M. Solow. *Linear Programming and Economic Analyses*, McGraw-Hill Book Co., New York, 1958.
127. Dreyfus, S. E. "An Appraisal of Some Shortest Path Algorithms," *Operations Research*, **17** (3), pp. 395–412, 1969.
128. Duffin, R. J. "Infinite Programs," in H. W. Kuhn and A. W. Tucker (eds.), *Linear Inequalities and Related Systems*, Annals of Mathematics Study No. 38, Princeton University Press, Princeton, N.J., 1956.
129. Duffin, R. J. The Extremal Length of a Network, Office of Technical Services, Document No. AD-253 665, 1961.
130. Duffin, R. J. "Dual Programs and Minimum Costs," *Journal of the Society for Industrial and Applied Mathematics*, **10**, pp. 119–123, 1962.
131. Duffin, R. J. "Convex Analysis Treated by Linear Programming," *Mathematical Programming* (Netherlands), **4** (2), pp. 125–143, April 1973.
132. Edmonds, J., and R. M. Karp. "Theoretical Improvements in Algorithmic Efficiency for Network Flow Problems," *ACM*, **19** (2), pp. 248–264, April, 1972.
133. Egerváry, E. On Combinatorial Properties of Matrices (1931), translated by H. W. Kuhn, Paper No. 4, George Washington University Logistics Research Project, 1954.
134. Eggleston, H. G. *Convexity*, Cambridge University Press, New York, 1958.
135. Eisemann, K. "The Primal-Dual Method for Bounded Variables," *Operations Research*, **12** (1), January–February 1964.
136. Elmaghraby, Salah E. "Sensitivity Analysis of Multiterminal Flow Networks," *Operations Research*, **12** (5), pp. 680–688, September–October 1964.
137. Elmaghraby, Salah E. "The Theory of Networks and Management Science. Part I," *Management Science*, **17** (1), pp. 1–34, September 1970.
138. Elmaghraby, Salah E. "The Theory of Networks and Management Science. Part II," *Management Science*, **17** (2), pp. B54–B71, October 1970.
139. Everett, H. "Generalized Lagrange Multiplier Method for Solving Problems of Optimum Allocation of Resources," *Operations Research*, **11**, pp. 399–417, 1963.
140. Falk, J. E. "Lagrange Multipliers and Nonlinear Programming," *Journal of Mathematical Analysis and Applications*, **19**, pp. 141–159, 1967.

141. Farbey, B. A., A. H. Lard, and J. D. Murchland. "The Cascade Algorithm for Finding All Shortest Distances in a Directed Graph," *Management Science*, **14** (1), pp. 19–28, September 1967.
142. Farkas, J. "Uber die Theorie der einfachen Ungleichungen," *Journal für die reine und angewandte Mathematik*, **124**, pp. 1–27, 1901.
143. Ferguson, Allen R., and G. B. Dantzig. Notes on Linear Programming: Part XVI–The Problem of Routing Aircraft–a Mathematical Solution, Research Memorandum RM-1369, also RAND Paper P-561, 1954, also *Aeronautical Engineering Review*, **14** (4), pp. 51–55, April 1955.
144. Ferguson, Allen R., and G. B. Dantzig. "The Allocation of Aircraft to Routes—An Example of Linear Programming Under Uncertain Demand," *Management Science*, **3** (1), pp. 45–73, October 1956.
145. Fisher, F. P. "Speed Up the Solution to Linear Programming Problems," *Journal of Industrial Engineering*, **12** (6), pp. 412–416, November–December 1961.
146. Flood, M. M. "On the Hitchcock Distribution Problem," *Pacific Journal of Mathematics*, **3** (2), 1953.
147. Flood, M. M. "Application of Transportation Theory to Scheduling a Military Tanker Fleet," *Operations Research*, **2** (2), 1954.
148. Flood, M. M. "An Alternative Proof of a Theorem of Konig as an Algorithm for the Hitchcock Distribution Problem," in R. Bellman and M. Hall (eds.), *Proceedings of Symposia in Applied Mathematics*, American Mathematical Society, Providence, R.I., pp. 299–307, 1960.
149. Flood, M. M. "A Transportation Algorithm and Code," *Naval Research Logistics Quarterly*, **8** (3), September 1961.
150. Florian, M., and P. Robert. "A Direct Search Method to Locate Negative Cycles in a Graph," *Management Science*, **17** (5), pp. 307–310, January 1971.
151. Floyd, R. W. "Algorithm 97: Shortest Path," *Communication of ACM*, **5** (6), p. 345, 1962.
152. Ford, L. R., and D. R. Fulkerson. "Maximal Flow Through a Network," *Canadian Journal of Mathematics*, **8** (3), pp. 399–404, 1956.
153. Ford, L. R., and D. R. Fulkerson. "A Simple Algorithm for Finding Maximal Network Flows and an Application to the Hitchcock Problem," *Canadian Journal of Mathematics*, **9** (2), pp. 210–218, 1957.
154. Ford, L. R., and D. R. Fulkerson. "Solving the Transportation Problem," *Management Science,* **3** (1), 1956.
155. Ford, L. R., and D. R. Fulkerson. "A Primal-Dual Algorithm for the Capacitated Hitchcock Problem," *Naval Research Logistics Quarterly*, **4** (1), 1957.
156. Ford. L. R., and D. R. Fulkerson. "Constructing Maximal Dynamic Flows from Static Flows," *Operations Research*, **6** (3), 1958.

157. Ford, L. R., and D. R. Fulkerson. "Suggested Computation of Maximal Multi-Commodity Network Flows," *Management Science*, **5** (1), pp. 97–101, October 1958.

158. Ford, L. R., and D. R. Fulkerson. *Flows in Networks*, Princeton University Press, Princeton, N.J., 1962.

159. Fox, Bennett. "Finding Minimal Cost-Time Ratio Circuits," *Operations Research*, **17** (3), pp. 546–551, May–June 1969.

160. Frank, H., and I. T. Frisch. *Communication, Transmission, and Transportation Networks*, Addison-Wesley, Reading, Mass., 1971.

161. Francis, R. L., and J. A. White. *Facility Layout and Location,* Prentice-Hall, Englewood Cliffs, N. J., 1974.

162. Frazer, Ronald J. *Applied Linear Programming*, Prentice-Hall, Englewood Cliffs, N.J., 1968.

163. Frisch, R. The Multiplex Method for Linear Programming, Memorandum of Social Economic Institute, University of Oslo, Oslo, Norway, 1955.

164. Fulkerson, D. R. "A Network Flow Feasibility Theorem and Combinatorial Applications," *Canadian Journal of Mathematics*, **11** (3), 1959.

165. Fulkerson, D. R. "Increasing the Capacity of a Network, the Parametric Budget Problem," *Management Science*, **5** (4), pp. 472–483, July 1959.

166. Fulkerson, D. R. On the Equivalence of the Capacity-Constrained Transshipment Problem and the Hitchcock Problem, Research Memorandum RM-2480, The Rand Corporation, Santa Monica, Calif., 1960.

167. Fulkerson, D. R. "An Out-of-Kilter Method for Minimal Cost Flow Problems," *Journal of the Society for Industrial and Applied Mathematics*, **9** (1), 1961.

168. Fulkerson, D. R. "A Network Flow Computation for Project Cost Curves," *Management Science*, **7** (2), pp. 167–178, January 1961.

169. Fulkerson, D. R. "Flow Networks and Combinatorial Operations Research," *The American Mathematical Monthly*, **73** (2), February 1966.

170. Fulkerson, D. R. "Networks, Frames, and Blocking Systems," *Mathematics of the Decision Sciences, Part* 1, Lectures in Applied Mathematics, vol. 11 (eds. G. B. Dantzig and A. F. Veinott), American Mathematical Society, pp. 304–334, 1968.

171. Fulkerson, D. R., and G. B. Dantzig. "Computations of Maximal Flows in Networks," *Naval Research Logistic Quarterly*, **2** (4), pp. 277–283, December 1955.

172. Gale, D. "Neighboring Vertices on a Convex Polyhedron," in H. W. Kuhn and A. W. Tucker (eds), *Linear Inequalities and Related Systems*, Annals of Mathematics Study No. 38, Princeton University Press, Princeton, N.J., pp. 255–263, 1956.

173. Gale, D. "A Theorem on Flows in Networks," *Pacific Journal of Mathematics*, **7**, *pp.* 1073–1082, 1957.

174. Gale, D. Transient Flows in Networks, Research Memorandum RM-2158, The Rand Corporation, Santa Monica, Calif., 1958.
175. Gale, D. *The Theory of Linear Economic Models*, McGraw-Hill Book Co., New York. 1960.
176. Gale, D. On the Number of Faces of a Convex Polytope, Technical Report No. 1, Department of Mathematics, Brown University, 1962.
177. Gale, D., H. W. Kuhn, and A. W. Tucker. "Linear Programming and the Theory of Games," chapter 19 of T. C. Koopmans (ed.), *Activity Analysis of Production and Allocation*, Cowles Commission Monograph 13, John Wiley & Sons, New York, 1951.
178. Garfinkel, R. S., and G. L. Nemhauser. *Integer Programming*, John Wiley & Sons, New York, 1972.
179. Garvin, W. W., H. W. Crandall, J. B. John, and R. A. Spellman. "Applications of Linear Programming in the Oil Industry," *Management Science*, **3** (4), pp. 407–430, July 1957.
180. Garvin, W. W. *Introduction to Linear Programming*, McGraw-Hill Book Co., New York, 1960.
181. Gass, S. I. "A First Feasible Solution to the Linear Programming Problem," in H. Antosiewicz (ed.), *Proceedings of the Second Symposium in Linear Programming*, vols. 1 and 2, DCS/Comptroller, Headquarters U. S. Air Force, Washington, D. C., pp. 495–508, January, 1955.
182. Gass. S. I. *Linear Programming: Methods and Applications*, McGraw-Hill Book Co., New York, 1958.
183. Gass, S. I. The Dualplex Method for Large-scale Linear Programs, ORC Report 66-15, Operations Research Center, University of California, Berkeley, Calif., June 1966.
184. Gass, S. I., and T. Saaty. "The Computational Algorithm for the Parametric Objective Function," *Naval Research Logistics Quarterly*, **2**, (1–2), 1955.
185. Gass, S. I., and T. L. Saaty. "Parametric Objective Function. Part II: Generalization," *Operations Research*, **3**, 1955.
186. Gassner, Betty J. "Cycling in the Transportation Problem," *Naval Research Logistics Quarterly*, **11** (1), March 1964.
187. Geary, R. C., and M. C. McCarthy. *Elements of Linear Programming, with Economic Applications*, Charles Griffin & Co., London, 1964.
188. Geoffrion, A. M. "Primal Resource-Directive Approaches for Optimizing Nonlinear Decomposable Systems," *Operations Research*, **18** (3), pp. 375–403, May-June 1970.
189. Geoffrion, A. M. "Duality in Nonlinear Programming: A Simplified Applications-Oriented Development," *SIAM Review*, **13**, pp. 1–37, 1971.
190. Gilmore, P. C., and R. E. Gomory. "A Linear Programming Approach to the Cutting Stock Problem—Part 1," *Operations Research*, **9**, pp. 849–859, 1961.

191. Gilmore, P. C., and R. E. Gomory. "A Linear Programming Approach to the Cutting Stock Problem—Part 2," *Operations Research*, **11** (6), pp. 863–887, 1963.

192. Glassey, C. Roger. "Nested Decomposition and Multi-Stage Linear Programs," *Management Science*, **20** (3), pp. 282–292, November 1973.

193. Glicksman, M. A. *Linear Programming and Theory of Games*, John Wiley & Sons, New York, 1963.

194. Glover, F. "A New Foundation for a Simplifed Primal Integer Programming Algorithm," *Operations Research*, **16** (4), pp. 727–740, July–August, 1968.

195. Glover, F., and D. Klingman. "Basic Dual Feasible Solutions for a Class of Generalized Networks," *Operations Research*, **20**, pp. 126–136, 1972.

196. Glover, F., D. Klingman, and R. S. Barr. An Improved Version of the Out-of-Kilter Method and a Comparative Study of Computer Codes, Report CS-102, Center for Cybernetic Studies, University of Texas, Austin, Texas, 1972.

197. Glover, F., D. Karney, and D. Klingman. "Implementation and Computational Comparisons of Primal, Dual and Primal-Dual Computer Codes for Minimum Cost Network Flow Problems," *Networks*, **4** (3), pp. 191–212, 1974.

198. Goldman, A. J. "Resolution and Separation Theorems for Polyhedral Convex Sets," in H. W. Kuhn and A. W. Tucker (eds), *Linear Inequalities and Related Systems*, Annals of Mathematics Study No. 38, Princeton University Press, Princeton, N.J., pp. 41–51, 1956.

199. Goldman, A. J., and A. W. Tucker. "Theory of Linear Programming," in H. W. Kuhn and A. W. Tucker (eds), *Linear Inequalities and Related Systems*, Annals of Mathematics Study No. 38, Princeton University Press, Princeton, N.J., pp. 53–98, 1956.

200. Goldman, A. J., and A. W. Tucker. "Polyhedral Convex Cones," in H. W. Kuhn and A. W. Tucker (eds), *Linear Inequalities and Related Systems*, Annals of Mathematics Study No. 38, Princeton University Press, Princeton, N.J., pp. 19–39, 1956.

201. Golomski, William A. Linear Programming in Food Blending, Annual Convention Transactions, 17th Annual Convention, American Society for Quality Control, pp. 147–152, 1963.

202. Gomory, R. E. An Algorithm for the Mixed Integer Problem, Research Memorandum RM-2597, The Rand Corporation, Santa Monica, Calif., 1960.

203. Gomory, R. E. "An Algorithm for Integer Solutions to Linear Programs," in R. L. Graves and P. Wolfe (eds), *Recent Advances in Mathematical Programming*, McGraw-Hill Book Co., New York, pp. 269–302, 1963.

204. Gomory, R. E. "All-Integer Integer Programming Algorithm," in J. F. Muth and G. L. Thompson (eds.), *Industrial Scheduling*, Prentice-Hall, Englewood Cliffs, N.J., pp. 193–206, 1963.

205. Gomory, R. E., and W. J. Baumol. "Integer Programming and Pricing," *Econometrica*, **28** (3), pp. 521–550, 1960.
206. Gomory, R. E., and T. C. Hu. "Multi-Terminal Network Flows," *SIAM*, **9** (4), 1961.
207. Gomory, R. E., and T. C. Hu. "An Application of Generalized Linear Programming to Network Flows," *SIAM*, **10** (2), 1962.
208. Gomory, R. E., and T. C. Hu. "Synthesis of a Communication Network," *SIAM*, **12** (2), June 1964.
209. Goncalves, A. S. "Basic Feasible Solutions and the Dantzig-Wolfe Decomposition Algorithm," *Operations Research Quarterly*, **19** (4), pp. 465–469, December 1968.
210. Gorham. William. "An Application of a Network Flow Model to Personnel Planning," *IEEE Transactions on Engineering Management*, **10** (3), pp. 121–123, September 1963.
211. Gould, F. J. "Proximate Linear Programming: A Variable Extreme Point Method," *Mathematical Programming*, **3** (3), pp. 326–338, December 1972.
212. Graves, R. L, and P. Wolfe (eds). *Recent Advances in Mathematical Programming*, McGraw-Hill Book Co., New York, 1963.
213. Greenberg, Harold. "Modification of the Primal-Dual Algorithm for Degenerate Problems," *Operations Research*, **16** (6), pp. 1227–1230, November–December 1968.
214. Grigoriadis, M. D. "A Dual Generalized Upper Bounding Technique," *Management Science*, **17** (5), pp. 269–284, January 1971.
215. Grigoriadis, M. D., and W. W. White. "A Partitioning Algorithm for the Multicommodity Network Flow Problem," *Mathematical Programming*, **3**, pp. 157–177, 1972.
216. Grigoriadis, M. D., and W. W. White. "Computational Experience with a Multicommodity Network Flow Algorithm," in R. Cottle and J. Krarup (eds.), *Optimization Methods for Resource Allocation*, English University Press, 1972.
217. Grinold, R. C. "A Multicommodity Max-Flow Algorithm," *Operations Research*, **16**, pp. 1234–1238, 1968.
218. Grinold, R. C. "A Note on Multicommodity Max-Flow Algorithm," *Operations Research*, **17**, p. 755, 1969.
219. Grinold, R. C. Calculating Maximal Flows in a Network with Positive Gains, working paper WP CP-337, Center for Research in Management Science, University of California, Berkeley, Calif., 1971.
220. Grinold, R. C. "Steepest Ascent for Large Scale Linear Programs," *SIAM Review*, **14**, pp. 447–464, 1972.
221. Gross, O. The Bottleneck Assignment Problem, Paper P-1630, The Rand Corporation, Santa Monica, Calif., March 1959.
222. Gross, Oliver. A Linear Program of Prager's. Notes on Linear Programming and Extensions, Part 60, Research Memorandum RM-2993-PR, The Rand Corporation, Santa Monica, Calif., April 1962.

223. Gunther, Paul. "Use of Linear Programming in Capital Budgeting," *Operagions Research*, **3** (2), pp. 219–224, May 1955.

224. Gutnik, L. A. "On the Problem of Cycling in Linear Programming," *Dokl. AN SSR* (USSR), **170** (1), pp. 53–56, 1966.

225. Hadley, G. *Linear Algebra*, Addison-Wesley, Reading, Mass., 1961.

226. Hadley, G. *Linear Programming*, Addison-Wesley, Reading, Mass., 1962.

227. Hadley, G., and M. A. Simonnard. A Simplified Two-Phase Technique for the Simplex Method," *Naval Research Logistics Quarterly*, **6** (3), 1959.

228. Hagelschuer, Paul B. *Theory of Linear Decomposition*, Springer-Verlag, New York, 1971.

229. Hakimi, S. L. On Simultaneous Flows in a Communication Network, Document No. AD-267 090, Office of Technical Services, 1961.

230. Haley, K. B. "The Existence of a Solution to the Multi-Index Problem" *Operational Research Quarterly*, **16** (4), pp. 471–474, December, 1965.

231. Halmos, P. R., and H. E. Vaughan. "The Marriage Problem," *American Journal of Mathematics*, **72**, (1), pp. 214–215, January 1950.

232. Harris, Milton Y. "A Mutual Primal-Dual Linear Programming Algorithm," *Naval Research Logistics Quarterly*, **17** (2), pp. 199–206, June 1970.

233. Hartman, J. K., and L. S. Lasdon. "A Generalized Upper Bounding Algorithm for Multicommmodity Network Flow Problems," *Networks* **1**, pp. 333–354, 1972.

234. Heady, E. O., and W. Candler. *Linear Programming Methods*, Iowa State College Press, Ames, Iowa, 1958.

235. Held, M., P. Wolfe, and H. D. Crowder. "Validation of Subgradient Optimization," *Mathematical Programming*, **6**, pp. 62–68, 1974.

236. Heller, I. "Constraint Matrices of Transportation-Type Problems," *Naval Research Logistics Quarterly*, **4**, pp. 73–78, 1957.

237. Heller, I. "On Linear Programs Equivalent to the Transportation Problem," *SIAM*, **12**, (1), pp. 31–42, March 1964.

238. Heller, I., and C. B. Tompkins. "An Extension of a Theorem of Dantzig's," in H. W. Kuhn and A. W. Tucker (eds.), *Linear Inequalities and Related Systems*, Annals of Mathematics Study No. 38, Princeton University Press, Princeton, N. J., pp. 247–254, 1956.

239. Himmelblau, David M. *Decomposition of Large Scale Problems*, North-Holland, Amsterdam, 1973.

240. Hitchcock, F. L. "Distribution of a Product from Several Sources to Numerous Localities," *Journal of Mathematical Physics*, **20**, pp. 224–230, 1941.

241. Hoffman, A. J. "How to Solve a Linear Program," in H. Antosiewicz (ed.), *Proceedings of the Second Symposium in Linear Programming*, vols. 1 and 2, DCS/Comptroller, Headquarters U. S. Air Force, Washington, D. C., pp. 397–424, January, 1955.

242. Hoffman, A. J. Cycling in the Simplex Algorithm, Report No. 2974,

National Bureau of Standards, Washington, D. C., 1953.

243. Hoffman, A. J., and J. B. Kruskal. "Integral Boundary Points of Convex Polyhedra," in H. W. Kuhn and A. W. Tucker (eds.), *Linear Inequalities and Related Systems*, Annals of Mathematics Study No. 38, Princeton University Press, Princeton, N. J., pp. 233–246, 1956.

244. Hu, T. C. "The Maximum Capacity Route Problem," *Operations Research*, **9** (6), pp. 898–900, November–December 1961.

245. Hu, T. C. "Multi-Commodity Network Flows," *Operations Research*, **11** (3), pp. 344–360, May–June 1963.

246. Hu, T. C. "On the Feasibility of Multicommodity Flows in a Network," *Operations Research*, **12**, pp. 359–360, 1964.

247. Hu, T. C. Multi-Terminal Shortest Paths, U.S.G.R. & D.R. Order AD-618 757 from Clearinghouse, Operations Research Center, University of California, Berkeley, Calif., April 1965.

248. Hu, T. C. "Minimum Convex Cost Flows," *Naval Research Logistics Quarterly*, **13** (1), pp. 1–9, March 1966.

249. Hu, T. C. "Revised Matrix Algorithms for Shortest Paths in a Network," *SIAM*, **15**, (1), pp. 207–218, January 1967.

250. Hu, T. C. "Laplace Equation and Network Flows," *Operations Research*, **15** (2), pp. 348–356, April 1967.

251. Hu, T. C. "Decomposition Algorithm for Shortest Paths in a Network," *Operations Research*, **16** (1), pp. 91–102, January–February 1968.

252. Hu, T. C. *Integer Programming and Network Flows*, Addison-Wesley, Reading, Mass., 1969.

253. Iri, M. "A New Method of Solving Transportation Network Problems," *Journal of the Operations Research Society of Japan*, **3**, (1–2), pp. 27–87, October 1960.

254. Iri, M. "An Extension of the Maximum-Flow Minimum-Cut Theorem to Multicommodity Networks," *Journal of the Operations Research Society of Japan*, **13**, 1971.

255. Jacobs, W. W. "The Caterer Problem," *Naval Research Logistics Quarterly*, **1** (2), 1954.

256. Jarvis, J. J. "On the Equivalence Between the Node-Arc and Arc-Chain Formulations for the Multicommodity Maximal Flow Problem," *Naval Research Logistics Quarterly*, **16**, pp. 515–529, 1969.

257. Jarvis, J. J., and A. M. Jezior. "Maximal Flow with Gains Through a Special Network," *Operations Research*, **20**, pp. 678–688, 1972.

258. Jarvis, J. J., and P. D. Keith. Multicommmodity Flows with Upper and Lower Bounds, Unpublished paper, School of Industrial and Systems Engineering, Georgia Institute of Technology, Atlanta, Ga., 1974.

259. Jarvis, J. J., and J. B. Tindall. "Minimum Disconnecting Sets in Directed Multi-Commodity Networks," *Naval Research Logistics Quarterly*, **19**, pp. 681–690, 1972.

260. Jewell, W. S. Optimal Flow Through Networks, Interim Technical Report No. 8, M. I. T. Project, Fundamental Investigations in Methods of Operations Research, 1958.

261. Jewell, W. S. "Optimal Flow Through Networks with Gains," *Operations Research*, **10** (4), 1962.

262. Jewell, W. S. A Primal-Dual Multicommodity Flow Algorithm, ORC Report 66–24, University of California, Berkeley, Calif., 1966.

263. Jewell, W. S. Multi-Commodity Network Solutions, Research Report ORC 66-23, Mathematical Science Division, Operations Research Center, University of California, Berkeley, Calif., September 1966.

264. Jewell, W. S., Complex: A Complementary Slackness, Out-of-Kilter Algorithm for Linear Programming, ORC 67-6, University of California, Berkeley, Calif., 1967.

265. John, F. "Extremum Problems with Inequalities as Subsidiary Conditions," in *Studies and Essays*, Wiley Interscience, New York, pp. 187–204, 1948.

266. Johnson, E. L. Programming in Networks and Graphs, Research Report ORC 65-1, University of California, Berkeley, Calif., 1965.

267. Johnson, E. L. "Networks and Basic Solutions," *Operations Research*, **14**, pp. 619–623, 1966.

268. Jones, W. G., and C. M. Rope. "Linear Programming Applied to Production Planning," *Operational Research Quarterly*, **15** (4), December 1964.

269. Kahle, Robert V. "Application of Linear Programming for Industrial Planning," *Proceedings of the American Institute of Industrial Engineers*, 1962.

270. Kalaba, R. E., and M. L. Juncosa. "Optimal Design and Utilization of Communication Networks," *Management Science*, **3** (1), pp. 33–44, 1956.

271. Kantorovich, L. *Mathematical Methods in the Organization and Planning of Production*, Publication House of The Leningrad State University, 1939. Translated in *Management Science*, **6**, pp. 366–422, 1958.

272. Kantorovich, L. "On the Translocation of Masses," *Compt. Rend. Academy of Sciences, U. R. S. S.*, **37**, pp. 199–201, 1942. Translated in *Management Science*, **5** (1), 1958.

273. Kantorovich, L. V., and M. K. Gavurin. "The Application of Mathematical Methods to Problems of Freight Flow Analysis," *Akademii Nauk SSSR*, 1949.

274. Kapur, J. N. "Linear Programming in Textile Industry," *Journal of National Productivity Council* (India), **4** (2), pp. 296–302, April–June 1963.

275. Karlin, S. *Mathematical Methods and Theory in Games, Programming and Economics*, **1–2**, Addison-Wesley, Reading, Mass., 1959.

276. Karush, W. Duality and Network Flow, Report TM-1042-201-00, System Development Corporation, Document no. AD-402 643, March 1963.

277. Kelley, J. E. "Parametric Programming and the Primal-Dual Algorithm," *Operations Research*, **7**, 1959.

278. Kelley, J. E. "The Cutting Plane Method for Solving Convex Programs," *SIAM*, **8** (4), pp. 703–712, December 1960.

279. Kelley, J. E. "Critical-Path Planning and Scheduling, Mathematical Basis," *Operations Research*, **9** (2), pp. 296–320, May 1961.

280. Kennington, J. L. Multicommodity Network Flows: A Survey, Technical report CP74015, Department of Computer Science and Operations Research, Southern Methodist University, 1974.

281. Klee, V. L. "A String Algorithm for Shortest Paths in a Directed Network," *Operations Research*, **12** (3), pp. 428–432, May–June 1964.

282. Klee, Victor, and George J. Minty. How Good is the Simplex Algorithm?, Mathematical Note no. 643, Mathematics Research Laboratory, Boeing Scientific Research Labs, February 1970.

283. Klein, M. "A Primal Method for Minimal Cost Flows," *Management Science*, **14** (3), pp. 205–220, November 1967.

284. Kleitman, D. J. "An Algorithm for Certain Multi-Commodity Flow Problems," *Networks*, **1**, 75–90, 1971.

285. Klingman, D., and R. Russell. "Solving Constrained Transportation Problems," *Operations Research*, **23**, pp. 91–106, 1975.

286. Kobayashi, Takashi. "On Maximal Flow Problem in a Transportation Network with a Bundle," *Journal of the Operations Research Society of Japan*, **10** (3–4), pp. 69–75, June 1968.

287. Koch, James V. "A Linear Programming Model of Resource Allocation in a University," *Decision Sciences*, **4** (4), pp. 494–504, October 1973.

288. Koenigsberg, E. "Some Industrial Applications of Linear Programming," *Operations Research Quarterly*, **12**, (2), pp. 105–114, June 1961.

289. Kondor, Y. "Linear Programming of Income Tax Rates," *Israel Journal of Technology* (Israel), **6** (5), pp. 341–354, November–December 1968.

290. Koopmans, T. C. "Optimum Utilization of the Transportation System," *Econometrica*, **17** (3–4), 1949.

291. Koopmans, T. C. (ed.). *Activity Analysis of Production and Allocation*, Cowles Commission Monograph 13, John Wiley & Sons, New York, 1951.

292. Koopmans, T. C., and S. Reiter. "A Model of Transportation," in T. C. Koopmans (ed.), *Activity Analysis of Production and Allocation*, John Wiley & Sons, New York, pp. 222–259, 1951.

293. Kruskal, J. B. "On the Shortest Spanning Subtree of a Graph and the Traveling Salesman Problem," *Proceedings of the American Mathematical Society*, **7**, pp. 48–50, 1956.

294. Kuhn, H. W. "The Hungarian Method for the Assignment Problem," *Naval Research Logistics Quarterly*, **2** (1–2), March–June 1955.

295. Kuhn, H. W., and A. W. Tucker. "Nonlinear Programming," in J. Neyman (ed.), *Proceedings of the Second Berkeley Symposium on Mathematical Statistics and Probability*, University of California Press, Berkeley, Calif., pp. 481–492, 1950.

296. Kuhn, H. W. and A. W. Tucker (eds). Linear Inequalities and Related Systems, *Annals of Mathematics Study* No. 38, Princeton University Press, Princeton, N. J., 1956.

297. Kunzi, H. P., and K. Kleibohm. "The Triplex Method," *Unternehmensforschung* (Germany), **12** (3), pp. 145–154, 1968.

298. Kunzi, H. P., and W. Krelle. *Nonlinear Programming*, Blaisdell, Waltham, Mass., 1966.

299. Lageman, J. J. "A Method for Solving the Transportation Problem," *Naval Research Logistics Quarterly*, **14** (1), March 1967.

300. Land, A. H., and A. G. Doig. "An Automatic Method for Solving Discrete Programming Problems," *Econometrica*, **28**, pp. 497–520, 1960.

301. Land, A. H., and S. W. Stairs. "The Extension of the Cascade Algorithm to Large Graphs," *Management Science*, **14** (1), pp. 29–33, September 1967.

302. Langley, R. W. Continuous and Integer Generalized Flow Problems, Ph.D. Dissertation, Georgia Institute of Technology, 1973.

303. Langley, R. W., and J. L. Kennington. The Transportation Problem: A Primal Approach, Presented to the 43rd ORSA National Meeting, May 1973.

304. Larionov, B. A. "Abridgement of the Number of Operations in the Solution of the Transportation Problem of Linear Programming," *Trudy Tashkentskogo Instituta Inzhenernogozheleznodorognogo transporta* (USSR), **29**, pp. 211–214, 1964.

305. Lasdon, L. S. *Optimization Theory for Large Systems*, Macmillan, New York, 1970.

306. Lasdon, L. S. "Duality and Decomposition in Mathematical Programming," *IEEE Transactions on Systems Science and Cybernetics*, **4** (2), pp. 86–100, 1968.

307. Lavallee, R. S. "The Application of Linear Programming to the Problem of Scheduling Traffic Signals," *Operations Research* **3** (4), 1955.

308. Lemke, C. E. "The Dual Method of Solving the Linear Programming Problem," *Naval Research Logistics Quarterly*, **1** (1), 1954.

309. Lemke, C. E. "The Constrained Gradient Method of Linear Programming," *SIAM* **9** (1), pp. 1–17, March 1961.

310. Lemke, C. E., and T. J. Powers. A Dual Decomposition Principle, Document No. AD-269 699, 1961.

311. Leontief, W. W. *The Structure of the American Economy*, 1919–1939, Oxford University Press, New York, 1951.

312. Liebling, Thomas M. "On the Number of Iterations of the Simplex Method," *Operations Research Verfahren* (Germany), **17**, pp. 248–264, 1973.

313. Llewellyn, Robert W. *Linear Programming*, Holt, Rinehart and Winston, New York, 1964.

314. Lombaers, H. J. M. *Project Planning by Network Analysis*, North-Holland Publishing Co., Amsterdam, 1969.

315. Lourie, Janice R. "Topology and Computation of the Generalized Trans-

portation Problem," *Management Science*, **11**, (1), September 1964.
316. Luenberger, D. G. *Introduction to Linear and Nonlinear Programming*, Addison-Wesley, Reading, Mass., 1973.
317. Maier, S. F. Maximal Flows Using Spanning Trees, Report 71-14, Operations Research House, Stanford University, Stanford, Calif., 1971.
318. Malek-Zavarei, M., and J. K. Aggarwal. "Optimal Flow in Networks with Gains and Costs," *Networks*, **1** (4), pp. 355–365, 1972.
319. Mangasarian, O. L. *Non-Linear Programming*, McGraw-Hill Book Co., New York, 1969.
320. Manne, A. S. Notes on Parametric Linear Programming, RAND Report P-468, The Rand Corporation, Santa Monica, Calif., 1953.
321. Manne, A. S. *Scheduling of Petroleum Refinery Operations*, Harvard Economic Studies No. 48, Harvard University Press, Cambridge, Mass., 1956.
322. Markland, R. E. Analyzing Multi-Commodity Production-Distribution Networks with Transformed Flows, unpublished paper, School of Business Administration, University of Missouri, St. Louis, 1972.
323. Markowitz, H. M. "The Elimination Form of the Inverse and its Application to Linear Programming," *Management Science*, **3** (3), 1957.
324. Markowitz, H. M., and A. S. Manne. "On the Solution of Discrete Programming Problems," *Econometrica,* **25** (1), p. 19, January 1957.
325. Marshall, Clifford W. *Applied Graph Theory*, Wiley-Interscience, New York, 1971.
326. Masse, Pierre, and R. Gibrat. "Applications of Linear Programming to Investments in the Electric Power Industry," *Management Science*, **3** (1), pp. 149–166, January 1957.
327. Maurras, J. F. "Optimization of the Flow Through Networks with Gains," *Mathematical Programming*, **3**, pp. 135–144, 1972.
328. Mayeda, W., and M. E. Van Valkenburg. Set of Cut Sets and Optimum Flow, U. S. G. R. & D. R. Order AD-625 200 from CFSTI, Coordinated Science Laboratory, Illinois University, Urbana, Ill., November 1965.
329. McIntosh, Peter T. "Initial Solutions to Sets of Related Transportation Problems," *Operational Research Quarterly*, **14** (1), pp. 65–69, March 1963.
330. Metzger, R. W., and R. Schwarzbeck. "A Linear Programming Application to Cupola Charging," *Journal of Industrial Engineering*, **12** (2), pp. 87–93, March–April 1961.
331. Miller, R. E. "Alternative Optima, Degeneracy and Imputed Values in Linear Programs," *Journal of Regional Science*, **5** (1), pp. 21–39, 1963.
332. Mills, G. "A Decomposition Algorithm for the Shortest-Route Problem," *Operations Research*, **14** (2), pp. 279–291, March–April 1966.
333. Minieka, E. "Optimal Flow in a Network with Gains," *INFOR*, **10**, pp. 171–178, 1972.
334. Minieka, E. "Parametric Network Flows," *Operations Research*, **20** (6), pp. 1162–1170, November–December 1972.
335. Minkowski, H. *Geometry der Zahlen*, Teubner, Leipzig, 2nd ed., 1910.

336. Minty, G. J. "On an Algorithm for Solving Some Network-Programming Problems," *Operations Research*, **10** (3), pp. 403–405, May–June 1962.

337. Moore, E. F. "The Shortest Path Through a Maze," *Proceedings of the International Symposium on the Theory of Switching*, Part II, April 2–5, 1957, The Annals of the Computation Laboratory of Harvard University, vol. 30, Harvard University Press, 1959.

338. Motzkin, T. S. Beiträge zur Theorie der Linearen Ungleichungen, Ph. D. Dissertation, University of Zurich, 1936.

339. Motzkin, T. S. "The Multi-index Transportation Problem," *Bulletin of American Mathematical Society*, **58** (4), p. 494, 1952.

340. Motzkin, T. S "The Assignment Problem," *Proceedings of the 6th Symposium in Applied Mathematics*, McGraw-Hill Book Co. New York, pp. 109–125, 1956.

341. Motzkin, T. S., and I.G. Schoenberg, "The Relaxation Method for Linear Inequalities," *Canadian Journal of Mathematics,* **6,** pp. 393–404, 1954.

342. Mueller, R. K., and L. Cooper. "A Comparison of the Primal Simplex and Primal-dual Algorithms in Linear Programming," *Communications of the ACM*, **18** (11), November 1965.

343. Mueller-Merbach, H. "An Approximation Method for Finding Good Initial Solutions for Transportation Problems," *Elektronische Datenverarbeitung* (Germany), **4** (6), pp. 255–261, November–December 1962.

344. Mueller-Merbach, H. Several Approximation Methods for Solving the Transportation Problem, IBM-Form 78 106, IBM-Fachbibliothek (Germany), November 1963.

345. Mueller-Merbach, H. "Optimal Acceleration of Projects by Parametric Linear Programming," *Elektronische Datenverarbeitung* (Germany), **9** (1), pp. 33–39, January 1967.

346. Mueller-Merbach, H. "The Method of Direct Decomposition in Linear Programming," *Ablauf- und Planungsforschung* (Germany), **6** (2), pp. 306–322, April–June 1965.

347. Mueller-Merbach, H. "An Improved Starting Algorithm for the Ford-Fulkerson Approach to the Transportation Problem," *Management Science,* **13** (1), pp. 97–104, September 1966.

348. Munkres, James. "Algorithms for the Assignment and Transportation Problems," *SIAM*, **5** (1), pp. 32–38, March 1957.

349. Murchland, J. D. A New Method for Finding All Elementary Paths in a Complete Directed Graph, Report LSE-TNT-22, Transport Network Theory Unit, London School of Economics, London, England, October 1965.

350. Murchland, J. D. The Once-Through Method of Finding All Shortest Distances in a Graph from a Single Origin, Report LBS-TNT-56, Transport Network Theory Unit, London School of Economics, London, England, August 1967.

351. Naniwada, M. "Multicommodity Flows in a Communication Network," *Electronics Communica ions of Japan*, **52**, pp. 34–40, 1969.

352. Nemhauser, G. L. "A Generalized Permanent Labeling Setting Algorithm

for the Shortest Path Between Specified Nodes," *Journal of Mathematical Analysis and Applications*, **38**, pp. 328–334, 1972.

353. Nicholson, T. A. J. "Finding the Shortest Route Between Two Points in a Network," *The Computer Journal*, **9** (3), pp. 275–280, November 1966.
354. Ohtsuka, Makoto. "Generalization of the Duality Theorem in the Theory of Linear Programming," *Journal of Science of the Hiroshima University*, Ser. A-I (Japan), **30** (1), pp. 31–39, July 1966.
355. Ohtuska, M. "Generalized Capacity and Duality Theorems in Linear Programming," *Journal of Science of the Hiroshima University*, Ser. A-I (Japan), **30** (1), pp. 45–56, July 1966.
356. Onaga, K. "Optimum Flows in General Communications Networks," *Journal of the Franklin Institute*, **283**, pp. 308–327, 1967.
357. Onaga, K. "A Multicommodity Flow Theorem," *Electronics Communications of Japan*, **53**, pp. 16–22, 1970.
358. Orchard-Hays, W. A Composite Simplex Algorithm-II, Research Memorandum RM-1275, The Rand Corporation, Santa Monica, Calif., May 1954.
359. Orchard-Hays, W. Background, Development and Extensions of the Revised Simplex Method, Research Memorandum RM-1433, The Rand Corporation, Santa Monica, Calif., 1954.
360. Orchard-Hays, W. Matrices, Elimination and the Simplex Method, Report, CEIR, Inc., Bethesda, Md., 1961.
361. Orden, A. A Procedure for Handling Degeneracy in the Transportation Problem, mimeograph, DCS/Comptroller, Headquarters U. S. Air Force, Washington, D. C., 1951.
362. Orden, A. "The Transshipment Problem," *Management Science*, **2** (3), 1956.
363. Paranjape, S. R. "The Simplex Method: Two Basic Variables Replacement," *Management Science*, **12** (1), September 1965.
364. Pierskalla, W. P. "The Multidimensional Assignment Problem," *Operations Research*, **16**, pp. 422–431, 1968.
365. Pollack, M., and G. Wallace. Methods for Determining Optimal Traffic Routes in Large Communication Networks, U. S. G. R. R. Document Number AD-434 856, Standord Research Institute, Menlo Park, Calif., June 1962.
366. Potts, R. B., and R. M. Oliver. *Flows in Transportation Networks*, Academic Press, New York, 1972.
367. Prager, W. "On the Caterer Problem," *Management Science*, **3** (1), pp. 15–23, October 1956.
368. Prager, W. "A Generalization of Hitchcock's Transportation Problem," *Journal of Mathematical Physics* (M. I. T.), **36** (2), pp. 99–106, July 1957.
369. Pshchenichnii, B. M. "The Connection Between Graph Theory and the Transportation Problem," *Dopovidi Akademii Nauk URSR* (USSR), **4**, pp. 427–430, 1963.
370. Quandt, R. E., and H. W. Kuhn. "On Upper Bounds for the Number of

Iterations in Solving Linear Programs," *Operations Research*, **12** (1), 1964.

371. Ravindran, A., and T. W. Hill. "A Comment on the Use of Simplex Method for Absolute Value Problems," *Management Science*, **19** (5), pp. 581–582, January 1973.
372. Reban, K. R. "Total Unimodularity and the Transportation Problem: A Generalization," *Linear Algebra and Its Applications*, **8**, pp. 11–24, 1974.
373. Riesco, A., and M. E. Thomas. "A Heuristic Solution Procedure for Linear Programming Problems with Special Structure," *AIIE Transactions*, **1** (2), pp. 157–163, June 1969.
374. Riley, V., and S. I. Gass. *Linear Programming and Associated Techniques: A Comprehensive Bibliography on Linear, Nonlinear and Dynamic Programming*, John Hopkins Press, Baltimore, Md., 1958.
375. Ritter, K. A Decomposition Method for Linear Programming Problems with Coupling Constraints and Variables, MRC Report no. 739, Mathematics Research Center, U. S. Army, University of Wisconsin, Madison, Wis., April 1967.
376. Robacker, J. T. Concerning Multicommodity Networks, Research Memorandum RM-1799, The Rand Corporation, Santa Monica, Calif., 1956.
377. Rockafeller, R. T. *Convex Analysis*, Princeton University Press, Princeton, N. J., 1970.
378. Rohde, F. V. "Bibliography on Linear Programming," *Operations Research*, **5** (1), 1957.
379. Rosen, J. B. "The Gradient Projection Method for Nonlinear Programming: Part I–Linear Constraints," *SIAM*, **8** (1), pp. 181–217, 1960.
380. Rosen, J. B. "The Gradient Projection Method for Nonlinear Programming: Part II," *SIAM*, **9** (4), 1961.
381. Rosen, J. B. "Primal Partition Programming for Block Diagonal Matrices," *Numerical Mathematics*, **6**, 250–260, 1964.
382. Rothfarb, B., N. P. Shein, and I. T. Frisch. "Common Terminal Multicommodity Flow," *Operations Research*, **16**, pp. 202–205, January–February 1968.
383. Rothfarb, B., and I. T. Frisch. "On the Three-Commodity Flow Problem," *SIAM*, **17**, pp. 46–58, 1969.
384. Rothchild, B., and A. Whinston. "On Two-Commodity Network Flows," *Operations Research*, **14** (3), pp. 377–388, May–June 1966.
385. Rothchild, B., and A. Whinston. "Feasibility of Two-Commodity Network Flows," *Operations Research*, **14**, pp. 1121–1129, 1966.
386. Rothchild, B., and A. Whinston. "Maximal Two-Way Flows," *SIAM*, **15**, (5), pp. 1228–1238, September 1967.
387. Roy, B. "Extremum Paths," *Gestion* (France), **5**, pp. 322–335, May 1966.
388. Russel, A. H. "Cash Flows in Networks," *Management Science*, **16** (5), pp. 357–373, January 1970.
389. Rutenberg, D. P. "Generalized Networks, Generalized Upper Bounding

and Decomposition of the Convex Simplex Method," *Management Science*, **16** (5), pp. 388–401, January 1970.

390. Saaty, T. L. "The Number of Vertices of a Polyhedron," *American Mathematical Monthly*, **62**, 1955.
391. Saaty, T. L. "Coefficient Perturbation of a Constrained Extremum," *Operatons Research*, **7**, May–June 1959.
392. Saaty, T. L., and S. I. Gass. "The Parametric Objective Function, Part I," *Operations Research*, **2**, 1954.
393. Saigal, R. Multicommodity Flows in Directed Networks, ORC Report 67–38, University of California, Berkeley, Calif., 1967.
394. Saigal, R. "A Constrained Shortest Route Problem," *Operations Research*, **16** (1), pp. 205–209, January–February 1968.
395. Saigal, R. "On the Modularity of a Matrix," *Linear Algebra and Its Applications*, **5**, pp. 39–48, 1972.
396. Sakarovitch, M. The Multi-Commodity Maximum Flow Problem, ORC Report 66-25, University of California, Berkeley, Calif., 1966.
397. Sakarovitch, M., and R. Saigal. "An Extension of Generalized Upper Bounding Techniques for Structured Linear Programs," *SIAM*, **15** (4), pp. 906–914, July 1967.
398. Sandor, P. E. "Some Problems of Ranging in Linear Programming," *Journal of the Canadian Operational Research Society*, **2** (1), pp. 26–31, June 1964.
399. Scoins, H. I. The Compact Representation of a Rooted Tree and the Transportation Problem, presented at the International Symposium on Mathematical Programming, London, 1964.
400. Seiffart, E. "An Algorithm for Solution of a Parametric Distribution Problem," *Ekonomicko-matematicky obzor* (Czechoslovakia), **2** (3), pp. 263–283, August 1966.
401. Sengupta, J. K., and T. K. Kumar. "An Application of Sensitivity Analysis to a Linear Programming Problem," *Unternehmensforschung* (Germany), **9** (1), pp. 18–36, 1965.
402. Seshu, S., and M. Reed. *Linear Graphs and Electrical Networks*, Addison-Wesley, Reading, Mass., 1961.
403. Shanno, D. F. and R. L. Weil. "Linear Programming With Absolute Value Functionals," *Management Science*, **16** (5), p. 408, January 1970.
404. Shapley, L. S. On Network Flow Functions, Research Memorandum RM-2338, The Rand Corporation, Santa Monica, Calif., 1959.
405. Shetty, C. M. "A Solution to the Transportation Problem with Nonlinear Costs," *Operations Research*, **7** (5), pp. 571–580, September–October 1959.
406. Shetty, C. M. "Solving Linear Programming Problems with Variable Parameters," *Journal of Industrial Engineering*, **10** (6), 1959.
407. Shetty, C. M. "On Analyses of the Solution to a Linear Programming Problem," *Operational Research Quarterly*, **12** (2), pp. 89–104, June 1961.

408. Simonnard, M. A. Transportation-Type Problems, Interim Technical Report No. 11, Massachusetts Institute of Technology, Cambridge, Mass., 1959.

409. Simonnard, M. A. "Structure des bases dans les problèmes de transport," *Rev. Fr. Rech. Op.*, **12**, 1959.

410. Simonnard, M. A. *Linear Programming*, Prentice-Hall, Englewood Cliffs, N. J., 1966.

411. Simonnard, M. A., and G. F. Hadley. "Maximum Number of Iterations in the Transportation Problem," *Naval Research Logistics Quarterly*, **6** (2), 1959.

412. Smith, C. W. Maximal Flow at Minimal Cost Through a Special Network with Gains, unpublished thesis, School of Industrial and Systems Engineering, Georgia Institute of Technology, Atlanta, Ga., 1971.

413. Smith, D. M., and W. Orchard-Hays. "Computational Efficiency in Product Form LP Codes," in R. Graves and P. Wolfe (eds.), *Recent Advances in Mathematical Programming*, McGraw-Hill Book Co., New York, 1963.

414. Spivey, W. A. *Linear Programming*, Macmillan, New York, 1963.

415. Srinivasan, V., and G. L. Thompson. "Accelerated Algorithms for Labeling and Relabeling of Trees with Applications to Distribution Problems," *Journal of the ACM*, **19**, pp. 712–726, 1972.

416. Staffurth, C. (ed.). *Project Cost Control Using Networks*, The Operational Research Society and the Institute of Cost and Works Accountants, London, 1969.

417. Stevens, B. H. "Linear Programming and Location Rent," *Journal of Regional Science*, **3** (2), pp. 15–26, 1961.

418. Stoinova-Penkrova, N. "The Simplex Method without Fractions," *Trudove Vissh Ikonomicheshi Institute Sofia* (Bulgaria), **1**, pp. 357–370, 1964.

419. Stokes, R. W. "A Geometric Theory of Solution of Linear Inequalities," *Transactions of the American Mathematical Society*, **33**, pp. 782–805, 1931.

420. Symonds, G. H. *Linear Programming: The Solution of Refinery Problems*, Esso Standard Oil Company, New York, 1955.

421. Stroup, J. W. "Allocation of Launch Vehicles to Space Missions: A Fixed-Cost Transportation Problem," *Operations Research*, **15** (6), pp. 1157–1163, November–December 1967.

422. Strum, J. E. *Introduction to Linear Programming*, Holden Day, San Francisco, 1972.

423. Swarup, K. "Duality for Transportation Problem in Fractional Programming," *CCERO* (Belgium), **10** (1), pp. 46–54, March 1968.

424. Takahashi, I. "Tree Algorithm for Solving Resource Allocation Problems," *Operations Research Society of Japan*, **8**, pp. 172–191, 1966.

425. Takahashi, I. "Tree Algorithm for Solving Network Transportation Problems," *Operations Research Society of Japan*, **8**, pp. 192–216, 1966.

426. Tan, S. T. "Contributions to the Decomposition of Linear Programs," *Unternehmensforschung* (Germany), **10**, (3–4), pp. 168–189, 247–268, 1966.

427. Tang, D. T. "Bipath Networks and Multicommodity Flows," *IEEE Transactions*, **11**, pp. 468–474, 1964.
428. Thompson, G. L., F. M. Tonge, and S. Zionts. "Techniques for Removing Nonbinding Constraints and Extraneous Variables from Linear Programming Problems," *Management Science*, **12** (7), pp. 588–608, March 1966.
429. Thrall, R. M. The Mutual Primal-Dual Simplex Algorithm, Report no. 6426–27, University of Michigan Engineering Summer Conferences on Operations Research, Summer 1964.
430. Tomizawa, N. "On Some Techniques Useful for Solution of Transportation Network Problems," *Networks*, **1**, pp. 173–194, 1971.
431. Tomlin, J. A. "Minimum-Cost Multi-Commodity Network Flows," *Operations Research*, **14** (1), pp. 45–51, February 1966.
432. Tompkins, C. B. "Projection Methods in Calculation," in H. A. Antosiewicz (ed.), *Proceedings of the Second Symposium in Linear Programming*, **2**, National Bureau of Standards, 1955.
433. Tompkins, C. B. "Some Methods of Computational Attack on Programming Problems Other Than the Simplex Method," *Naval Research Logistics Quarterly*, **4** (1), pp. 95–96, March 1957.
434. Torng, H. C. "Optimization of Discrete Control Systems Through Linear Programming," *Journal of the Franklin Institute*, **278** (1), pp. 28–44, July 1964.
435. Tucker, A. W. "Linear Programming and Theory of Games," *Econometrica*, **18** (2), p. 189, April 1950.
436. Tucker, A. W. "Linear Inequalities and Convex Polyhedral Sets," in H. A. Antosiewicz (ed.), *Proceedings of the 2nd Symposium in Linear Programming*, **2**, National Bureau of Standards, pp. 569–602, 1955.
437. Tucker, A. W. "Linear and Nonlinear Programming," *Operations Research*, **5** (2), pp. 244–257, April 1957.
438. Tucker, A. W. "Dual Systems of Homogeneous Linear Relations," in H. W. Kuhn and A. W. Tucker (eds.), *Linear Inequalities and Related Systems*, Annals of Mathematics Study No. 38, Princeton University, Princeton, N. J., pp. 3–18, 1960.
439. Tucker, A. W. On Directed Graphs and Integer Programs, Technical Report, IBM Mathematical Research Project, Princeton University, Princeton, N. J., 1960.
440. Tucker, A. W. "Solving a Matrix Game by Linear Programming," *IBM Journal of Research and Development*, **4** (5), pp. 507–517, November 1960.
441. Tucker, A. W. "Combinatorial Theory Underlying Linear Programs," in R. Graves and P. Wolfe (eds.), *Recent Advances in Mathematical Programming*, McGraw-Hill Book Co., New York, 1963.
442. Tyndall, W. F. "An Extended Duality Theorem for Continuous Linear Programming Problems," *SIAM*, **15** (5), pp. 1294–1298, September 1967.
443. Vajda, S. *The Theory of Games and Linear Programming*, John Wiley & Sons, New York, 1956.

444. Vajda, S. *Readings in Linear Programming*, John Wiley & Sons, New York, 1958.
445. Vajda, S. *Mathematical Programming*, Addison-Wesley, Reading, Mass., 1961.
446. Van de Panne, C., and A. Whinston. "The Simplex and Dual Method for Quadratic Programming," *Operational Research Quarterly*, **15** (4), December 1964.
447. Van de Panne, C., and A. Whinston. An Alternative Interpretation of the Primal-Dual Method and Some Related Parametric Methods, U. S. G. R. & D. R. Order AD-624 499, University of Virginia, Charlottesville, Va., August 1965.
448. Van de Panne, C. *Linear Programming and Related Techniques*, North-Holland/American Elsevier, New York, 1971.
449. Veinott, A. F., and H. M. Wagner. "Optimum Capacity Scheduling, I and II," *Operations Research*, **10**, pp. 518–546, 1962.
450. Verkhovskii, B. S. "The Existence of a Solution of a Multi-Index Linear Programming Problem," *Doklady Akademii Nauk SSSR* (USSR), **158** (4), pp. 763–766, 1964.
451. Von Neumann, J. "Uber ein Okonomisches Gleichungssystem und ein Verallgemeinerung des Brouwerschen Fixpunktsatzes," *Ergebnisse eines Mathematischen Kolloguims*, **8**, 1937.
452. Von Neumann, J. On a Maximization Problem, manuscript, Institute for Advanced Studies, Princeton, N. J., 1947.
453. Von Neumann, J. "A Certain Zero-Sum Two-Person Game Equivalent to the Optimal Assignment Problem" in H. W. Kuhn and A. W. Tucker (eds.), *Contributions to the Theory of Games*, **2**, Annals of Mathematics Study No. 28, Princeton University Press, Princeton, N. J., pp. 12–15, 1953.
454. Von Neumann, J., and O. Morgenstern. *Theory of Games and Economic Behavior*, Princeton University Press, Princeton, N. J., 1944.
455. Votaw, D. F., and A. Orden. "Personnel Assignment Problem," in A. Orden and L. Goldstein (eds.), *Symposium on Linear Inequalities and Programming*, **10**, Planning Research Division, Director of Management Analysis Service, Comptroller, U. S. Air Force, Washington, D. C., pp. 155–163, 1952.
456. Wagner, H. M. A Linear Programming Solution to Dynammic Leontief Type Models, Research Memorandum RM-1343, The Rand Corporation, Santa Monica, Calif., 1954.
457. Wagner, H. M. "A Two-Phase Method for the Simplex Tableau," *Operations Research*, **4** (4), 1956.
458. Wagner, H. M. "A Comparison of the Original and Revised Simplex Methods," *Operations Research*, **5** (3), 1957.
459. Wagner, H. M. "A Supplementary Bibliography on Linear Programming," *Operations Research*, **5** (4), 1957.

460. Wagner, H. M. "On the Distribution of Solutions in Linear Programming Problems," *Journal of the American Statistical Association*, **53**, pp. 161–163, 1958.

461. Wagner, H. M. "The Dual Simplex Algorithm for Bounded Variables," *Naval Research Logistics Quarterly*, **5** (3), September 1958.

462. Wagner, H. M. On the Capacitated Hitchcock Problem, Technical Report no. 54, Stanford University, Stanford, Calif., 1958.

463. Wagner, H. M. "Linear Programming Techniques for Regression Analysis," *Journal of the American Statistical Association*, March 1959.

464. Walker, W. E. "A Method for Obtaining the Optimal Dual Solution to a Linear Program Using the Dantzig-Wolfe Decomposition," *Operations Research*, **17** (2), pp. 368–370, March–April 1969.

465. Waugh, F. V. "The Minimum Cost Dairy Feed," *Journal of Farm Economics*, **33** (3), pp. 299–310, August 1951.

466. White, D. J. "A Linear Programming Analogue, a Duality Theorem, and a Dynamic Algorithm," *Management Science*, **21** (1), pp. 47–59, September 1974.

467. White, W. C., M. B. Shapiro, and A. W. Pratt. "Linear Programming Applied to Ultraviolet Absorption Spectroscopy," *Communications of the ACM*, **6** (2), pp. 66–67, February 1963.

468. Williams, A. C. "A Treatment of Transportation Problems by Decomposition," *SIAM*, **10** (1), pp. 35–48, January–March 1962.

469. Williams, A. C. "Marginal Values in Linear Programming," *SIAM*, **11** (1), pp. 82–94, March 1963.

470. Williams, A. C. "Complementary Theorems for Linear Programming," *SIAM Review*, **12** (1), January 1970.

471. Williamson, J. "Determinants Whose Elements are 0 and 1," *American Mathematical Monthly*, **53**, pp. 427–434, 1946.

472. Wolfe, P. An Extended Composite Algorithm for Linear Programming, Paper P-2373, The Rand Corporation, Santa Monica, Calif., 1961.

473. Wolfe, P. A Technique for Resolving Degeneracy in Linear Programming, Research Memorandum RM-2995-PR, The Rand Corporation, Santa Monica, Calif., 1962.

474. Wolfe, P. "The Composite Simplex Algorithm," *SIAM Review*, **7** (1), 1965.

475. Wolfe, P. The Product Form of the Simplex Method, U. S. G. R. & D. R. Order AD-612 381, The Rand Corporation, Santa Monica, Calif., 1965.

476. Wolfe, P., and G. B. Dantzig. "Linear Programming in a Markov Chain," *Operations Research*, **10**, pp. 702–710, 1962.

477. Wollmer R. "Removing Arcs ffom a Network," *Operations Research*, **12** (6), November–December 1964.

478. Wollmer, R. "Maximizing Flow Through a Network with Node and Arc Capacities," *Transportation Science*, **2** (3), pp. 213–232, August 1968.

479. Wollmer, R. The Dantzig-Wolfe Decomposition Principle and Minimum

Cost Multicommodity Network Flows, Paper P-4191, The Rand Corporation, Santa Monica, Calif., 1969.

480. Wollmer, R. Multicommodity Networks with Resource Constraints: The General Multicommodity Flow Problem, Report TM393-50, Jet Propulsion Laboratory, California Institute of Technology, Pasadena, Calif., 1971.

481. Yaspan, A. "On Finding a Maximal Assignment," *Operations Research*, **14** (4), pp. 646–651, July–August 1966.

482. Yoshida, M. "Some Examples Related to the Duality Theorem in Linear Programming," *Journal of Science of the Hiroshima University*, Series A-I (Japan), **30** (1), pp. 41–43, July 1966.

483. Young, R. D. "A Primal (All-Integer) Integer Programming Algorithm," *Journal of Research*—National Bureau of Standards: B, Mathematics and Mathematical Physics, **69** (3), pp. 213–250, July–September 1965.

484. Young, R. D. "A Simplified Primal (All Integer) Integer Programming Algorithm," *Operations Research*, **16** (4), pp. 750–782, July–August 1968.

485. Zangwill, W. I. "The Convex Simplex Method," *Management Science*, **14** (3), pp. 221–238, November 1967.

486. Zangwill, W. I. *Nonlinear Programming: A Unified Approach*, Prentice-Hall, Englewood Cliffs, N. J., 1969.

487. Zangwill, W. I. "Minimum Convex Cost Flows in Certain Networks," *Management Science*, **14** (7), pp. 429–450, March 1968.

488. Zionts, S. "The Criss-Cross Method for Solving Linear Programming Problems," *Management Science*, **15** (7), pp. 426–445, March 1969.

489. Zionts, S. *Linear and Integer Programming*, Prentice-Hall, Englewood Cliffs, N. J., 1974.

490. Zoutendijk, G. *Methods of Feasible Directions*, Elsevier, New York, 1960.

INDEX

N

O

P